Ecosystem Modeling: Theory and Practice

Ecosystem Modeling: Theory and Practice

Edited by Rosemary Charles

SYRAWOOD
PUBLISHING HOUSE
New York

Published by Syrawood Publishing House,
750 Third Avenue, 9th Floor,
New York, NY 10017, USA
www.syrawoodpublishinghouse.com

Ecosystem Modeling: Theory and Practice
Edited by Rosemary Charles

International Standard Book Number: 978-1-68286-425-8 (Hardback)

Cataloging-in-publication Data

Ecosystem modeling : theory and practice / edited by Rosemary Charles.
p. cm.
Includes bibliographical references and index.
ISBN 978-1-68286-425-8
1. Biotic communities. 2. Ecosystem management. 3. Ecology. I. Charles, Rosemary.
QH541 .E36 2017
577--dc23

Printed in China.

TABLE OF CONTENTS

PREFACE

Ecosystem modeling is a method whereby entire ecosystems are studied through mathematical models. Newer concepts are generated on the basis of ecosystem modeling as it facilitates a better understanding of the ecosystem. This book on ecosystem modeling presents interdisciplinary viewpoints on the potential benefits and processes of ecosystem modeling. Different approaches, evaluations, methodologies and advanced studies have been included in this text. It covers in details some existence theories and innovative concepts revolving around ecosystem modeling. This book would prove useful for professionals and students in the fields of ecosystem engineering, restoration ecology and computational biology. It aims to serve as a resource guide for students and experts alike and contribute to the growth of the discipline.

The researches compiled throughout the book are authentic and of high quality, combining several disciplines and from very diverse regions from around the world. Drawing on the contributions of many researchers from diverse countries, the book's objective is to provide the readers with the latest achievements in the area of research. This book will surely be a source of knowledge to all interested and researching the field.

In the end, I would like to express my deep sense of gratitude to all the authors for meeting the set deadlines in completing and submitting their research chapters. I would also like to thank the publisher for the support offered to us throughout the course of the book. Finally, I extend my sincere thanks to my family for being a constant source of inspiration and encouragement.

Editor

1

Exploring Trade-Offs between Fisheries and Conservation of the Vaquita Porpoise (*Phocoena sinus*) Using an Atlantis Ecosystem Model

Hem Nalini Morzaria-Luna[1]*, Cameron H. Ainsworth[1¤], Isaac C. Kaplan[2], Phillip S. Levin[2], Elizabeth A. Fulton[3]

1 Marine Resources Assessment Group Americas Incorporated, Seattle, Washington, United States of America, **2** Conservation Biology Division, Northwest Fisheries Science Center, National Marine Fisheries Service, National Oceanic and Atmospheric Administration, Seattle, Washington, United States of America, **3** Division of Marine and Atmospheric Research, The Commonwealth Scientific and Industrial Research Organisation Wealth from Oceans Flagship, Hobart, Tasmania, Australia

Abstract

Background: Minimizing fishery bycatch threats might involve trade-offs between maintaining viable populations and economic benefits. Understanding these trade-offs can help managers reconcile conflicting goals. An example is a set of bycatch reduction measures for the Critically Endangered vaquita porpoise (*Phocoena sinus*), in the Northern Gulf of California, Mexico. The vaquita is an endemic species threatened with extinction by artisanal net bycatch within its limited range; in this area fisheries are the chief source of economic productivity.

Methodology/Principal Findings: We analyze trade-offs between conservation of the vaquita and fisheries, using an end-to-end Atlantis ecosystem model for the Northern Gulf of California. Atlantis is a spatially-explicit model intended as a strategic tool to test alternative management strategies. We simulated increasingly restrictive fisheries regulations contained in the vaquita conservation plan: implementing progressively larger spatial management areas that exclude gillnets, shrimp driftnets and introduce a fishing gear that has no vaquita bycatch. We found that only the most extensive spatial management scenarios recovered the vaquita population above the threshold necessary to downlist the species from Critically Endangered. The scenario that excludes existing net gear from the 2008 area of vaquita distribution led to moderate decrease in net present value (US$ 42 million) relative to the best-performing scenario and a two-fold increase in the abundance of adult vaquita over the course of 30 years.

Conclusions/Significance: Extended spatial management resulted in the highest recovery of the vaquita population. The economic cost of proposed management actions was unequally divided between fishing fleets; the loss of value from finfish gillnet fisheries was never recovered. Our analysis shows that managers will have to confront difficult trade-offs between management scenarios for vaquita conservation.

Editor: Konstantinos I. Stergiou, Aristotle University of Thessaloniki, Greece

Competing Interests: The authors have the following conflicts: Dr. Morzaria-Luna is a government contractor employed through a commercial company, Marine Resources Assessment Group (MRAG) Americas Inc. Dr. Ainsworth was previously a government contractor employed through this same company. This does not alter the authors' adherence to all the PLoS ONE policies on sharing data and materials.

* E-mail: HemNalini.MorzariaLuna@noaa.gov

¤ Current address: College of Marine Science, University of South Florida, Saint Petersburg, Florida, United States of America

Introduction

The incidental capture or entanglement of non-target species ("bycatch") is a ubiquitous feature of marine fisheries, and is a major factor underlying the declines of marine megafuna worldwide [1–3]. In some situations eliminating all mortality associated with bycatch is an imperative to avert extinction [3]. This may take the form of modifying fishing gears or in some cases closing or limiting fisheries. However, many such restrictions on fisheries will result in at least short-term reductions in fishers' employment and incomes. Such economic impacts potentially lead to costly and socially divisive conflicts, particularly in regions that rely heavily on fishing and/or areas where alternative employment opportunities are poor [4]. Understanding the trade-offs between conservation benefits and economic costs is thus crucial for developing well-informed policies and management strategies [5].

Vaquita, *Phocoena sinus*, is a small porpoise endemic to the Upper Gulf of California, Mexico. The vaquita population is small (245 individuals in 2008; 95% CI 68–884; [6]), with a localized distribution that is coincident with large and highly profitable industrial and artisanal fisheries, mainly for blue shrimp (*Litopenaeus stylirostris*) [7]. Vaquita are entangled in gillnets that target finfish, and driftnets targeting shrimp; these gears account for the majority of bycatch mortality [8,9]. Industrial shrimp trawling is also known to catch vaquita [10] and may disrupt vaquita behavior [8]. The combination of low fecundity and late maturity with high bycatch mortality [8] has resulted in vaquita being considered the world's most Critically Endangered cetacean [11].

Throughout the distribution of vaquita, the Mexican government is employing a set of economic incentives (Table 1) to eliminate shrimp driftnets and finfish gillnets [8,9,12] from the vaquita distribution area (see Text S1 for details), and has also reduced industrial trawl effort [13,14]. These incentives would need to compensate for the current profitability of fishing activity; in the Upper Gulf catch in artisanal fisheries is valued at over US$ 10 million [7]. Economic analysis of the plan to eliminate nets from the vaquita distribution area estimated a 93% reduction in the total value of catch of shrimp driftnet and finfish gillnets fisheries, although it did not consider catch revenue from other gears or alternate economic activities [15].

Ortiz [16] and Gerrodette and Rojas Bracho [17] analyzed the possibility of recovery of the vaquita using single-species models; they found that eliminating gillnets from the vaquita distribution area gave the most complete protection to vaquita, reducing the probability of extinction to 3% [16] and 2% respectively [17]. Likewise, applications of Ecopath with Ecosim ecosystem models [18,19], which consider trophic interactions and dynamic modeling to explore the impacts of fishing [20], found negative impacts on vaquita from the high incidental mortality. However, there has been no concurrent analysis of the effects of management actions on vaquita and Upper Gulf fisheries [15].

An important advance in marine conservation has been the rise of 'end-to-end" ecosystem models that integrate submodels of physical, biological and socio-economic components of the system [21–23]. These models have been particularly useful for exploring the ecological and conservation consequences of fisheries management scenarios, such as bycatch reduction programs. For instance, Hutton and colleagues [24] used an end-to-end model, built in the Atlantis modeling framework [21,22], to explore the effects of bycatch quotas and bycatch penalties on marine mammals. Atlantis sets single species dynamics in the context of trophic interactions, multiple fleets, and spatial and temporal variability [22]. However, ecosystem models such as Atlantis are not meant to fully capture statistical uncertainty in all parameters, as would a single species model. For this reason, these models are meant for strategic evaluations (e.g., ranking performance of alternative policy options) and not as tactical management aids (e.g., for setting total allowable catch or gear restrictions) [21].

Here, we use an Atlantis ecosystem model of the Northern Gulf of California to evaluate alternate management actions outlined in the vaquita recovery plan [9] and proposed by the International Committee for the Recovery of the Vaquita in 2012 [25]. We evaluate effects on vaquita abundance and on the net economic benefit of fisheries catch. Changes in biodiversity and other ecological processes derived from these management actions are analyzed in a separate manuscript (Morzaria et al., *unpublished information*). Simulating the effect of management on vaquita can yield valuable insights, as experiments are costly and intensive monitoring is required to detect whether and at what rate the population is recovering [6,26]. Our results illustrate potential trade-offs between fisheries and conservation. We found that only extensive spatial management scenarios recovered the vaquita population, at the cost of decreasing the net present value of fisheries.

Methods

Atlantis ecosystem model

We adapted the Atlantis model for the Northern Gulf of California, developed by Ainsworth et al. [27,28]. Technical specifications of the Atlantis code base and a review of existing applications are detailed elsewhere (e.g. [22,29]). In brief, Atlantis incorporates multiple submodels that simulate oceanography, biogeochemistry, food web interactions and human impacts, especially fisheries [30]. Species of particular ecological, management or conservation importance are represented with enough detail to evaluate direct effects of fishing, while other species are aggregated into functional groups with sufficient resolution to capture human, trophic, and climate impacts on the ecosystem [22].

The Northern Gulf of California Atlantis model was built to test ecosystem-based fisheries management questions and ecological hypotheses. The initial conditions of the model represent ecosystem structure and function for 2008 and provide a comprehensive representation of the Northern Gulf's oceanography, historical fishing patterns, migration and movement of key species, and variability in diet compositions, among other features. Ecosystem dynamics are simulated in 12-hour time steps. Model dynamics were calibrated against historical time series of abundance and catch from fisheries statistics, field data, and stock assessments available for the period from 1985–2008 (see Text S2 and [27] for details). Stock productivity and resilience were tested by a series of simulations ranging across fishing mortality rates. Unexploited biomasses of functional groups corresponded well to estimates by previous authors.

The model extends over 57800 km^2, from the Colorado River Delta south to the northern tip of Baja California Sur (Fig. 1). The model area is divided into 66 polygons based on ecologically important gradients and boundaries, stock assessments, catch data,

Table 1. Economic incentives within the vaquita recovery program for small-scale fishers and fishing cooperatives.

Option		Requirements	Benefits
Rentout	Annual payment to stop fishing	Agree to stop fishing in the Vaquita refuge	~$ 3, 500 USD‡
Switchout	Substitution of finfish gillnets and shrimp driftnets for gears with no vaquita bycatch	Turn in nets. Agree to use alternative gear.	~$ 25, 000 USD†. New fishing permit.
Buyout	Fishermen stop fishing and are given a payment destined specifically for an alternate business or economic activity	Turn in boat, nets and permits	~$ 25–35, 000 USD† (depending on number of permits)

Program is open to those based in the communities of San Felipe, in the state of Baja California, and in Golfo de Santa Clara and Puerto Peñasco, in the state of Sonora, Mexico, that own finfish gillnet and shrimp driftnet permits. Payments and guidelines for fiscal year 2011 [60].
‡Yearly payment,
†One-time payment.

and spatial management boundaries. Individual polygons include one sediment layer and up to six water depth layers. Biological, chemical and physical processes replicated within each spatial cell drive the model. Fluxes of water, heat and salt are forced by a Regional Oceanographic Model System (ROMS) [31]. Water flux drives the advection of plankton, nutrients and waste cycling; temperature affects growth, consumption and primary production rates. The spatial distribution of abundance for the 63 functional groups is defined per model polygon and depth layer.

Functional group dynamics are driven by formulations that describe biological processes, including consumption, reproduction, waste cycling, predation, recruitment, habitat dependency, mortality, growth, and movement (both foraging behavior and horizontal/vertical migration). Some key assumptions include density dependent movement, where predators move according to abundance of prey items; reproduction based on Beverton-Holt stock recruitment relationships (for fish except elasmobranchs) and fixed reproduction per female for elasmobranchs, mammals, birds, and turtles; predation calculated by a Holling Type II functional response, where diet compositions vary through time and feeding rates vary dynamically according to gape limitation; and dynamic weight at-age, where dynamic consumption rates result in variable weight at-age. For vertebrates, we track abundance and size-at-age per age class. For invertebrates, we track biomass pools. Unlike in single-species models, in Atlantis birth and death rates are not fixed, but rather evolve as predation, growth and fecundity change through time.

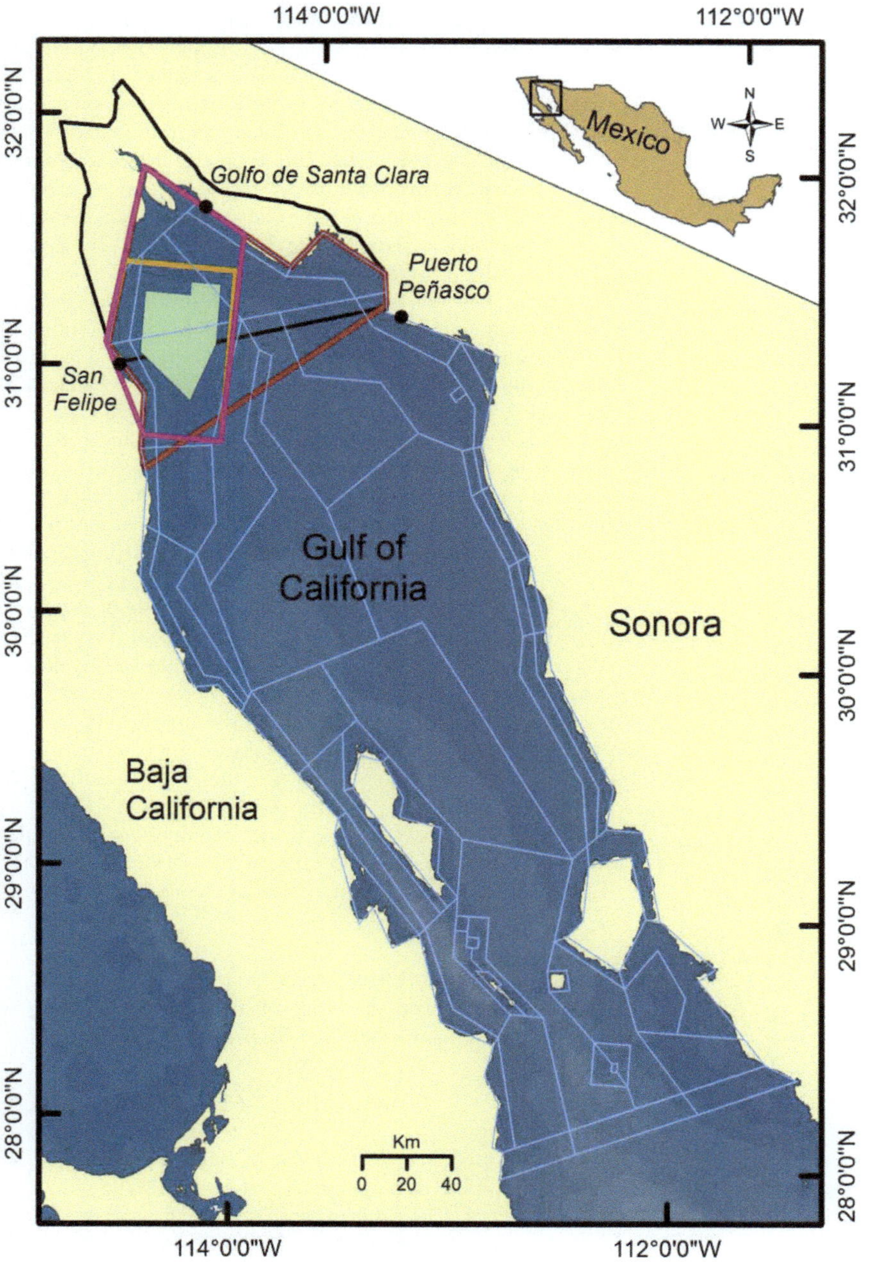

Figure 1. Northern Gulf of California and Atlantis model extent. Atlantis polygon geometry (blue lines), Biosphere Reserve (dark black line), vaquita refuge and core (green), extended refuge (orange) and fishing communities in the vaquita distribution area (red).

Vaquita age structure is divided in ten classes, with a maximum longevity of 20 years; they are considered sexually mature at age ten [32]. Average body weight was 22.3 kg vaquita^{-1} [33]. Diet composition for vaquita was based on averaged values from previous ecosystem models for the Northern Gulf [19,34], and is dominated by small pelagic fish (Table S1), although modeled diets will vary through time and space based on prey availability. Diet for juvenile vaquita was assumed similar to adults. Large pelagic sharks are considered the only predators on vaquita [10]. The model uses a fixed recruitment relationship for vaquita, in which each female is assumed to birth 0.26 viable offspring per year [27]. The initial distribution of vaquita spatial abundance follows Gerrodette et al. [6,35] (Fig. S1). Vaquita newborns and all older age classes are subject to fishing mortality by all fleets in the model [36], insufficient data was available to specify age selectivity by gears (but see [36,37]). We set initial vaquita mortality at 0.15 year^{-1} under No management (see below *Scenarios*) the median estimate in 2007 [17], before the implementation of bycatch reduction measures. When vaquita dynamics in Atlantis are tested against the historical vaquita abundance trajectory for the period 1993–2008, model results are within the 95% confidence intervals for published abundance estimates [6,38,39] (see Text S2 for details).

Scenarios

We projected five management scenarios over 30 years, from 2008 to 2038. The scenarios described here began with the same initial parameterization of ecology and oceanography; thus, the differences in outcomes result from the direct and indirect effects of fishing. Fishing mortality is imposed by the fishing fleets onto all relevant target and bycatch functional groups.

We first simulated a reference scenario (No management scenario) that did not include any management actions for vaquita protection. This scenario includes the current degree of compliance with existing fisheries restrictions (see Ainsworth et al. [28] for details); it incorporates a 30% reduction in trawl effort within the Upper Gulf Biosphere Reserve (Fig. 1) implemented in 2008 [13]. We based initial (2008) catches of species other than vaquita on the average of 2000–2007 catches, assembled from official fishery statistics, port-level surveys, and fisher log books (Table S2). We used data on fleet and gear catch composition to assign a proportion of the catch to thirty-three fleets in the model (Tables S3 and S4); each of these fleets has specific fishing areas (see Ainsworth et al. [28] for details). We set vaquita mortality at 0.15 year^{-1}, the median estimate for 2007 [17], before the implementation of bycatch reduction measures. Although industrial shrimp trawlers have been reported to catch vaquitas, there are no fleet estimates of bycatch mortality [10], therefore we allocated all vaquita bycatch to shrimp driftnet and finfish gillnet fleets. Allocation of vaquita bycatch between gillnet fleets was based on the number of fishing permits in 2008 per fleet [40] as a measure of effort, as there appears to be no difference in bycatch mortality between different fisheries and mesh sizes [37].

The No management scenario was then compared to four scenarios that simulate management actions in the species recovery plan [9], mitigation measures in the Environmental Impact Assessment for industrial shrimp trawling within the Upper Gulf Biosphere Reserve [13,14] and the latest recommendations by the International Committee for the Recovery of the Vaquita [25]. The scenarios combine spatial closures for industrial shrimp trawls and gillnets while allowing the shrimp driftnet fleet to switch to a light trawl instead of being excluded. The spatial closures occur in the area where vaquita sightings are concentrated [38]. We began all scenarios using the most recent estimate of vaquita abundance (245; [6]), and complemented these with scenarios initialized at the upper and lower limits of the 95% confidence interval (68–884). We used this approach to estimate uncertainty derived from initial population sizes since Atlantis is a deterministic model and computational limits prevent extensive stochastic realizations.

The four scenarios (Fig. 1) include a 1264-km^2 spatial closure to industrial shrimp trawls within the current refuge [13,14]. The scenarios also include progressively larger spatial closures for shrimp driftnets and finfish gillnets: 1) The Vaquita refuge scenario includes a 1264-km^2 spatial closure [9,13], representing the 2010 status quo (see Supporting Information for details). 2) The Extended refuge scenario is a 3579-km^2 spatial closure representing an option from the species recovery plan [9]. 3) The Primary area scenario excludes nets from a refuge encompassing the vaquita distribution as of 2008 (5339 km^2) [6], corresponding to the recommendation of the International Committee for the Recovery of the Vaquita in 2012 [25]. 4) The Distribution area scenario closes off the entire known vaquita range (8432 km^2) and is equivalent to the 2012 target in the species recovery plan [9]. The closures were simulated as partial or complete spatial closures to model cells affected for the corresponding fleets, with fishing effort reduced proportionally to area closed. The reconversion program being implemented to eliminate vaquita bycatch is designed to minimize redistribution of fishing effort (Table 1); we did not consider possible increases in illegal fishing.

In each of these four scenarios a new light shrimp trawl fleet that eliminates vaquita bycatch [41] was allowed to operate within the spatial closure area. We assumed that as the area closed to shrimp driftnets and finfish gillnets increases, adoption of the light trawl will increase, as many fishers want to continue fishing [7], shrimp are profitable [42], and removal of driftnets and gillnets facilitates the use of light trawl by reducing chances of entanglement. This is consistent with a recent analysis which found that the fishers which opted for the buyout were those close to retirement and that no fishers have opted to leave the fishery since 2010 [12]. Our analysis does not consider participation in the rentout option (Table 1). Shrimp catch rates for the light trawl may be dependent on the level of training and skill by fishers [25,43], varying from reductions of 13% [43] to 56% [44] relative to the shrimp driftnet. Therefore, we ran each spatial closure scenario considering 10%, 20%, 30%, 40% and 50% reductions in shrimp catch, to assess the variation in net profits. Based on results from recent trials we set the shrimp:bycatch ratio of 1.37 for the light trawl fleet [43] compared to a ratio of 1.23 for the shrimp driftnet [45] (bycatch is primarily fish, and not vaquita). Bycatch composition of the light trawl fleet also differed from that of the shrimp driftnet fleet ([41,45]; Table S5). Fishing mortality of vaquita was set to 0 for this fleet.

Assessment of economic costs

To assess the economic cost of management actions, we calculated net economic benefit (NB) per functional group as,

$$NB = GBt[1 - C]$$

where GB is gross benefit (i.e. value of catch) for year t and C is cost rate for fishing. We assumed a cost rate of fishing of 32% for artisanal fleets [7] and a base cost of 96% for industrial fleets (Table S6). Prices were determined from the average 2005–2010 value, dollars tonne^{-1}, by functional group (Table S7).

We then calculated the net present value (NPV) of profits to commercial fisheries by discounting the net benefit over time with a standardized rate [46]. Discounting implies that $1 received at some time in the future is perceived to be worth less to a person than $1 received now, and reflects uncertainty and lost opportunity costs. By calculating NPV, a common metric in cost-benefit analysis, the economic benefits of all scenarios are standardized and are comparable against other alternative investments that a fisher could receive, such as earning bank interest. NPV was calculated as

$$NPV = \sum_{t=0}^{T} d^t \cdot NB_t$$

where NB is the sum of net benefits accruing in year t for all functional groups and d is the discount factor,

$$d = \frac{1}{1+\delta}$$

As a discount rate (δ), we used the 2000–2010 Mexican government bond real interest rate of 3.6% (Center for the Study of Public Finance, House of Representatives).

For industrial shrimp trawl fleets, we only included shrimp and blue crab catch in calculation of value and assumed all other groups were discarded [47]. We present net benefit separately for industrial and artisanal fleets, and by gear (grouped fleets), to make it easier to explain the effects of the simulated management actions. Results for the end of the 30-year simulations are presented as the average of the last five years, 2034–2038 (± SE) to emphasize differences amongst scenarios rather than temporal trends which are dominated by underlying biomass dynamics.

Results

Vaquita recovery

Mature vaquita abundance trajectories for each scenario, across the 30-year simulations, are shown in Fig. 2. In the No management scenario and in the Vaquita refuge scenario that included a 1264 km^2 spatial closure, abundance showed a

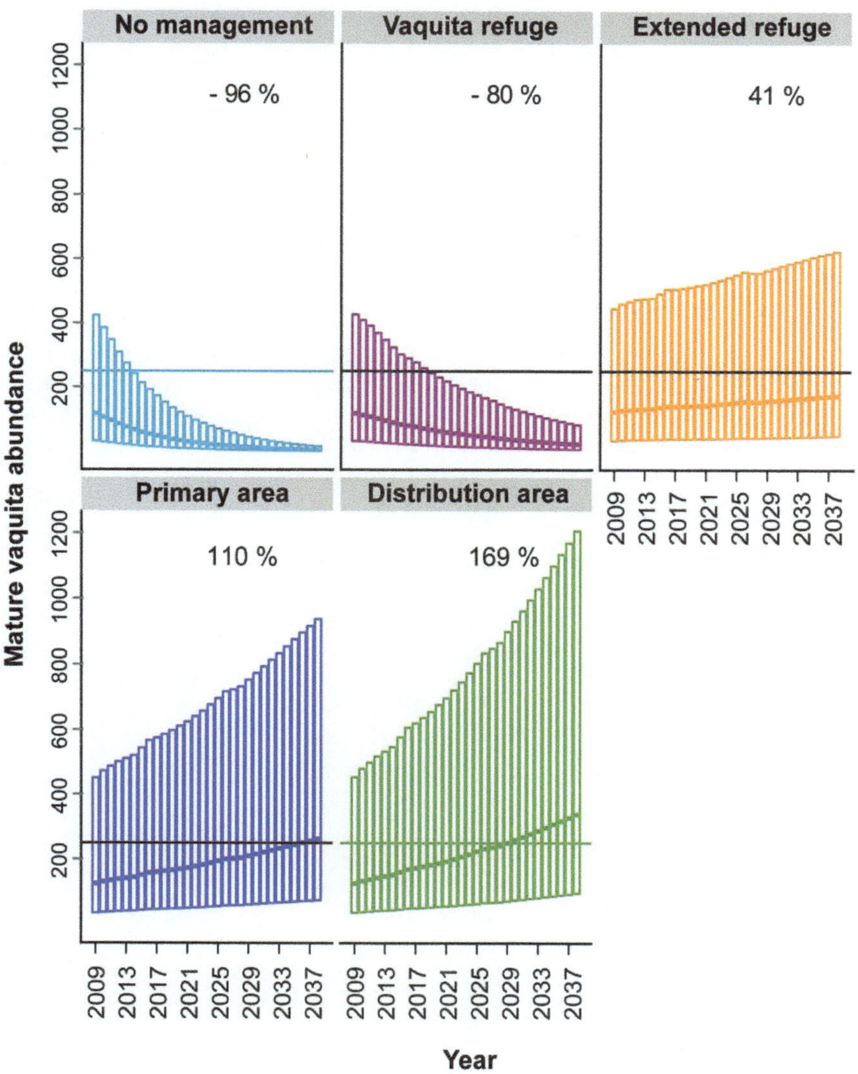

Figure 2. Mature vaquita abundance trajectories from 2009 to 2038 for each management scenario. Dark lines are abundance assuming an initial vaquita population of 245 individuals [6]. Bars are simulation results for 95% CI interval estimates for vaquita abundance, top of bars is an initial 884 vaquitas and bottom of bars is an initial 68 vaquitas. Numbers are percent change in abundance relative to 2009, considering an initial abundance of 245 vaquitas. Solid line indicates 250 mature individuals, a criterion for downlisting from the Critically Endangered category of the IUCN Red List of Threatened Species™ [11].

downward trend regardless of the initial vaquita abundance. For these scenarios, mature vaquita abundance in year 30 decreased significantly (−96%, No management; −80%, Vaquita refuge) relative to the start of the simulations. Given the strategic nature of Atlantis and the lack of demographic stochasticity in the model, the species can be considered functionally extinct in simulations where the population decreased and remained at low numbers for long periods. Atlantis does not include a population viability threshold but single-species models have found that persistent vaquita bycatch mortality would drive the population extinct [16,17].

Only scenarios with large spatial closures led to a sustained increase in vaquita. The highest increase occurred in the Distribution area scenario, where mature vaquita abundance in year 30 was nearly 3 times higher than 2008, considering an initial abundance of 245 vaquitas. Even in these scenarios that reduce vaquita bycatch, mature vaquita abundance remained under the population threshold necessary to downlist the species from critically endangered [11] for the first 20 years of the simulations except when considering an initial abundance of 884 individuals (upper 95% CI on 2008 abundance; [17]); this pattern emphasizes that current rates of bycatch mortality are unsustainable [8].

We constructed a vaquita abundance equilibrium curve (Fig. 3) to further analyze population dynamics under varying rates of bycatch mortality. This equilibrium curve assumed deterministic population behavior in growth, recruitment and mortality; however it also accounted for species interactions, which can affect population dynamics in unexpected ways [48]. As expected, abundance of vaquita was highest under zero fishing effort. The model predicted an ecological carrying capacity of 773 individuals. This predicted carrying capacity is on the low range of a few thousand individuals suggested by the genetic analysis [49]. We found that the population could withstand bycatch mortality of 0.03 $year^{-1}$, without vaquita abundance decreasing relative to initial abundance over the 30-year simulation. Bycatch mortality under the No management scenario, 0.17 $year^{-1}$, was more than five times this level of bycatch mortality that sustains the population at the current level.

In addition to changes in abundance, vaquita prey composition changed across management scenarios. Although vaquita still consumed the same 15 prey groups (Table S1), by year 30 diet proportions had changed (Fig. 4). In the most restrictive scenario, Distribution area, there was a decrease in mojarra (−8%) consumption relative to No management and increased predation by vaquita on small pelagic fish (132%), scorpionfish (71%), and flatfish (57%). Abundance of the groups that comprise less than 1% of vaquita diet also increased significantly in the Distribution area scenario relative to No management, including crabs and lobsters (323%), squid (142%), and totoaba (102%). Biomass of these prey groups increases as a result of decreased fishing pressure within spatial management areas.

Fishery value

We first show results for net benefit, considering a 10% reduction in catch of the shrimp light trawl fleet relative to the driftnet fleet. We also analyze the effect of the range of reductions (10–50%) when examining NPV.

Temporal trends in the undiscounted annual net benefit of fisheries were dominated by underlying biomass dynamics (Fig. 5). For artisanal net fleets (finfish gillnets, shrimp driftnets and shrimp light trawl), which are directly affected by simulated management actions, net benefit increased the most in the No management scenario over the 30-year simulation (17% relative to 2009), to an average of US$72±0.8 million annually over the last five years of the simulation. However, the Vaquita refuge was the highest-value scenario by the end of the simulation, 4% higher than No management (equivalent to US$3 million). In this scenario, both the finfish gillnet and shrimp driftnet fleets still operate over a large area while the shrimp light trawl is being introduced. Nonetheless, as the area of spatial restrictions over gillnets and driftnet increases, net benefit decreases. In the most restrictive scenario, at the end of the simulation, average annual net benefit is US$23 million lower than No-management (Distribution area; US$49±0.5 million). Fishery gains from spatial management are

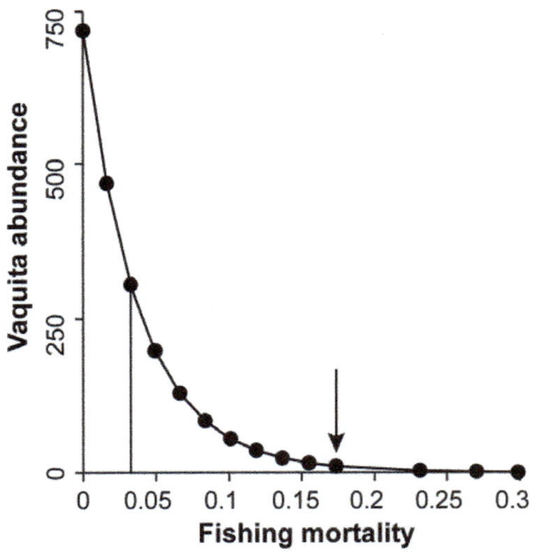

Figure 3. Vaquita abundance equilibrium curve. Vaquita abundance as a function of absolute bycatch mortality. Each bycatch mortality point in the figure is the ratio of average catch and biomass for the last 5 years (2033–2038) of a simulation run under a specific bycatch mortality value. Solid vertical line indicates maximum sustainable bycatch mortality. Arrow indicates bycatch mortality for the No management scenario.

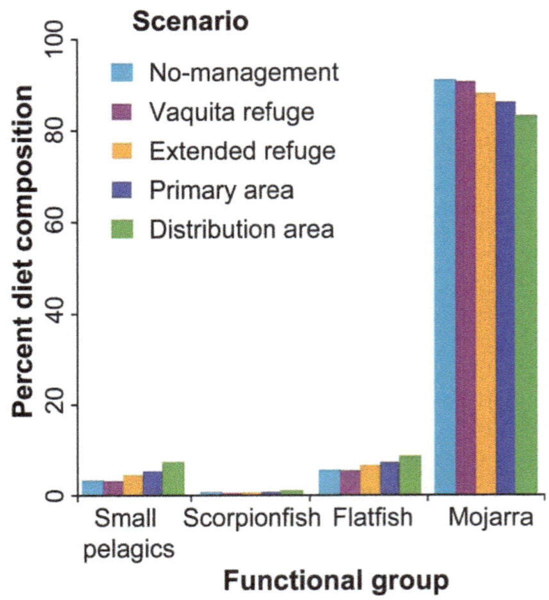

Figure 4. Vaquita prey composition for each management scenario. Prey that compose 99% of vaquita diet are shown.

Artisanal

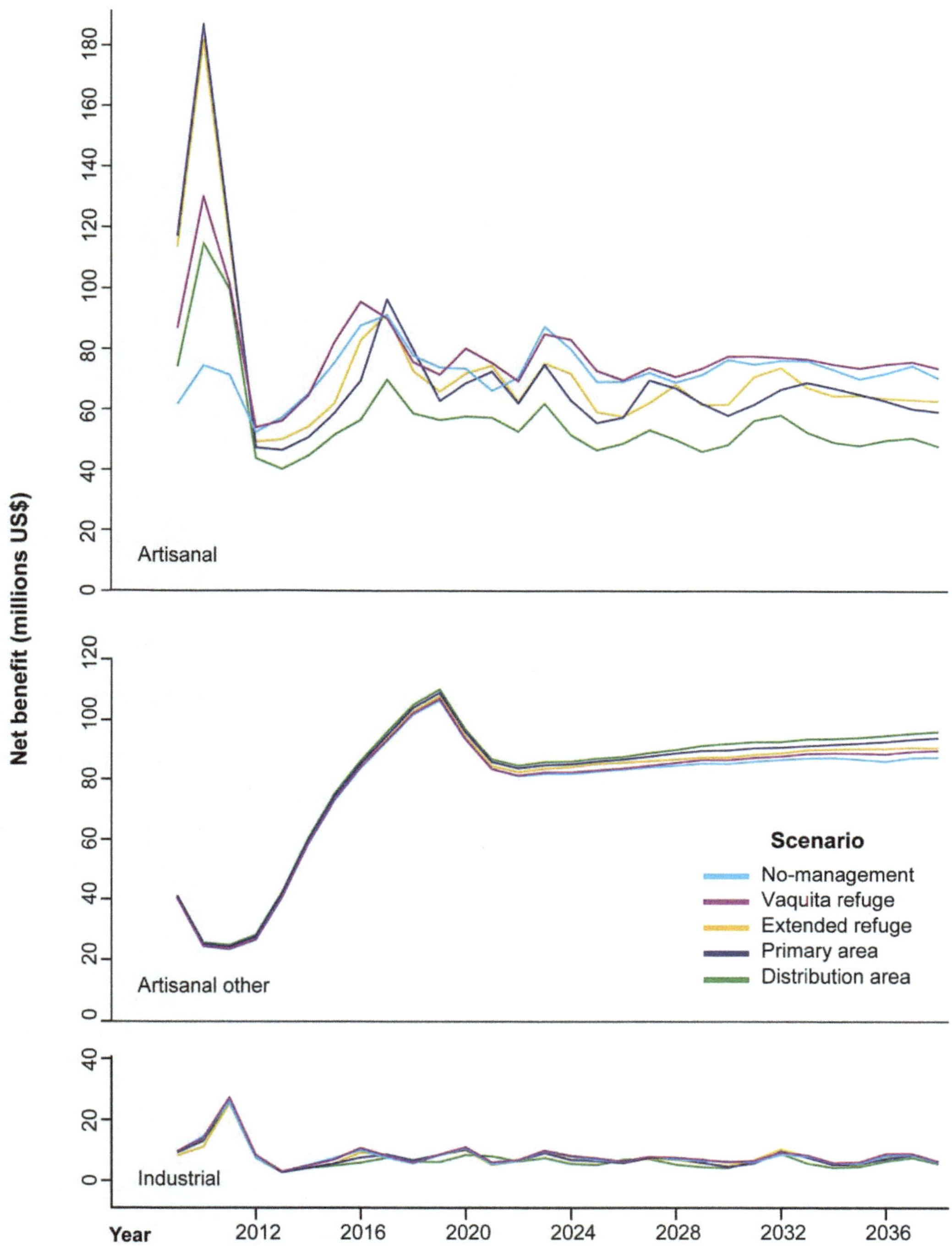

Figure 5. Undiscounted net benefit trajectories from 2009 to 2038 for each management scenario, considering a 10% reduction in shrimp catch for the light trawl fleet relative to the shrimp driftnet.

modest because model-wide net benefit is driven by abundant finfish (over 65% of net benefit across scenarios) rather than harvest of sedentary species and overfished species more likely to benefit from reserves (Fig. S2). However, the contribution of different functional groups to net benefit changes with spatial management; target groups that decrease relative to No-management include Gulf grouper, large pelagics, and drums and croakers (Fig. S3). A greater proportion of catch value under spatial management comes from herbivorous fish and sharks.

Other artisanal fleets, including longline, handline, traps, and diving, benefit from spatial management. For all scenarios, average net benefit to these other artisanal fleets at the end of the simulations outperformed No management by US$ 2–8 million (Fig. 5). The highest-value scenario by the end of the simulation was the Distribution area (US$95±0.4 million). These gains in net benefit are a result of higher catch from groups that experience decreased fishing pressure from gillnet fleets through spatial restrictions, including herbivorous fish, sharks, Amarillo snapper and drums and croakers (Fig. S3).

There was a modest decrease in the net benefit of industrial fleets with spatial management, <US$1 million annually in every case. On average, Net benefit of industrial fisheries in the last five

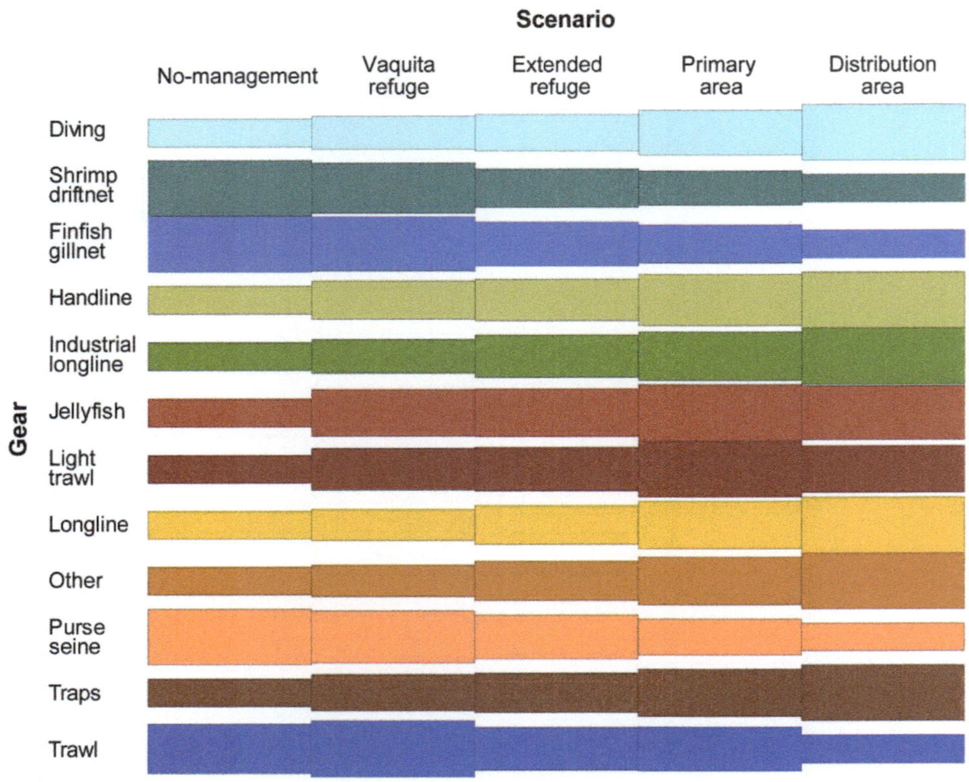

Figure 6. Comparison of undiscounted net benefit for gears between management scenarios. The tile plots shows average net benefit (value of catch minus costs) for gears (aggregated fleets) for the last five years of the 30 year simulation for each management scenario. The net benefit for each gear is scaled between the scenario with the highest and lowest value of catch.

years of the simulation was highest under the Vaquita refuge scenario (US\$7.1±0.7 million); but this was only 5% higher than No management. The losses that industrial fleets experience under spatial management are due to trophic effects. Net benefit (profits) from harvesting main target groups including shrimp (−23%) and small pelagics (−53%) decrease under spatial management relative to No management (Fig. S3), as a result of increased predation pressure (Fig. S4).

We assessed how the distribution of average undiscounted net benefits among gears (grouped fleets) for the last five years of the simulations compared amongst scenarios (Fig. 6). Net benefit for all gears except the shrimp driftnet, gillnets, and purse seine, was lower in No management than in one or more other scenarios that allowed additional stocks to rebuild and led to subsequent increases in catch. Several gears including diving, handline, traps, and longline had highest value of catch under the restrictive Distribution area scenario.

Summing over all fleets, there was an inverse relationship between NPV and the abundance of mature vaquita. In scenarios without extensive spatial management (No management and Vaquita refuge), mature vaquita abundance remained well below the population threshold necessary to downlist the species from Critically Endangered [11]. NPV was lower in No management than all scenarios with spatial management except Distribution area (Fig. 7). The maximum NPV from fisheries, US\$ 3billion, was achieved under the Vaquita refuge scenario. This value represents the discounted profit stream over the next 30 years. As the area under spatial management increased relative to the Vaquita refuge, NPV decreased 1–12% primarily due to declines in catch from net fleets. The distribution of NPV values suggests there may be convex relationship between NPV and mature vaquita abundance, possibly indicating a Pareto efficiency where joint benefits are maximized. The Primary area scenario, which represents the current recommendation of the International Committee for the Recovery of the Vaquita [25], could be close to this maximum. In this scenario NPV was US\$ 2.96 billion and mature vaquita abundance was above the population threshold considering an initial abundance of 245 vaquitas.

We found that fishers' experience and training in operating the shrimp light trawl could affect NPV [25,43].The 50% reduction in shrimp catch of the light trawl fleet, relative to the shrimp driftnet fleet, could account for US\$193 million (Primary area) to US\$ 98 million (Vaquita refuge) in lost NPV, relative to the best-case scenario (10% reduction; Table 2).

Discussion

Our study illustrates potential trade-offs between economic and conservation objectives when reducing bycatch threats for an endangered cetacean. Trade-offs are inherent in conservation, as any actions intended to reduce pressures from development or resource extraction will involve costs to local inhabitants that rely on those activities [50]; nonetheless trade-offs are rarely recognized or debated explicitly [51]. The rhetoric of win-win solutions underlies many popular conservation programs, although in practice gains for both biodiversity and human well-being are difficult to attain [52]. For example, the harvest of non-timber forest products was long thought to be both more profitable than other forest uses such as logging or agriculture and less ecologically destructive, but recent analyses show that there is uncertainty on the impact of non-timber forest product harvesting and on income generation [50].

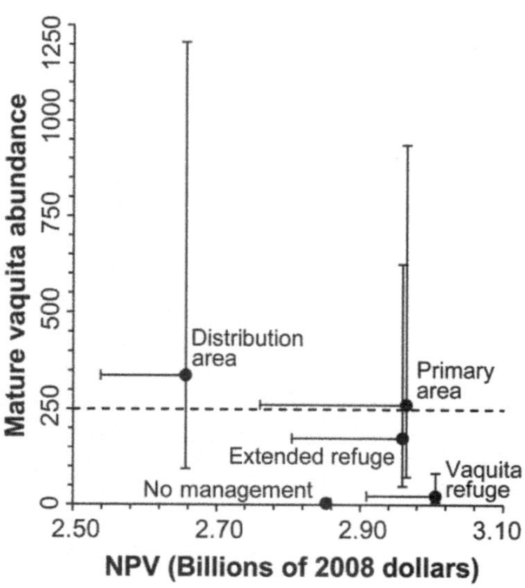

Figure 7. Net present value of fisheries catch for each management scenario, plotted against mature vaquita abundance at the end of the 30 year simulation. Net present value is discounted benefit over time with an alternate rate of return from bank investment (δ = 3.6%). Markers are results for initial vaquita abundance of 245 individuals and a 10% reduction in shrimp catch in the light trawl relative to the shrimp driftnet. Vertical confidence intervals are mature vaquita abundance results for initial abundance of 68 and 884 and horizontal intervals show 50% reductions in shrimp catch. Dashed line indicates 250 mature individuals, a criterion for downlisting from the Critically Endangered category of the IUCN Red List of Threatened Species^{TM} [11].

In marine environments, there is empirical [53] and theoretical evidence [54] that spatial closures can result in win-win solutions, with both gains for conservation and increased economic benefits from fishery enhancement, as fish spillover from reserves to harvested areas. We found a more complex result. Our model results suggest that the Primary area scenario, which excludes finfish gillnets and shrimp driftnets from the vaquita distribution area as of 2008 (5339 km^2) [6] could be a good compromise between vaquita conservation and fisheries. Under this scenario mature vaquita increase two-fold relative to 2008 while bycatch mortality remained at 0.03 year^{-1}, equal to the maximum that the population could withstand without decreasing abundance over the 30 year simulation. From a fisheries perspective, NPV in the Primary area scenario decreased moderately (US$ 42 million) relative to the best-performing scenario. Nonetheless, the outcomes for individual fisheries varied. Finfish gillnet fisheries experienced large losses in net benefit under spatial management, while longline, handline, traps, and dive fisheries benefited from spatial management and net benefit of industrial fisheries decreased slightly. Therefore, the economic cost of proposed management actions will be unequally partitioned within the fishing community. Fishers will likely respond in changing fishing tactics and strategies, within (or outside) the regulatory framework, according to the type of alternate fishery available, fisher experience, risk perception and opportunity cost [55].

Our results agree with previous authors [16,17,19,56] who concluded that preventing extinction of vaquita would require eliminating all nets and trawls from its distribution area or known range; this promises to be an expensive, politically unpopular process that has not been tried elsewhere on this scale [3], and would likely have high social and cultural costs [7]. Vaquita recovery will therefore be contingent on implementation of management actions, particularly economic alternatives provided to fishers [7,57]. In a policy context, the value of those alternative income streams can be measured against the expected benefits (NPVs) provided here.

Table 2. Net present value (NPV; billions of 2008 dollars per year) for spatial management scenarios under various reductions of shrimp catch.

Scenario	Reduction in shrimp catch				
	10%	20%	30%	40%	50%
Vaquita refuge	3.006	2.982	2.958	2.934	2.909
Extended refuge	2.960	2.922	2.882	2.844	2.805
Primary area	2.965	2.913	2.867	2.820	2.761
Distribution area	2.657	2.628	2.599	2.570	2.538

We applied a range of reductions in shrimp catch in the light trawl fleet relative to the shrimp driftnet fleet (10%–50%), as shrimp catch in the light trawl will be dependent on fishers' experience and training [25,43]. NPV for the No management scenario was 2.853.

As with vaquita, many conservation projects around the world center on endangered and threatened species [1]; where preventing extinction of one species is costly and the outcome is uncertain. Nonetheless there are examples, such as the Steller sea lion in the North Pacific [58] where the risk of extinction for the population has declined following the implementation of management measures that restricted human activities. Incorporating an evaluation of associated ecosystem-level benefits, the probability of success, and the opportunity costs of conservation actions (i.e. alternative goals that could be achieved with the same resources) would help managers reconcile potentially conflicting benefits and values [59]. Ultimately, as in this case, society may often be required to confront the hard choices between protection of charismatic species and local livelihoods; end-to-end models such as the one applied here will be useful to evaluate the trade-offs.

Supporting Information

Figure S1 Initial vaquita spatial abundance distribution considering 245 individuals. Grey lines are Atlantis polygon geometry. Abundance per polygon changes dynamically during simulations dependent on predator-prey and local habitat influences.

Figure S2 Undiscounted net benefit for each gear and management scenario. Values are averages for the last 5 years of the 30-year simulations (± SE). (*) indicates the scenario with the highest value for each gear. See text for calculation of net benefit.

Figure S3 Percent change in undiscounted net benefit deriving from harvest of selected functional groups across scenarios, relative to No management. Net benefit is average for the last 5 years of the 30-year simulations.

Figure S4 Percent increase in predation for selected target groups across scenarios relative to No management, in the last year of the 30-year simulation.

Text S1 Management actions for vaquita conservation.

Text S2 Historical vaquita abundance trajectory (1993–2008).

Table S1 Vaquita diet based on averaged values from the ecosystem models developed by Morales-Zárate [1] and Lozano [2]; based on data analyzed by [3,4]. Diet for juveniles was assumed similar to adults. Values are proportion of total diet for each functional group used at the start of the simulations, realized diets will vary through time and space based on prey availability.

Table S2 Catch by functional group used as baseline in the No management scenario. Catches are the average of the 2000–2007 model catch series from Ainsworth et al. [1] summed for all fleets and modified as described in Ainsworth et al. [2]. Vaquita mortality was set at 0.15 year-1, the median estimate prior to 2007 [3]. Since the publication of Ainsworth et al. [1], the model has been simplified to only include catch for a generic Penaeid shrimp group rather than for separate shrimp groups. Ainsworth et al. [1] provide species composition for each functional group.

Table S3 Fishery fleets in the Atlantis Northern Gulf of California model. Fleets are defined based on gears, targets, bycatch, home ports and fishery utilization areas. Table indicates ports, functional groups targeted out of 63 in the model and gears. See Ainsworth et al. [1] and Ainsworth [2] for more details on fleets and functional groups used.

Table S4 Catch per fleet used as a baseline. Asterisk indicates group is bycatch; fleet name is in bold. Vaquita catch was updated considering a mortality rate of 0.15 year-1. Bycatch composition for the Upper Gulf shrimp driftnet fleet was updated using more monitoring data [1]. Otherwise catch per fleet values are unmodified from Ainsworth et al. [2,3].

Table S5 Catch for the shrimp light trawl fleet. Bycatch composition was based on INAPESCA & NMFS [1] and considering a reduction in shrimp catch of 10% [1] and an increase in the ratio of shrimp to bycatch (other than vaquita) of 11% [1,2]. Vaquita bycatch was set to 0 for this fleet.

Table S6 Sources for estimation of industrial fleets' cost rate of in the Northern Gulf of California. Average cost rate of 0.956 was used to calculate net present value.

Table S7 Price matrix for Atlantis functional groups. Values are dollars tonne-1 for 2010 or the most recent year for which data was available. For Penaeid shrimp, prices were set by fleet, weighted by the amount of blue, brown and Japanese shrimp caught. Information from National statistics for Sonora and Baja California, (Anuarios Estadísticos www.inegi.org.mx), state statistics for Sonora (www.oeidrus-sonora.gob.mx/), and port-level data for both states (unpublished data, A. Cinti, The University of Arizona, acinti@email.arizona.edu). Values were converted from Mexican pesos to dollars using the exchange rate from 2005–2010 (www.x-rates.com).

Acknowledgments

We acknowledge researchers from the PANGAS research initiative (http://www.pangas.arizona.edu) who participated in development of the Atlantis model for the Northern Gulf of California. We thank D. Holland who advised us on the use of interest rates, S. Perez-Valencia whose input helped with scenario development, A. Munguia-Vega who provided information on vaquita population sizes and R. Estrella and personnel at the "Gerardo Cornejo Murrieta" library for providing access to bibliographic material. The manuscript was substantially improved by the constructive comments of T. Gerrodette, C. Harvey, M. Ford and one anonymous reviewer.

Author Contributions

Conceived and designed the experiments: HML CHA ICK PSL. Performed the experiments: HML CHA ICK. Analyzed the data: HML CHA ICK. Contributed reagents/materials/analysis tools: EAF CHA HML ICK. Wrote the paper: HML CHA ICK PSL EAF. Built the Ecosystem model: CHA ICK EAF.

References

1. Hooker SK, Gerber LR (2004) Marine reserves as a tool for ecosystem-based management: The potential importance of megafauna. Bioscience 54: 27–39.
2. Lewison RL, Crowder LB, Read AJ, Freeman SA (2004) Understanding impacts of fisheries bycatch on marine megafauna. Trends Ecol Evol 19: 598–604. doi:10.1016/j.tree.2004.09.004.
3. Read AJ (2008) The looming crisis: interactions between marine mammals and fisheries. J Mammal 89: 541–548. doi:10.1644/07-MAMM-S-315R1.1.
4. Mardle S, Pascoe S (2002) Modelling the effects of trade-offs between long and short-term objectives in fisheries management. J Environ Manag 65: 49–62. doi:10.1006/jema.2001.0518.
5. Cheung WWL, Sumaila UR (2008) Trade-offs between conservation and socio-economic objectives in managing a tropical marine ecosystem. Ecol Econ 66: 193–210. doi:10.1016/j.ecolecon.2007.09.001.
6. Gerrodette T, Taylor BL, Swift R, Rankin S, Jaramillo-Legorreta AM, et al. (2011) A combined visual and acoustic estimate of 2008 abundance, and change in abundance since 1997, for the vaquita, *Phocoena sinus*). Mar Mammal Sci 27: E97–E100. doi:10.1111/j.1748-7692.2010.00438.x.
7. Aragón-Noriega EA, Rodríguez-Quiroz G, Cisneros-Mata MA, Ortega-Rubio A (2010) Managing a protected marine area for the conservation of critically endangered vaquita (*Phocoena sinus* Norris, 1958) in the Upper Gulf of California. Int J Sust Dev World 17: 410. doi:10.1080/13504509.2010.500823.
8. Rojas-Bracho L, Reeves RR, Jaramillo-Legorreta A (2006) Conservation of the vaquita *Phocoena sinus*. Mammal Rev 36: 179–216.
9. SEMARNAT (2008) Programa de Acción para la Conservación de la Especie: Vaquita (*Phocoena sinus*). Estrategia integral para el manejo sustentable de los recursos marinos y costeros del Alto Golfo de California. Mexico, D.F.: Secretaría del Medio Ambiente y Recursos Naturales. 106 p.
10. Vidal O (1995) Population biology & incidental mortality of the Vaquita, *Phocoena sinus*. R Int Whal Com (Spec Issue) 16: 247–272.
11. IUCN (2010) IUCN Red List of Threatened Species. Version 2010.3. Available:www.iucnredlist.org. Accessed 2 September 2010.
12. Avila-Forcada S, Martínez-Cruz AL, Muñoz-Piña C (2012) Conservation of vaquita marina in the Northern Gulf of California. Mar Policy 36: 613–622. doi:10.1016/j.marpol.2011.10.012.
13. SGPA (2009) Resolutivo a la Manifestacion de Impacto Modalidad regional. Pesca de camarón en el Alto Golfo de California. 26SO2009P0006. Promovente Armadores Unidos de Puerto Peñasco, S.A. de C.V. Subsecretaria de Gestión para la Protección Ambiental. Dirección General de Impacto y Riesgo Ambiental. S.G.P.A. – DGIRA - DG. 6690.09. 89 p.
14. SGPA (2010) Resolutivo a la Manifestacion de Impacto Modalidad regional. Pesca sustentable del camarón azul y cafe y reconversión productiva, en el Alto Golfo de California. 26SO2010PD053. Promovente Federacion de Armadores del Piedron del Alto Golfo S.C. de R.L.de C.V. Subsecretaria de Gestión para la Protección Ambiental. Dirección General de Impacto y Riesgo Ambiental. S.G.P.A. – DGIRA - DG. 9150.10. 58 p.
15. INE WWF (2006) Conservación de la Vaquita y su hábitat en el Alto Golfo de California. Fondo de compensación económica para la pesca con redes de enmalle y el fomento de alternativas a la pesca en el Alto Golfo de California. Instituto Nacional de Ecología. Dirección General de Investigación en Política y Economía Ambiental. WWF programa México. 46 p.
16. Ortiz I (2002) Impacts of fishing and habitat alteration on the population dynamics of the vaquita, (*Phocoena sinus*) [M.S. Thesis]. Seattle, WA: University of Washington, School of Aquatic and Fishery Sciences. 72 p.

17. Gerrodette T, Rojas-Bracho L (2011) Estimating the success of protected areas for the vaquita, *Phocoena sinus*. Mar Mammal Sci 27: E101–E125. doi:10.1111/j.1748-7692.2010.00449.x.
18. Díaz-Uribe JG, Arreguín-Sánchez F, Lercari-Bernier D, Cruz-Escalona VH, Zetina-Rejón MJ, et al. (2012) An integrated ecosystem trophic model for the North and Central Gulf of California: An alternative view for endemic species conservation. Ecol Model 230: 73–91. doi:10.1016/j.ecolmodel.2012.01.009.
19. Lozano H (2006) Historical ecosystem modelling of the Upper Gulf of California (Mexico): Following 50 years of change. [PhD dissertation]. Vancouver, British Columbia: The University of British Columbia, The Faculty of Graduate Studies. 266 p.
20. Christensen V, Walters CJ (2004) Ecopath with Ecosim: methods, capabilities and limitations. Ecol Model 172: 109–139. doi:10.1016/j.ecolmodel.2003.09.003.
21. Fulton EA, Smith ADM, Smith DC (2007) Alternative management strategies for Southeast Australian Commonwealth Fisheries: Stage 2: Quantitative Management Strategy Evaluation. Final report June 2007. Canberra, AU: Australian Fisheries Management Authority. Fisheries Research and Development Corporation. 398 p.
22. Fulton EA, Link JS, Kaplan IC, Savina-Rolland M, Johnson P, et al. (2011) Lessons in modelling and management of marine ecosystems: The Atlantis experience. Fish Fisheries 12: 171–188.
23. Rose KA, Allen JI, Artioli Y, Barange M, Blackford J, et al. (2010) End-to-end models for the analysis of marine ecosystems: challenges, issues, and next steps. Mar Coast Fish Dyn Mgmt Ecosys Sci 2: 115. doi:10.1577/C09-059.1.
24. Hutton T, Thebaud O, Fulton EA, Pascoe S, Innes J, et al. (2010) Use of economic incentives to manage fisheries bycatch. An application to key sectors in Australia's Southern and Eastern scalefish and shark fisheries. Final report August 2010 Australia: Commonwealth Scientific and Industrial Research Organisation (CSIRO). 398 p.
25. CIRVA (2012) Report of the fourth meeting of the International Committee for the Recovery of the Vaquita (CIRVA). Final report 21–24 February 2012. Ensenada, Baja California, México: International Committee for the Recovery of the Vaquita. 36 p.
26. Taylor BL, Gerrodette T (1993) The uses of statistical power in conservation biology: The Vaquita and Northern Spotted Owl. Conserv Biol 7: 489–500.
27. Ainsworth C, Kaplan IC, Levin PS, Cudney-Bueno R, Fulton EA, et al. (2011) Atlantis model development for the Northern Gulf of California. May 2011 Seattle, WA: U.S. Department of Commerce. National Oceanic and Atmospheric Administration. National Marine Fisheries Service. NOAA Technical Memorandum NMFS-NWFSC-110. 293 p.
28. Ainsworth CH, Morzaria-Luna H, Kaplan IC, Levin PS, Fulton EA (2012) Full compliance with harvest regulations yields ecological benefits: Northern Gulf of California case study. J Appl Ecol 49: 63–72. doi:10.1111/j.1365-2664.2011.02064.x.
29. Fulton EA (2010) Approaches to end-to-end ecosystem models. J Marine Syst 81: 171–183. doi:10.1016/j.jmarsys.2009.12.012.
30. Fulton EA, Parslow JS, Smith ADM, Johnson CR (2004) Biogeochemical marine ecosystem models. 2. The effect of physiological data on model performance. Ecol Model 173: 371–406.
31. Lluch-Cota SE, Parés-Sierra A, Magaña-Rueda VO, Arreguín-Sánchez F, Bazzino G, et al. (2010) Changing climate in the Gulf of California. Prog Oceanogr 87: 114–126. doi:10.1016/j.pocean.2010.09.007.
32. Rojas-Bracho L, Reeves RR, Taylor BL, Jaramillo-Legorreta A (2008) *Phocoena sinus*. IUCN 2010. IUCN Red List of Threatened Species. Version 2010.3. Available:<www.iucnredlist.org>. Downloaded on 24 September 2010.
33. Trites A, Pauly D (1998) Estimating mean body masses of marine mammals from maximum body lengths. Can J Zool 76: 886–896.
34. Morales-Zarate MV, Arreguin-Sanchez F, Lopez-Martinez J, Lluch-Cota SE (2004) Ecosystem trophic structure and energy flux in the Northern Gulf of California, Mexico. Ecol Model 174: 331–345.
35. Gerrodette T, Fleischer LA, Pérez-Cortés H, Villa-Ramírez B (1995) Distribution of the vaquita, *Phocoena sinus*, based on sightings from systematic surveys. R Int Whal Com (Spec Issue) 16: 273–281.
36. Hohn AA, Read AJ, Fernandez S, Vidal O, Findley LT (1996) Life history of the vaquita, *Phocoena sinus* (Phocoenidae, Cetacea). J Zoo 239: 235–251. doi:10.1111/j.1469-7998.1996.tb05450.x.
37. D'Agrosa C, Vidal O, Graham WC (1995) Mortality of the Vaquita (*Phocoena sinus*) in Gillnet Fisheries Durin 1993–94. R Int Whal Com (Spec Issue) Special Issue 16.
38. Jaramillo-Legorreta A, Rojas-Bracho L, Gerrodette T (1999) A new abundance estimate for vaquitas: first step for recovery. Mar Mammal Sci 15: 957–973.
39. Barlow J, Gerrodette T, Silber G (1997) First estimates of vaquita abundance. Mar Mammal Sci 13: 44–55. doi:10.1111/j.1748-7692.1997.tb00611.x.
40. Acuacorp de Hidalgo (2009) Pesca ribereña multiespecífica. Manifestación de Impacto Ambiental modalidad regional. Promovente: Federación de Sociedades Cooperativas de Producción Pesquera Ribereña del Puerto de San Felipe, S. de R.L. de C.V.. Clave 02BC2009P0005. 337 p.
41. INAPESCA (2011) Evaluación biotecnológica de la red de arrastre prototipo "RS-INP-MEX" para captura de camarón en el Alto Golfo de California. Final report April 2011. México: SAGARPA. INAPESCA. 22 p.
42. Lluch-Cota SE, Aragón-Noriega EA, Arreguín-Sánchez F, Aurioles-Gamboa D, Baustista-Romero JJ, et al. (2007) The Gulf of California: review of ecosystem status and sustainability challenges. Prog Oceanogr 73: 1–26.
43. INAPESCA NMFS (2012) Fishing technology research collaboration: Efficiency and selectivity of two trawl nets design to fish blue shrimp (*Litopenaeus stylirostris*) in the artisanal shrimp fishery of the Upper Gulf of California. 9 p.
44. PRONATURA (2010) Pesca experimental de camarón con la Red de Arrastre Prototipo RS-INP-MEX en el Alto Golfo de California: Programa de observadores de pesca en acciones de desarrollo tecnológico PACE VAQUITA 2009. Final report. February 2010: PRONATURA Noroeste. A.C. CEC-CCA-CCE. CEDO Intercultural. NOS Noroeste Sustentable. 31 p.
45. Pérez-Valencia S, Gorostieta-Monjaraz M, Castañeda-Fernández de Lara V, Loaiza-Villanueva R, Turk-Boyer P, et al. (2011) Manifestación de Impacto Ambiental para la pesca ribereña responsable en la Reserva de la Biosfera Alto Golfo de California y Delta del Río Colorado: Costa Este. Puerto Peñasco, Sonora: Centro Intercultural de Estudios de Desiertos y Océanos, A.C. 221 p.
46. Ainsworth CH, Sumaila UR (2005) Intergenerational valuation of fisheries resources can justify long-term conservation: a case study in Atlantic cod (*Gadus morhua*). Can J Fish Aquat Sci 62: 1104–1110. doi:10.1139/f05-010.
47. Arreguín-Sánchez F, Beléndez-Moreno L, Gómez-Humarán IM, Solana Sansores R, Rangel Dávalos C (2006) Sustentabilidad y pesca responsable en Mexico; Evaluación y manejo. Secretaria de Agricultura, Ganadería, Desarrollo Rural, Pesca y Alimentación. 560 p.
48. Ainsworth CH, Varkey DA, Pitcher TJ (2008) Ecosystem simulations supporting ecosystem-based fisheries management in the Coral Triangle, Indonesia. Ecol Model 214: 361–374. doi:10.1016/j.ecolmodel.2008.02.039.
49. Munguia-Vega A, Esquer-Garrigos Y, Rojas-Bracho L, Vazquez-Juarez R, Castro-Prieto A, et al. (2007) Genetic drift vs. natural selection in a long-term small isolated population: major histocompatibility complex class II variation in the Gulf of California endemic porpoise (*Phocoena sinus*). Mol Ecol 16: 4051–4065. doi:10.1111/j.1365-294X.2007.03319.x.
50. Chan KMA, Pringle RM, Ranganathan J, Boggs CL, Chan YL, et al. (2007) When agendas collide: Human welfare and biological conservation. Conserv Biol 21: 59–68. doi:10.1111/j.1523-1739.2006.00570.x.
51. Leader-Williams N, Adams WM, Smith RJ (2010) Trade-offs in Conservation: Deciding What to Save. Oxford: John Wiley and Sons. 429 p.
52. McShane TO, Hirsch PD, Trung TC, Songorwa AN, Kinzig A, et al. (2011) Hard choices: Making trade-offs between biodiversity conservation and human well-being. Biol Conserv 144: 966–972. doi:16/j.biocon.2010.04.038.
53. Levin PS, Kaplan I, Grober-Dunsmore R, Chittaro PM, Oyamada S, et al. (2009) A framework for assessing the biodiversity and fishery aspects of marine reserves. J Appl Ecol 46: 735–742. doi:10.1111/j.1365-2664.2009.01667.x.
54. Sanchirico JN, Malvadkar U, Hastings A, Wilen JE (2006) When are no-take zones an economically optimal fishery management strategy? Ecol Appl 16: 1643–1659.
55. Salas S, Gaertner D (2004) The behavioural dynamics of fishers: management implications. Fish Fisheries 5: 153–167. doi:10.1111/j.1467-2979.2004.00146.x.
56. Jaramillo-Legorreta A, Rojas-Bracho L, Brownell RL, Read AJ, Reeves RR, et al. (2007) Saving the vaquita: Immediate action, not more data. Conserv Biol 21: 1653–1655.
57. Bobadilla M, Alvarez-Borrego S, Avila-Foucat S, Lara-Valencia F, Espejel I (2011) Evolution of environmental policy instruments implemented for the protection of totoaba and the vaquita porpoise in the Upper Gulf of California. Environ Sci Policy 14: 935–1220. doi:16/j.envsci.2011.06.003.
58. Boyd IL (2010) Assessing the effectiveness of conservation measures: Resolving the "wicked" problem of the Steller sea lion. Biol Conserv 143: 1664–1674. doi:10.1016/j.biocon.2010.04.006.
59. Bottrill MC, Joseph LN, Carwardine J, Bode M, Cook C, et al. (2008) Is conservation triage just smart decision making? Trends Ecol Evol 23: 649–654. doi:16/j.tree.2008.07.007.
60. SEMARNAT, CONANP (2010) Lineamientos para el otorgamiento de apoyos del Programa de Acción para la Conservación de la Especie: Vaquita (*Phocoena sinus*). Ejercicio fiscal 2010. Available:http://www.conanp.gob.mx/contenido/pdf/LINEAMIENTOS%20PACE%20VAQUIITA%2020100001__.pdf. Accessed July 25, 2012.

2

Estimating Carbon Flux Phenology with Satellite-Derived Land Surface Phenology and Climate Drivers for Different Biomes: A Synthesis of AmeriFlux Observations

Wenquan Zhu[1,2]*, Guangsheng Chen[3], Nan Jiang[2], Jianhong Liu[2], Minjie Mou[2]

1 State Key Laboratory of Earth Surface Processes and Resource Ecology, Beijing Normal University, Beijing, China, **2** College of Resources Science and Technology, Beijing Normal University, Beijing, China, **3** Environmental Sciences Division, Oak Ridge National Laboratory, Oak Ridge, Tennessee, United States of America

Abstract

Carbon Flux Phenology (CFP) can affect the interannual variation in Net Ecosystem Exchange (NEE) of carbon between terrestrial ecosystems and the atmosphere. In this study, we proposed a methodology to estimate CFP metrics with satellite-derived Land Surface Phenology (LSP) metrics and climate drivers for 4 biomes (i.e., deciduous broadleaf forest, evergreen needleleaf forest, grasslands and croplands), using 159 site-years of NEE and climate data from 32 AmeriFlux sites and MODIS vegetation index time-series data. LSP metrics combined with optimal climate drivers can explain the variability in Start of Carbon Uptake (SCU) by more than 70% and End of Carbon Uptake (ECU) by more than 60%. The Root Mean Square Error (RMSE) of the estimations was within 8.5 days for both SCU and ECU. The estimation performance for this methodology was primarily dependent on the optimal combination of the LSP retrieval methods, the explanatory climate drivers, the biome types, and the specific CFP metric. This methodology has a potential for allowing extrapolation of CFP metrics for biomes with a distinct and detectable seasonal cycle over large areas, based on synoptic multi-temporal optical satellite data and climate data.

Editor: Bruno Hérault, Cirad, France

Funding: This work was supported by the National Natural Science Foundation of China (Grant No. 41371389), the State Key Laboratory of Earth Surface Processes and Resource Ecology (Grant No. 2013-ZY-14), and the Fundamental Research Funds for the Central University. The funders had no role in study design, data collection and analysis, decision to publish, or preparation of the manuscript.

Competing Interests: The authors have declared that no competing interests exist.

* E-mail: zhuwq75@bnu.edu.cn

Introduction

Vegetation phenology plays an important role in adjusting the annual Net Ecosystem Exchange (NEE) (see Acronym S1 in supporting information for a list of acronyms and definitions used in this paper) of carbon between terrestrial ecosystems and the atmosphere [1–5]. The interannual variation in ecosystem productivity caused by vegetation phenology shifts was widely investigated by field studies [6–9] and ecosystem models [10–14]. An earlier start or/and a later end of vegetation growing season can extend the period of photosynthesis, and thus increased primary productivity is expected. Indeed, some previous studies have shown a positive effect of Growing Season Length (GSL) on net productivity (e.g., 5.9 g C•m^{-2}•d^{-1} in a deciduous forest [15] and around 4 g C•m^{-2}•d^{-1} in a subtropical forest stand [16]). Moreover, the length of Carbon Uptake Period (CUP) has much predictive power about the spatial variation of annual NEE. For example, the length of CUP can explain 80% of the spatial variance in annual NEE for deciduous forests across a latitudinal and continental gradient [17].

There are currently numerous data sources available for estimating the timing of recurrent vegetation phenology transitions, such as the ground-, satellite- and eddy covariance flux-based data sources [18]. Land Surface Phenology (LSP) is defined as the study of the timing of recurring seasonal pattern of variation in vegetated land surfaces observed from synoptic sensors [19,20]. Satellite-based LSP is characterized by the Start (SOS) and End (EOS) of growing Season, which are closely related to vegetation growth or photosynthesis. Carbon Flux Phenology (CFP) is defined as the detrended zero-crossing timing of NEE from a source to a sink in spring and *vice versa* in autumn [3,4,18,19]. CFP is characterized by the Start (SCU) and End (ECU) of Carbon Uptake, which are closely related to the difference between growth and respiration. LSP allows the determination of GSL or the duration of canopy coverage from the difference between EOS and SOS, while CFP allows the determination of CUP from the difference between ECU and SCU. The CUP is controlled by GSL, but is not identical because growth will typically commence and terminate some time before and after the NEE changes sign in spring and autumn, respectively [19,21]. White & Nemani [13] found that there was a strong relationship between NEE and CUP, but a very weak relationship between NEE and GSL for deciduous forests. Thus, CUP is a potentially useful indicator of annual carbon sequestration [3]. However, the application of CUP is hindered by the limited number of flux towers and the distribution and footprint of these flux towers [3,19,21]. Although more than 500 tower sites from approximately 30 regional networks across 5 continents are currently operating on a long-term basis, these globally distributed eddy flux sites sample only a small subset of the Earth's biomes, disturbance regimes, and land management systems. Thus, estimation of CUP over large areas remains challenging [19,21,22].

Some limited attempts have been made to estimate CFP dates beyond the footprints of flux towers [18,19,21,22]. Using over 30 site-years of data from 12 eddy flux sites, Baldocchi *et al.* [22] found that 64% of variance in SCU can be explained by the date when soil temperature matched the mean annual air temperature. Remote sensing provides spatially comprehensive measures of ecosystem activity and therefore is a potentially powerful tool to allow extrapolation of CUP over large areas. To test the capabilities of remote observations in estimating CUP, Churkina *et al.* [21] related the GSL from remotely sensed data to the CUP from eddy flux tower measurements and found a strong relationship between them. However, a comparison of multiple phenology data sources indicated that no single source of phenological data was able to accurately describe annual patterns of flux phenology [18]. Therefore, Gonsamo et al. [19] combined LSP dates with the mean monthly surface temperature derived from remote sensing observations to predict CUP. Their results indicated that remote sensing-derived multiple surface variables can explain CUP variability by more than 70% in spring and autumn. However, this CUP determination approach is just based on four selected temperate and boreal deciduous forest CO_2 flux tower sites. A more comprehensive analysis, based on multi-year data from eddy flux sites across large areas for various biome types, is still expected. Moreover, improved estimation of LSP dates combined with optimal climate drivers may further enhance the CUP estimation performance.

Using data from a large number of AmeriFlux sites, this study aims to estimate CFP metrics with satellite-derived LSP metrics and climate drivers for different biomes, including deciduous broadleaf forest, evergreen needleleaf forest, grasslands and croplands. We first evaluated different LSP retrieval methods and Vegetation Index (VI) products based on the observed CFP dates and selected the best performing method and VI product to retrieve LSP dates as the explanatory variables in estimating both SCU and ECU. Then, we carried out a sensitive analysis to search the optimal explanatory climate drivers for the estimation of SCU and ECU. Finally, the estimated LSP dates and the selected optimal explanatory climate drivers were combined to estimate CFP dates, and a comprehensive discussion was given to highlight the limitations and potentials of the proposed methodology.

Data and methods

Data and pre-processing

Site carbon flux and meteorological data. The daily NEE ($g\ C{\bullet}m^{-2}{\bullet}d^{-1}$), air temperature (°C) and precipitation (mm) data used in this study were derived from the post-processed Level 4 product (available at: http://daac.ornl.gov/FLUXNET/fluxnet.shtml) of the AmeriFlux sites (Figure 1, Dataset S1). The covered period for the product was generally from 1995 to 2007 but depending on the specific site. For example, the acquired NEE and meteorological data were from 1998 to 2007 for the Niwot Ridge site, while they were from 1995 to 1999 for the Walker Branch site in the United States. The same years for having both NEE and Moderate Resolution Imaging Spectroradiometer (MODIS) VI data were used for analysis. Therefore, our analysis only focused on the period of 2000–2007 since the overlay time period for both data sets only covered from February 2000 (Start date for MODIS VI data) to December 2007 (End date for available NEE data). For each biome type, we first excluded the sites with more than 60 days deviations from the average SCU and ECU. We regarded each year for each flux tower site as one site-year and excluded the site-years whose daily NEE values were missing for the carbon source-sink or sink-source transition period. Moreover, only the biomes with at least 10 site-years were included for analysis in order to get robust estimations for CFP dates. Therefore, we got 32 eddy flux sites, which covered 4 biome types according to the International Geosphere-Biosphere Program (IGBP) classification system (Figure 1). There were totally 73 site-years involved in the spring carbon source-sink transition period and 86 site-years involved in the autumn carbon sink-source transition period.

Remotely sensed data. The Terra's MODIS 250 m 16-day composited VI products (MOD13Q1, V005) for the 32 flux tower sites were used in this analysis (available at: http://daac.ornl.gov/MODIS/). The first VI product was the standard Normalized Difference Vegetation Index (NDVI), and the second was the Enhance Vegetation Index (EVI). The VI time series for the pixel located at the center of the flux tower was used to retrieve land surface phenological metrics. The covered period for the VI data was the same as the NEE data for a given flux site. Much noise existed in the VI time series because of cloud contamination, atmospheric variability and sun-sensor-surface viewing geometries [23,24]. A filtering process was needed before using VI to retrieve phenological metrics [25]. We used the Savitzky-Golay filter method to remove the noise in the VI time series [26].

Methods

Retrieving CFP dates from NEE data. The SCU and ECU were retrieved based on the method proposed by Baldocchi *et al.* [22]. The original method is based on visual interpretation of the daily NEE time series. We developed this method to retrieve SCU and ECU automatically through fitting a regression equation between the daily NEE and the Julian Day of Year (DOY), using subsets of NEE data from spring source-sink and autumn sink-source transition periods, respectively (Figure 2). Specifically, SCU and ECU were automatically retrieved by the following three steps: (1) the original daily NEE was smoothed with a moving average of a 15-day width; (2) based on the smoothed daily NEE, a 10-day width window with the first 5 elements greater than zero and the last 5 elements less than zero was selected in the spring/summer period to predict SCU, and another 10-day width window with the first 5 elements less than zero and the last 5 elements greater than zero was selected in the autumn/winter period to predict ECU; and (3) the smoothed daily NEE in the two selected windows was linearly regressed to predict SCU and ECU at the zero intersection.

Retrieving LSP dates from remotely sensed data. A number of methods have been developed to retrieve land surface phenology metrics using satellite VI time series [27–29]. These methods can be classified into 3 types: the threshold method (i.e., a global absolute threshold value or a local relative threshold value defined as a fraction of the annual amplitude) [30–33], the autoregressive moving average method [27,34] and the function fitting method [28,35–39]. Almost all the methods mentioned above have been proven to be consistent with their given references (e.g., ground observed phenology events, model simulated vegetation phenology or eddy covariance flux tower-derived phenological metrics), but it was very difficult to give the ordinal rank of SOS methods because they varied geographically [27]. Therefore, this study first investigated these 3 types of satellite methods (including 6 specific retrieval methods) based on the first MODIS VI product (i.e., NDVI), and selected the one with the best performance to retrieve LSP dates. Then, a comparison between the two MODIS VI products (i.e., NDVI and EVI) was carried out based on the best-performing LSP retrieval method, and the more suitable VI product was selected to

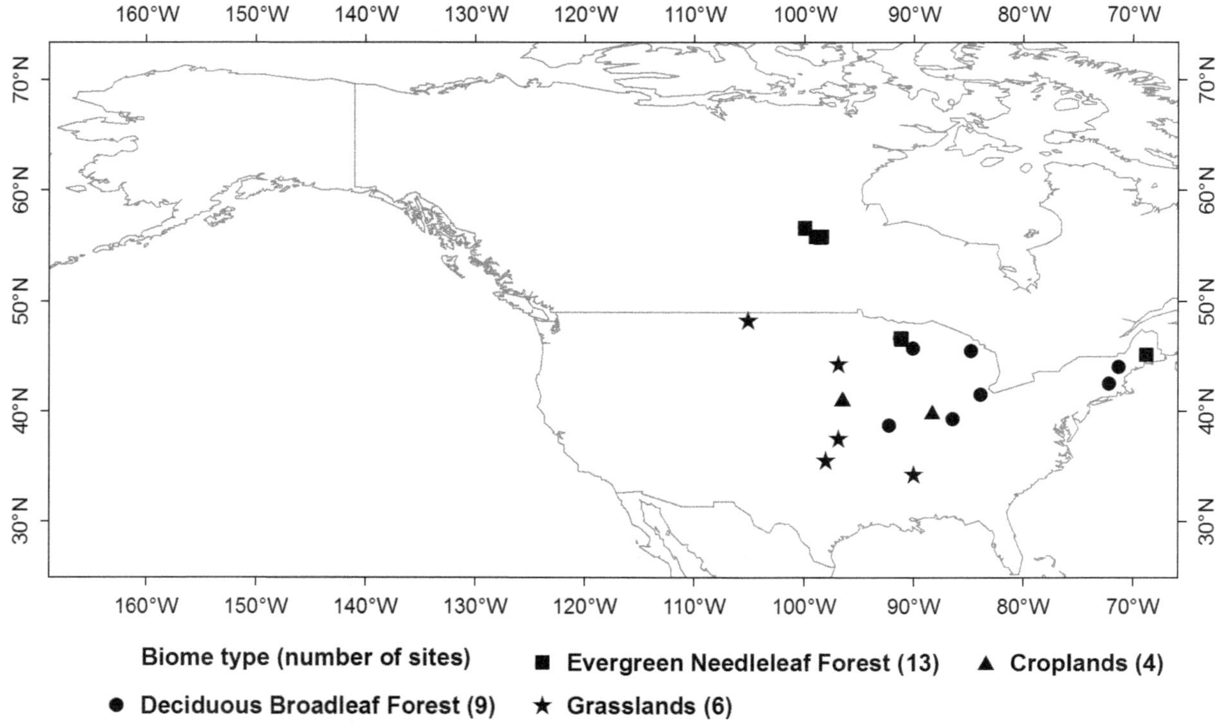

Figure 1. Distribution of eddy flux towers and their corresponding biome types.

retrieve LSP dates as the explanatory variables in estimating CFP dates. The detailed descriptions about these 6 retrieval methods and the evaluation process were given in Text S1.

Identifying the explanatory climate drivers. The LSP dates derived from the more suitable VI product with the best satellite retrieval method, the cumulative daily air temperature (above 0°C) and total precipitation were used to identify the explanatory climate drivers. Previous studies indicated that SCU usually occurs 0–20 days later than SOS while ECU usually occurs 0–60 days earlier than EOS [4,19,21]. Therefore, we restricted the impact period of climate drivers on SCU/ECU in the range from 60 days before SOS/EOS to 20 days after SOS/EOS. To identify the optimal impact period for each climate driver, different impact periods were tested according to the distance (in days) from SOS/EOS, 10-day after SOS/EOS and 20-day after SOS/EOS with a step of 10 days (Figure 3). Therefore, we got 18 candidate impact periods for each climate driver (i.e., cumulative daily air temperature (above 0°C) or total precipitation). The coefficient of determination (R^2) between observed SCU/ECU and each climate driver with different candidate impact periods was used to select the best explanatory climate drivers. Only the climate driver in a given impact period with the highest R^2 in its group (i.e., 18 candidate cumulative temperature or total precipitation data for each biome type and each phenological metric) and with a statistical significance at the 0.05 level will be selected to estimate SCU/ECU.

Estimating CFP dates. Using the least-squares linear regression model, the CFP dates (i.e., SCU and ECU) can be estimated with the LSP dates (i.e., SOS and EOS) and the selected explanatory climate drivers. The estimating performance of the linear regression models was evaluated with coefficient of determination (R^2), Root Mean Square Error (RMSE) and the leave-one-out cross-validation approach [19,40]. Significance test for these linear regression models was conducted by F-test with the standard 0.05 cutoff indicating statistical significance (i.e., $P<0.05$).

Results

The relationship between CFP and LSP dates

Our evaluation about the different LSP retrieval methods and MODIS VI products indicated that the NDVI-derived LSP dates with the local mean midpoint threshold method were more consistent with the observed CFP dates (see details in Text S2). Table 1 showed the relationship between CFP and LSP dates. SOS explained the SCU variance by 43.1%–78.4% for different biomes. The RMSE between SCU and SOS ranged from 2.7 to 7.6 days, which was far smaller than the temporal resolution of the satellite data (~16 days). This indicated that the SCU can be estimated with SOS to the accuracy that was comparable to the 16-day composited temporal resolution of satellite sensor. Comparing with the SCU, lower performance was found for estimating ECU based on EOS, with relatively lower explanatory variances and higher RMSE for different biomes. However, this RMSE was still comparable with the 16-day composited temporal sampling resolution of satellite data.

Different biomes showed distinctive CFP dates as estimated based on the LSP dates (Table 1). For example, evergreen needleleaf forest had the highest explanatory variance (78.4%) in estimating SCU based on SOS while grasslands had the lowest (43.1%). On the contrary, evergreen needleleaf forest showed the poorest performance in estimating ECU based on EOS while grasslands showed the best performance (67.1%). In general, the performance in estimating CFP dates for a single biome was better than multiple biomes.

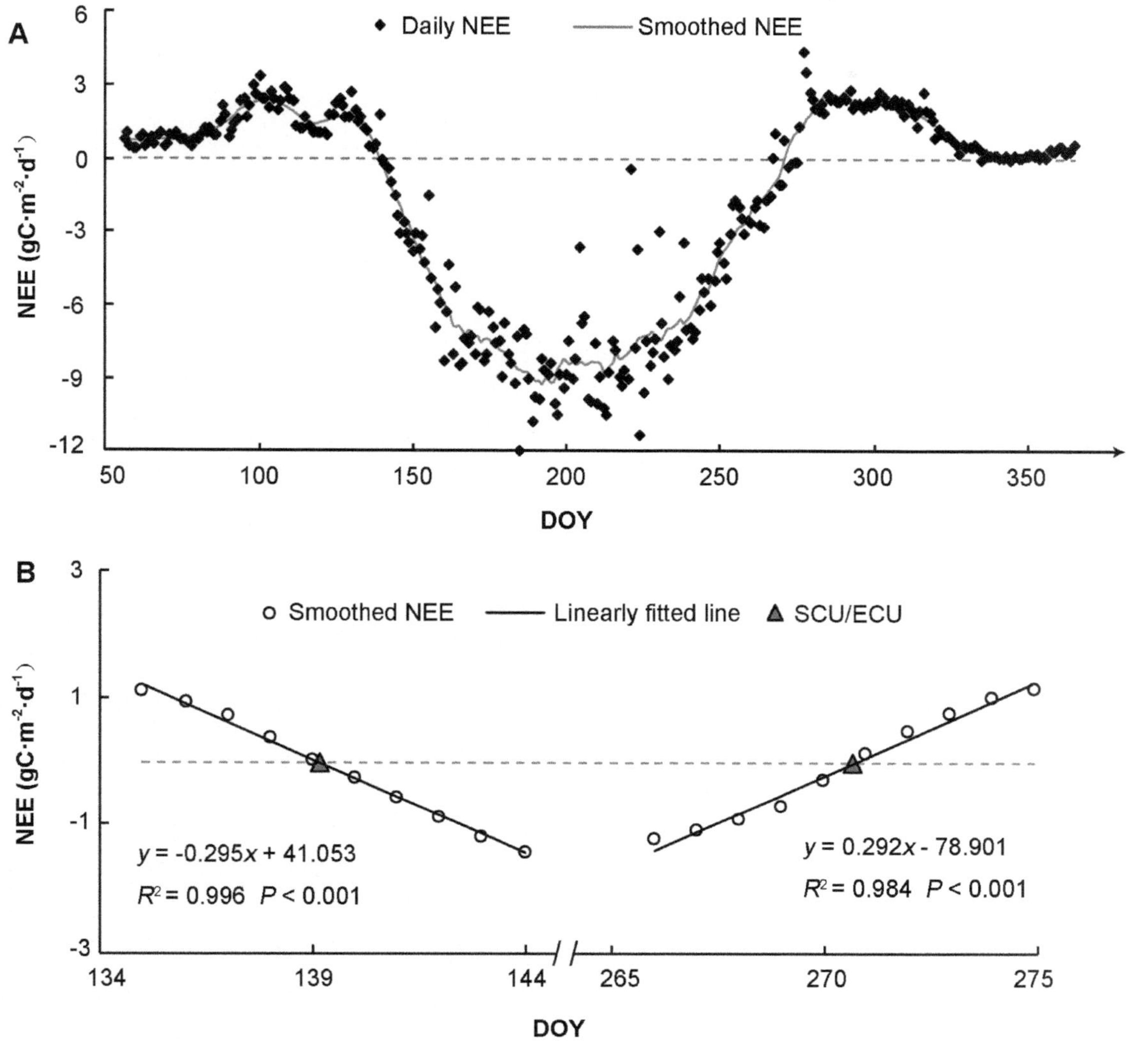

Figure 2. A schematic demonstration of the retrieval method for carbon flux phenology dates. A) The original and smoothed 15-day mean Net Ecosystem Exchange (NEE) of carbon, and B) the two selected transition periods for spring source-sink and autumn sink-source for identifying linear regressions between NEE and the Julian Day of Year (DOY). Start/End of Carbon Uptake (SCU/ECU) is estimated at the zero intersection.

The relationship between CFP dates and climate drivers

Figure 4 showed the coefficient of determination (R^2) between CFP dates and climate drivers during different impact periods. For SCU explanatory variances by the cumulative daily air temperature (above 0°C) (Figure 4A), evergreen needleleaf forest demonstrated better performance than other 3 biomes with consistently the highest R^2 values (ranging from 88.6% to 95.7% among the 18 tested impact periods). Considering its high explained variances, we selected the cumulative daily air temperature above 0°C during 20 days before NDVI-derived SOS as one of the explanatory variables in estimating SCU. Similarly, the total precipitation during the period of 30-day before SOS and 10-day after SOS was also selected for evergreen needleleaf forest in estimating SCU (Figure 4B). All the total precipitation variables during different impact periods were not significant at the 0.05 level for both deciduous broadleaf forest and croplands (Figure 4B). Therefore, no precipitation variables were selected for these two biomes to estimate SCU. In summary, the optimal impact periods for different climate drivers (i.e., cumulative air temperature and total precipitation) and different CFP metrics (i.e., SCU and ECU) were marked with stars in Figure 4. Only the climate drivers being significant during their impact periods were selected as the explanatory variables to estimate CFP dates.

The sensitivity of CFP metrics to climate drivers varied among different biomes (Figure 4). For SCU, evergreen needleleaf forest showed higher sensitivities to both cumulative temperature (above 0°C) and total precipitation, while deciduous broadleaf forest and cropland were only sensitive to cumulative temperature (Figure 4 A, B). Grassland had a higher sensitivity to total precipitation during the period of 20-day before SOS and 10-day after SOS but a lower sensitivity to cumulative temperature for the 6 selected grassland sites. For ECU, herbaceous biomes (i.e., grasslands and croplands) showed a higher sensitivity to cumulative temperature than woody biomes (e.g., deciduous broadleaf and evergreen needleleaf forest) (Figure 4 C). Deciduous broadleaf forest showed higher explained variances by total precipitation variables in different impact periods for ECU, while evergreen needleleaf forest demonstrated lower explained variances (Figure 4 D).

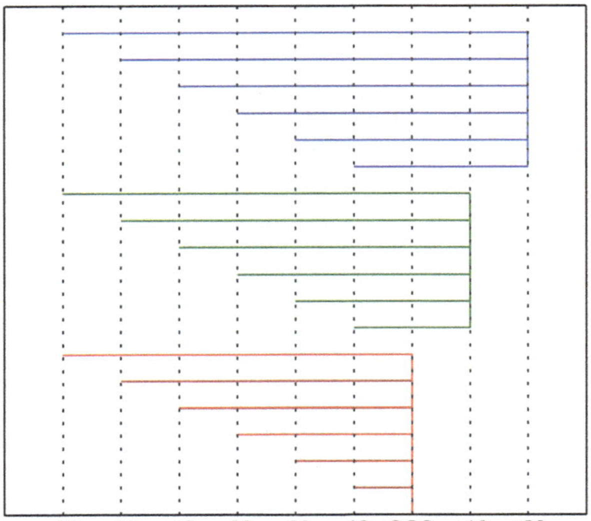

Figure 3. A schematic representation of the different impact periods for climate drivers. The different impact periods of climate drivers on carbon flux phenology dates were determined in terms of the distance (in days) from satellite-derived Start/End of Season (SOS/EOS), 10-day after SOS/EOS and 20-day after SOS/EOS with a step of 10 days. There were totally 18 candidate impact periods for each climate driver. Negative values indicate the days before SOS/EOS and positive values indicates the days after SOS/EOS.

Estimation of CFP dates with LSP dates and climate drivers

Figure 5 showed the relationships between observed and estimated CFP dates based on the linear regression (Figure 5 A, C, E, G) and the leave-one-out cross-validation (Figure 5 B, D, F, H) approaches in terms of the best performing explanatory variables. All of the estimation performances were statistically significant ($P<0.05$). The explained variance for SCU ranged from 71% to 97%, and the RMSE for SCU ranged from 2.6 to 5.2 days (Figure 5 A, C, E, G). Comparing with SCU, the ECU estimation showed a relatively lower performance, with a relatively lower R^2 (60%–84%) and a higher RMSE (5.3–8.5 days) (Figure 5 A, C, E, G).

The CFP estimation performance varied among different biomes. The estimation RMSE for both SCU and ECU was less than 9 days for deciduous broadleaf forest (Figure 5 A, B). Evergreen needleleaf forest had a lower RMSE for SCU but a higher RMSE for ECU (Figure 5 C, D). Grasslands had the highest R^2 but also the highest RMSE for ECU (Figure 5 E, F), while croplands had the lowest RMSE for both SCU and ECU (Figure 5 G, H).

Discussion

The optimal explanatory variables in estimating CFP dates

The temporal and spatial variation in CFP metrics (i.e., SCU and ECU) is controlled by many factors, including the biome type, canopy structures, species compositions, soil type, forest age and meteorological factors (e.g., temperature, precipitation, etc.) [3,18,19,41]. Wu *et al.* [3] demonstrated that the interannual variation in NEE and phenological indicators at a study site could be mainly resulted from the meteorological factors, while differences of canopy structures and species compositions had no significant impacts. On the contrary, when the spatial variation was considered, the primary controlling factors may be site-specific differences in canopy structures, soil prosperities and biome types. Therefore, they suggested a separated analysis for spatial and temporal variation in the response of annual NEE to CUP and its transitions. In fact, the most challenging aspect in estimating CFP dates is to identify the optimal explanatory variables whether or not the temporal and spatial variation in CFP metrics is separately considered.

The satellite-derived phenological metrics reflect an integrated signal of a group of species (individuals) in a pixel because remote sensing can capture the spectral characteristics of green leaf and vegetation canopy structures at moderate to coarse spatial resolutions [19,42,43]. Our results showed high explained SCU/ECU variances with SOS/EOS (Table 1), which have been also found by Gonsamo *et al.* [19]. This suggested that satellite-derived LSP dates can effectively reflect the spatial and temporal variations in CFP dates and should be selected as one of the primary explanatory variables in estimating CFP dates, especially for the large-scale (e.g., regional or continental) studies. However, we should also note that large discrepancies exist in different LSP retrieval methods in terms of the CFP estimating performances (Text S2). Because of the different SOS/EOS definitions (Text S1), the satellite-derived LSP metrics with different methods do

Table 1. The coefficient of determination (R^2), Root Mean Square Error (RMSE) and Bias between Net Ecosystem Exchange (NEE)-derived carbon flux phenology dates and Normalized Difference Vegetation Index (NDVI)-derived land surface phenology dates based on the best performing retrieval method (i.e., the local mean midpoint threshold method) for different biomes.

Biome type	SOS vs. SCU†				EOS vs. ECU†			
	Samples	R^2 (%)	RMSE	Bias	Samples	R^2 (%)	RMSE	Bias
Deciduous broadleaf forest	24	74.3*	7.5	−10.5	20	51.4*	6.3	4.6
Evergreen needleleaf forest	16	78.4*	7.6	15.8	30	43.5*	13.1	35.6
Grasslands	16	43.1*	6.5	3.3	14	67.1*	10.3	9.4
Croplands	17	68.8*	2.7	−0.2	22	65.0*	5.6	14.7
All biomes	73	49.6*	17.1	0.7	86	43.5*	14.6	18.8

†SOS = Start of Season derived from satellite data, SCU = Start of Carbon Uptake derived from carbon flux data, EOS = End of Season derived from satellite data, ECU = End of Carbon Uptake derived from carbon flux data.
*Statistically significant at the 0.05 level.

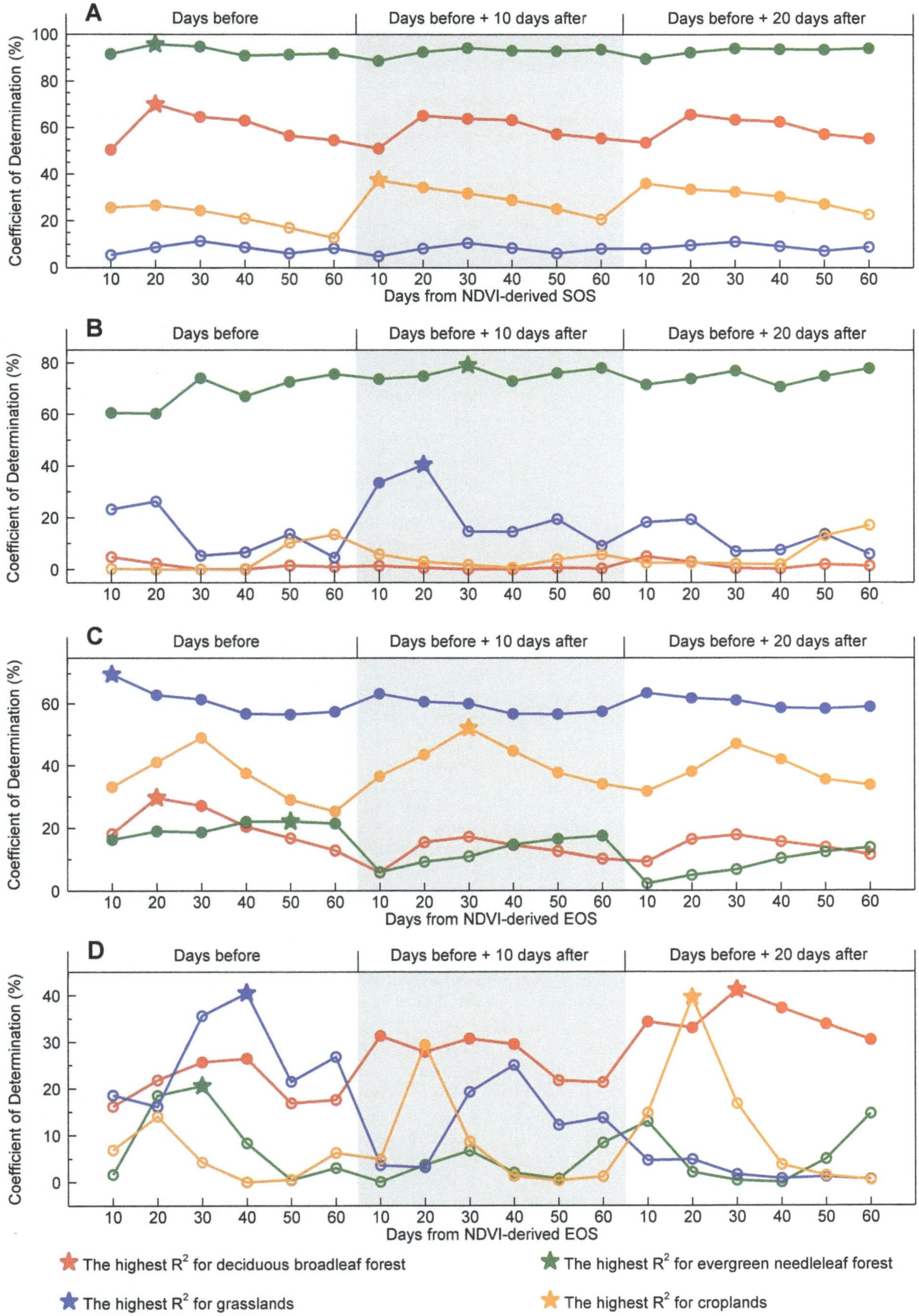
A
Days before
Days before + 10 days after
Days before + 20 days after
Coefficient of Determination (%)
Days from NDVI-derived SOS
B
C
Days from NDVI-derived EOS
D
The highest R² for deciduous broadleaf forest
The highest R² for evergreen needleleaf forest
The highest R² for grasslands
The highest R² for croplands

Figure 4. The relationships between carbon flux phenology dates and climate drivers in different impact periods. A) The coefficient of determination (R^2) between Net Ecosystem Exchange (NEE)-derived Start of Carbon Uptake (SCU) and the cumulative daily air temperature (above 0°C) for different periods around Normalized Difference Vegetation Index (NDVI)-derived Start of Season (SOS). B) The R^2 between NEE-derived SCU and the total precipitation for different periods around NDVI-derived SOS. C) The R^2 between NEE-derived End of Carbon Uptake (ECU) and the cumulative daily air temperature (above 0°C) for different periods around NDVI-derived End of Season (EOS). D) The R^2 between NEE-derived ECU and the total precipitation for different periods around NDVI-derived EOS. Red colored line: Deciduous broadleaf forest; green: evergreen needleleaf forest; blue: grassland; orange: cropland. Stars indicate the locations with the highest R^2 for each biome and with a statistical significance at the 0.05 level. Solid circles indicate statistically significant R^2 at the 0.05 level, and hollow circles indicate statistically non-significant R^2.

not actually measure the same phenological traits [44]. Our results demonstrated that the SCU/ECU is much closer to and more consistent with the time when satellite-derived vegetation index reaches its midpoint in a growing season for the 4 tested biomes.

Meteorological factors are important candidate explanatory variables in estimating CFP dates. Previous results [18,19] and our results (Table 1, Figure 5) all indicated that combining optimal climate drivers with LSP dates can obviously improve the estimation robustness when the established regression models are applied in a wide spatial and temporal range. Our results also demonstrated the differences in the optimal impact periods of climate drivers on SCU/ECU for different biomes (Figure 4). Therefore, the key question is how to select the optimal explanatory climate drivers for different biomes. Gonsamo *et al.* [19] conducted a sensitivity analysis to search the optimal impact period based on human calendar month (e.g., the mean air temperature in April and May for SCU). Rather than applying the human calendar month, we used the distance (in days) from the LSP dates to identify the optimal impact periods. This is because the LSP dates retrieved with the best performing method are close to the CFP dates and vary simultaneously with CFP dates when across regions and/or biomes, while the unified human calendar month cannot reflect the variations in vegetation phenology across large heterogeneous areas and thus may fail to describe the actual impact period of climate drivers on SCU/ECU.

The difference in CFP estimation performance among different biomes

Large discrepancies existed in the CFP estimation performance for different biomes. The CFP dates are relatively easy to be estimated for deciduous broadleaf forest because of its distinct seasonal variation in canopy structure and carbon flux which can be effectively captured by remote sensing and eddy covariance system. Our estimated CFP dates based on both LSP dates and climate drivers for deciduous broadleaf forest was comparable with that based on observed carbon flux and meteorology data (5.20 vs. 5.12 for SCU, 5.30 vs. 6.65 for ECU in RMSE) and slightly better than that solely based on satellite data (5.20 vs. 6.98 for SCU, 5.30 vs. 8.88 for ECU in RMSE) from Gonsamo *et al.* [19].

Theoretically, it is difficult to define CFP metrics for conifers because NEE may be negative throughout the year. The transition from net positive to net negative NEE is more flexible in coniferous than in deciduous forest because the seasonality of coniferous forest is not related to changes in canopy structure [45]. However, in high latitude snow-dominated coniferous forests, the annual cycle from near-total snow cover to a mature canopy provides a distinct and detectable VI and NEE cycle [27]. The flux sites for evergreen needleleaf forest in this study are distributed above 45°N and are fully covered by snow in winter. The satellite-derived LSP dates are closely related to the timing of spring snowmelt and winter snow, which substantially reflects the change in air temperature. Therefore, the CFP dates estimation performance with LSP dates and climate drivers for evergreen needleleaf forest in this study was relatively high, especially in estimating SCU. It should be noted that the high estimation performance for evergreen needleleaf forest may be only suitable for snow-dominated ecosystems and cannot be extrapolated to other coniferous forests, such as temperate and subtropical evergreen needleleaf forest.

Out results also demonstrated a high estimation performance for CFP dates of herbaceous biomes (i.e., grassland, crop). The high estimation performance may result from the primary control of satellite-derived LSP dates on estimating CFP dates (Table 1). For example, the phenological development for the 6 selected grassland sites was mainly precipitation-driven (Figure 4 A, B) and their leaf-out and NEE transition was usually occurred in a short period. Moreover, herbaceous biomes do not have understory plants that could confound the spectral signal. Therefore, LSP dates showed a significant variance explanatory rate in estimating CFP dates for herbaceous biomes.

The difference in estimation performance between SCU and ECU

The SCU estimation performance was generally better than ECU (Figure 5). The explained variance by LSP dates (Table 1) and climate drivers (Figure 4) for ECU was generally lower than that for SCU, implying that the satellite-derived EOS and climate drivers had relatively weak relationships with ECU. During the greenup phase, increasing greenness is closely related to chlorophyll, leaf area and changes in canopy structure, which scale rather well with photosynthesis and respiration [46,47]. However, during the leaf senescence phase, changes in leaf color, environmental stress (e.g., drought stress), and meteorological conditions (e.g., cooler air temperature) may complicate the relationships between canopy structure-based phenology metrics and carbon fluxes, and in general make the detection of senescence events more difficult [41,48]. In fact, the factors controlling senescence and dormancy are not well-documented in all biomes [41]. A mechanistic understanding of the drivers controlling senescence and dormancy is urgently needed.

Potentials of the optimization method in estimating CFP metrics

The optimization method proposed in this study can be used to extrapolate regional CFP metrics through extending the footprints of flux towers. As for a given biome over large heterogeneous areas, a synoptic train of thought is first to classify the biome to smaller ecoregions, or use existing ecoregion maps (e.g., the terrestrial ecoregions compiled by the World Wildlife Fund (WWF) (available at: http://worldwildlife.org/biome-categories/terrestrial-ecoregions)), since vegetation phenology may differ significantly even within the same biome. Then for each ecoregion, time-series satellite data, climate data and NEE data for the involved eddy flux sites can be used to build an optimized empirical model to predict CFP metrics beyond the footprints of flux towers.

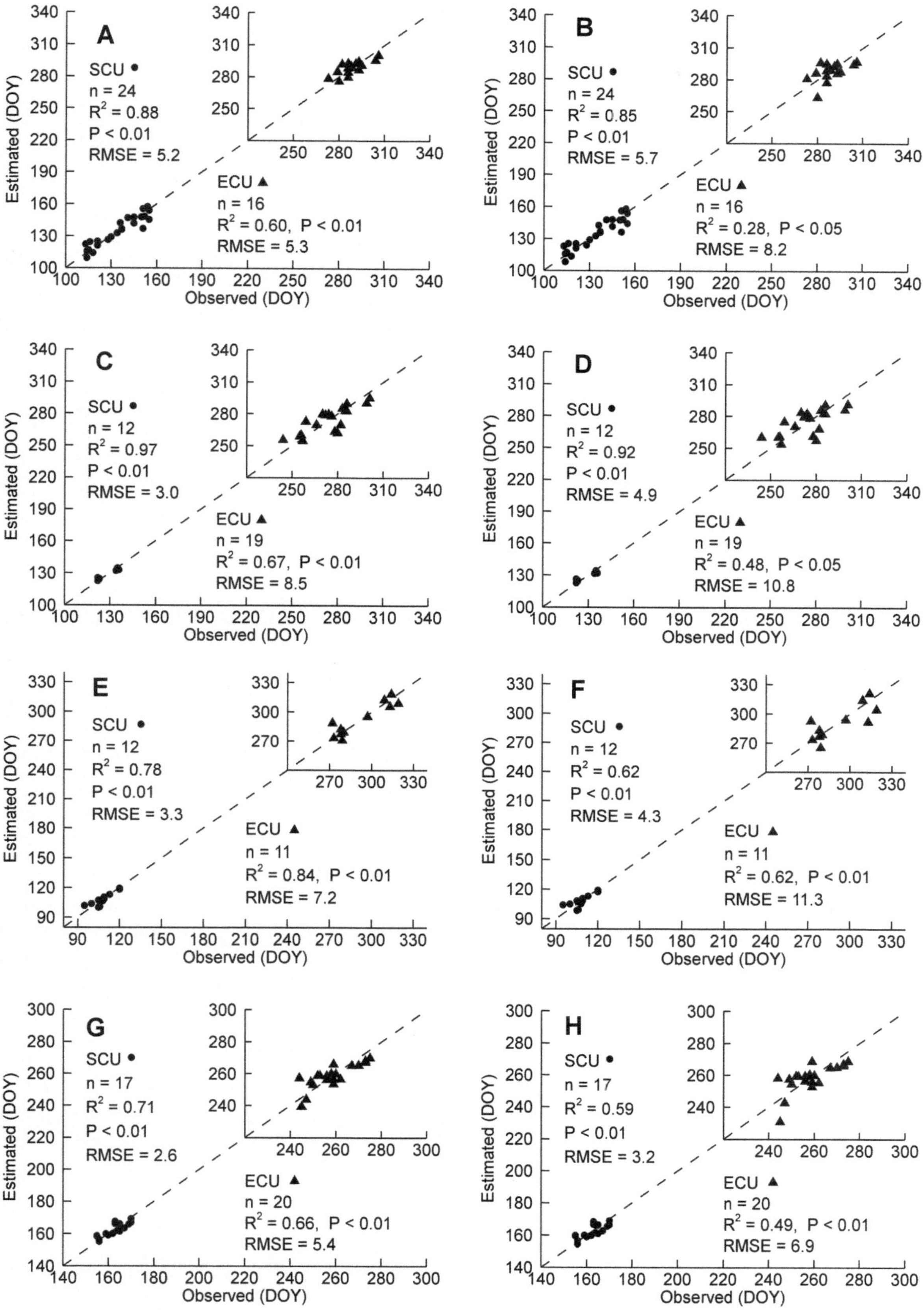

Figure 5. Relationships between observed and estimated carbon flux phenology dates for different biomes. A) and B) Deciduous broadleaf forest, C) and D) Evergreen needleleaf forest, E) and F) Grasslands, and G) and H) Croplands. The left panel (i.e., A, C, E and G) indicates the relationships between observed Start/End of Carbon Uptake (SCU/ECU) in Julian Day of Year (DOY) and estimated with the best performing explanatory variables given in Figure 4, and the right panel (i.e., B, D, F and H) indicates the relationships between observed SCU/ECU and estimated with the best performing explanatory variables based on the leave-one-out cross-validation approach.

Conclusions

This study provided a methodology to estimate CFP metrics with satellite-derived LSP metrics and climate drivers for different biomes through a synthesis of AmeriFlux observations. LSP metrics combined with optimal climate drivers can explain SCU variability by more than 70% (ranging from 71% to 97% for different biomes) and ECU variability by more than 60% (ranging from 60% to 84% for different biomes). The RMSE of the estimations ranged from 2.6 to 5.2 days for SCU and from 5.3 to 8.5 days for ECU. The results of our study highlighted the relative strengths and weaknesses of LSP metrics derived from different methods and climate drivers in different impact periods in estimating a specific CFP metric for different biomes. The estimation performance for the methodology was primarily dependent on the optimal combination of the LSP retrieval methods, the explanatory climate drivers, the biome types, and the specific CFP metric. Although the proposed methodology showed high performance in estimating CFP metrics for biomes with a distinct and detectable VI and NEE cycle, a better mechanistic understanding of the drivers controlling vegetation phenology is urgently needed in order to improve the estimation performance, especially for senescence and dormancy phenology.

Acknowledgments

Special thanks to the data archive at the Carbon Dioxide Information Analysis Center (CDIAC) for providing flux NEE data, and the Oak Ridge National Laboratory for the National Aeronautics and Space Administration for providing MODIS data. We also appreciate the three anonymous reviewers for providing very useful comments and suggestions, which led to a major improvement of this paper.

Author Contributions

Conceived and designed the experiments: WQZ. Performed the experiments: WQZ MJM NJ JHL. Analyzed the data: WQZ GSC. Wrote the paper: WQZ GSC. Edited manuscript: WQZ GSC..

References

1. Richardson AD, Black TA, Ciais P, Delbart N, Friedl MA, et al. (2010) Influence of spring and autumn phenological transitions on forest ecosystem productivity. Philosophical Transactions of the Royal Society B-Biological Sciences 365: 3227–3246.
2. Dragoni D, Schmid HP, Wayson CA, Potter H, Grimmond C, et al. (2011) Evidence of increased net ecosystem productivity associated with a longer vegetated season in a deciduous forest in south-central Indiana, USA. Global Change Biology 17: 886–897.
3. Wu CY, Gonsamo A, Chen JM, Kurz WA, Price DT, et al. (2012) Interannual and spatial impacts of phenological transitions, growing season length, and spring and autumn temperatures on carbon sequestration: A North America flux data synthesis. Global and Planetary Change 92–93: 179–190.
4. Wu CY, Chen JM, Gonsamo A, Price DT, Black TA, et al. (2012) Interannual variability of net carbon exchange is related to the lag between the end-dates of net carbon uptake and photosynthesis: Evidence from long records at two contrasting forest stands. Agricultural and Forest Meteorology 164: 29–38.
5. Wu C, Chen JM, Black TA, Price DT, Kurz WA, et al. (2013) Interannual variability of net ecosystem productivity in forests is explained by carbon flux phenology in autumn. Global Ecology and Biogeography 22: 994–1006.
6. Chen WJ, Black TA, Yang PC, Barr AG, Neumann HH, et al. (1999) Effects of climatic variability on the annual carbon sequestration by a boreal aspen forest. Global Change Biology 5: 41–53.
7. Black TA, Chen WJ, Barr AG, Arain MA, Chen Z, et al. (2000) Increased carbon sequestration by a boreal deciduous forest in years with a warm spring. Geophysical Research Letters 27: 1271–1274.
8. Bergeron O, Margolis HA, Black TA, Coursolle C, Dunn AL, et al. (2007) Comparison of carbon dioxide fluxes over three boreal black spruce forests in Canada. Global Change Biology 13: 89–107.
9. Richardson AD, Hollinger DY, Dail DB, Lee JT, Munger JW, et al. (2009) Influence of spring phenology on seasonal and annual carbon balance in two contrasting New England forests. Tree Physiology 29: 321–331.
10. Kramer K (1995) Modeling comparison to evaluate the importance of phenology for the effects of climate-change on growth of temperate-zone deciduous trees. Climate Research 5: 119–130.
11. White MA, Running SW, Thornton PE (1999) The impact of growing-season length variability on carbon assimilation and evapotranspiration over 88 years in the eastern US deciduous forest. International Journal of Biometeorology 42: 139–145.
12. Cramer W, Bondeau A, Woodward FI, Prentice IC, Betts RA, et al. (2001) Global response of terrestrial ecosystem structure and function to CO_2 and climate change: results from six dynamic global vegetation models. Global Change Biology 7: 357–373.
13. White MA, Nemani AR (2003) Canopy duration has little influence on annual carbon storage in the deciduous broad leaf forest. Global Change Biology 9: 967–972.
14. Piao SL, Ciais P, Friedlingstein P, Peylin P, Reichstein M, et al. (2008) Net carbon dioxide losses of northern ecosystems in response to autumn warming. Nature 451: 49–52.
15. Baldocchi DD, Wilson KB (2001) Modeling CO_2 and water vapor exchange of a temperate broadleaved forest across hourly to decadal time scales. Ecological Modelling 142: 155–184.
16. Zhang WJ, Wang HM, Yang FT, Yi YH, Wen XF, et al. (2011) Underestimated effects of low temperature during early growing season on carbon sequestration of a subtropical coniferous plantation. Biogeosciences 8: 1667–1678.
17. Baldocchi D, Falge E, Gu LH, Olson R, Hollinger D, et al. (2001) FLUXNET: A new tool to study the temporal and spatial variability of ecosystem-scale carbon dioxide, water vapor, and energy flux densities. Bulletin of the American Meteorological Society 82: 2415–2434.
18. Garrity SR, Bohrer G, Maurer KD, Mueller KL, Vogel CS, et al. (2011) A comparison of multiple phenology data sources for estimating seasonal transitions in deciduous forest carbon exchange. Agricultural and Forest Meteorology 151: 1741–1752.
19. Gonsamo A, Chen JM, Wu CY, Dragoni D (2012) Predicting deciduous forest carbon uptake phenology by upscaling FLUXNET measurements using remote sensing data. Agricultural and Forest Meteorology 165: 127–135.
20. Gonsamo A, Chen JM, Price DT, Kurz WA, Wu CY (2012) Land surface phenology from optical satellite measurement and CO_2 eddy covariance technique. Journal of Geophysical Research-Biogeosciences 117, G03032.
21. Churkina G, Schimel D, Braswell BH, Xiao XM (2005) Spatial analysis of growing season length control over net ecosystem exchange. Global Change Biology 11: 1777–1787.
22. Baldocchi DD, Black TA, Curtis PS, Falge E, Fuentes JD, et al. (2005) Predicting the onset of net carbon uptake by deciduous forests with soil temperature and climate data: A synthesis of FLUXNET data. International Journal of Biometeorology 49: 377–387.
23. Holben B, Fraser RS (1984) Red and near-infrared sensor response to off-nadiir viewing. International Journal of Remote Sensing 5: 145–160.
24. Kobayashi H, Dye DG (2005) Atmospheric conditions for monitoring the long-term vegetation dynamics in the Amazon using normalized difference vegetation index. Remote Sensing of Environment 97: 519–525.
25. Hird JN, McDermid GJ (2009) Noise reduction of NDVI time series: An empirical comparison of selected techniques. Remote Sensing of Environment 113: 248–258.
26. Chen J, Jönsson P, Tamura M, Gu Z, Matsushita B, et al. (2004) A simple method for reconstructing a high-quality NDVI time-series data set based on the Savitzky-Golay filter. Remote Sensing of Environment 91: 332–344.
27. White MA, Beurs KM, Didan K, Inouye DW, Richardson AD, et al. (2009) Intercomparison, interpretation, and assessment of spring phenology in North America estimated from remote sensing for 1982–2006. Global Change Biology 15: 2335–2359.
28. Zhang XY, Friedl MA, Schaaf CB, Strahler AH, Hodges J, et al. (2003) Monitoring vegetation phenology using MODIS. Remote Sensing of Environment 84: 471–475.
29. Cong N, Wang T, Nan HJ, Ma YC, Wang XH, et al. (2013) Changes in satellite-derived spring vegetation green-up date and its linkage to climate in China from 1982 to 2010: a multimethod analysis. Global Change Biology 19: 881–891.

30. Lloyd D (1990) A phenological classification of terrestrial vegetation cover using shortwave vegetation index imagery. International Journal of Remote Sensing 11: 2269–2279.
31. Fischer A (1994) A model for the seasonal variations of vegetation indices in coarse resolution data and its inversion to extract crop parameters. Remote Sensing of Environment 48: 220–230.
32. Markon CJ, Fleming MD, Binnian EF (1995) Characteristics of vegetation phenology over the Alaskan landscape using AVHRR time-series data. Polar Record 31: 179–190.
33. White MA, Thornton PE, Running SW (1997) A continental phenology model for monitoring vegetation responses to interannual climatic variability. Global Biogeochemical Cycles 11: 217–234.
34. Reed BC, Brown JF, VanderZee D, Loveland TR, Merchant JW, et al. (1994) Measuring phenological variability from satellite imagery. Journal of Vegetation Science 5: 703–714.
35. Beck PSA, Atzberger C, Hogda KA, Johansen B, Skidmore AK (2006) Improved monitoring of vegetation dynamics at very high latitudes: A new method using MODIS NDVI. Remote Sensing of Environment 100: 321–334.
36. Jönsson P, Eklundh L (2002) Seasonality extraction by function fitting to time-series of satellite sensor data. IEEE Transactions on Geoscience and Remote Sensing 40: 1824–1832.
37. Piao SL, Fang JY, Zhou LM, Ciais P, Zhu B (2006) Variations in satellite-derived phenology in China's temperate vegetation. Global Change Biology 12: 672–685.
38. Zhang XY, Friedl MA, Schaaf CB, Strahler AH (2004) Climate controls on vegetation phenological patterns in northern mid- and high latitudes inferred from MODIS data. Global Change Biology 10: 1133–1145.
39. Zhu W, Tian H, Xu X, Pan Y, Chen G, et al. (2012) Extension of the growing season due to delayed autumn over mid and high latitudes in North America during 1982–2006. Global Ecology and Biogeography 21: 260–271.
40. Shao J (1993) Linear-model selection by cross-validation. Journal of the American Statistical Association 88: 486–494.
41. Richardson AD, Keenan TF, Migliavacca M, Ryu Y, Sonnentag O, et al. (2013) Climate change, phenology, and phenological control of vegetation feedbacks to the climate system. Agricultural and Forest Meteorology 169: 156–173.
42. Campbell JB (2007) Introduction to Remote Sensing, 4th edn., The Guilford Press, New York.
43. Myneni RB, Maggion S, Iaquinto J, Privette JL, Gobron N, et al. (1995) Optical remote-sensing of vegetation: modeling, caveats, and algorithms. Remote Sensing of Environment 51: 169–188.
44. Badeck FW, Bondeau A, Bottcher K, Doktor D, Lucht W, et al. (2004) Responses of spring phenology to climate change. New Phytologist 162: 295–309.
45. Tanja S, Berninger F, Vesala T, Markkanen T, Hari P, et al. (2003) Air temperature triggers the recovery of evergreen boreal forest photosynthesis in spring. Global Change Biology 9: 1410–1426.
46. Lindroth A, Lagergren F, Aurela M, Bjarnadottir B, Christensen T, et al. (2008) Leaf area index is the principal scaling parameter for both gross photosynthesis and ecosystem respiration of Northern deciduous and coniferous forests. Tellus Series B-Chemical and Physical Meteorology 60: 129–142.
47. Peng Y, Gitelson AA, Keydan G, Rundquist DC, Moses W (2011) Remote estimation of gross primary production in maize and support for a new paradigm based on total crop chlorophyll content. Remote Sensing of Environment 115: 978–989.
48. Sparks TH, Menzel A (2002) Observed changes in seasons: An overview. International Journal of Climatology 22: 1715–1725.

Fishery-Induced Changes in the Subtropical Pacific Pelagic Ecosystem Size Structure: Observations and Theory

Jeffrey J. Polovina*, Phoebe A. Woodworth-Jefcoats

Pacific Islands Fisheries Science Center, NOAA Fisheries, Honolulu, Hawaii, United States of America

Abstract

We analyzed a 16-year (1996–2011) time series of catch and effort data for 23 species with mean weights ranging from 0.8 kg to 224 kg, recorded by observers in the Hawaii-based deep-set longline fishery. Over this time period, domestic fishing effort, as numbers of hooks set in the core Hawaii-based fishing ground, has increased fourfold. The standardized aggregated annual catch rate for 9 small (<15 kg) species increased about 25% while for 14 large species (>15 kg) it decreased about 50% over the 16-year period. A size-based ecosystem model for the subtropical Pacific captures this pattern well as a response to increased fishing effort. Further, the model projects a decline in the abundance of fishes larger than 15 kg results in an increase in abundance of animals from 0.1 to 15 kg but with minimal subsequent cascade to sizes smaller than 0.1 kg. These results suggest that size-based predation plays a key role in structuring the subtropical ecosystem. These changes in ecosystem size structure show up in the fishery in various ways. The non-commercial species lancetfish (mean weight 7 kg) has now surpassed the target species, bigeye tuna, as the species with the highest annual catch rate. Based on the increase in snake mackerel (mean weight 0.8 kg) and lancetfish catches, the discards in the fishery are estimated to have increased from 30 to 40% of the total catch.

Editor: David Hyrenbach, Hawaii Pacific University, United States of America

Funding: These authors have no support or funding to report.

Competing Interests: The authors have declared that no competing interests exist.

* E-mail: jeffrey.polovina@noaa.gov

Introduction

The North Pacific subtropical gyre is a large oceanic gyre bounded on the south by the North Equatorial Current, on the west by the Kuroshio Current, on the north by the Kuroshio Extension Current and the North Pacific Current, and on the east by the California Current [1]. Although low in primary productivity, the warm, vertically stratified oligotrophic waters of the subtropical gyre contain a highly diverse food web populated by tunas, sharks, and billfishes at the top trophic levels [2,3]. Since the 1950s, large-scale fisheries have targeted the tunas, billfishes, and other large predators in this ecosystem. Several studies have suggested possible ecosystem impacts from fishing [3–5]. A comparison of catch, size, and species composition between a research longline survey in the 1950s and observer data from commercial longliners in the 1990s suggested a substantial decline in the abundance of large predators, the mean size of these predators, and some evidence of an increased abundance of formerly rare species [5]. Models of the North Pacific subtropical gyre were generated with Ecopath with Ecosim (EwE) to investigate whether the ecosystem contained any keystone species [3,4]. The results suggested that there was not any single species group that functioned as a keystone, but that a broad reduction of apex predators as a result of fishing might result in an increase in prey in response to a decreased predation [3,4]. In effect the fishing fleet is the keystone predator [4]. However, another modeling effort using an EwE model that incorporated some size-class structure found that while fishing decreased predator abundance there was limited evidence of trophic cascades or other ecosystem impacts based on the decline in predators [6].

A more recent analysis of catch rates for the 13 most abundant species caught in the deep-set Hawaii-based longline fishery over the past decade (1996–2006) provided evidence of a top-down response of the North Pacific subtropical ecosystem. Catch rates for apex predators such as blue shark (*Prionace glauca*), bigeye (*Thunnus obesus*) and albacore (*Thunnus alalunga*) tunas, shortbill spearfish (*Tetrapturus angustirostris*), and striped marlin (*Tetrapturus audax*) declined from 3 to 9% per year while catch rates for 4 mid-trophic species, mahimahi (*Coryphaena hippurus*), sickle pomfret (*Taractichthys steindachneri*), escolar (*Lepidocybium flavobrunneum*), and snake mackerel (*Gempylus serpens*), increased by 6 to 18% per year [7].

Ecosystem food webs and models for the central North Pacific subtropical pelagic ecosystem have traditionally been built from a species-specific perspective [2–4,6]. Recent analysis of the temporal ecosystem dynamics used trophic levels derived from species-specific models [7]. However, recent applications of size-based ecosystem models across various ecosystems show they are emerging as a powerful tool, particularly in pelagic environments where predation is more strongly driven by body size than species' taxonomic identity [8–13]. A key advantage to size-based models is that they are based on broad ecological and physiological relationships requiring few region-specific parameter estimates apart from sea surface temperature (SST), the size structure at the base of the food web, and fishing gear selectivity.

In this paper we further examine ecosystem changes in the subtropical pelagic system from a size-based perspective and compare observations from the Hawaii-based longline fishery with simulations from a dynamic size-based ecosystem model.

Materials and Methods

Catch and fishing effort data from the Hawaii-based longline fishery are collected in two ways. Federally mandated logbooks are required from all fishers licensed in that fishery. The logbooks report daily records of fishing activity including location, catch by species, number of hooks per set, and other data on the fishing operation. While logbook data provides complete coverage from all vessels, it is most reliable for the landed portion of the catch while catches of discarded species including sharks and fishes with low or no economic value are often not recorded. Fishery observers are placed on a subset of all vessels to record all the species caught, fishing effort, and various operational aspects and since 2006 they have recorded the length of every third fish caught. Over the period 1996–2011, 16% of the deep-set trips had observer coverage. However, even with this relatively modest observer coverage, observer and logbook catch rates for the commercial species were highly correlated. For example, over the period 1996–2006, annual catch rates from observer and logbook data had correlations ranging from over 0.93 for albacore, striped marlin, shortbill spearfish bigeye tuna, and pomfret, between 0.80 and 0.89 for mahimahi, ono, and yellofin tuna, 0.78 for skipjack, and 0.76 for blue shark [7].

The Hawaii-based longline fishery consists of two components: the daytime deepset fishery targeting bigeye tuna, and the nighttime shallow-set fishery targeting swordfish (*Xiphias gladius*). The deep-set fishery typically sets hooks between 100 m and 400 m with the median hook depth at about 250 m while the median depth of the deepest hook in the shallow-set fishery is 60 m [14]. Deep sets and shallow sets can be identified based on a very strong bimodal distribution of the number of hooks between floats. Shallow sets use 2–6 hooks per float while deep sets use 20–32 hooks per float [14]. For our analysis we identified deep sets as those with 10 or more hooks per float and shallow sets as those with fewer than 10 hooks per float. The shallow-set fishery occurs primarily in the winter and spring within a narrow band of 28°–32° N latitude. The shallow-set fishery was closed for several years to reduce interactions with sea turtles. This paper focuses exclusively on the deepset fishery occurring throughout the year over a broad geographic region and provides an uninterrupted observed catch and effort time series from 1996. Restricting our analysis to the deep-set fishery provides a relatively standardized depth range and method of gear deployment. This analysis was further restricted to data that were obtained from the core region of the fishing ground defined as bounded by 12°–27° N latitude. In some years, the fishery made excursions as far south as the equator and as far north as 32° N latitude; however, fishing in these areas was inconsistent over the study period.

Over the past two decades fishing effort in the Hawaii-based longline fishing ground and fishing mortality over the basin has increased about fourfold. For example, from 1996 to 2008 the number of hooks set in the Hawaii-based longline fishery increased from 10 million to 40 million. Recent stock assessments for yellowfin and bigeye tuna in the central and western Pacific estimated an increase in fishing mortality from 0.1 in 1990 to 0.3–0.4 in 2010 [15]. Unfortunately, we do not have any estimates of fishing mortality for either target species or the ecosystem that apply specifically to the central North Pacific, the area covered by the Hawaii deepset fishery.

We based this study on the catch and effort data for 23 species defined as those that have a mean catch per unit effort (CPUE) of at least 0.05 fish per 1000 hooks set over the 1996–2011 period. For those 23 species we estimated the species' mean weights from published length-weight equations by using lengths recorded by observers pooled over the 2006–2011 period (Table S1). Although some species had sufficient lengths to compute annual weights we chose to use mean sizes pooled over the entire time period for all species. The reason was our focus on changes in the ecosystem rather than within-species size structure, and many of our 23 species did not have sufficient length data for finer temporal resolution. An analysis of temporal changes in length for the most abundant tunas and billfishes in the catch is presented in Gilman *et al.* [16].

As a robust indicator of ecosystem size structure, we computed annual combined catch for small species (those with mean weights less than 15 kg) and large species (those with mean weights equal to or greater than 15 kg). The value of 15 kg was determined from the species-specific regressions from Table 1 as discussed in the results section. A generalized additive model (GAM) was fit to these two time series to estimate a standardized CPUE time series for small and large species.

Static size-based models, based on metabolic theory and empirical relationships between body size and trophic level, have been applied to investigate unexploited production and biomass of larger marine animals in the global oceans based on current environmental conditions [17]. Dynamic size-spectrum models can extend this approach by considering the time-dependent and continuous growth and mortality processes that result from size-structured feeding, representative of pelagic ecosystems [8,12]. They can be used to predict the consequences of fishing mortality and changes in primary production as well as temperature effects on dynamic changes in the community size spectrum [13]. A key attribute of these models is that the probability of a predator of size M eating an encountered prey of smaller size m is given by a lognormal probability density function, with a mean value representing the preferred predator–prey mass ratio and a standard deviation that represents the breadth of the relative prey mass. A key strength of this approach is that realized predator–prey mass ratios in fish communities do not appear to vary systematically with temperature or primary production in the world's oceans [18].

A size-based ecosystem model based on the pelagic component of the model detailed in Blanchard *et al.* [19] and adapted for the subtropical pelagic ecosystem [20] was used to simulate the response of the size structure to fishing pressure. Input for the model consists of small (<5 μm) and large (>5 μm) phytoplankton densities, SST, size of entry to the fishery, and gear selectivity as a function of size. We used phytoplankton densities and SST output by the NOAA Geophysical Fluid Dynamics Laboratory prototype Earth System Model 2.1 [20,21], averaged both spatially (12°–27° N, 180°–140°W) and temporally (1996–2011) so the only variable input to the size-based model is fishing mortality.

Results

Linear regressions fit to the annual CPUE time series for each of the 23 species found 12 species had statistically significant positive or negative slopes (Table 1). Eight species 16.4 kg or larger had declining CPUE trends while the remaining 6 species 16.4 kg or larger had no significant trend (Table 1). Three species 12.1 kg or

Table 1. Change in catch rate estimated from statistically significant ($P<0.01$) linear regressions over 1996–2011, in order by by fish size.

Species	% Annual Change in CPUE[a] (*P*-value)	Mean Weight in kg (*N*)
Blue Marlin (*Makaira nigricans*)	−5.0 (0.005)	224.0 (1295)
Blue Shark (*Prionace glauca*)	−3.7 (0.004)	106.4 (22856)
Striped Marlin (*Tetrapturus audax*)	−5.0 (0.004)	93.5 (3800)
Shortbill Spearfish (*Tetrapturus angustirostris*)	−4.2 (0.008)	75.7 (4078)
Shortfin Mako Shark (*Isurus oxyrinchus*)	0	48.3 (624)
Swordfish (*Xiphias gladius*)	0	42.0 (1509)
Yellowfin Tuna (*Thunnus albacares*)	0	33.5 (9224)
Opah (*Lampris guttatus*)	−4.1 (0.008)	30.2 (3923)
Bigeye thresher Shark (*Alopias superciliosus*)	0	24.0 (1922)
Unidentified Tuna	0	24.0 (49)
Bigeye Tuna (*Thunnus obesus*)	−2.1 (0.005)	22.5 (41456)
Oceanic White-tip Shark (*Carcharinus longimanus*)	−6.9 (<0.0001)	19.0 (277)
Albacore Tuna (*Thunnus alalunga*)	−6.9 (<0.0001)	17.1 (4718)
Wahoo (*Acanthocybium solandri*)	0	16.4 (4172)
Escolar (*Lepidocybium flavobrunneum*)	12.1 (<0.0001)	12.1 (9817)
Mola (*Ranzania laevis* and *Mola mola*)	0	8.8 (521)
Skipjack Tuna (*Katsuwonus pelamis*)	0	7.9 (9352)
Mahi Mahi (*Coryphaena hippurus*)	0	7.4 (19346)
Lancetfish (*Alepisaurus ferox*)	2.2 (0.026)	7.1 (34186)
Great Barracuda (*Sphyraena jello*)	0	5.9 (1198)
Pomfrets (*Taractichthys steindachneri* and *Brama japonica*)	0	4.9 (14898)
Pelagic Stingray (*Pteroplatytrygon violacea*)	−5.4 (<0.0001)	3.0 (4165)
Snake Mackerel (*Gempylus serpens*)	15.1 (<0.0001)	0.8 (15371)

From left, columns indicate species, annual percent change in CPUE based on linear regression (*P*-values for significant trends in parentheses, insignificant fits denoted by a 0% change), mean species weight as determined from length-weight conversion (number of fish used in length-weight conversion indicated in parentheses).
[a]from linear fit.

smaller had increasing CPUE trends while one, pelagic stingray, had a declining trend and the remaining 5 had no trend (Table 1). The declining species included billfishes, sharks, and tunas, with linear CPUE trend declines ranging from 2 to 7% annually over the 16-year time period. The species with increasing linear CPUE trends were escolar and two noncommercial species, snake mackerel and lancetfish, with increases of about 12, 15, and 2% annually, respectively (Table 1).

To more rigorously and robustly examine the different CPUE time trends for large and small fishes, we split the catch data into 2 groups consisting of 9 species with mean weights <15 kg, termed small fishes and 14 species with mean weights ≥15 kg, termed large fishes (Table 1). The value of 15 kg used to classify large and small fishes was based on the individual CPUE trends in Table 1 showing that 8 of the 9 species with declining CPUE trends had mean weights of 16.4 kg or larger and all of the species with declining CPUE trends had mean weights 12.1 kg or smaller. A GAM was fit to the catch in numbers per longline set for each size group using independent variables: hooks per set, set latitude, set longitude, SST at set location recorded by the observers, and year (all were significant with $p<2\times10^{-16}$). The GAM was then used to estimate annual CPUE for each size group by first using the model to estimate catch for each set, then cumulating the estimated catch for each year, and dividing annual estimated catch by annual observed effort to obtain annual estimated CPUE. The resultant estimated annual CPUE time series fits the observed annual CPUE time series quite well with correlations between the observed and estimated CPUEs of 0. 97 for large fishes and 0.86 for small fishes (Fig. 1). From the GAM, standardized annual CPUE time series for large and small fishes are computed as a function of year by replacing the set SST, latitude, and longitude by mean SST, longitude, and latitude from the 16 year period in the GAM and following the steps outlined previously to obtain annual CPUE. This standardized CPUE time series, computed with mean SST, latitude and longitude, is standardized to eliminate any trends in CPUE due to changes in these variables over the 16-year time period. The temporal trend in standardized annual CPUE for small fishes increased about 25% while it decreased by about 50% for large fishes over the time period (Fig. 2a). To check how robust the observer CPUE pattern was we computed standardized CPUEs for large and small fishes from the logbook data. The results, presented as supplemental material (Table S2 and Fig. S1) show the same pattern of about a 50% decrease in large fish CPUE and about a 33% increase in small fish CPUE (Fig. S1) very similar to that seen with the observer data, although the logbook data reports only the commercially valuable species omitting lancetfish and snake mackerel in the small size group and sharks in the large size groups.

The mean weights of fishes caught in the Hawaii-based deep-set fishery range from 0.8 kg for snake mackerel to 224 kg for blue marlin (Table 1). Thus for the size at entry to the fishery we use 1.0 kg. To estimate the selectivity of the gear we examine the

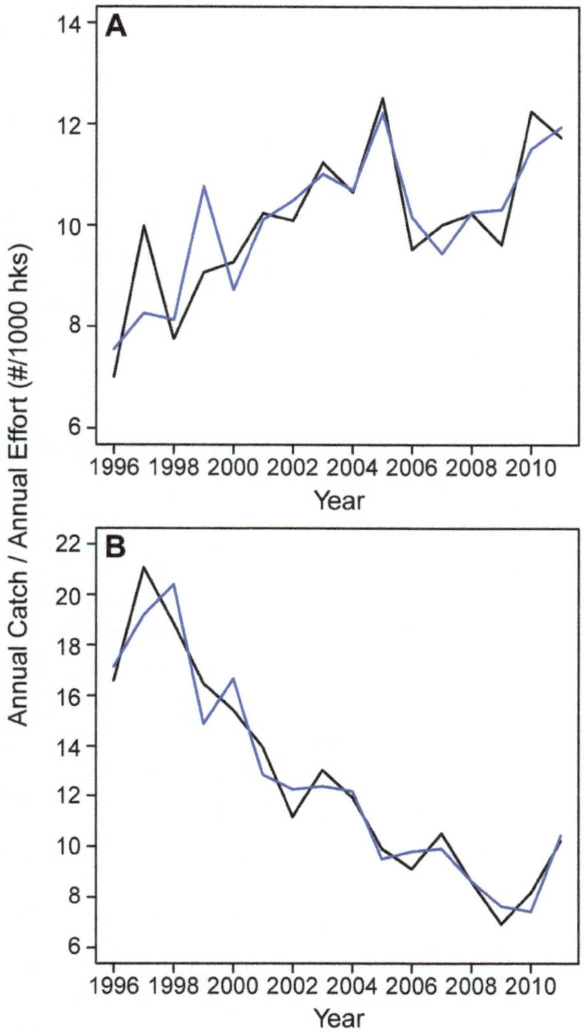

Figure 1. The annual observer and generalized additive model CPUE (# fish per 1000 hooks). Panels indicate (**A**) fishes <15 kg and (**B**) fishes ≥15 kg. In both panels black line represents CPUE from observer data, blue line represents CPUE estimated from the generalized additive model.

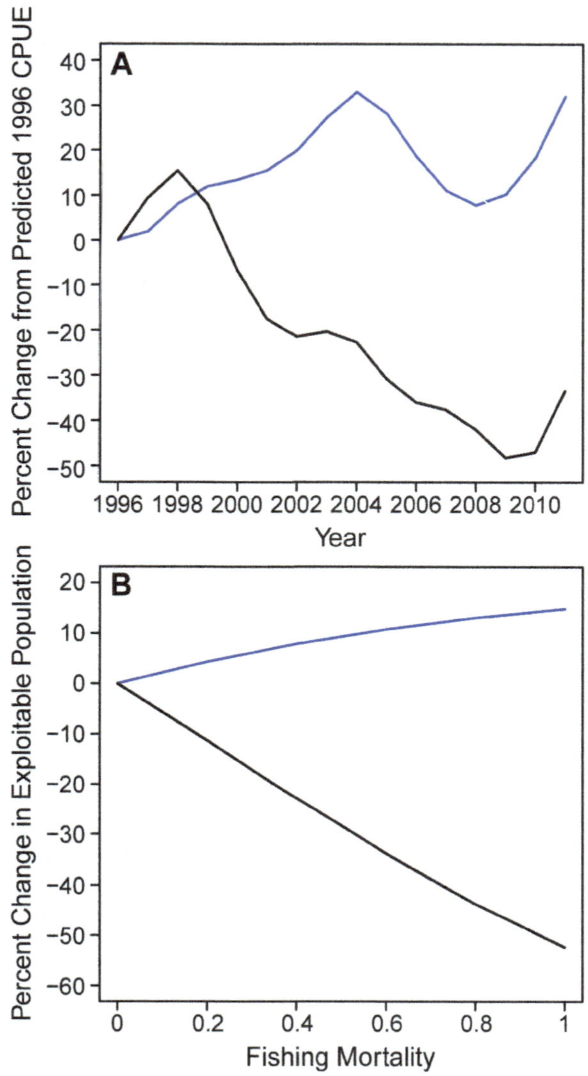

Figure 2. Percent change in standardized CPUE and population size for small and large fishes. Panels indicate (**A**) change in generalized additive model (GAM) standardized CPUE and (**B**) change in size-based model estimated population size for fishes <15 kg (blue) and fishes ≥15 kg (black).

weight-frequency distribution. The weight-frequency distribution of the catch pooled over the 16-year period shows a typical exponential frequency decline with weight above about 15 kg suggesting that fish above this size are largely fully exploited while this is not the case for smaller fishes (Fig. 3). To further define a selectivity function our pelagic size-based model was run with fishing mortality (F) for $F = 0.4$ and $F = 0.6$ to generate catch size distributions. We found a simple size selectivity function where fishes greater than 15 kg experience the full level of F while for fishes in the range of 1–15 kg, F is one fourth the level for the larger fishes generated catch size structures similar to that in Fig. 3.

Using this size selectivity function we compute population abundance as a function of F with the size-based model. As F increases from 0 to 1.0, the population of large fishes declines about 60%, relative to $F = 0$, while the population of small fishes increases about 20%, relative to $F = 0$ (Fig. 2b). The standardized fishery CPUE derived from the GAM shows the same pattern of a decline in large fish CPUE and concurrent increase in small fish CPUE (Fig. 2a). Further, both observed and model trends show the decline in large fishes is substantially greater than the increase in small fishes (Fig. 2).

Next we use the size-based model to examine the change in the entire ecosystem (fished and unfished) size structure in response to various fishing levels. We express the change in ecosystem size structure as a percent change in the size frequency distribution in the absence of fishing relative to that of a fished ecosystem for various levels of F. The change in ecosystem size structure as a function of F shows a decline in the abundance of fishes larger than15 kg, the size of full recruitment to the fishery, and an increase in abundance of fishes within the size range 0.1–15 kg (Fig. 4). For any level of F, the magnitude of the increase in the 0.1–15 kg size range is less than the decline above 15 kg (Fig. 4). A portion of the small size class (1–15 kg) is also fished so this group is responding to both fishing and top-down impacts. For organisms weighing less than 0.1 kg, there is a very slight decrease in abundance but essentially the top-down or size-based cascade has only one cascade with declines for fishes greater than 15 kg resulting in increases for fishes between 0.1–15 kg (Fig. 4).

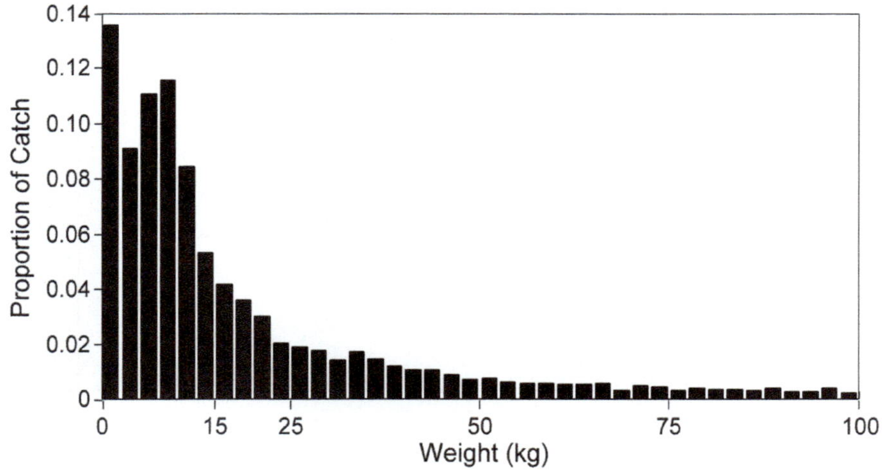

Figure 3. Longline catch weight-frequency distribution. The estimated longline catch weight-frequency distribution for the 23 species listed in Table 1 from the observer data 1996–2011. Distribution truncated above 100 kg.

Size-based indicators derived from catch data including the mean size of the catch or the proportion of large fishes have been proposed as useful indicators to monitor fishery trends and ecosystem impacts. In the presence of a size-based cascade, indicators based only on catch data will necessarily underestimate the impact from fishing on the ecosystem size structure. For example, from the observer data we can monitor the change in the size structure of the catch as the proportion of the catch greater than 15 kg. In 1996 about 70% of the catch was greater than 15 kg but this proportion has declined over time to about 45% by 2011, roughly a 25% decline (Fig. 5a). The size-based model predicts that as F increases from 0.01 to 1.0 the proportion of fishes larger than 15 kg in the catch will decline about 30% (Fig. 5b). This is similar to the decline seen in the observer data. However, we have seen from the size-based model that fishing impacts the ecosystem size structure down to 0.1 kg. Thus a more complete measure of the change in the size structure due to fishing would be to compute the ratio of the population of fish larger than

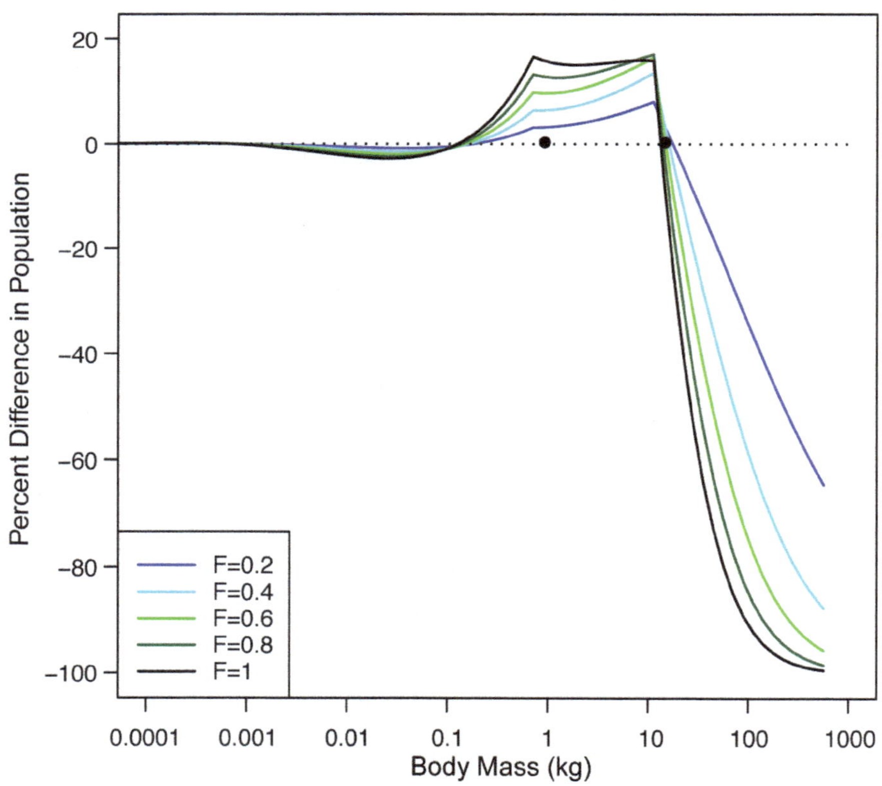

Figure 4. Change in fished ecosystem size structure relative to unfished ecosystem size structure. The percent change in ecosystem abundance by size between the unfished size structure and the fished size structure for F ranging from 0.2 to 1.0. The dots are located at 1 and 15 kg to indicate the size at entry to the fishery and the size of full recruitment.

15 kg relative to all fish larger than 0.1 kg. While we can't do this with catch data, we can with the size-based model. This ratio declines, as F goes from 0.01 to 1.0, by about 55% or almost twice the decline seen in the model catch data since not only does the proportion of large fish decrease but smaller ones increase (Fig. 5b). Thus the change in ecosystem size structure due to fishing may be underestimated if computed from fisheries catch data alone.

The rise in catch rates of noncommercial snake mackerel and lancetfish has an impact on the discards in the fishery. The major components of discards in the fishery are the two non-commercial fishes, lancetfish and snake mackerel, which are 100% discarded, and sharks of which about 95% are discarded based on logbook records. Pelagic stingrays are also discarded but their contribution to the total discards is minimal (Fig. 6). Using these discard proportions and the catch for these 3 species from the observer data, we estimate that in 1996 about 30% of the total catch was discarded while by 2011 this proportion had increased to nearly 40% with about one third of the total catch consisting of snake mackerel and lancetfish (Fig. 6).

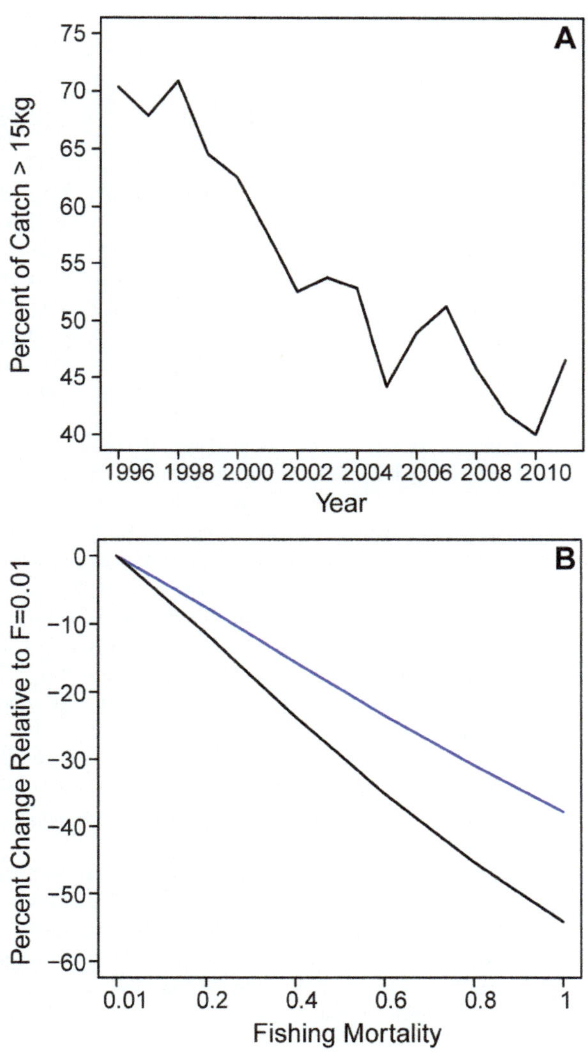

Figure 5. Percent change in catch and biomass. Panels indicate (**A**) the percent of the observed catch >15 kg, by year and (**B**) the percent change in the size-based model large fish catch and biomass relative to F = 0.01 for F ranging from 0.2 to 1.0 for the proportion of catch ≥15 kg (blue) and the proportion of exploitable biomass >0.1 kg that is ≥15 kg (black).

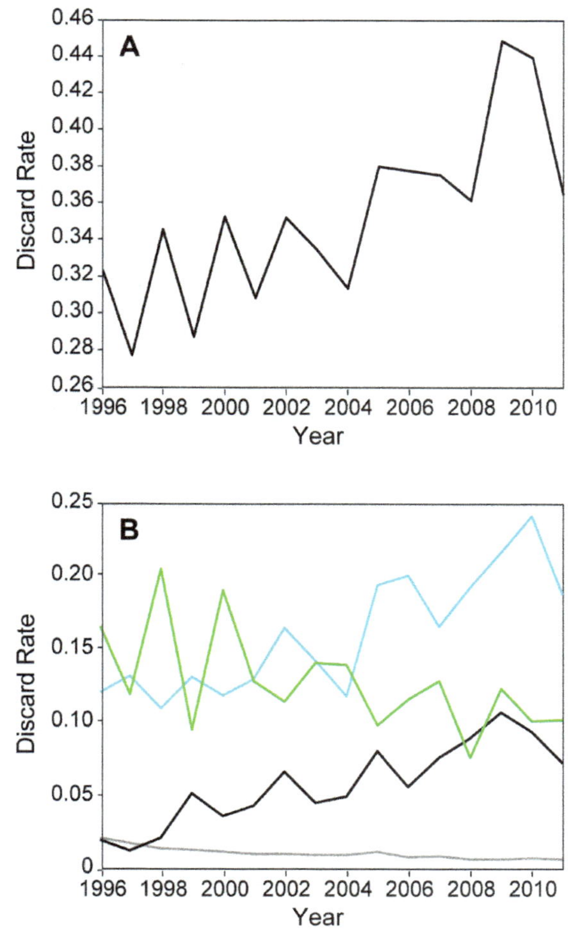

Figure 6. The proportion of the total observed catch estimated to be discarded. Panels indicate (**A**) the combined ratio of estimated discards consisting of the catches of pelagic stingray, snake mackerel, lancetfish, and 95% of the shark catch to total catch and (**B**) the ratio by species of Lancetfish (blue), snake mackerel (black), pelagic stingray (grey), and 95% of the shark catch (green) to total catch.

Discussion

Evidence of fishing impacts to marine ecosystems, especially reductions in trophic structure (fishing down the food web), has been widely reported [22]. Further, decreases in the catch rates of large fishes and increases in the catch rates of small fishes based on longline catches in the central North Pacific between 1950s and 1990s has been previously reported [5]. Our results indicate this trend has continued through 2011. Further, we show that the temporal change in the species and size composition seen over the past 16 years in the Hawaii-based longline observer data is consistent with and can be explained by a size-based model. This suggests that size-based predation is the dominant mechanism in structuring the subtropical pelagic ecosystem, at least the upper trophic levels caught in the deep-set fishery. Earlier work [7] used a species-based model (EwE) to explain this temporal trend as a top-down response using estimated trophic level instead of size. Our current results and the previous ones are consistent in describing the ecosystem change as a top-down response. The fact that two different model approaches reach the same conclusion is seen as positive. We see the current results as a step forward in that a conceptually (size-based predation) and operationally (requiring many fewer estimated input parameters) simpler model explains the observed ecosystem changes.

One implication of this result is that we have a model-based description of the impact of fishing on the entire ecosystem size structure. The model describes a one-step size-based cascade where a reduction of fishes above the size that is fully exploited by the fishery increases the abundance of organisms from about the size of full entry to the fishery down to about 2 orders of magnitude in size but results in little impact on smaller micronekton and plankton. The reason the top-down impact reaches sizes as small as 0.1 kg is predators have a mean prey size that is $1/100^{th}$ their weight with a lognormal distribution around that mean. The prey of fishes above 15 kg would have a mean size 0.15 kg and larger. The reason the size-based cascade, expressed as a percent of ecosystem abundance, diminishes with declining size is the ecosystem abundance increases exponentially as size decreases. Thus predation impacts from a fixed larger size represent a smaller and smaller fraction of the prey population as size decreases.

A key result of this work is the observation that the impacts to size structure extends to sizes below those caught in the fishery, and hence catch-based indicators will underestimate the impact of fishing on the ecosystem size structure. Further, unless there is a targeted sampling for the sizes below the size at entry to the fishery these changes will not be recognized. Size-based models can help to more fully represent the full ecosystem impact of fishing on size structure.

One exception to the pattern of a decline in CPUE of large fishes and concurrent increase for small fishes is the pelagic stingray (mean weight 3 kg) that exhibited a 5.4% annual decline. However, an earlier study [5] found that this species increased in the central Pacific longline catches between 1950s and 1990s and attributed this change to a reduction in predation due to declining shark abundance. Fishing effort since the 1990s has continued to increase and while the rays are not retained they are often severely damaged in the release process [23]. Thus their increased bycatch mortality may exceed the decreased predation mortality resulting in a population decline. Pelagic stingrays have been characterized as one of two elasmobranchs with the lowest risk of extinction due to their resilient life history characteristics [23] but our data suggest this may not be the case.

A recent analysis of temporal trends in catch rates of tunas, sharks, and billfishes based on a GAM using observer data the Hawaii longline fishery documented strong general declining trends in standardized catch rates for bigeye, yellowfin and albacore tunas, blue and oceanic white tip sharks, shortbill spearfish and striped marlin [16]. Our results for these species, just based on annual CPUEs but from a more geographically restricted region and over a slightly different time period, also showed significant declining trends in these species with the exception of yellowfin tuna where our estimated linear trend was not statistically significant. Additionally, Gilman *et al.* [16] looked at trends in length for tunas and billfishes and found the lengths significantly increased over time due to the distributions of length classes having shifted towards larger fish. The authors suggest reasons for this shift may include operational changes in the fishery and/or increased catches of juveniles in the purse seine fishery [16]. Initially, this shift to larger fish seems contrary to our finding of an ecosystem shift to small-sized fishes but the difference is a within-species vs. between-species comparison. We used mean weights averaged over the entire time series for each of the 23 species and described the shift in the ecosystem size structure as the shift in the relative abundance of small and large species. We did not examine temporal size trends within species, as many of our 23 species did not have sufficient length data. However, looking at the modest within species length changes presented in Gilman *et al.* [16] relative to the pretty substantial changes in the proportions of large and small fishes in the catch data we conclude that the main change to ecosystem size structure comes from changes in relative abundance between large and small species and not the smaller changes in size within species.

An ecosystem approach to fisheries management looks at fishery impacts to the entire ecosystem. Clearly the longline fishery is changing the subtropical ecosystem size structure. Time series of CPUEs computed separately for the pooled small and large fishes represent an informative ecosystem indicator of this trend and should be monitored and reported in any analysis of the fishery. Current reporting in the fishery shows only catch and catch rates in numbers of fishes so managers are not as likely to be aware of the greater decline in weight per effort compared to numbers per effort, and the former may be more closely related to economics of the fishery. Lastly this work shows the value of observer data, which unlike the more commonly collected logbook data, provides information on bycatch and discarded species that contributes to a more complete understanding of ecosystem dynamics.

Our size-based model does not suggest any obvious threshold in changes to an ecosystem size structure that could serve as a management target. A recent meta-analysis of global fisheries explores tradeoffs between multispecies maximum sustainable yield (MMSY) and the collapse of individual stocks [24]. Their model finds that for a wide range of exploitation rates ranging from 0.25 to 0.60 the resultant catches equal or exceed 90% of the MMSY, but with an exploitation rate of 0.60 almost half the species in the ecosystem are expected to collapse, while with an exploitation rate of 0.25 less than 10% of the species are expected to collapse [24]. Thus in a multispecies context, taking into consideration aspects of ecosystem structure and function, the exploitation rate that achieves maximum sustainable yield should be considered an upper limit rather than a management target [24]. Unfortunately our observer data represents only a small portion of the Pacific pelagic fishery so estimating MMSY and the corresponding F is problematic. This analysis needs to be conducted on the basin-scale by the appropriate regional fisheries management organizations. The sharp decline of stingrays and oceanic white-tip shark presents concern of collapse for these species. Currently management of the longline fishery is based on single species basin-wide quotas for yellowfin tuna, bigeye tuna, and striped marlin set by the Western and Central Pacific Fisheries Commission. To the extent that these quotas cap fishing effort and mortality on all species they could prevent further ecosystems impacts. Further, ways to reduce the estimated 40% discard rate in the fishery should be a management focus as well. Lastly, this work highlights the critical importance of observer data in monitoring ecosystem changes.

While this paper has focused on changes in ecosystem structure it is clear that with increases in escolar and snake mackerel CPUEs of 12 and 15% per year respectively and declines in pelagic stingray and oceanic white-tip CPUEs of 5.4 and 6.9% per year respectively we are also seeing changes in the ecosystem composition with potential significant impacts on ecosystem function.

Lastly, while we have seen evidence of changes at the base of the ecosystem in the subtropical Pacific over the past decade, they have been modest relative to the substantial increase in fishing effort [25,26]. However, going forward the impact of climate change has been projected to increase its ecosystem impact and shift the subtropical ecosystem size structure toward smaller sizes even if fishing effort remains constant [20,27]. Thus the combined impacts of increased fishing effort and future climate change are projected to be additive and accelerate a shift of ecosystem size structure to smaller sizes. The time series of CPUE for our small fishes group shows considerably more interannual variation than the large fishes group. Many of these small fishes have faster growth rates and shorter life spans than the larger fishes and hence may be more responsive to interannual environmental changes. Thus a shift to smaller fishes may result in greater interannual variation in the longline fishery CPUE.

Supporting Information

Figure S1 The annual logbook and generalized additive model CPUE (# fish per 1000 hooks). Panels indicate (**A**) fishes <15 kg and (**B**) fishes ≥15 kg. In both panels black line represents CPUE from logbook data, blue line represents CPUE estimated from the generalized additive model.

Table S1 Mean species weight, length, and length-weight conversion factors. From left, columns indicate species, mean species weight as determined form length-weight conversions, mean length from those recorded by observers measuring every third fish from 2006–2011, *a*, *b*, and reference listing length-weight conversion factors. To convert length (*L*) in cm to weight (*W*) in g, the equation $aL^b = W$ was used.

Table S2 Change in logbook catch rate estimate from statistically significant ($P<0.01$) linear regressions over 1996–2011, ordered by fish size. From left, columns indicate species, annual percent change in CPUE based on linear regression (*P*-values for significant trends in parentheses, insignificant fits denoted by a 0% change), and mean species weight as determined form length-weight conversion.

Acknowledgments

We thank Dr. Julia Blanchard who provided code and assistance to greatly facilitate our application of the size-based model. We also thank all the scientific observers employed thorough the Pacific Islands Region for their collection of observer data from the longline fishery.

Author Contributions

Analyzed the data: PWJ JJP. Wrote the paper: JJP.

References

1. Pickard GL, Emery WJ (1990) Descriptive physical oceanography: an introduction, 5th ed. Oxford: Pergamon Press. 336 p.
2. Seki MP, Polovina JJ (2001) Ocean gyre ecosystems. In: Steele JH et al., editors. The Encyclopedia of Ocean Sciences, vol. 4. San Diego, CA: Academic Press. 1959–1964.
3. Kitchell JF, Essington TE, Boggs CH, Schindler DE, Walters CJ (2002) The role of sharks and longline fisheries in a pelagic ecosystem of the central Pacific. Ecosystems 5: 202–216.
4. Kitchell JF, Boggs C, He X, Walters CJ (1999) Keystone predators in the Central Pacific. In: Ecosystem approaches to fisheries management. Univ. Alaska Sea Grant. AL-SG-99-01. 665–683.
5. Ward P, Myers RA (2005) Shifts in open-ocean fish communities coinciding with the commencement of commercial fishing. Ecology 86: 835–847.
6. Cox SP, Essington TE, Kitchell JF, Martell SJD, Walters CJ, et al. (2002) Reconstructing ecosystem dynamics in the central Pacific Ocean, 1952–1998. II. A preliminary assessment of the trophic impacts of fishing and effects on tuna dynamics. Canadian Journal of Fisheries and Aquatic Sciences 59: 1736–1747.
7. Polovina JJ, Abecassis M, Howell EA, Woodworth P (2009) Increases in the relative abundance of mid-trophic level fishes concurrent with declines in apex predators in the central North Pacific subtropical gyre, 1996–2006. Fishery Bulletin 107: 523–531.
8. Benoit E, Rochet M (2004) A continuous model of biomass size spectra governed by predation and the effects of fishing on them. Journal of Theoretical Biology 226: 9–21.
9. Blanchard JL, Jennings S, Law R, Castle MD, McCloghrie P, et al. (2009) How does abundance scale with body size in coupled size-structured food webs? Journal of Animal Ecology, 78: 270–280.
10. Blanchard JL, Law R, Castle MD, Jennings S (2011) Coupled energy pathways and the resilience of size-structured food webs. Theoretical Ecology 4: 289–300.
11. Jennings S, Pinnegar JK, Polunin NVC, Boon T (2001) Weak cross-species relationships between body size and trophic level belie powerful size-based trophic structuring in fish communities. Journal of Animal Ecology 70: 934–944.
12. Law R, Plank MJ, James A, Blanchard JL (2009) Size-spectra dynamics from stochastic predation and growth of individuals. Ecology 90: 802–11.
13. Maury O, Shin Y, Faugeras B, Benari T, Marsac F (2007) Modeling environmental effects on the size-structured energy flow through marine ecosystems. Part 2: Simulations. Progress in Oceanography 74: 500–514.
14. Bigelow KA, Musyl MK, Poisson F, Kleiber P (2006) Pelagic longline gear depth and shoaling. Fisheries Research 77: 173–183.
15. Harley SJ, Williams P, Nicol S, Hampton J (2011) The western and central Pacific tuna fishery: 2010 overview and status of stocks. Tuna Fisheries Assessment Report 11. Noumea, New Caledonia: Secretariat of the Pacific Community.
16. Gilman E, Chaloupka M, Read A, Dalzell P, Holetschek J, et al. (2012) Hawaii longline tuna fishery temporal trends in standardized catch rates and length distributions and effects on pelagic and seamount ecosystems. Aquatic Conservation: Marine and Freshwater Ecosystems 22: 446–488.
17. Jennings S, Mélin F, Blanchard JL, Forster RM, Dulvy NK, et al. (2008) Global-scale predictions of community and ecosystem properties from simple ecological theory. Proceedings of the Royal Society of London B: Biological Sciences 275: 1375–1383. doi:10.1098/rspb.2008.0192.
18. Barnes C, Maxwell D, Reuman DC, Jennings S (2010) Global patterns in predator–prey size relationships reveal size dependency of trophic transfer efficiency. Ecology 91: 222–232. doi:10.1890/08-2061.1.
19. Blanchard JL, Jennings S, Holmes R, Harle J, Merino G, et al. (2012) Potential consequence of climate change on primary production and fish production in 28 large marine ecosystems. Philosophical Transactions of the Royal Society of London B: Biological Sciences 367: 2979–2989. doi: 10.1098/rstb.2012.0231.
20. Woodworth-Jefcoats PA, Polovina JJ, Dunne JP, Blanchard JL (2012) Ecosystem size structure response to 21st century climate projection: large fish abundance decreases in the central North Pacific and increases in the California Current. Global Change Biology 19(3): 724–733.
21. Delworth TL, Broccoli AJ, Rosati A, Stouffer RJ, Balaji V, et al. (2006) GFDL's CM2 global coupled climate models. Part I: formulation and simulation characteristics. Journal of Climate 19: 643–374.
22. Pauly D, Palomares ML (2005) Fishing down the food web: It is far more pervasive than we thought. Bulletin of Marine Science 76:197–211.
23. Dulvy NK, Baum JK, Clarke S, Compagno LJV, Cortés E, et al. (2008) You can swim but you can't hide: the global status and conservation of oceanic pelagic sharks and rays. Aquatic Conserv: Mar. Freshw. Ecosyst. 18: 459–482.
24. Worm B, Hilborn R, Baum JK, Branch TA, Collie JS, et al. (2009) Rebuilding global fisheries. Science 325: 578–585.
25. Polovina JJ, Howell EA, Abecassis M (2008) The ocean's least productive waters are expanding. Geophysical Research Letters 35: L03618. doi:10.1029/2007GL031745.
26. Polovina JJ, Woodworth PA (2012) Declines in phytoplankton cell size in the subtropical oceans estimated from satellite remotely-sensed temperature and chlorophyll, 1998–2007. Deep-Sea Research II 77–80: 82–88.
27. Howell EA, Wabnitz CCC, Dunne JP, Polovina JJ (2012) Climate-induced primary productivity change and fishing impacts on the central North Pacific ecosystem and Hawaii-based pelagic longline fishery. Climatic Change. doi 10.1007/s10584-012-0597-z.

4

Role and Value of Nitrogen Regulation Provided by Oysters (*Crassostrea virginica*) in the Mission-Aransas Estuary, Texas, USA

Jennifer Beseres Pollack[1]*, David Yoskowitz[2], Hae-Cheol Kim[3], Paul A. Montagna[2]

1 Department of Life Sciences, Texas A&M University-Corpus Christi, Corpus Christi, Texas, United States of America, **2** Harte Research Institute, Texas A&M University-Corpus Christi, Corpus Christi, Texas, United States of America, **3** I.M. Systems Group at National Centers for Environmental Prediction, National Oceanic Atmospheric Administration, College Park, Maryland, United States of America

Abstract

Suspension-feeding activities of oysters impart a potentially significant benefit to estuarine ecosystems via reduction of water column nutrients, plankton and seston biomass, and primary productivity which can have a significant impact on human well-being. This study considered nitrogen regulation by eastern oysters *Crassostrea virginica* in the Mission-Aransas Estuary, Texas, USA, as a function of denitrification, burial, and physical transport from the system via harvest. Oyster reefs were estimated to remove 502.5 kg N km^{-2} through denitrification of biodeposits and 251.3 kg N km^{-2} in burial of biodeposits to sediments. Nitrogen is also physically transported out of the estuary via harvest of oysters. Commercial harvest of oysters in the Mission-Aransas Estuary can remove approximately 21,665 kg N per year via physical transport from the system. We developed a transferable method to value the service of nitrogen regulation by oysters, where the potential cost equivalent value of nitrogen regulation is quantified via cost estimates for a constructed biological nutrient removal (BNR) supplement to a wastewater treatment plant. The potential annual engineered cost equivalent of the service of nitrogen regulation and removal provided by reefs in the Mission-Aransas Estuary is \$293,993 yr^{-1}. Monetizing ecosystem services can help increase awareness at the stakeholder level of the importance of oysters beyond commercial fishery values alone.

Editor: Simon Thrush, National Institute of Water & Atmospheric Research, New Zealand

Funding: This study was partially supported by grant number NA09NMF4720179 from the National Oceanic and Atmospheric Administration under the Comparative Assessment of Marine Ecosystem (CAMEO) program, and by grant number 12-142 from the Texas General Land Office Coastal Management Program. Support was also provided by Texas A&M University-Corpus Christi and the Harte Research Institute. The funders had no role in study design, data collection and analysis, decision to publish, or preparation of the manuscript.

Competing Interests: The authors have declared that no competing interests exist.

* E-mail: jennifer.pollack@tamucc.edu

Introduction

Oysters have long been acknowledged for their economic importance as a commercially and recreationally harvested species. Within the United States, Texas produces the second-largest oyster harvest, with an estimated \$19 million generated in 2010 [1]. Additionally, oysters offer a suite of critically important ecosystem services that benefit human well-being. Oyster reef structures provide essential refuge and foraging habitat for fish [2–4] and invertebrate species [5]. Filter feeding imparts a potentially tremendous benefit to the ecosystem by influencing nutrient cycling, gas regulation, and water quality [6–8]. Furthermore, the presence of reefs can positively impact the growth of submerged aquatic vegetation (SAV) either through removing suspended solids or acting as a breakwater to reduce sediment re-suspension, both which increase light penetration [9], [10].

Shellfish (including oyster) reefs, are the most imperiled marine habitats on earth, with an estimated 85% lost in relation to historic levels [11]. In response to the drastic decline in oyster populations, there has been an increased focus on the loss of associated ecosystem services that contribute to human well-being (Table 1) [12–15]. Suspension feeding and nutrient regulation activities of oysters have been of particular interest, as they can ameliorate some of the negative effects of nutrient overenrichment, including enhanced primary production and algal blooms [16]. Using a dynamic ecosystem model, Cerco and Noel [17] estimated water quality improvements resulting from different oyster biomass scenarios in Chesapeake Bay. Results suggested that a ten-fold increase in oyster biomass could potentially remove 30,000 kg d^{-1} nitrogen from the system. Newell et al. [18] estimated the value of oyster-mediated nitrogen removal of 13,080 kg N yr^{-1} from the upper Choptank estuary at \$314,836. Using an average value for nitrogen removal, Kasperski and Wieland [19] estimated an ecosystem service value of \$18,135.69 per million oysters in Chesapeake Bay.

The goal of this study was to build on previous research, primarily from the Atlantic U.S. coast, to estimate the role and value of nutrient regulation provided by eastern oysters within a Gulf of Mexico estuary using the replacement cost method. Quantitative valuation of ecosystem services provided by oysters is important for increasing awareness of stakeholders, and is particularly critical for justifying restoration and recovery dollars post-disturbance that exceed the calculated value as a fishery.

Table 1. Ecosystem services of oysters, using functions and services modified from Faber et al. (2006).

Service	Description	Oyster Example
Habitat	Physical place where organisms reside	Fish and invertebrate habitat
Gas Regulation	Regulation of the chemical composition of the atmosphere and oceans	Potential sequestration of atmospheric carbon dioxide
Disturbance Regulation	Dampening environmental fluctuations and disturbance	Shoreline protection, buffering wave energies, erosion prevention
Water Regulation	Flow and purification of water	Influences circulation patterns
Soil Retention	Erosion control and sediment retention	Sediment stabilization, creation of shell hash and sand as shells break down
Nutrient Regulation	Maintenance of major nutrients within acceptable bounds	Reduce water column nutrients, phytoplankton and primary productivity; clean water
Food	Provisioning of edible plants and animals for human consumption	Commercial and subsistence harvesting
Raw materials	Building, manufacturing, fuel, soil, fertilizer	Road base, chicken calcium supplement, cosmetics
Ornamental Resources	Resources for fashion, handicraft, jewelry, etc.	Belt buckles, ornamental construction
Recreation	Opportunities for rest, refreshment, recreation	Fishing, birdwatching
Science and Education	Use of natural areas for science and educational enhancement	Research about oysters, natural reefs provide metrics for restoration
Spiritual and historic	Spiritual and historical information	Middens, oystermen, seafood festivals

Methods

Study Location

Our study site is the Mission-Aransas estuary, Texas, USA. The shallow, bar-built estuary is approximately 540 km^2 with an average depth of 2 m at mid-tide level (Figure 1) [20]. The estuary comprises several bays, the largest of which are Copano Bay, Aransas Bay and Mesquite Bay. Oysters occur primarily as large subtidal reefs in low- to moderate-salinity regions of the estuary [21]. Vertical relief of the reefs ranges from ~0.3 to 1.8 m in height, and the areal extent of is approximately 18.11 km^2 (1811 ha).

Field Data

Oysters and environmental variables were collected monthly throughout 2007 and 2008 by oyster dredge at randomly selected locations on known reefs in Copano Bay and Aransas Bay in cooperation with the Texas Parks and Wildlife Department's (TPWD) Coastal Fisheries Division as described [22]. All necessary collecting permits were obtained from TPWD. Up to 20 live oysters per location were randomly selected and measured for shell height and biomass parameters (wet weight, dry weight, ash-free dry weight). Water samples were collected and hydrographic characteristics (salinity, temperature, dissolved oxygen, pH) were measured approximately 0.2 m above the top of the oyster reef using a YSI series 6 data sonde. Chlorophyll-*a* (Chl-*a*) samples were filtered onto glass fiber filters and placed on ice (<4.0°C). Chl-*a* was extracted overnight and read fluorometrically on a Turner Model 10-AU using a non-acidification technique (EPA method 445.0) [23]. Total suspended solids were measured using EPA method 160.2.

Nitrogen Regulation by Oysters

Nitrogen can be removed from an estuary via physical transport, denitrification, or burial [24]. Bivalves remove significant amounts of planktonic nitrogen from the water column, which is assimilated at rates that vary seasonally [25]. Eastern oysters can assimilate an estimated 50% of the particulate organic nitrogen (PON) filtered, with the remainder voided as biodeposits [26]. Assimilated nitrogen can be removed from the estuarine system via oyster harvest or temporarily sequestered in shell and tissue and available to tertiary consumers. Oyster biodeposits may undergo burial and/or denitrification [18], also resulting in nitrogen removal from the system [27], [28] [29].

Seasonal oyster filtration rates were empirically derived [30] using the general form:

$$FR = \frac{(L^{0.96}T^{0.95})}{2.95} \tag{1}$$

where *FR* is filtration rate (ml ind^{-1} min^{-1}), *L* is oyster shell height (mm), and *T* is temperature (°C).

This equation for filtration rate is modified as a function of salinity [29]:

$$If: \ S \geq 7.5\ psu; \qquad Then: \ FR_S = FR \tag{2}$$

$$If: \ 3.5\ psu < S < 7.5\ psu; \quad Then: \ FR_S = \frac{FR\ x\ (S-3.5)}{4.0} \tag{3}$$

$$If: \ S \leq 3.5\ psu; \qquad Then: \ FR_S = 0.0 \tag{4}$$

where *S* is salinity and FR_S is the salinity effect-corrected filtration rate (ml ind^{-1} min^{-1}).

Filtration rate is also modified as a function of suspended solids in the water column [30] as:

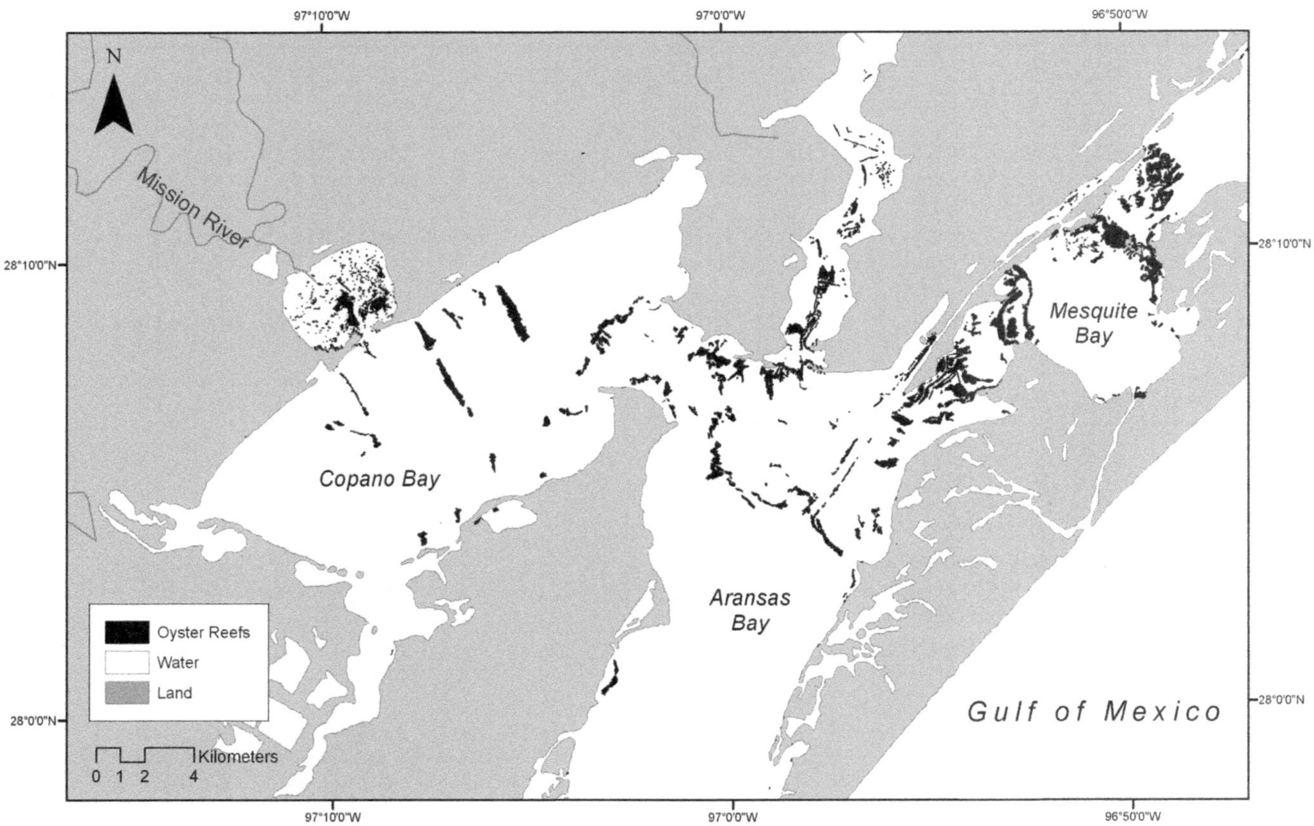

Figure 1. Mission-Aransas Estuary, Texas, USA, showing locations of oyster reefs.

$$FR_{\tau} = FR_S\left[1 - 0.01\left(\frac{\log_{10}\tau + 3.38}{0.418}\right)\right] \tag{5}$$

where τ is total particulate content (g l^{-1}).

Seasons were designated as follows: winter = Dec, Jan, Feb; spring = Mar, Apr, May; summer = Jun, Jul, Aug; fall = Sep, Oct, Nov.

The amount of food available to oyster populations is a function of the amount of food present and the characteristics of water flow delivering the food particles [31], [32]. Chlorophyll-*a* (Chl-*a*) concentration can be used as an index of available food [33], [32] using:

$$Food = 0.085\ x\ Chl \tag{6}$$

where *Chl* is Chl-*a* concentration in μg l^{-1}. Chl-*a* was then converted to nitrogen using an established Chl-*a*:N ratio from the literature [34].

Nitrogen uptake is a function of clearance rate and assimilation efficiency. Seasonal clearance rates were then calculated as [30]:

$$CR = FR\ x\ Food\ x\ 0.001\ l\ ml^{-1} \tag{7}$$

where *CR* is clearance rate (μg N ind^{-1} min^{-1}), *FR* is filtration rate (ml ind^{-1} min^{-1}), and *Food* is available food converted from Chl-*a* in nitrogen units (μg N l^{-1}).

Existing field and laboratory measurements were used to estimate nitrogen removal. To determine the amount of undigested PON that was transferred to sediments in biodeposits, an average nitrogen assimilation efficiency of 50% was applied, as derived from physiological data for eastern oysters [26]. We estimated the amount of PON from biodeposits that is removed through coupled nitrification-denitrification by applying a denitrification rate of 20%, from laboratory derived estimates for eastern oysters [27]. We also applied an average nitrogen burial rate of 10% to the seasonal biodeposition data, based on field measurements from the Choptank River [29]. This method and series of assumptions follow those used by Newell et al. [18] to derive the 33% annual loss estimated for natural intertidal oyster reefs [35].

In order to make our estimates comparable to other systems, we scaled up from individual oyster physiology to standardized units of kg N km^{-2} yr^{-1}. We first calculated nitrogen regulation per km^2 in the Mission-Aransas estuary per year by applying the average number of oysters (408 m^{-2}) and the areal extent of oyster reefs within the estuary (18.11 km^2). A 25% live-to-dead ratio was applied (Beseres Pollack, unpublished data), to adjust for dead oysters that are not producing biodeposits.

Valuation

A straight forward approach to assessing the cost equivalent value of the nutrient regulating service that oysters provide with regards to nitrogen, is to look at the replacement costs to provide an engineered solution via wastewater treatment and specifically a biological nitrogen removal (BNR) processes. The engineered BNR process removes total nitrogen and total phosphorus from wastewater through the use of microorganisms under different environmental conditions [36].

Replacement costs can be a suitable measure of economic value if three conditions are met: i) the engineered system provides the same service as the natural system, ii) the alternative is the least cost, and iii) the services are demanded by society [37]. We estimated the amount of nitrogen removed through an engineered process using the Back River Wastewater Treatment Plant in Maryland as the processing example of the BNR method [36]. The Back River plant, which can process 180 million-gallons-day (mgd), was used because it had relevant data available appropriate to demonstrate the method described here. The dollar cost equivalent of nitrogen regulation via an engineered process was calculated given capital, maintenance, and operation costs for the entire Mission-Aransas estuary.

Results

Environmental conditions were variable within the Mission-Aransas Estuary (Figure 2). Seasonal temperatures ranged from (mean ± SD) 14.4±3.4°C in winter to 30.4±1.0°C in summer. Salinities ranged from (mean ± SD) 15.1±8.4 in winter to 29.6±6.8 in summer. Total suspended solids ranged from (mean ± SD) 0.0277±0.02 g l^{-1} in winter to 0.0365±0.02 g l^{-1} in spring. Seasonal Chl-*a* concentrations ranged from (mean ± SD) 2.69±2.60 ug l^{-1} in summer to 6.31±6.01 ug l^{-1} in fall. Average oyster sizes (shell heights, ± SD) were 69.6±21.7 mm in winter, 73.8±20.8 mm in spring, 70.6±22.3 mm in summer, and 62.5±23.3 mm in fall.

Average oyster filtration rates were calculated seasonally for the Mission-Aransas estuary and ranged from (mean ± SD) 14.5±5.2 ml ind^{-1} min^{-1} in winter to 23.52±7.53 ml ind^{-1} min^{-1} in spring (Table 2). After adjusting for food availability, Chl-*a* was converted to nitrogen at 1 μg Chl-*a*:14 μg N [34]. Seasonal clearance rates were then calculated, ranging from (mean ± SD) 0.063±0.02 μg ind^{-1} min^{-1} in winter to 0.124±0.09 μg ind^{-1} min^{-1} in fall (Table 2).

Oysters assimilate nitrogen with a wide range of efficiencies, due in part to sources of particulate organic matter [38] and season [39]. We applied an average assimilation efficiency of 50% as reported [26] for *Crassostrea virginica* feeding on natural seston at comparable concentrations (4 to 20 mg L^{-1}) in order to estimate the total amount of undigested PON transferred to the sediments in biodeposits. Scaling up to seasonal and annual averages, nitrogen removal from biodeposits associated with coupled nitrification-denitrification ranged from 85.2 kg N km^{-2} in winter to 165.7 kg N km^{-2} in fall, with an annual total of 502.5 kg N km^{-2}, or 9100.28 kg N for 18.11 km^2 of oyster reef in the Mission-Aransas Estuary. Seasonal average nitrogen removal due to PON in biodeposits that is buried in sediments ranged from 42.6 kg N km^{-2} in winter to 82.9 kg N km^{-2} in fall, with an annual total of 251.3 kg N km^{-2} or 4550.14 kg N for 18.11 km^2 of oyster reef in the Mission-Aransas Estuary.

Nitrogen is also physically transported out of the estuary via harvest of oysters. Approximately 8,497,910 kg of oysters (wet meat weight 450,647 kg) were harvested in 2007 and 3,862,997 kg of oysters (wet meat weight 204,856 kg) were harvested in 2008 from the Mission-Aransas Estuary (TPWD commercial landings data). Eastern oyster tissue and shell both comprise nitrogen – at approximately 7% and 0.3% respectively on a dry weight basis [16] – which can be permanently removed from the system as a result of harvest. The allometric relationship between wet and dry weight for oysters in the Mission-Aransas Estuary [22] was then applied:

$$W_{dry} = 0.179 W_{wet} - 0.113 \quad (8)$$

where W_{dry} and W_{wet} are dry weight (g) and wet weight (g) of oyster tissue, respectively. Applying this relationship, approximately 80,666 kg of tissue and 36,669 kg of tissue (on a dry-weight basis) were harvested in 2007 and 2008, respectively. Therefore, approximately 29,789 kg N was removed at harvest in 2007, and 13,541 kg N was removed at harvest in 2008.

How much total nitrogen is removed through an engineered process depends upon the influent and technology used to remove it. At the Back River Wastewater Treatment Plant in Maryland, which uses the Modified Ludzack-Ettinger process (MLE), influent into the BNR process contains 30 mg/l total nitrogen [36]. After treatment, it is 7.6 mg/l, close to a 75% reduction or 22.4 mg/l. Therefore, a facility like the Back River Wastewater Treatment Plant theoretically could remove:

$$\begin{aligned} TN\ removed &= 0.0001869\ (lbs\ gal^{-1})\ x\ 180\ mgd \\ &= 33{,}642\ lbs\ day^{-1}\ (15{,}260\ kg\ day^{-1}) \\ &= 12{,}279{,}330\ lbs\ yr^{-1}\ (5{,}569{,}810\ kg\ yr^{-1)} \end{aligned} \quad (9)$$

Capital, and maintenance and operations costs vary with the type of technology employed as well as the design capacity of the plant. We have calculated that the oyster reefs in the Mission-Aransas system can remove approximately 21,665 kg N per year (average 2007–2008) via physical transport from the system through harvest. This permanent removal equates to ~0.4% of the of the Back River BNR facility while the denitrification and burial equates to ~0.25% of Back River. Therefore, oysters that are harvested can provide the nutrient regulation service, removal in this case through harvest, roughly equivalent to the size of a 0.7 mgd wastewater treatment plant, which has a calculated capital cost of:

$$\begin{aligned} &0.7\ mgd\ x\ \$2{,}975{,}631\ (mgd^{-1}\ unit\ capital\ \cos t) \\ &= \$2{,}082{,}941 \end{aligned} \quad (10)$$

Where the unit capital cost is calculated from the relevant MLE plants in [36]. The annualized value based upon a 15 year life span of a BNR system [40] straight-line depreciation schedule with no scrap value would be $138,863. Straight-line depreciation is the expensing of the original cost of the asset in equal increments over the useful life of the asset. In our example this technique is used to allocate the cost equivalent for each year: $2,082,941/ 15 yrs = $138,863. Assuming conservative maintenance and operation expenses of 2% cost of capital annually [41], the potential annual engineered cost equivalent of the service of nitrogen regulation and removal provided by harvest of oysters in the Mission-Aransas system is:

$$\$138{,}863 + (\$2{,}082{,}941\ x\ 0.02) = \$180{,}522\ yr^{-1} \quad (11)$$

Utilizing the same approach described above, the capital cost equivalent for nitrogen removal via:

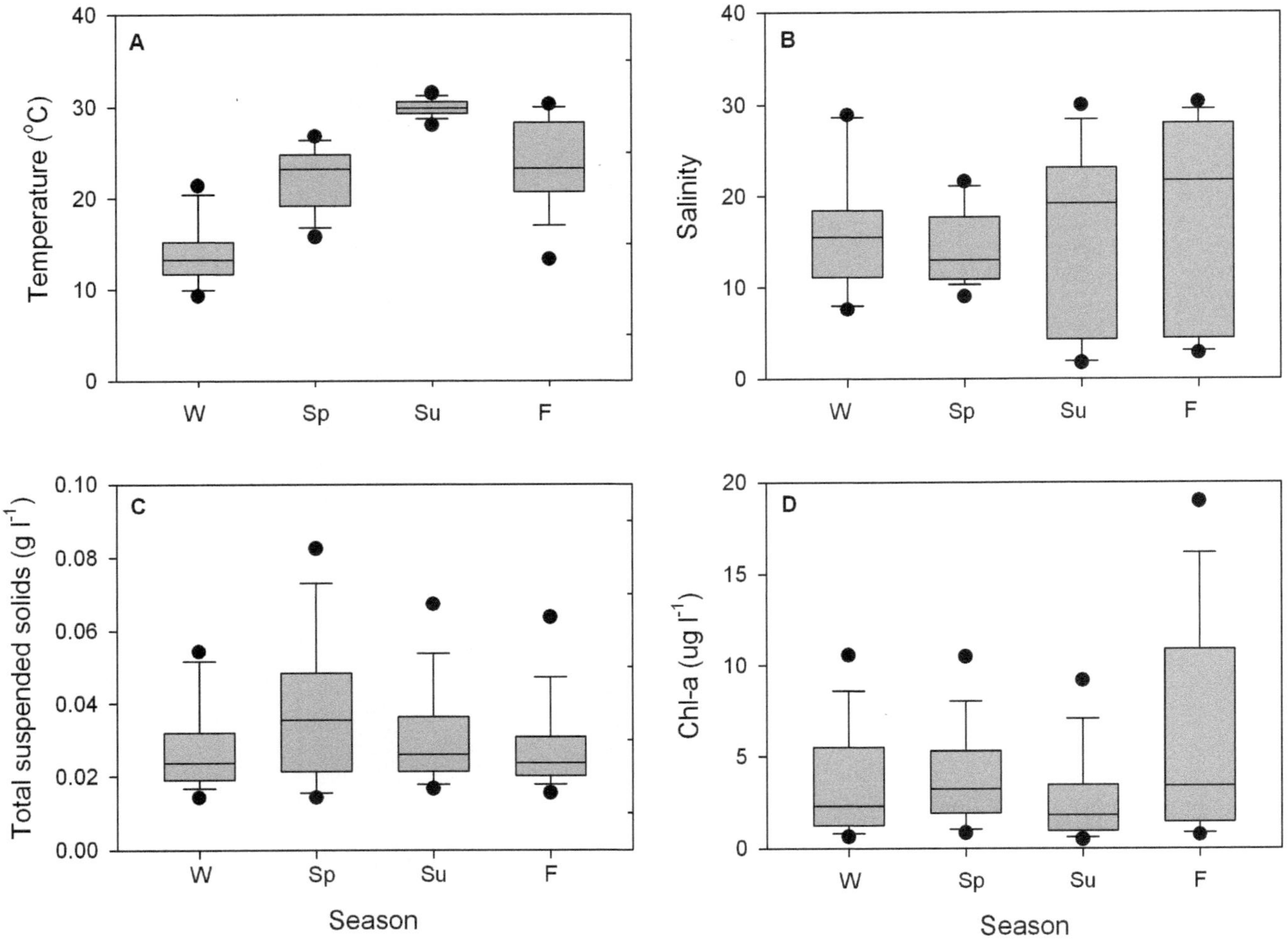

Figure 2. Seasonal environmental variables measured in the Mission-Aransas Estuary, Texas. (W = winter, Sp = spring, Su = summer, F = fall).

Denitrification.

$$0.29\,mgd\ x\ \$2{,}975{,}631\,(mgd^{-1}\ unit\ capital\ \cos t) = \$862{,}933 \quad (12)$$

Burial.

$$0.15\ mgd\ x\ \$2{,}975{,}631\,(mgd^{-1}\ unit\ capital\ \cos t) = \$446{,}345 \quad (13)$$

The annualized cost of a BNR system, once again assuming a 15 year life span and no scrap value, is $57,529 for denitrification and $29,756 for burial. With an annual maintenance and operation cost 2% cost of capital the engineered cost equivalent of the nitrogen regulation and removal service is:

Denitrification

$$\$57{,}529 + (\$862{,}933\ x\ 0.02) = \$74{,}788\,yr^{-1} \quad (14)$$

Burial

$$\$29{,}756 + (\$446{,}345\ x\ 0.02) = \$38{,}683\,mgd^{-1} \quad (15)$$

The combined value from oysters in place, aiding in denitrification and burial, is $113,471 yr^{-1} (see Table 3).

Discussion

Nutrient overenrichment of coastal areas is an ongoing global issue. The most common strategy for dealing with eutrophication is reduction of anthropogenic nutrient loads [42], [43]. Habitat restoration has also been suggested as a potential mechanism for improved nitrogen removal [17], [44], [45]. Suspension feeding activities of oysters can improve water quality in estuarine systems by exerting top-down control on phytoplankton populations [46], [16]. Eastern oysters are unique among other suspension-feeding bivalves in that they can maintain high clearance rates even at high seston concentrations [18]. Oysters transfer water column nitrogen to the sediments in biodeposits, enhancing removal via burial and denitrification [18], [28], [47]. Oysters can also sequester and assimilate nitrogen from the water column into shell

Table 2. Monthly average water temperature (°C), seston concentrations (measured as TSS (mg L^{-1})), and phytoplankton Chl-*a* (μg L^{-1}) in the Mission Aransas Estuary.

					Seasonal nutrient removal	
	Water temp (°C)	Seston (mg L^{-1})	Chl a (μg L^{-1})	Clearance rate (μg ind^{-1} min^{-1})	kg N denitrified km^{-2}	kg N buried km^{-2}
Winter	14.4	27.7	3.68	0.063	85.20	42.60
Spring	20.7	36.5	4.12	0.115	154.71	77.36
Summer	30.4	29.8	2.69	0.072	96.86	48.43
Fall	23.1	28.4	6.31	0.124	165.73	82.87
Annual Total					502.50	251.25

See text for details of calculations of clearance rate and N removal.

and tissue, which can be removed from the system via direct harvest.

We calculated that the 18.11 km^2 oyster reefs in the Mission-Aransas Estuary can annually remove 9100 kg N via coupled denitrification on biodeposits and 4550 kg N via burial of biodeposits in sediments. We also calculated that harvested oysters in the Mission-Aransas system can remove approximately 21,665 kg N annually via physical transport from the system. Although harvested oysters can act as a vehicle for one-time permanent transport of N away from the estuary, unharvested oysters continue to provide nutrient regulation services over their life span [18]. The loss of filtration capacity concurrent with oyster population declines in the Chesapeake Bay over the past century [48] provides a particularly compelling illustration of this point.

The importance of nitrogen regulation services provided by oyster reefs has been demonstrated in several studies. Oyster restoration has been proposed as a management strategy for improving water quality in Chesapeake Bay via top-down control of phytoplankton populations, with estimates of approximately 753 kg N yr^{-1} removed per million oysters [18]. Use of habitat restoration to increase oyster biomass 10-fold in Chesapeake Bay is predicted to reduce Chl-*a* concentrations by 1 mg m^{-3} and remove 30,000 kg N d^{-1} through denitrification [17]. Denitrification rates commonly show seasonal differences, with higher rates in warmer months, concurrent with higher metabolism [49], [50], [28], [47]. Piehler and Smyth [28] measured denitrification rates in Bogue Sound, NC, and estimated the annual value of nitrogen removal by oysters as approximately $3,000 per acre per year. The amount of nitrogen pollution removed by oysters in the Choptank River would otherwise cost over $300,000 yr^{-1} to remove via combination of methods [18]. In order to have a meaningful effect, the population of oysters needs to be large enough to both remove new nitrogen being introduced to the system while also decreasing the existing concentration of nitrogen within the estuary [19].

Commercial and recreational harvest of oysters and other bivalves and extractive aquaculture provide direct mechanisms for nitrogen removal from coastal systems. Commercial mussel farming in Sweden has been predicted to reduce the net transport of nitrogen at the mouth of the Gullmar Fjord by 20%, providing a significant service to Swedish society [51]. The potential harvest of 10^6 cultured oysters yr^{-1} from Chesapeake Bay is estimated to remove 132 kg N yr^{-1}, and could offer financial incentives in the form of nutrient reduction payments to help scale-up aquaculture production activities [52], [53].

There is a clear trade-off between the one-time value of harvesting oysters and leaving them within the estuary to provide ongoing ecosystem services [18]. For example, in Chesapeake and Delaware Bays, severe declines in eastern oyster populations, due in part to habitat destruction and overfishing, corresponded with rapid ecosystem change to planktonic food-web dominant systems [48], [54]. When oysters are harvested, it potentially diminishes the pool of available suspension feeders and reduces their overall capacity to filter the water column [48].

Nutrient regulation is not the only service potentially supplied by oyster reefs. In fact, as Table 1 illustrates there are numerous services that only add to the value of nutrient regulation, so any monetary estimate that focuses only on one ecosystem service will undervalue a reef. For example, three-dimensional oyster reef structures provide nursery, refuge, and foraging habitat for fish and macroinvertebrates [2], [3]. Reefs can also provide sediment stabilization and shoreline protection from both human (wake from passing boats) and natural disturbance (storms) [55]. Cultural services such as recreational fishing over and around reefs [56], [57], science and education opportunities that are available when research is conducted on oysters [7], [18], food [58], and spiritual and historic services such as the tradition of oystermen [59] are also extremely important and valued by society.

In the current study, we demonstrate a process in which one ecosystem service can be monetized. The value is derived via a cost equivalent of an engineered approach using a BNR process. This ecosystem service is provided at no direct charge to the public but can have a significant impact on human well-being. For example, humans do not need to pay increased tax or utility rates for a wastewater treatment system that would include nutrient removal. When utilizing the replacement cost approach it should be done under the notion that the "costs" will not necessarily equate to "benefits" except under very restricted circumstances [60]. What is demonstrated here is simply a cost equivalent of the service based upon the BNR process.

Table 3. Nitrogen removal and cost equivalent values provided by 18.11 km^2 of oyster reef in the Mission-Aransas Estuary.

	Harvest	Denitrification	Burial	Total
Annual nitrogen removal (kg)	21,665	9,100	4,550	35,315
Cost equivalent value (*yr^{-1}*)	$180,522	$74,788	$38,683	**$293,993**

While there is currently no regulatory need for nutrient regulation in the Mission-Aransas estuary, this pilot study demonstrates a valuation process that can be transferable to nutrient enriched systems. In eutrophic estuaries, oyster restoration is recommended as a complement, not a substitute, to load reduction [17]. The results of this study are sensitive to the assimilation rate and the capital cost. The assimilation rate varies based upon oyster size, available food, and environmental conditions [61]. The capital costs utilized in this study were the best available estimations on average and are not site specific, which would produce better estimates of the replacement cost value of nitrogen regulation. This study utilizes a novel approach, the replacement cost method, for estimating the ecosystem service of nutrient regulation provided by oysters. Previous studies most commonly associated with waste or nutrient regulation and storm protection are linked to the services provided by wetlands [62–64].

Other techniques exist for estimating the value of ecosystem services such as willingness-to-pay (WTP), attribute based stated-choice, hedonic pricing, and travel cost. Utilizing different techniques will generate different values so it is critical to employ the appropriate technique for the question being asked. For example, Piehler and Smyth [28] utilized values derived from the North Carolina nutrient offset trading program of \$13 per kg of nitrogen removed in order to quantify the denitrification value of various estuarine habitats. Newell et al. [18] employ an average monetary value of \$24.07 kg^{-1} for removing nitrogen from an early study and apply it specifically to their study site. Including the combined value for denitrification and burial calculated in this study of \$8.33 kg^{-1} using a replacement cost approach, the values generated by these three studies are reasonably similar given their different approaches.

Effective management of oyster reefs and the services they provide requires that both natural and social scientists collaborate to develop tools that can be utilized by stakeholders. Monetizing ecosystem services can help increase awareness at the stakeholder level of the importance of oysters beyond commercial fishery values alone. While discussion of potential reef conservation and restoration mechanisms are beyond the scope of this paper, serious consideration should be given to the process of "operationalizing" ecosystem services of oyster reefs for restoration so that both public and private entities have the opportunity to support restoration and at the same time receive credit for enhanced services, such as reducing nutrient loads, creating habitat, or protecting shorelines. An increase in a multidisciplinary effort, including funding, to quantify coastal and marine ecosystem services both in monetary and non-monetary terms is critical to help close the information gap that exists to help make effective management decisions.

Acknowledgments

We thank Texas Parks and Wildlife Department's Coastal Fisheries Division in Rockport, Texas, especially Nick Ahrens, Willy Cupit, Mark Fisher, Christopher Mace, Karen Meador, and Luis Uballe. We also thank Carlota Santos from the Harte Research Institute for her assistance.

Author Contributions

Conceived and designed the experiments: DY JBP HK. Performed the experiments: DY JBP HK. Analyzed the data: DY JBP HK PAM. Wrote the paper: DY JBP.

References

1. NOAA Fisheries (2011) Office of Science and Technology, annual commercial landing statistics. Available: http://www.st.nmfs.noaa.gov/st1/commercial/landings/annual_landings.html. Accessed 2011 Oct 12.
2. Coen LD, Luckenbach M, Breitburg D (1999) The role of oyster reefs as essential fish habitat: a review of current knowledge and some new perspectives. In Fish habitat: essential and rehabilitation, Symposium 438–454. Bethesda, MD: American Fisheries Society.
3. Reese Robillard M, Stunz G, Simons J (2010) Relative value of deep subtidal oyster reefs to other estuarine habitat types using a novel sampling method. Jrnl of Shellfish Res 29: 291–302.
4. Stunz G, Minello T, Rozas L (2010) Relative value of oyster reef as habitat for estuarine nekton in Galveston Bay, Texas. Mar Ecol Prog Ser 406: 147–159.
5. Wells HW (1961) The fauna of oyster beds with special reference to the salinity factor. Ecol Mono 31: 239–266.
6. French McCay D, Peterson C, DeAlteris J, Catena J (2003) Restoration that targets function as opposed to structure: replacing lost bivalve production and filtration. Mar Ecol Prog Ser 264: 197–212.
7. Grizzle R, Greene J, Coen LD (2008) Seston removal by natural and constructed intertidal Eastern Oyster (Crassostrea virginica) Reefs: A comparison with previous laboratory studies, and the value of in situ methods. Estuaries Coasts, 31: 1208–1220.
8. Nelson K, Leonard L, Posey M, Alphin T, Mallin M (2004) Using transplanted oyster (*Crassostrea virginica*) beds to improve water quality in small tidal creeks: a pilot study. J Exp Mar Biol Ecol 298: 347–368.
9. Dennison W, Orth R, Moore K, Stevenson J, Carter V, et al. (1993) Assessing water quality with submersed aquatic vegetation. BioScience 43: 86–94.
10. Smith KA, North EW, Shi F, Chen S, Hood RR, et al. (2009) Modeling the effects of oyster reefs and breakwaters on seagrass growth. Estuaries Coasts 32: 748–757.
11. Beck MW, Brumbaugh RD, Airoldi L, Carranza A, Coen LD, et al. (2011) Oyster reefs at risk and recommendations for conservation, restoration, and management. Bioscience 61: 107–116.
12. Peterson C, Lipcius R (2003) Conceptual progress towards predicting quantitative ecosystem benefits of ecological restorations. Mar Ecol Prog Ser 264: 297–307.
13. Brumbaugh RD, Toropova C (2008) Economic valuation of ecosystem services: A new impetus for shellfish restoration? Basins Coasts 2: 8–15.
14. Coen LD, Brumbaugh RD, Bushek D, Grizzle R, Luckenbach M, et al. (2007) Ecosystem services related to oyster restoration. Mar Ecol Prog Ser 341: 303–307.
15. Grabowski J, Peterson C (2007) Restoring oyster reefs to recover ecosystem services. In Ecosystem Engineers - Plants to Protists, ed. Cuddington K, Byers J, Wilson J, Hastings WA Elsevier-Academic Press, Amsterdam, p 281–298.
16. Newell RIE (2004) Ecosystem influences of natural and cultivated populations of suspension-feeding bivalve molluscs: A review. J Shellfish Res 23: 51–61.
17. Cerco CF, Noel MR (2007) Can oyster restoration reverse cultural eutrophication in Chesapeake Bay? Estuaries Coasts 30: 331–343.
18. Newell RIE, Fisher TR, Holyoke RR, Cornwell JC (2005) Influence of eastern oysters on nitrogen and phosphorus regeneration in Chesapeake Bay, USA. In: The comparative Roles of Suspension Feeders in Ecosystems. R Dame and S. Olenin (Eds.) Vol 47 NATO Science Series: IV - Earth and Environmental Sciences. Springer, Netherlands. P 93–120.
19. Kasperski S, Wieland R (2009) When is it optimal to delay harvesting? The role of ecological services in the northern Chesapeake Bay oyster fishery. Mar Resour Econ 24: 361–385.
20. Finkbeiner M, Simons J, Robinson C, Wood J, Summers A, et al. (2009) Atlas of Shallow-Water Benthic Habitats of Coastal Texas: Espiritu Santo Bay to Lower Laguna Madre, 2004 and 2007. NOAA Coastal Services Center, Charleston, SC.
21. Beseres Pollack J, Cleveland A, Palmer TA, Reisinger AS, Montagna PA (2012) A restoration suitability index model for the Eastern oyster (*Crassostrea virginica*) in the Mission-Aransas Estuary, TX, USA. PLoS ONE 7(7): e40839. doi:10.1371/journal.pone.0040839.
22. Beseres Pollack J, Kim HC, Morgan E, Montagna P (2011) Role of flood disturbance in natural oyster (*Crassostrea virginica*) population maintenance in an estuary in South Texas, USA. Estuaries Coasts 34: 187–197.
23. Welschmeyer NA (1994) Fluorometric analysis of chlorophyll a in the presence of chlorophyll b and pheopigments. Limnol Oceanogr 39: 1985–1992.
24. Vitousek PM, Ader J, Howarth RW, Likens GE, Matson PA, et al. (1997) Human alteration of the global nitrogen cycle: sources and consequences. Ecol Appl 7(3): 737–750.
25. Kreeger DA, Newell RIE (2001) Seasonal utilization of different seston carbon sources by the ribbed mussel, *Geukensia demissa* (Dillwyn) in a mid-Atlantic salt marsh. J Exp Mar Biol Ecol 260: 71–91.
26. Newell RIE, Jordan SJ (1983) Preferential ingestion of organic material by the American Oyster *Crassostrea virginica* Mar Ecol Prog Ser 13: 47–537.
27. Newell RIE, Cornwell JC, Owens MS (2002) Influence of simulated bivalve biodeposition and microphytobenthos on sediment nitrogen dynamics: A laboratory study. Limnol Oceanogr 47: 1367–1379.
28. Piehler MF, Smyth AR (2011) Habitat-specific distinctions in estuarine denitrification affect both ecosystem function and services. Ecosphere 2: art 12.

29. Boynton WR, Garber JH, Summers R, Kemp WM (1995) Input, transformations, and transport of nitrogen and phosphorus in Chesapeake Bay and selected tributaries. Estuaries 18: 285–314.
30. Powell E, Klinck J, Hofmann E, Wilson-Ormond E, Ellis M (1995) Modeling oyster populations. V. Declining phytoplankton stocks and the population dynamics of the American oyster (Crassostrea virginica) populations. Fish Res 24: 199–222.
31. Soniat TM, Ray SM, Jeffrey LM (1984) Components of the seston and possible available food for oysters in Galveston Bay, Texas. Contrib Mar Sci 27: 127–141.
32. Wilson-Ormond EA, Powell EN, Ray SM (1997) Short-term and small-scale variation in food availability to natural oyster populations: food, flow, and flux. P.S.Z.N.: Mar Ecol 18: 1–34.
33. Hyun KH, Pang IC, Klinck JM, Choi KS, Lee JB, et al. (2001) The effect of food composition on Pacific oyster *Crassostrea gigas* (Thunberg) growth in Korea: a modeling study. Aquaculture 199: 41–62.
34. Parsons TR, Takahashi M, Hargrave B (1984) Biological Oceanographic Processes. Pergamon Press, New York, 330 p.
35. Dame RF, Spurrier JD, Wolaver TG (1989) Carbon, nitrogen and phosphorus processing by an oyster reef. Mar Ecol Prog Ser 54: 249–256.
36. Environmental Protection Agency, Office of Water (2007) Biological Nutrient Removal Processes and Costs. EPA-823-R-07–002, Washington D.C.
37. Shabman L, Batie S (1978) Economic value of natural coastal wetlands: A critique. Coast Manage 4: 231–247.
38. Romberger HP, Epifanio CE (1981) Comparative effects of diets consisting of one or two algal species upon assimilation efficiencies and growth of juvenile oysters, *Crassostrea virginica* (Gmelin). Aquaculture 25(1): 77–87.
39. Bayne BL, Newell RC (1983) Physiological energetics of marine molluscs. In: The Mollusca, Volume 4. Physiology, Part 1, ASM Saleuddin and KM Wilbur (Eds), Academic Press, New York, USA, p 407–515.
40. Foley J, de Hass D, Hartley K, Lant P (2007) Life cycle assessment of biological nutrient removal wastewater treatment plants. 3rd International Conference on Life Cycle Management. Zurich Switzerland.
41. Scranton Sewer Authority (2006) Long Term Control Plan Report. Available: http://www.scrantonsewer.org/docs/LTCP.pdf.
42. D'Elia CF, Harding LW, Leffler M, Mackiernan GB (1992) The role and control of nutrients in Chesapeake Bay. Water Sci Technol 26: 2635–2644.
43. National Research Council (NRC) (2000) Clean Coastal Waters. National Academy Press, Washington, D.C.
44. Byers JE, Cuddington K, Jones CG, Talley TS, Hastings A, et al. (2006) Using ecosystem engineers to restore ecological systems. Trends Ecol Evol 21: 493–500.
45. Craig LS, Palmer MA, Richardson DC, Filoso S, Bernhardt ES, et al. (2008) Stream restoration strategies for reducing river nitrogen loads. Front Ecol Environ 6: 529–538.
46. Dame RF (1996) Ecology of marine bivalves: an ecosystem approach. CRC Press, Boca Raton, FL.
47. Smyth AR, Thompson SP, Siporin KN, Gardner WS, McCarthy MJ, et al. (2013) Assessing nitrogen dynamics throughout the estuarine landscape. Estuaries Coasts 36: 44–55.
48. Newell RIE (1988) Ecological changes in Chesapeake Bay: are they the result of overharvesting the American oyster, *Crassostrea virginica*? In: Lynch MP, Krome EC (eds). Understanding the estuary: advances in Chesapeake Bay research. Chesapeake Research Consortium Publication 129, Baltimore, MD, 536–546.
49. Yoon WB, Benner R (1992) Denitrification and oxygen consumption in sediments of two south Texas estuaries. Mar Ecol Prog Ser 90: 157–167.
50. Thompson S, Paerl H, Go M (1995) Seasonal patterns of nitrification and denitrification in a natural and restored salt marsh. Estuaries Coasts 18: 399–408.
51. Lindahl O, Hart R, Hernroth B, Kollberg S, Loo L, et al. (2005) Improving marine water quality by mussel farming: A profitable solution for Swedish society. AMBIO: J Human Environ 34: 131–138.
52. Shumway SE, Davis C, Downey R, Karney R, Kraeuter J, et al. (2003) Shellfish aquaculture - in praise of sustainable economies and environments. World Aquaculture 34: 15–17.
53. Higgins C, Stephenson K, Brown B (2011) Nutrient bioassimilation capacity of aquacultured oysters: Quantification of an ecosystem service. J Environ Qual 40: 271–277.
54. Rothschild B, Ault J, Goulletquer P, Heral M (1994) Decline of the Chesapeake Bay oyster population: A century of habitat destruction and overfishing. Mar Ecol Pro Ser 111: 29–39.
55. Piazza B, Banks P, La Peyre M (2005) The potential for created oyster shell reefs as a sustainable shoreline protection strategy in Louisiana. Restor Ecol 13: 499–506.
56. Isaacs J, Keithly W, Lavergne D (2004) The Value of Louisiana Oyster Reefs to Recreational Fishermen, in: Louisiana's Oyster Shell Recovery Pilot Project. Final report prepared by Louisiana Department of Wildlife and Fisheries, Socioeconomics Research and Development Section and Marine Fisheries Division. Submitted to the National Marine Fisheries Service under grant number NA96FK0188.
57. Hicks R (2004) Recreational Fishing and the Benefits of Oyster Reef Restoration in the Chesapeake Bay (No. #1), Working Papers. Department of Economics, College of William and Mary, Williamsburg, VA.
58. Lipton D (2008) Economic benefits of a restored oyster fishery in Chesapeake Bay. J Shellfish Res 27: 619–623.
59. Gaillard F (2007) After the storms: Tradition and change in Bayou La Batre. J Am Hist 94: 856–862.
60. Barbier E (2007) Valuing ecosystem services as productive inputs. Econ Policy. 22: 177–229.
61. Dame RF (1972) The ecological energies of growth, respiration and assimilation in the intertidal American oyster, *Crassostrea virginica*. Mar Biol 17: 243–250.
62. Breaux A, Farber S, Day J (1995) Using natural coastal wetlands systems for wastewater treatment: an economic benefit analysis. J Environ Manage 44: 285–291.
63. Byström O (2000) The replacement value of wetlands in Sweden. Environ Resour Econ 16: 347–362.
64. Gunawardena M, Rowan J (2005) Economic valuation of a mangrove ecosystem threatened by shrimp aquaculture in Sri Lanka. J Environ Manage 36: 535–550.

5

Warming Reduces Carbon Losses from Grassland Exposed to Elevated Atmospheric Carbon Dioxide

Elise Pendall[1]*, Jana L. Heisler-White[2,3], David G. Williams[3], Feike A. Dijkstra[2,4], Yolima Carrillo[1,4], Jack A. Morgan[2], Daniel R. LeCain[2]

1 Department of Botany and Program in Ecology, University of Wyoming, Laramie, Wyoming, United States of America, **2** United States Department of Agriculture – Agricultural Research Service, Rangeland Resources Research Unit and Northern Plains Area, Fort Collins, Colorado, United States of America, **3** Departments of Botany; Ecosystem Science and Management, and Program in Ecology, University of Wyoming, Laramie, Wyoming, United States of America, **4** Faculty of Agriculture and Environment, University of Sydney, Sydney, New South Wales, Australia

Abstract

The flux of carbon dioxide (CO_2) between terrestrial ecosystems and the atmosphere may ameliorate or exacerbate climate change, depending on the relative responses of ecosystem photosynthesis and respiration to warming temperatures, rising atmospheric CO_2, and altered precipitation. The combined effect of these global change factors is especially uncertain because of their potential for interactions and indirectly mediated conditions such as soil moisture. Here, we present observations of CO_2 fluxes from a multi-factor experiment in semi-arid grassland that suggests a potentially strong climate – carbon cycle feedback under combined elevated [CO_2] and warming. Elevated [CO_2] alone, and in combination with warming, enhanced ecosystem respiration to a greater extent than photosynthesis, resulting in net C loss over four years. The effect of warming was to reduce respiration especially during years of below-average precipitation, by partially offsetting the effect of elevated [CO_2] on soil moisture and C cycling. Carbon losses were explained partly by stimulated decomposition of soil organic matter with elevated [CO_2]. The climate – carbon cycle feedback observed in this semiarid grassland was mediated by soil water content, which was reduced by warming and increased by elevated [CO_2]. Ecosystem models should incorporate direct and indirect effects of climate change on soil water content in order to accurately predict terrestrial feedbacks and long-term storage of C in soil.

Editor: Carl J. Bernacchi, University of Illinois, United States of America

Funding: This project was supported by the United States Department of Agriculture Agricultural Research Service Climate Change, Soils & Emissions Program, USDA-CSREES Soil Processes Program (Grant no. 2008-35107-18655), United States Department of Energy's Office of Science (BER), through the Terrestrial Ecosystem Science program and the Western Regional Center of the National Institute for Climatic Change Research at Northern Arizona University, and by NSF (DEB# 1021559). The funders had no role in study design, data collection and analysis, decision to publish, or preparation of the manuscript.

Competing Interests: The authors have declared that no competing interests exist.

* E-mail: Pendall@uwyo.edu

Introduction

Models predict declining C sequestration in the coming century [1,2,3] as ecosystem respiration (R_{eco}) is preferentially stimulated over ecosystem photosynthesis (P_{eco}), but experimental tests for these predictions are lacking [4]. Experimental manipulations of single global change factors have greatly improved our understanding of ecological processes that regulate C exchange [5,6]. Photosynthesis is stimulated by elevated [CO_2] due to increased biochemical forcing and improved water use efficiency [7], but the magnitude of these mechanisms varies both within and across ecosystems [8]. Warming has been shown to increase, decrease, and have no effect on C assimilation [6] – responses that are tied to enzymatic reaction rates, plant photosynthetic acclimation, potential changes in growing season length, and resource availability. On a physiological level, R_{eco}, composed of both autotrophic (R_a) and heterotrophic (R_h) respiration, responds more strongly to temperature than does photosynthesis [9], underpinning model predictions that ecosystem C storage will gradually decrease in a future warmer world [1,10]. Multifactor climate change experiments are needed to test the predictions of global C cycle models and identify the strength of interactive effects on ecosystem C uptake and loss [4].

Recent modeling and meta-analysis studies suggest that the combination of warming and elevated [CO_2] will increase biomass and soil respiration in grassland [11,12], but this does not address what mechanisms underlie the responses. Furthermore, only six experiments in the meta-analysis combined these treatments in natural ecosystems, and of those, just one reported measurements of R_h, which is a key determinant of long-term climate – C cycle feedbacks. Biomass and respiration responses to elevated [CO_2] are known to be mediated by indirect effects of soil water enhancement [8,13], whereas warming-induced drying may counteract the effects of moisture on C cycling. These emergent, ecosystem-level properties are not well represented in meta-analyses or simulation models, because virtually no experimental data exists to validate the models.

Since 2006, we have conducted a global change experiment in a temperate semi-arid native grassland in southeastern Wyoming, USA to study the combined impacts of elevated [CO_2] and warming on ecosystem C dynamics and C balance. The **P**rairie **H**eating **A**nd **C**O_2 **E**nrichment (**PHACE**) experiment combines a full-factorial manipulation of these global change conditions and a supplemental irrigation treatment across 25 replicate plots (n = 5 per treatment type). Previous research at PHACE demonstrated that elevated [CO_2] and warming together enhanced net primary production (NPP), especially in C4 grasses [14], but the gross and net CO_2 fluxes, and therefore feedbacks to climate change, are still in question.

We hypothesized that soil water availability (SWC) would mediate treatment effects on gross CO_2 fluxes and R_h, such that 1) elevated [CO_2] alone (referred to as the Ct treatment) would stimulate R_{eco} more than P_{eco} due to enhanced R_h, leading to net C losses, compared to the ambient (ct) treatment [13,15]; 2) warming alone (cT treatment) would stimulate P_{eco} and suppress R_{eco} relative to ambient conditions, as soil moisture limits R_h, leading to net C gains relative to ambient conditions [6]; and that 3) elevated [CO_2] plus warming (CT treatment) would lead to no stimulation or suppression of R_{eco} or P_{eco} relative to ambient conditions, because SWC is the same in this treatment as ambient [14]. We further hypothesized that CO_2 fluxes in the irrigated (ct-i) treatment would follow a similar pattern to the elevated [CO_2] treatment because SWC was manipulated to match that of the Ct treatment.

Materials and Methods

Experimental Manipulation and Field Site

The Prairie Heating and CO_2 Enrichment (PHACE) experiment is located in a northern mixed grass prairie (NMP) ecosystem at the United States Department of Agriculture Agricultural Research Service (USDA-ARS) High Plains Grasslands Research Station in Cheyenne, Wyoming, USA, with full approval by HPGRS management. The vegetation is dominated by the C_3 grass *Pascopyrum smithii* (Rydb.) A Love and the C_4 grass *Bouteloua gracilis* (H.B.K) Lag, and other abundant species include the C_3 grass *Hesperostipa comata* Trin and Rupr., the sedge *Carex eleocharis* L. Bailey, and the sub-shrub *Artemisia frigida* Willd. No protected species were sampled during this research. Mean annual precipitation is 384 mm and mean maximum and minimum air temperatures are 17.5°C in July and −2.5°C in January, respectively. The soil is a fine-loamy, mixed, mesic Aridic Argiustoll.

In 2005, 25 circular plots were established ca. 3.4 m in diameter and surrounded by an impermeable barrier that was buried to 60 cm soil depth. Free-Air CO_2 Enrichment [16] began in 2006 and elevates [CO_2] to 600 μmol mol^{-1}, and warming began in 2007 with infrared heaters that elevate plant canopy temperatures 1.5 and 3.0°C during the day and night, respectively [17], in a full factorial design with 5 replicates for each of the 4 combinations (ct, ambient CO_2 and ambient temperature; cT, ambient CO_2 and elevated temperature; Ct, elevated CO_2 and ambient temperature; and CT, elevated CO_2 and elevated temperature). Five plots were exposed to ambient CO_2 and temperature and received periodic irrigations to maintain soil water content (SWC) similar to that in elevated CO_2 plots (referred to as ct-i in figures). These plots received 20-mm irrigations five times in 2007 (7 June, 20 June, 11 July, 21 September, and 15 November), three times in 2008 (26 June, 18 July, and 19 September), three times in 2009 (17 July, 10 August, and 28 September), and three times in 2010 (1 July, 22 July, and 20 August).

Continuous Measurement of Soil Moisture

Within each plot, volumetric soil moisture was measured at 10 and 20 cm soil depths (EnviroSMART probe; Sentek Sensor Technologies, Stepney, Australia) and logged (via CR10X data loggers; Campbell Scientific, Logan, Utah, USA) hourly from 2007–2010. We calculated soil water content (SWC; cm H_2O) for the 0–10 cm and 10–20 cm soil depths and summed the amounts to arrive at SWC for the upper 0–20 cm of the soil profile.

Ecosystem C Flux Measurements

We used a static chamber method [18] to measure ecosystem C fluxes on ca. 50 days between May 2006 and October 2010 through a combination of mid-day and diurnal sampling campaigns. Flux measurements occurred every 2–4 weeks during the growing season, with diurnal sampling campaigns at approximately 6 week intervals and midday campaigns during the intervening periods. Both net ecosystem exchange (NEE) and ecosystem respiration (R_{eco}) were measured, and ecosystem photosynthesis (P_{eco}) was calculated according to $P_{eco} = NEE - R_{eco}$. For each diurnal, NEE and R_{eco} were measured 5 times over the course of 24 hours (at ca. 0400, 0900, 1300, 1600, and 2100 hours). We used a Lexan polycarbonate (GE plastics, Pittsfield, MA) chamber fitted with 2 circulating fans, a Q190 photosynthetically active radiation sensor (LI-COR, Lincoln, NE), and an open-path LI-7500 infrared gas analyzer (LI-COR) for measuring [CO_2]. Two chambers were used in tandem so that the sampling time period was constrained to ≤2 hours. NEE was measured for 2 minutes, and then an opaque cover was placed over the chamber to block light and eliminate photosynthesis to measure R_{eco} for the next 2 minutes.

Ecosystem CO_2 fluxes were calculated after applying a correction for water vapor dilution [19]. Comparisons of CO_2 fluxes between the two chambers on a subset of the plots revealed no significant offset from a 1:1 line, with a correlation coefficient of 0.96, indicating no chamber bias. For each diurnal field campaign, integrations of daily NEE, P_{eco}, and R_{eco} were calculated using linear interpolation between measurement points to calculate hourly C balance, summed over 24-hours and presented as g C m^{-2} day^{-1}.

We used simple linear regression to establish scaling relationships between mid-day and daily C fluxes [18] (Figures S1, S2). This allowed us to scale frequent mid-day measurements to daily sums of C uptake or loss (g m^{-2} day^{-1}). Because daily NEE cannot be directly estimated from mid-day NEE, we used the equation $NEE_{daily} = P_{ecodaily} + R_{ecodaily}$ (where $P_{ecodaily}$ and $R_{ecodaily}$ were first estimated by linear regression). We arrived at growing season (May – October) P_{eco}, R_{eco} and net ecosystem production (NEP) values by using linear interpolation to estimate daily net C fluxes for all days between measurement dates and then summing all daily data for each season. This simple gap-filling method to estimate seasonal carbon fluxes allowed us to make nearly simultaneous measurements at 25 plots during 50 campaigns. Our measurements were representative of climatic conditions encountered within the field site. This temporal variability did not interact significantly with the climate change treatments [18], so the summation method does not affect the outcome of statistical tests.

Heterotrophic Respiration and Soil C

In early May of 2008, we established root exclusion plots, a standard method for separating soil respiration into root and microbial components [20]. We installed root barriers to 25 cm depth and applied glyphosate, a broad-spectrum herbicide, to a small area of each plot. PVC rings 25-cm in diameter were inserted 8-cm deep into the soil and a standard static chamber method [21] was used to analyze CO_2 efflux from headspace samples collected at weekly to biweekly intervals. Syringes were used to collect headspace air three-four times over a 45 minute period, which were analyzed for CO_2 by gas chromatography

(Varian 3800 gas chromatograph equipped with thermal conductivity and flame ionization detectors, Varian Instruments, Sunnyvale, CA, USA). Soil C concentrations were determined on root-free, acidified (1 N H_3PO_4) soil samples collected by coring plots in mid-July with a Costech elemental analyzer (Cernusco, Italy).

Methodological Limitations

Ecosystem-scale flux measurements are required to quantify net C storage on land, and partitioning the net CO_2 flux into its gross components of ecosystem respiration and photosynthesis demonstrates the physiological control over C storage. We acknowledge the potential for artifacts associated with our chamber techniques, including short-term light-enhanced dark respiration [22] and offsets between light and dark microbial respiration [23]. While increased recognition of these leaf-level phenomena demonstrates a role in ecosystem-scale C cycling, their contribution to fluxes reported here is estimated to be <10%, and within the error of the measurements, because aboveground biomass is only about 20% of total biomass [14], and ecosystem respiration is contributed by approximately equal parts of soil organic matter decomposition and root respiration.

Statistical Analyses

We used a repeated measures mixed effects model with CO_2 level, temperature level, and year to test for main and interactive effects of global change treatments on cumulative annual carbon fluxes (P_{eco}, R_{eco}, and NEP). To determine whether C fluxes in response to elevated [CO_2] and irrigation were similar, a single factor ANOVA with irrigation as the main effect was used. When necessary, data were log-transformed to meet assumptions of normality and equal variance (soil C). Throughout the text, we characterize significant results according to $P \leq 0.05$ and numerical values are presented with standard error of the mean (SEM).

Results

Under present ambient conditions, our data suggest that the semi-arid grassland of the PHACE experiment site ranged from being a slight C sink to a slight source to the atmosphere over 2006–2010; with growing season (April–October) net ecosystem production (NEP, the sum of NEE) losses averaging +33 g m^{-2} over this period (Fig. 1C; positive fluxes indicate mass transfer from the ecosystem to the atmosphere, and negative fluxes indicate mass transfer from the atmosphere to the ecosystem). Elevated [CO_2] reduced net C uptake in 2006, and caused significantly greater net C losses than ambient in 2009 and 2010 (Fig. 1C; ANOVAR $P<0.01$ in all 3 years; $n=5$). No significant elevated [CO_2] effects on NEP were observed in 2007–2008. Warming alone never affected NEP, but when combined with elevated [CO_2], led to significant net C loss in 2007 (Fig. 1C; ANOVAR $P=0.02$; $n=5$); consistent net C losses were observed from CT treatment from 2007 through 2010.

Elevated [CO_2] enhanced C cycling by stimulating both gross fluxes, P_{eco} and R_{eco}, from 2006–2008 (Fig. 1), years in which both average and below-average annual precipitation were experienced [14]. During these first three years of the experiment, P_{eco} was stimulated by 19–40%, and R_{eco} by 13–42% (Table 1). Beginning in 2009, however, elevated [CO_2] ceased to stimulate P_{eco} but continued to stimulate R_{eco}. This continued stimulation of R_{eco} led to NEP losses that averaged 93 g m^{-2} per growing season (Fig. 1; Table 1). These cumulative CO_2 losses did not lead to measurable net changes in soil C storage or concentrations during 2006–2010.

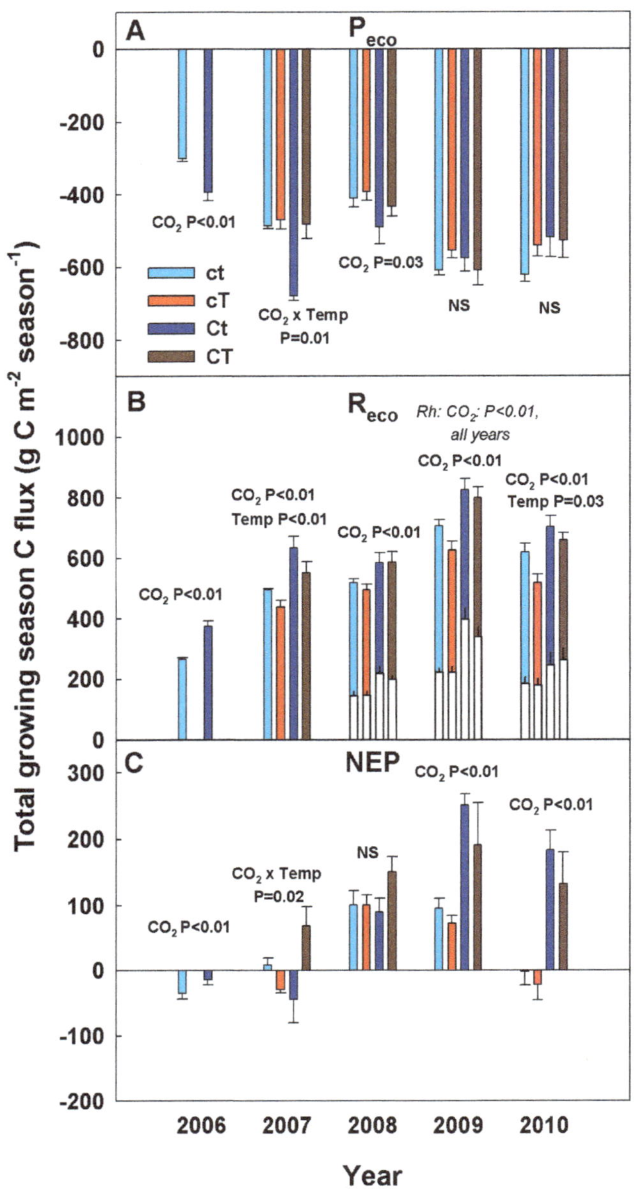

Figure 1. Growing season carbon fluxes in response to global changes. Growing season sums (April–October, 2006–2010) for **A)** gross ecosystem production (P_{eco}), **B)** ecosystem respiration (R_{eco}) and heterotrophic respiration (Rh) inset white bars, and **C)** net ecosystem production (NEP) for control and global change treatments at the Prairie Heating and CO_2 Enrichment Experiment in Cheyenne, WY USA. Negative (−) values indicate C uptake and positive (+) values indicate C efflux. Treatment codes are: ct = ambient [CO_2] and temperature, cT = ambient [CO_2] and warming, Ct = elevated [CO_2] and ambient temperature, and CT = elevated [CO_2] and warming. Statistically significant main and interactive treatment effects (within a given year) along with p-values are indicated (n = 5 for all measurements).

Throughout the experiment, soil water content (SWC) was increased in elevated [CO_2] plots on average by 21% during the growing season (Fig. 2), owing to reduced stomatal conductance and transpiration under elevated [CO_2] [8]. We evaluated the influence of this indirect SWC effect on C fluxes by frequently irrigating an additional set of non-CO_2 treated plots (n = 5) and maintaining SWC close to that observed under elevated [CO_2] (Fig. 2). Both the seasonal pattern of daily P_{eco} (Fig. 3) and the

Table 1. Treatment effects on ecosystem CO_2 fluxes.

Treatment	Flux	2006	2007	2008	2009	2010
Warming	P_{eco}	NA	0.97	0.96	0.91	0.87
Elevated [CO_2]		1.31*	1.40*	1.19*	0.95	0.84
Elevated [CO_2]×Warming		NA	0.99	1.05*	1.00	0.85
Irrigation		NA	*1.38*	*1.21*	*0.93*	*0.84*
Warming	R_{eco}	NA	0.89*	0.95	0.89	0.84*
Elevated [CO_2]		1.42*	1.28*	1.13*	1.17*	1.14*
Elevated [CO_2]×Warming		NA	1.12*	1.13*	1.13*	1.07*
Irrigation		NA	*1.18*	*1.11*	*0.97*§	*0.90*§

Ecosystem photosynthesis (P_{eco}) and respiration (R_{eco}) were measured at the Prairie Heating and CO_2 Enrichment Experiment during 2006–2010, and are presented as response ratios. Data reflect ratios of a given treatment relative to ambient conditions and are based on cumulative growing season fluxes. Irrigation effects are shown in italics. Statistically significant differences ($P<0.05$, ANOVA) between core treatments (Warming, Elevated [CO_2], and Elevated [CO_2]×Warming) and the ambient treatment within a given year are indicated by an asterisk (*). Significant differences ($P<0.05$, ANOVA) between Irrigation effects and Elevated [CO_2] effects are indicated by (§).

annual P_{eco} response were nearly identical in elevated [CO_2] and irrigated plots in all years of the experiment (Fig. 3; Table 1). Further, the addition of 60 mm of water to irrigated plots did not stimulate P_{eco} in 2009 and 2010 (Table 1, Fig. 3), indicating that water availability under ambient conditions did not limit P_{eco} in these two wetter years. This suggests that stimulation of P_{eco} by elevated [CO_2] was primarily due to enhanced soil water availability rather than biochemical forcing, and that elevated [CO_2] alleviates water limitations that might otherwise constrain P_{eco}.

Seasonal trends for R_{eco} under elevated [CO_2] were similar to the shallow irrigation treatment in 2007–2008 (Fig. 3), suggesting that SWC was important in stimulating R_{eco} during those years of average moisture. However, in 2009–2010, the seasonal trends and cumulative fluxes diverged between elevated [CO_2] and irrigation treatments (Fig. 3; Table 1), suggesting that other mechanisms, in addition to SWC, drove the R_{eco} response to elevated [CO_2].

Warming suppressed C cycling by reducing R_{eco}, especially during dry years of 2007 ($P<0.01$) and 2010 ($P=0.03$) (Fig. 1; Table 1). In 2007, the suppression of R_{eco} by warming alone was sufficient to enhance net C uptake, but when warming was combined with elevated [CO_2], net C losses were enhanced (Fig. 1; [CO_2]×temperature interaction, $P=0.02$). The warming treatment decreased growing season SWC by 15% on average during 2007–2010 compared to ambient (Fig. 2). Cumulative P_{eco} was not affected by warming (Fig. 1A; Table 1), although it was stimulated early in the wet growing season of 2010 (Fig. 3) when SWC was highest (Fig. 2). This stimulatory effect of warming, however, was quickly reversed as soil water was depleted below levels observed in ambient treatment plots. These opposing responses, within a single growing season, emphasize the role of water availability in mediating C assimilation responses to warming.

Elevated [CO_2] led to both gains and losses of C (depending on the year), but when elevated [CO_2] was combined with warming, only net C loss was observed (Fig. 1C). This was driven by consistent enhancement of R_{eco} with elevated [CO_2] (Fig. 1B) combined with a neutral response of P_{eco} (Fig. 1A). From 2007–2010, R_{eco} was stimulated by an average of 111%, resulting in a net efflux of 259 g C m^{-2} over 4 years under elevated [CO_2] plus warming, compared to ambient conditions. Compared to NEP under ambient conditions, 545 g m^{-2} of C (237% increase) were lost from the ecosystem with elevated [CO_2] plus warming during 2007–2010 (Fig. 1C).

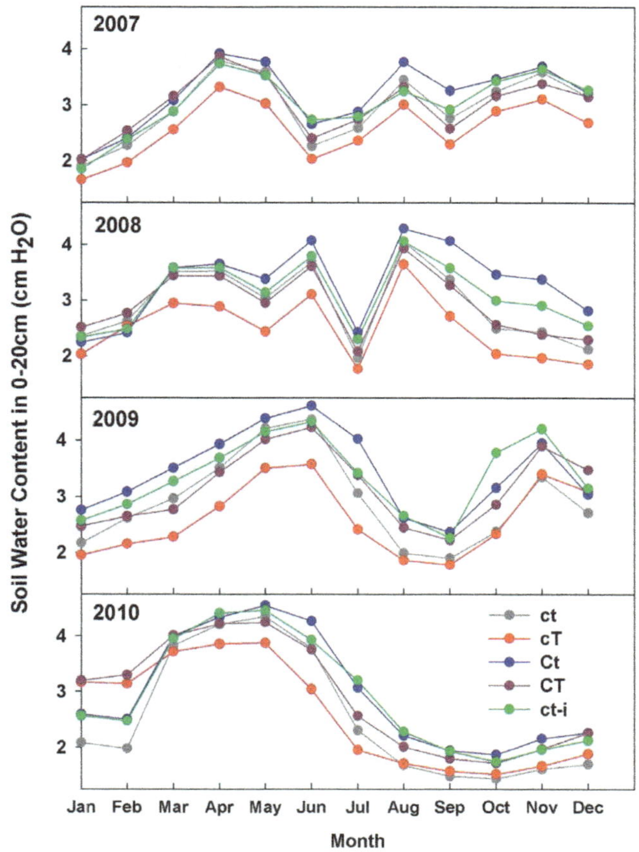

Figure 2. Monthly average soil water content in response to global changes. Mean monthly soil water content (SWC, 0–20 cm) for ambient and global change treatments (2007–2010). Treatment codes are ct = ambient temperature and ambient [CO_2]; cT = elevated temperature and ambient [CO_2]; Ct = ambient temperature and elevated [CO_2]; and CT = elevated temperature and elevated [CO_2]; ct-i = ambient temperature and ambient [CO_2]+irrigation.

We tested for the possibility of enhanced decomposition at PHACE by measuring respiration from root exclusion plots beginning in 2008. Elevated [CO_2] was observed to stimulate decomposition, or R_h, by 145% averaged over the last three years of the experiment (Fig. 1B, white bars), but R_a (estimated as the difference between R_{eco} and R_h) was not significantly affected by any treatment. Soil moisture in root exclusion plots was not affected by elevated [CO_2], and was less affected by warming compared to plots with intact vegetation [24]. Therefore the stimulation of R_h by elevated [CO_2] may have mainly been driven by increased C substrate in the form of dissolved organic C [25] and/or fine root biomass [14].

Discussion

Our measurements of net CO_2 exchange (NEE) and its components of ecosystem CO_2 uptake (P_{eco}) and release (R_{eco}) in future climate conditions indicated consistent, inter-annual net CO_2 loss in a semi-arid grassland in response to the combined global changes of elevated [CO_2] and warming. This result is in contrast to our third hypothesis that the combined treatment

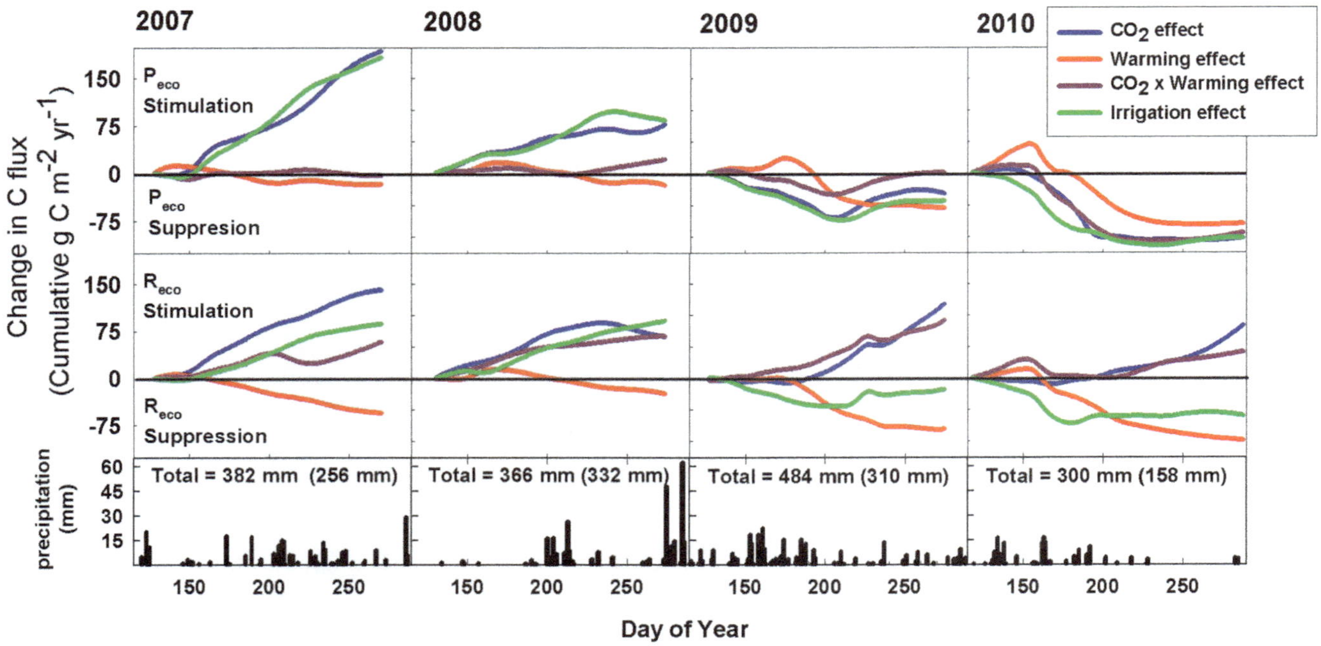

Figure 3. Seasonal patterns of global change effects on C fluxes. Cumulative effects of global changes on gross ecosystem production (P_{eco}; g m^{-2} yr^{-1}) and ecosystem respiration (R_{eco}; g m^{-2} yr^{-1}) in 2007–2010 at the Prairie Heating and CO_2 Enrichment Experiment (PHACE) in Cheyenne, WY USA. Treatment effects are differences from ambient values for each year of the experiment with increases in fluxes expressed as "stimulation" and decreases in fluxes expressed as "suppression." Data are expressed as cumulative across the growing season with the CO_2 effect in blue, warming effect in red, CO_2×warming interactive effect in purple, and irrigation effect in green. Daily precipitation amounts are depicted in the bottom panel. Both annual and growing season precipitation totals (in parentheses) are included for each year.

would not be different than ambient conditions, and to a recent modeling study indicating net C uptake with elevated [CO_2] and warming in other grasslands [12]. This positive feedback to climate change arose due to the prolonged stimulation of R_{eco} and the absence of consistent P_{eco} stimulation by elevated [CO_2], in agreement with our first hypothesis. The main effect of warming was to dampen the CO_2 loss, particularly in dry years, but P_{eco} was never stimulated by warming, and net C gain never occurred, in opposition to our second hypothesis. While it has been suggested that part of the terrestrial C sink is due to recent warming and lengthening of the growing season [26] our results agree with evidence from climate and remote sensing data [27], suggesting that these changes do not necessarily lead to greater C uptake and assimilation when integrated over the growing season.

Our results indicate that indirect effects of both warming and elevated [CO_2] on soil moisture strongly affected R_{eco}. Soil water availability is an important driver of soil and ecosystem respiration – which are comparable in our ecosystem due to the short vegetation stature – especially in water limited environments [28]. A modeling study suggested that precipitation was the dominant environmental influence over R_s in drylands [29], which contrasts somewhat with our findings that added irrigation sometimes stimulated and sometimes suppressed R_{eco} (Table 1; Figure 3). The same modeling study suggests that in simulated future climate, warming will increase the effect of elevated [CO_2] by additional stimulation of R_s [29]. This positive effect of combined warming and elevated [CO_2] was observed in a mesic herbaceous ecosystem [30]. By contrast, our results suggest warming suppressed the effect of elevated [CO_2] on R_{eco}, probably due to soil drying (Table 1; Figure 2).

The importance of substrate availability in the enhancement of respiration by elevated [CO_2] was inferred from the larger stimulation of R_{eco} than in the irrigated plots (Figure 3). Stimulation of P_{eco} by elevated [CO_2] in the initial years of this experiment, and subsequently greater above- and belowground biomass production [14], may have increased substrate availability [25,31] to stimulate decomposition by an enhanced priming effect [32]. This is consistent with findings from a CO_2 enrichment experiment in a similar grassland where increased belowground C availability was much more important than soil moisture in stimulating soil respiration [33]. Further, increased labile soil C has often been associated with increased R_s in elevated [CO_2] experiments [13,34,35]. If labile C inputs lead to priming of soil organic matter decomposition (R_h) [13], C losses may be increased with elevated [CO_2] [15,36]. In our experiment, increased supply of C substrates belowground from a larger root system also stimulated microbial activity [25], contributing to the enhanced R_h associated with elevated [CO_2]. Our work demonstrates that the considerable stocks of C stored in grassland soils will be vulnerable to future global changes if R_h is broadly stimulated by elevated [CO_2] (Figure 1). However, soil C contents did not change over the 5-year duration of our study.

Warming at PHACE stimulates soil microbial temperature sensitivity at optimal moisture conditions [37], indicating that both elevated [CO_2] and warming can have the potential to diminish C sequestration in grassland soils. However, drier soils in the field may offset the enhanced temperature sensitivity we observed in the laboratory. The drying effect of warming clearly should be taken into account in modeling studies, some of which still predict enhanced R_h with warming, even in dryland ecosystems [29].

We expected to observe more consistent stimulation of C uptake (P_{eco}) by elevated [CO_2], although ample soil moisture availability can suppress the stimulatory effect of elevated [CO_2] on biomass growth [14]. Variable C uptake is expected in grassland ecosystems, which are characteristically water-limited and experience high interannual variability in precipitation [38]. Warming suppressed the enhancement of P_{eco} by elevated [CO_2] in 2007 with a significant interaction (Figure 1), probably due to reduction of moisture availability (Figure 3). Our experiment indicated relatively small effects of climate change on gross C uptake in comparison to gross C losses, indicating a continued need to improve understanding of respiratory process responses to climate change [39]. These findings underscore the need for continued measurements of interacting climate change factors and moisture mediated responses of ecosystem C metabolism to elevated [CO_2] and warming [12,40].

Supporting Information

Figure S1 Relationship between midday and daily measurements of ecosystem respiration. Individual data points reflect pairs of midday measurements and daily sums for a given treatment during diurnal field campaigns conducted in 2007–2010.

Figure S2 Relationship between midday and daily measurements of ecosystem photosynthesis. Individual data points reflect pairs of midday measurements and daily sums for a given treatment during diurnal field campaigns conducted in 2007–2010.

Acknowledgments

We thank D. Smith for installation and operation of the PHACE experiment, E. Hardy for assistance in installation, F. Miglietta for advice and help on the installation of the FACE system, and B. Kimball for advice and help on the installation of the infrared heating system. Technical assistance in the field and/or laboratory was provided by S. Bachman, A. Brennan, J. Newcomb, H. Munn, M. Wood, B. Mitra, D. Bronson, M. Taylor, E. Bladyka, J. Chen, K. Tylka, A. Eden, A. Jacobs and M. Smith. E. Ainsworth and C. Bernacchi provided feedback on the manuscript. Mention of commercial products is solely for the purpose of providing specific information and does not imply recommendation or endorsement by the USDA.

Author Contributions

Conceived and designed the experiments: JAM EP DGW. Performed the experiments: JHW EP DGW DRL FAD YC. Analyzed the data: JHW EP. Wrote the paper: EP JHW DGW JAM FAD YC.

References

1. Cox PM, Betts RA, Jones CD, Spall SA, Totterdell IJ (2000) Acceleration of global warming due to carbon-cycle feedbacks in a coupled climate model. Nature 408: 750–750.
2. Piao SL, Ciais P, Friedlingstein P, de Noblet-Ducoudre N, Cadule P, et al. (2009) Spatiotemporal patterns of terrestrial carbon cycle during the 20th century. Global Biogeochemical Cycles 23.
3. Davidson EA, Janssens IA (2006) Temperature sensitivity of soil carbon decomposition and feedbacks to climate change. Nature 440: 165–173.
4. Luo Y, Melillo J, Niu S, Beier C, Clark JS, et al. (2011) Coordinated approaches to quantify long-term ecosystem dynamics in response to global change. Global Change Biology 17: 843–854.
5. Pataki DE, Ellsworth DS, D ER, Gonzalez-Meler M, King J, et al. (2003) Tracing changes in ecosystem function under elevated carbon dioxide conditions. BioScience 53: 805–818.
6. Luo YQ (2007) Terrestrial carbon-cycle feedback to climate warming. Annual Review of Ecology Evolution and Systematics 38: 683–712.
7. Nowak RS, Ellsworth DS, Smith SD (2004) Functional responses of plants to elevated atmospheric CO2 - do photosynthetic and productivity data from FACE experiments support early predictions? New Phytologist 162: 253–280.
8. Morgan JA, Pataki DE, Korner C, Clark H, Del Grosso S, et al. (2004b) Water relations in grassland and desert ecosystems exposed to elevated atmospheric CO2. Oecologia 140: 11–25.
9. Kirschbaum MUF (2000) Will changes in soil organic carbon act as a positive or negative feedback on global warming? Biogeochemistry 48: 21–51.
10. Heimann M, Reichstein M (2008) Terrestrial ecosystem carbon dynamics and climate feedbacks. Nature 451: 289–292.
11. Dieleman WIJ, Vicca S, Dijkstra FA, Hagedorn F, Hovenden MJ, et al. (2012) Simple additive effects are rare: a quantitative review of plant biomass and soil process responses to combined manipulations of CO2 and temperature. Global Change Biology 18: 2681–2693.
12. Luo Y, Gerten D, Le Maire G, Parton WJ, Weng E, et al. (2008) Modeled interactive effects of precipitation, temperature, and CO2 on ecosystem carbon and water dynamics in different climatic zones. Global Change Biology 14: 1986–1999.
13. Pendall E, Del Grosso S, King JY, LeCain DR, Milchunas DG, et al. (2003) Elevated atmospheric CO2 effects and soil water feedbacks on soil respiration components in a Colorado grassland. Global Biogeochemical Cycles 17.
14. Morgan JA, LeCain DR, Pendall E, Blumenthal DM, Kimball BA, et al. (2011) C(4) grasses prosper as carbon dioxide eliminates desiccation in warmed semi-arid grassland. Nature 476: 202–U101.
15. Carney KM, Hungate BA, Drake BG, Megonigal JP (2007) Altered soil microbial community at elevated CO2 leads to loss of soil carbon. Proceedings of the National Academy of Sciences of the United States of America 104: 4990–4995.
16. Miglietta F, Hoosbeek MR, Foot J, Gigon F, Hassinen A, et al. (2001) Spatial and temporal performance of the MiniFACE (Free Air CO2 Enrichment) system on bog ecosystems in northern and central Europe. Environmental Monitoring and Assessment 66: 107–127.
17. Kimball BA (2005) Theory and performance of an infrared heater for ecosystem warming. Global Change Biology 11: 2041–2056.
18. Bachman S, Heisler-White JL, Pendall E, Williams DG, Morgan JA, et al. (2010) Elevated carbon dioxide alters impacts of precipitation pulses on ecosystem photosynthesis and respiration in a semi-arid grassland. Oecologia 162: 791–802.
19. Jasoni RL, Smith SD, Arnone JA (2005) Net ecosystem CO2 exchange in Mojave Desert shrublands during the eighth year of exposure to elevated CO2. Global Change Biology 11: 749–756.
20. Hanson PJ, Edwards NT, Garten CT, Andrews JA (2000) Separating root and soil microbial contributions to soil respiration: A review of methods and observations. Biogeochemistry 48: 115–146.
21. Livingston GP, Hutchinson GL, Spartalian K (2006) Trace gas emission in chambers: A non-steady-state diffusion model. Soil Science Society of America Journal 70: 1459–1469.
22. Atkin OK, Evans JR, Siebke K (1998) Relationship between the inhibition of leaf respiration by light and enhancement of leaf dark respiration following light treatment. Australian Journal of Plant Physiology 25: 437–443.
23. Ayub G, Smith RA, Tissue DT, Atkin OK (2011) Impacts of drought on leaf respiration in darkness and light in Eucalyptus saligna exposed to industrial-age atmospheric CO2 and growth temperature. New Phytologist 190: 1003–1018.
24. Carrillo Y, Dijkstra FA, Pendall E, Morgan JA, Blumenthal DM (2012) Controls over Soil Nitrogen Pools in a Semiarid Grassland Under Elevated CO2 and Warming. Ecosystems 15: 761–774.
25. Carrillo Y, Pendall E, Dijkstra FA, Morgan JA, Newcomb JM (2011) Response of soil organic matter pools to elevated CO(2) and warming in a semi-arid grassland. Plant and Soil 347: 339–350.
26. Nemani RR, Keeling CD, Hashimoto H, Jolly WM, Piper SC, et al. (2003) Climate-driven increases in global terrestrial net primary production from 1982 to 1999. Science 300: 1560–1563.
27. Angert A, Biraud S, Bonfils C, Henning CC, Buermann W, et al. (2005) Drier summers cancel out the CO2 uptake enhancement induced by warmer springs. Proceedings of the National Academy of Sciences of the United States of America 102: 10823–10827.
28. Parton W, Morgan J, Smith D, Del Grosso S, Prihodko L, et al. (2012) Impact of precipitation dynamics on net ecosystem productivity. Global Change Biology 18: 915–927.
29. Shen WJ, Reynolds JF, Hui DF (2009) Responses of dryland soil respiration and soil carbon pool size to abrupt vs. gradual and individual vs. combined changes in soil temperature, precipitation, and atmospheric [CO2]: a simulation analysis. Global Change Biology 15: 2274–2294.
30. Wan S, Norby RJ, Ledford J, Weltzin JF (2007) Responses of soil respiration to elevated CO2, air warming, and changing soil water availability in a model old-field grassland. Global Change Biology 13: 2411–2424.
31. Fontaine S, Barot S, Barre P, Bdioui N, Mary B, et al. (2007) Stability of organic carbon in deep soil layers controlled by fresh carbon supply. Nature 450: 277–U210.
32. Kuzyakov Y (2010) Priming effects: Interactions between living and dead organic matter. Soil Biology & Biochemistry 42: 1363–1371.
33. Adair EC, Reich PB, Trost JJ, Hobbie SE (2011) Elevated CO2 stimulates grassland soil respiration by increasing carbon inputs rather than by enhancing soil moisture. Global Change Biology 17: 3546–3563.
34. Pendall E, King JY (2007) Soil organic matter dynamics in grassland soils under elevated CO2: Insights from long-term incubations and stable isotopes. Soil Biology & Biochemistry 39: 2628–2639.

35. Dijkstra FA, Hobbie SE, Reich PB, Knops JMH (2005) Divergent effects of elevated CO2, N fertilizattion, and plant diversity on soil C and N dynamics in a grassland field experiment. Plant and Soil 272: 41–52.
36. Hungate BA, Holland EA, Jackson RB, Chapin FSI, Mooney HA, et al. (1997) The fate of carbon in grasslands under carbon dioxide enrichment. Nature 388: 576–579.
37. Nie M, Pendall E, Bell CW, Gasch C, Raut S, et al. (2012) Positive climate feedbacks of soil microbial communities in a semi-arid grassland. Ecology Letters 16: 234–241.
38. Knapp AK, Smith MD (2001) Variation among biomes in temporal dynamics of aboveground primary production. Science 291: 481–484.
39. Pendall E, Bridgham S, Hanson P, Hungate B, Kicklighter D, et al. (2004a) Belowground Process Responses to Elevated CO_2 and Temperature: A Discussion of Observations, Measurement Methods, and Models. New Phytologist 162: 311–322.
40. Dijkstra FA, Blumenthal D, Morgan JA, Pendall E, Carrillo Y, et al. (2010) Contrasting effects of elevated CO(2) and warming on nitrogen cycling in a semiarid grassland. New Phytologist 187: 426–437.

6

Acclimation of Foliar Respiration and Photosynthesis in Response to Experimental Warming in a Temperate Steppe in Northern China

Yonggang Chi[1,2], Ming Xu[1,3]*, Ruichang Shen[1,2], Qingpeng Yang[1,4], Bingru Huang[5], Shiqiang Wan[6]

1 Key Laboratory of Ecosystem Network Observation and Modeling, Institute of Geographic Sciences and Natural Resources Research, Chinese Academy of Sciences, Beijing, China, **2** University of Chinese Academy of Sciences, Beijing, China, **3** Department of Ecology, Evolution and Natural Resources, Center for Remote Sensing and Spatial Analysis, Rutgers University, New Brunswick, New Jersey, United States of America, **4** Huitong Experimental Station of Forest Ecology, State Key Laboratory of Forest and Soil Ecology, Institute of Applied Ecology, Chinese Academy of Sciences, Shenyang, China, **5** Department of Plant Biology and Pathology, Rutgers University, New Brunswick, New Jersey, United States of America, **6** Key Laboratory of Plant Stress Biology, College of Life Sciences, Henan University, Kaifeng, Henan, China

Abstract

Background: Thermal acclimation of foliar respiration and photosynthesis is critical for projection of changes in carbon exchange of terrestrial ecosystems under global warming.

Methodology/Principal Findings: A field manipulative experiment was conducted to elevate foliar temperature (T_{leaf}) by 2.07°C in a temperate steppe in northern China. R_d/T_{leaf} curves (responses of dark respiration to T_{leaf}), A_n/T_{leaf} curves (responses of light-saturated net CO_2 assimilation rates to T_{leaf}), responses of biochemical limitations and diffusion limitations in gross CO_2 assimilation rates (A_g) to T_{leaf}, and foliar nitrogen (N) concentration in *Stipa krylovii* Roshev. were measured in 2010 (a dry year) and 2011 (a wet year). Significant thermal acclimation of R_d to 6-year experimental warming was found. However, A_n had a limited ability to acclimate to a warmer climate regime. Thermal acclimation of R_d was associated with not only the direct effects of warming, but also the changes in foliar N concentration induced by warming.

Conclusions/Significance: Warming decreased the temperature sensitivity (Q_{10}) of the response of R_d/A_g ratio to T_{leaf}. Our findings may have important implications for improving ecosystem models in simulating carbon cycles and advancing understanding on the interactions between climate change and ecosystem functions.

Editor: Gil Bohrer, The Ohio State University, United States of America

Funding: Financial support came from the National Key Research and Development Program of China (2010CB833500, http://www.973.gov.cn), the 100 Talents Program of the Chinese Academy of Sciences supported M. Xu's work (http://www.cas.cn) and National Natural Science Foundation (30925009, http://www.nsfc.gov.cn). The funders had no role in study design, data collection and analysis, decision to publish, or preparation of the manuscript.

Competing Interests: The authors have declared that no competing interests exist.

* E-mail: mingxu@igsnrr.ac.cn

Introduction

The balance between respiration and photosynthesis is critical to the exchange of carbon between the atmosphere and the terrestrial biosphere [1–3]. Instantaneous increases in foliar temperature (T_{leaf}) typically result in an increase in respiration/photosynthesis (R/A) ratio because the response of respiration to T_{leaf} normally follows an approximate exponential-type curve (at moderate temperatures) while the response of photosynthesis to T_{leaf} often bears a bell-shaped curve [i.e. the thermal optimum (T_{opt}) of respiration is higher than that of photosynthesis] [4,5]. In contrast, long-term warming experiments have suggested that R/A ratio is often conservative to changes in growth temperature (T_{growth}) through acclimation, the metabolic adjustment for compensating changes in T_{growth} [6–8]. Acclimation could occur via suppression of respiration in response to changes in foliar carbohydrate supplies [4,9]. The thermal acclimation of respiration and photosynthesis is associated with multitudes of signal cascades and networks, which involves the reallocation of resources to achieve and maintain not only optimal R/A ratio but also protective strategies under sustained warming as projected by global climate models [10–12]. However, the mechanisms of thermal acclimation of respiration and photosynthesis to climate warming are far from clear, especially in natural ecosystems.

The acclimation of foliar respiration to warmer T_{growth} has been found in numerous studies [8,13–17], which may also be associated with plant developmental stage and other abiotic factors, such as drought and nutrient availability [18–21]. Thermal acclimation of respiration might occur via changes in the temperature sensitivity, Q_{10}, or the basal respiration, R_{10} (respiration at a reference temperature, such as 10°C) [11]. Altered Q_{10} partially reflects temperature-mediated changes in energy demand and/or available substrates [1,17,20] whereas changes in R_{10} may be associated with temperature-mediated changes in respiratory capacity, reflecting changes in mitochondrial abundance, structure and/or protein composition [22–24]. As a result, thermal acclimation of respiration may enhance plant net carbon assimilation by reducing carbon loss under warmer T_{growth} while maintaining basal rates of respiration in colder T_{growth} for subsequent recovery [12,20,25,26].

Figure 1. Layout of the experiment plots in a temperate steppe in northern China. Infrared heaters were suspended as the heating sources at the warming plots while 'dummy' heaters were suspended to simulate shading effects of the infrared heater at the control plots.

The thermal acclimation of the foliar net CO_2 assimilation rate (A_n) may involve three primary sets of processes that control the A_n/T_{leaf} curves (response of A_n versus T_{leaf}), namely respiratory, biochemical and stomatal processes [27]. First, A_n is the difference between gross CO_2 assimilation rate (A_g) and foliar dark respiration (R_d), $A_n = A_g - R_d$, which requires the decoupling of the two processes because A_g and R_d feature different thermal dynamic properties and thus involve different thermal acclimation processes [28]. This could result in a shift in T_{opt} and a change in the shape of the A_n/T_{leaf} curve. Therefore, R_d must be evaluated separately and factored out to understand the acclimation mechanisms of A_g in response to global warming [3,18,29]. Second, the acclimation of A_g to warmer T_{growth} deals with the changes in Rubisco activity [29–33] and electronic transport processes [34] where T_{growth} affects the thermal dependence of various enzymes in the dark and light reactions [35,36]. Therefore, the temperature sensitivity of the maximum rate of Rubisco carboxylation (V_{cmax}) and the maximum rate of photosynthetic electron transport (J_{max}) are associated with the acclimation of A_g [36,37]. In addition, the change in the balance between carboxylation and regeneration of RuBP, indicated by J_{max}/V_{cmax} ratio, may also result in the shift of T_{opt} of A_g due to nitrogen (N) partitioning in the photosynthetic apparatus [3,31,38,39]. Finally, the temperature-dependent diffusion processes of CO_2 to chloroplasts, such as stomatal conductance (g_s) and mesophyll conductance (g_m), can also affect the thermal acclimation of photosynthesis [36,40]. Kirschbaum and Farquhar [41] showed that higher conductance could cause an increase of CO_2 concentrations in the carboxylation site (C_c) and then resulted in a shift in limitation of A_g from Rubisco to electron transport capacity. Since T_{opt} of electron transport-limited A_g is higher than that of Rubisco-limited A_g, T_{opt} of A_g was increased (0.05°C per 1 μmol mol^{-1} CO_2) [36].

Stipa krylovii Roshev. is a keystone species in the temperate steppe in northern China [42,43]. Climate models predict this region will be 4°C warmer by 2100, which may have severe impacts on *Stipa krylovii* Roshev. [44]. Examining the respiration and photosynthesis of this species is critical to the steppe productivity and the carbon cycle of the ecosystem. The objectives of the current study are to examine: (1) the acclimation capacity of respiration and photosynthesis to experimental warming under field conditions, and (2) the homeostasis of respiration/photosynthesis ratio in response to experimental warming in the steppe ecosystem.

Materials and Methods

Site Description

The research site (42°02′ N, 116°17′ E, 1324 m a.s.l.) is a typical temperate steppe located in Duolun County, Inner Mongolia Autonomous Region, China. The experiment has received the permits for the field study from the land owner, Institution of Botany, Chinese Academy of Sciences. The mean

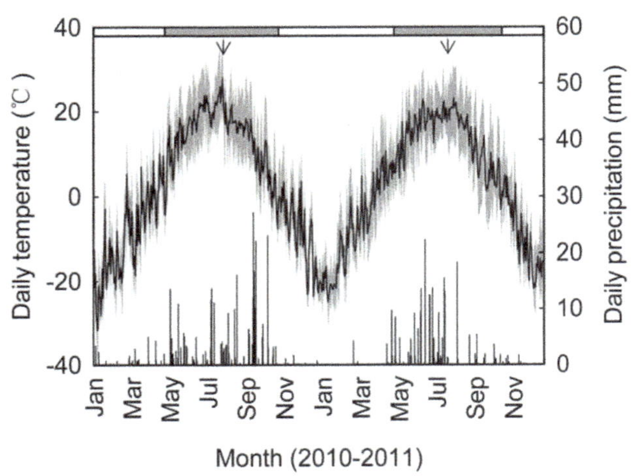

Figure 2. Daily maximum, minimum and mean air temperature (lines) and precipitation (bars) at the study site in 2010 and 2011. The filled rectangles on the top of figure indicate the growing season (May to October) and the open rectangles for the non-growing season (November to April). The arrows mark the timing of field campaigns when the gas exchange measurements were initiated.

annual temperature (MAT) is 2.1°C, with monthly mean temperature ranging from −17.5°C in January to 18.9°C in July. The mean annual precipitation (MAP) is approximately 385 mm with approximately 85% falling from May to September. The soils are chestnut (Chinese classification system) or Haplic Calcisols (FAO classification system), with 62.8% sand, 20.3% silt, and 17.0% clay respectively. The soils are characterized as sandy, slightly alkaline and nutrient poor with pH values around 7.7 and bulk density of 1.3 g cm^{-3} and soil total organic C and N concentrations of 16.1 and 1.5 g kg^{-1} respectively. The plant communities in the temperate steppe are dominated by *Stipa krylovii* Roshev., *Artemisia frigid* Willd., *Potentilla acaulis* L., *Cleistogenes squarrosa* (Trin.) Keng., *Allium bidentatum* Fisch. ex Prokh., and *Agropyroncristatum* (L.) Gaertn.

Warming Experiment

The warming experiment was initiated in April 2006 with infrared heaters (MSR-2420, Kalglo Electronics Inc., USA; radiation output is approximately 1600 W) as the heating source (Fig. 1). Briefly, an infrared heater of 1.65 m in length was suspended at 2.25 m above the ground in each warming plot which features a dimension of 3×4 m. A reflector associated with the heater can be adjusted so as to generate an evenly distributed radiant input to the plant canopy. In the control plots, a 'dummy' heater with the same shape and size was suspended at the same height to simulate shading effects of the infrared radiator. The effects of warming on T_{leaf} were measured using a portable infrared thermometer (FLUKE 574, Fluke Inc., USA). The mean daytime T_{leaf} in the warming plots was increased by 2.07°C compared to the control plots. The warming experiment was designed for long-term simulation of global change and it featured a complete random block design with multiple treatments (day warming, night warming, diel warming, and N addition) and six replicates. We took advantage of this multi-factor experiment by selecting the diel warming and control plots with all the other factors kept at control levels. The details of the experiment can be found in Wan *et al.* [44] and Xia *et al.* [45].

Gas Exchange Measurements

We measured foliar gas exchange using a portable photosynthesis system (LI-6400, LI-COR Inc., USA) in the middle of the growing seasons (late July to early August) in 2010 and 2011 (Fig. 2) to remove the effect of seasonal changes in photosynthetic and respiratory acclimation in *Stipa krylovii* Roshev. [19]. Four individuals (one individual per plot) were measured in each treatment. Eight days were required to complete all field measurements each year. Light, T_{leaf}, humidity, and CO_2

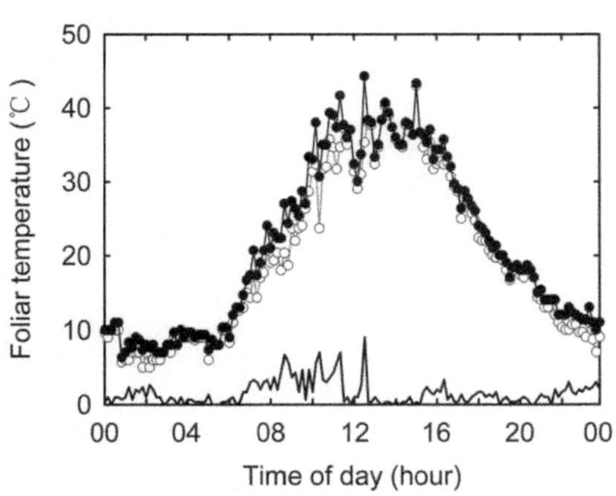

Figure 3. Representative 24-h foliar temperature (T_{leaf}) profiles from *Stipa krylovii* Roshev. grown in the control (open) and warming (filled) plots during the field measurement campaigns. Thick solid line indicates warming-induced changes in T_{leaf} between control and warming plots.

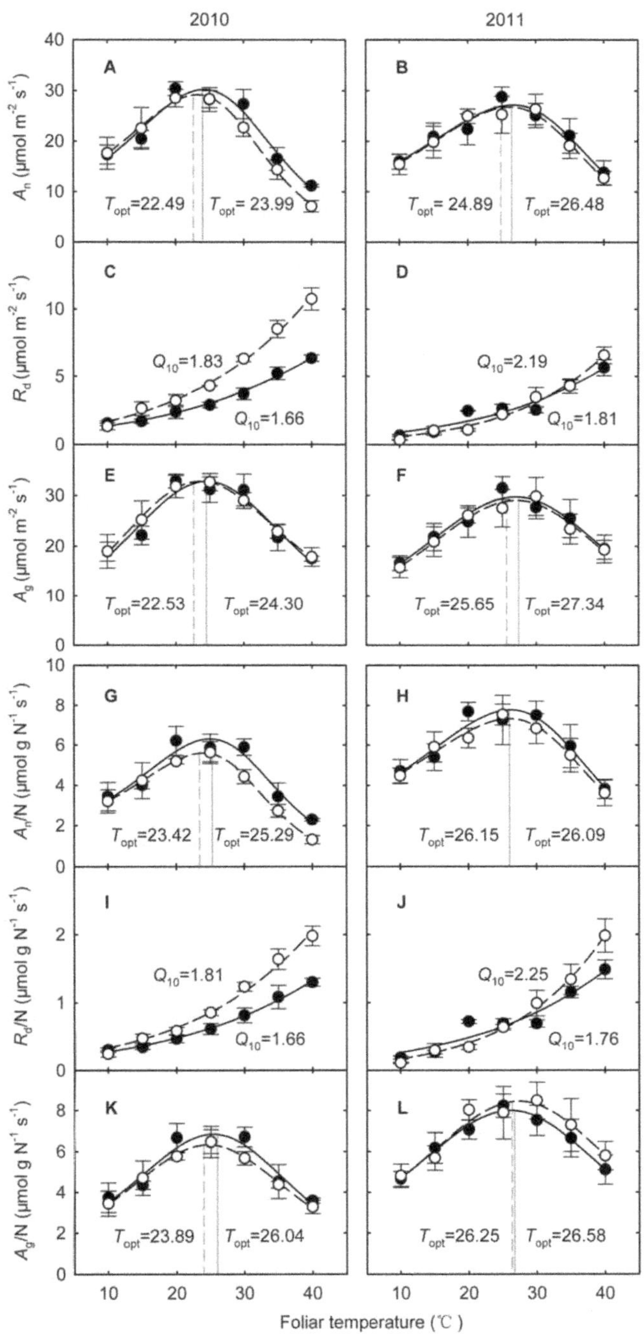

Figure 4. Warming effects on the responses of photosynthesis and respiration to foliar temperature (T_{leaf}) in 2010 (left panels) and 2011 (right panels. The filled circles indicate the warming plots and the open circles for the control plots. (A) to (F) foliar area based: (A) and (B) net CO_2 assimilation (A_n); (C) and (D) dark respiration (R_d); (E) and (F) gross CO_2 assimilation (A_g); (G) to (L) foliar nitrogen based: (G) and (H) A_n ; (I) and (J) R_d; (K) and (L) A_g. Each data point is the average of 4 replicates.

Table 1. Results (*P*-values) of one-way ANOVA on the effects of warming on the responses of A_n (the net CO_2 assimilation rate), R_d (dark respiration), A_g (the gross CO_2 assimilation rate), V_{cmax} (the maximum rate of Rubisco carboxylation) and J_{max} (the maximum rate of photosynthetic electron transport) expressed per unit foliar area and nitrogen to instantaneous change (10–40°C within a 5 h period) in T_{leaf} (foliar temperature) in 2010 and 2011.

Year	Parameters	*c*	ΔH_a	ΔH_d	ΔS	T_{opt}	Q_{10}	ref_{10}
2010	A_n	0.836	0.844	0.735	0.727	0.310	/	0.816
	R_d	**0.027**	**0.046**	/	/	/	**0.049**	0.090
	A_g	0.292	0.300	0.979	0.913	0.328	/	0.839
	V_{cmax}	0.055	0.064	/	/	/	0.062	0.784
	J_{max}	0.879	0.842	0.757	0.772	0.520	/	0.181
	A_n/N	0.726	0.732	0.575	0.612	0.323	/	0.955
	R_d/N	0.071	0.095	/	/	/	0.094	0.094
	A_g/N	0.977	0.976	0.138	0.178	0.302	/	0.996
	V_{cmax}/N	0.150	0.158	/	/	/	0.142	0.546
	J_{max}/N	0.474	0.468	0.678	0.646	0.874	/	0.381
2Δ011	A_n	0.474	0.472	0.986	0.923	0.619	/	0.865
	R_d	**0.042**	**0.040**	/	/	/	**0.042**	0.050
	A_g	0.403	0.400	0.600	0.529	0.637	/	0.758
	V_{cmax}	0.723	0.712	/	/	/	0.779	0.656
	J_{max}	0.166	0.167	0.325	0.369	0.317	/	0.487
	A_n/N	0.354	0.354	0.470	0.453	0.989	/	0.886
	R_d/N	**0.010**	**0.010**	/	/	/	**0.011**	**0.026**
	A_g/N	0.306	0.305	0.604	0.554	0.933	/	0.698
	V_{cmax}/N	0.074	0.079	/	/	/	0.093	0.703
	J_{max}/N	0.463	0.468	0.215	0.223	0.115	/	0.657

c is a scaling constant, ΔH_a is the activation energy, ΔH_d is a term for deactivation, ΔS is an entropy term, T_{opt} is the thermal optimum, Q_{10} is the temperature sensitivity and ref_{10} is the estimated basal rate at the reference temperature of 10°C. Significant values ($P<0.05$) are shown bold.

concentration were independently controlled in a 2×3 cm cuvette. Given the T_{leaf} control capacity is limited (within ±6°C) with the factory setup of the LI-6400 system, we modified the T_{leaf} control system by adding metal blocks with water channels to heat or cool the peltiers, thermoelectric cooling elements. The water channels were connected to a heating/cooling water bath whose temperature was controlled by adding hot water or ice. This modification allows holding T_{leaf} at any level between 10 and 40°C during the summer growing season in the steppe.

The photosynthetically active photon flux density (PPFD) was provided by the red/blue LED light source built in the foliar cuvette calibrated against an internal photodiode (LI-6400-02B, LI-COR Inc.). The vapor pressure deficit (VPD) in the foliar cuvette was controlled by passing the air entering the cuvette through either anhydrous calcium sulfate for the lower T_{leaf} when humidity was high or bubbling air via water at higher T_{leaf} when the air was dry. CO_2 concentrations in the cuvette were controlled using an injector system (LI-6400-01, LI-COR Inc.) which functions with a CO_2 mixer and compressed CO_2 cartridges. Cuvette was sealed with plasticine to prevent leakage. Potential leakage of CO_2 out and into the empty cuvette was determined for each concentration and used to correct the measured foliar fluxes with the equations provided by von Caemmerer and Farquhar [46] and Galmés *et al.* [47]. The gas exchange system was zeroed using H_2O and CO_2 free air every day.

Typical A_n/C_i curves (A_n versus calculated intercellular CO_2 concentrations, C_i) were measured at T_{leaf} changing from 10 to 40°C with 5°C increments each. We started with the A_n/C_i curves at low T_{leaf} (10°C) in the morning around 7:00 am finished at high T_{leaf} around noon. As to the problem of co-variance between the daily cycle and temperature, Luo *et al.* [48] and Way and Sage [3] suggested that the observed responses in the biochemical parameters resulted mainly from changes in temperature rather than changes in time of day. It usually took *c.* 5 min for T_{leaf} to reach stability at each step change in temperature. Photosynthesis was induced for 10 min in saturating PPFD (1500 µmol photons $m^{-2}\ s^{-1}$) and at ambient CO_2 concentration (C_a) of 380 ppmv. Measurements were made at saturating light (1500 µmol photons $m^{-2}\ s^{-1}$), and a leaf VPD between 0.5 and 2.0 kPa, except for 40°C where the VPD was 4.5±0.05 kPa. A_n was measured at cuvette CO_2 partial pressures between 50 and 1200 ppmv CO_2. The C_a was lowered stepwise from 380 to 50 ppmv and then increased again from 380 to 1200 ppmv with the total of 9 points. In total, 112 A_n/C_i curves were measured and used for the analysis of physiological parameters in this study. A_n/T_{leaf} curves (response of light-saturated A_n at 380 ppmv versus T_{leaf}) were obtained based on the A_n/C_i curves measured from 10 to 40°C.

R_d was measured by turning off the LED light source for at least 5 minutes in the cuvette after each A_n/C_i curve was accomplished [49]. All other conditions were the same as A_n/C_i curve measurements. Measurements of R_d on previously illuminated leaves were performed after a period of darkness in order to avoid light-enhanced dark respiration (LEDR) [13,18]. Five data points of R_d were logged at a 30 s interval and averaged for R_d at a given T_{leaf}. A_g was calculated by adding R_d to A_n at each T_{leaf}.

Table 2. Warming effects on the responses of A_n (the net CO_2 assimilation rate), R_d (dark respiration), A_g (the gross CO_2 assimilation rate), V_{cmax} (the maximum rate of Rubisco carboxylation) and J_{max} (the maximum rate of photosynthetic electron transport) expressed per unit foliar area and nitrogen to instantaneous change (10–40°C within a 5 h period) in T_{leaf} (foliar temperature) in the dry growing season (2010).

Parameters	Treatment	c	ΔH_a	ΔH_d	ΔS	T_{opt}	Q_{10}	ref_{10}
A_n (μmol m^{-2} s^{-1})	Control	24.24±5.86	50.48±14.38	166.73±19.47	0.56±0.07	22.49±1.04	/	17.33±3.63
	Warming	25.73±3.69	53.96±8.92	158.17±14.21	0.53±0.04	23.99±0.87	/	16.34±1.87
R_d (μmol m^{-2} s^{-1})	Control	20.03±0.92	45.93±2.27	/	/	/	1.83±0.05	1.70±0.12
	Warming	16.59±0.75	38.37±1.97	/	/	/	1.66±0.04	1.35±0.13
A_g (μmol m^{-2} s^{-1})	Control	38.57±6.03	83.57±14.11	134.98±14.15	0.46±0.05	22.53±1.38	/	18.44±3.52
	Warming	29.77±4.66	63.18±11.13	135.46±9.63	0.45±0.03	24.30±0.91	/	17.59±1.88
V_{cmax} (μmol m^{-2} s^{-1})	Control	24.68±1.26	49.08±3.08	/	/	/	1.91±0.07	46.38±2.43
	Warming	21.64±0.25	41.97±0.62	/	/	/	1.74±0.01	45.33±2.73
J_{max} (μmol m^{-2} s^{-1})	Control	34.22±4.47	68.79±10.39	126.60±17.24	0.43±0.05	25.50±0.90	/	126.47±15.86
	Warming	35.24±4.58	71.89±10.69	132.40±4.80	0.44±0.01	26.55±1.24	/	101.15±5.31
A_n/N (μmol g N^{-1} s^{-1})	Control	22.15±8.29	49.63±20.16	189.72±13.56	0.63±0.05	23.42±0.78	/	3.18±0.73
	Warming	27.95±13.45	63.21±32.05	201.66±14.95	0.67±0.06	25.29±1.55	/	3.13±0.58
R_d/N (μmol g N^{-1} s^{-1})	Control	18.04±0.83	45.10±2.03	/	/	/	1.81±0.05	0.33±0.02
	Warming	15.06±1.08	38.51±2.64	/	/	/	1.66±0.06	0.28±0.02
A_g/N (μmol g N^{-1} s^{-1})	Control	28.54±7.08	64.31±17.24	131.93±19.92	0.44±0.07	23.89±0.76	/	3.38±0.72
	Warming	29.01±13.69	65.48±32.55	176.38±16.66	0.58±0.06	26.04±1.75	/	3.39±0.60
V_{cmax}/N (μmol g N^{-1} s^{-1})	Control	22.47±1.15	47.69±2.81	/	/	/	1.87±0.07	9.18±0.49
	Warming	19.79±1.15	41.26±2.83	/	/	/	1.72±0.06	9.68±0.61
J_{max}/N (μmol g N^{-1} s^{-1})	Control	26.79±5.46	55.64±13.06	151.52±16.73	0.50±0.05	27.17±1.47	/	23.28±3.43
	Warming	35.83±10.49	77.39±24.84	165.48±27.26	0.55±0.09	27.58±1.98	/	19.04±2.88

c is a scaling constant, ΔH_a is the activation energy, ΔH_d is a term for deactivation, ΔS is an entropy term, T_{opt} is the thermal optimum, Q_{10} is the temperature sensitivity and ref_{10} is the estimated basal rate at the reference temperature of 10°C. Values are means (n = 4, ± SE).

Estimation of V_{cmax}, J_{max}, *TPU* and g_m

A_n/C_c curves (A_n versus chloroplastic CO_2 concentration) were fitted to estimate V_{cmax}, J_{max}, *TPU* (triose-phosphate utilization) and g_m. The spreadsheet-based software of Sharkey *et al.* [50] was modified (Appendix S1) to fit the A_n/C_c curve by fixing the R_d value which was measured following the A_n/C_i curve. This modification will improve the model performance by reducing the number of estimated parameters and thus decreasing the degree of freedom in fitting the model. As in the original software the optimum of V_{cmax}, J_{max}, *TPU* and g_m was obtained by minimizing the root mean square error (RMSE) of each curve [51,52].

Estimation of Dependence of Reaction Rates on Temperature

The responses of R_d and V_{cmax} to T_{leaf} were fitted to a non-peaked model, following Harley *et al.* [53], due to the fact that the deactivation of R_d and V_{cmax} was not observed in our study:

$$\text{Parameter}(R_d, V_{cmax}) = e^{(c\text{-}\Delta Ha/RT_k)} \quad (1)$$

where c is a scaling constant, ΔH_a is the activation energy, R is the molar gas constant (0.008314 kJ K^{-1} mol^{-1}) and T_k is the absolute T_{leaf} (K) [54]. Q_{10} of R_d and V_{cmax} were modeled using the following general function:

$$\text{Parameter}(R_d, V_{cmax}) = ref_{10} Q_{10}{}^{[(T_{lead}\text{-}10)/10]} \quad (2)$$

where ref_{10} is the estimated basal rate at the reference temperature of 10°C, and T_{leaf} is the leaf temperature (°C). The responses of A_n, A_g and J_{max} to T_{leaf} were fitted using a peak model in view that the deactivation at high T_{leaf} was substantial:

$$\text{Paramater}(A_n, A_g, J_{max}) = \frac{e^{(c\text{-}\Delta Ha/RT_k)}}{1 + e^{[(\Delta ST_k\text{-}\Delta H_d)/RT_k]}} \quad (3)$$

where ΔH_d is a term for deactivation and ΔS is an entropy term [54,55]. The second derivative of Eqn 3 shows that Topt can be calculated [56] as follows if the parameter includes a peak:

Table 3. Warming effects on the responses of A_n, R_d, A_g, V_{cmax} and J_{max} expressed per unit foliar area and nitrogen to instantaneous change (10–40°C within a 5 h period) in T_{leaf} in the wet growing season (2011).

Parameters	Treatment	c	ΔH_a	ΔH_d	ΔS	T_{opt}	Q_{10}	ref_{10}
A_n (μmol m^{-2} s^{-1})	Control	30.91±9.71	66.17±22.71	175.99±18.90	0.59±0.06	24.89±2.47	/	15.07±2.49
	Warming	21.82±6.90	44.81±16.09	175.39±26.86	0.58±0.08	26.48±1.76	/	15.56±1.29
R_d (μmol m^{-2} s^{-1})	Control	24.93±1.23	60.02±3.13	/	/	/	2.19±0.09	0.58±0.09
	Warming	18.95±1.97	44.85±4.88	/	/	/	1.81±0.11	0.92±0.11
A_g (μmol m^{-2} s^{-1})	Control	39.19±13.54	85.44±31.59	167.09±24.98	0.56±0.08	25.65±2.77	/	15.27±2.45
	Warming	25.00±8.06	52.17±18.80	148.67±22.07	0.49±0.07	27.34±1.99	/	16.16±1.26
V_{cmax} (μmol m^{-2} s^{-1})	Control	23.98±0.80	47.78±2.07	/	/	/	1.87±0.05	40.65±4.82
	Warming	23.25±1.81	45.87±4.47	/	/	/	1.83±0.11	43.85±4.84
J_{max} (μmol m^{-2} s^{-1})	Control	27.81±5.52	54.61±13.31	167.70±27.42	0.54±0.08	31.56±1.01	/	102.85±15.77
	Warming	41.44±6.65	86.59±15.41	136.76±8.98	0.46±0.02	30.03±0.97	/	90.79±4.06
A_n/N (μmol g N^{-1} s^{-1})	Control	28.91±9.48	64.42±22.35	277.95±128.81	0.91±0.41	26.15±3.34	/	4.38±0.71
	Warming	18.26±4.78	39.35±11.09	175.08±35.01	0.57±0.11	26.09±1.89	/	4.50±0.30
R_d/N (μmol g N^{-1} s^{-1})	Control	24.49±1.52	62.01±3.83	/	/	/	2.25±0.11	0.16±0.03
	Warming	16.71±1.44	42.52±3.54	/	/	/	1.76±0.08	0.26±0.02
A_g/N (μmol g N^{-1} s^{-1})	Control	37.55±11.17	84.51±26.22	193.11±55.95	0.64±0.17	26.25±3.03	/	4.37±0.64
	Warming	22.93±6.77	50.17±15.76	157.41±33.58	0.52±0.10	26.58±2.18	/	4.65±0.29
V_{cmax}/N (μmol g N^{-1} s^{-1})	Control	23.24±0.69	48.97±1.84	/	/	/	1.90±0.05	11.67±1.37
	Warming	21.11±0.70	43.81±1.61	/	/	/	1.78±0.04	12.26±0.58
J_{max}/N (μmol g N^{-1} s^{-1})	Control	27.40±9.59	56.67±22.90	218.93±62.97	0.71±0.20	32.46±1.55	/	28.09±3.70
	Warming	36.34±6.18	77.67±14.53	131.16±7.25	0.44±0.02	29.26±0.79	/	26.24±1.40

Values are means (n = 4, ± SE). See Table 2 for abbreviations defined.

$$T_{opt} = \frac{\Delta H_d}{\Delta S - R\ln[\Delta H_a/(\Delta H_d - \Delta H_a)]} \quad (4)$$

Estimation of Biochemical Limitations to Photosynthesis

Temperature dependence of A_g limited by RuBP carboxylation (A_c), RuBP regeneration (A_j) and *TPU* (A_p) were reconstructed as follows:

$$A_c = \frac{V_{cmax}(C_c - \Gamma^*)}{C_c + K_c(1 + O/K_o)} \quad (5)$$

$$A_j = \frac{J_{max}(C_c - \Gamma^*)}{4C_c + 8\Gamma*} \quad (6)$$

$$A_p = 3TPU \quad (7)$$

where V_{cmax}, J_{max} and *TPU* were derived from fitted kinetic parameters (c, ΔH_a, ΔH_d and ΔS) in our study, K_c, K_o and ?[?] were derived from a general set of kinetic parameters in Sharkey *et al.* [50]. C_c was set at 250.8 ppmv in view that the mean C_c/C_a ratio was 0.66 at ambient CO_2 concentration (380 ppmv) for all the A_n/C_i curves measured in the current study, O was the partial pressure of oxygen at Rubisco.

Foliar Characteristics

Foliar N concentration on an area basis was determined using the foliage covered in the cuvette during the gas exchange measurements. The foliage samples were first used to measure the leaf area with an area meter (Li-3100, Li-Cor Inc.) and then biomass where the samples were dried at 65°C for 48 h. Then the dry samples were ground to powder for measuring the total C and N concentrations with a CN analyzer (NA Series 2, CE Inc., Germany).

Data Analyses

The raw data from the gas exchange measurements were cleaned and processed in Excel spreadsheets where the non-linear A_n/C_c curve fitting was performed as in Sharkey *et al.* [50]. The fitting was improved by fixing R_d with the measured value (Appendix S1). Further statistical analyses were conducted using SPSS (version 17.0, SPSS Inc., USA). One-way ANOVA was used to analyze the effects of warming on (1) the foliar chemical properties (C, N, and C/N ratio) and (2) the thermal dynamic

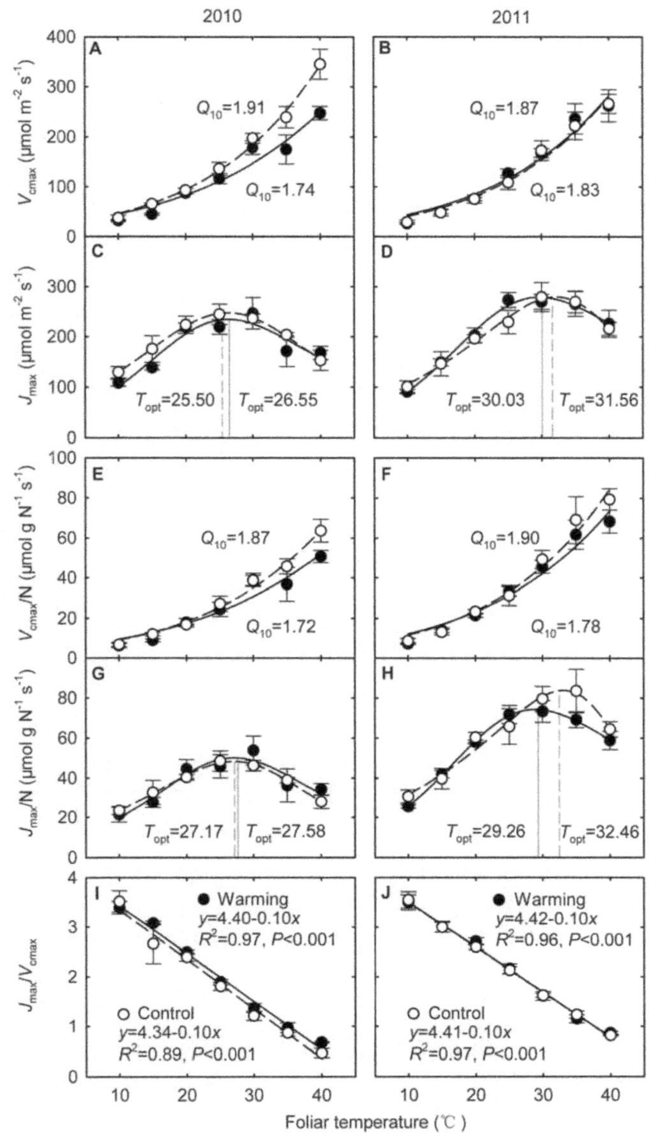

Figure 5. Warming effects on the responses of the maximum rate of Rubisco carboxylation (V_{cmax}), the maximum rate of photosynthetic electron transport (J_{max}) and the J_{max}/V_{cmax} ratio to foliar temperature (T_{leaf}) in 2010 (left panels) and 2011 (right panels). The filled circles indicate the warming plots and the open circles for the control plots. (A) and (B) area-based V_{cmax}; (C) and (D) area-based J_{max}; (E) and (F) N-based V_{cmax}; (G) and (H) N-based J_{max}; (I) and (J) the J_{max}/V_{cmax} ratio. Each data point is the average of 4 replicates.

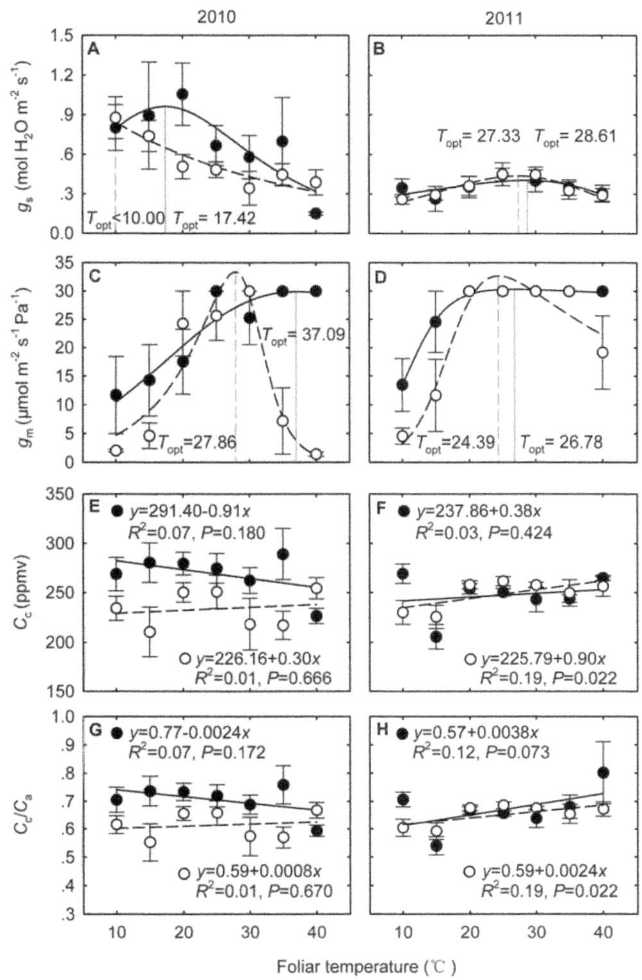

Figure 6. Warming effects on the responses of stomatal conductance (g_s) (A, B), mesophyll conductance (g_m) (C, D), carboxylation site CO_2 concentrations (C_c) (E, F), and C_c/C_a ratio (G, H) to foliar temperature (T_{leaf}) in 2010 (left panels) and 2011 (right panels). The filled circles indicate the warming plots and the open circles for the control plots. Each data point is the average of 4 replicates. **Note:** g_m is constrained to be 30 (μmol m^{-2} s^{-1} Pa^{-1}) or less.

properties (c, ΔH_a, ΔH_d, ΔS, Q_{10}, T_{opt} and ref_{10}) of foliar gas exchange (A_n, R_d and A_g) and photosynthetic metabolism (V_{cmax} and J_{max}). Differences were considered statistically significant at $P<0.05$. Linear regression was employed to examine relationships between foliar properties and climate (i.e. T_{growth}). T_{growth} in the control plots was an average for daytime T_{air} during the 5 d prior to gas exchange measurements in each plot. This choice was based on: (1) our observation that the bulk of individual foliar development by *Stipa krylovii* Roshev. species typically required 4–6 d; and (2) published results indicating that adjustments of foliar metabolism to climate change can occur rapidly (e.g. in a span of 1–5 d following a shift in T_{growth} [13,15,57–61]); (3) Gunderson *et al.* [60] found that T_{opt} for photosynthesis was

Table 4. Foliar characteristics of *Stipa krylovii* Roshev. grown in the control and warming plots.

Year	Treatment	N concentration	C concentration	C/N ratio
2010	Control	5.34±0.07	86.66±1.48	16.22±0.16
	Warming	5.02±0.16	92.91±3.42	18.48±0.26
	P value	0.063	0.100	**<0.001**
2011	Control	3.41±0.05	78.12±1.35	22.92±0.15
	Warming	3.68±0.06	79.70±1.44	21.65±0.11
	P value	**0.002**	0.426	**<0.001**

Warming effects on foliar nitrogen concentrations (g N m^{-2}), carbon concentrations (g C m^{-2}) and C/N ratio (g g^{-1}) were analyzed using one-way ANOVA for each year. Significant values ($P<0.05$) are shown bold (Mean ± SE, N = 28).

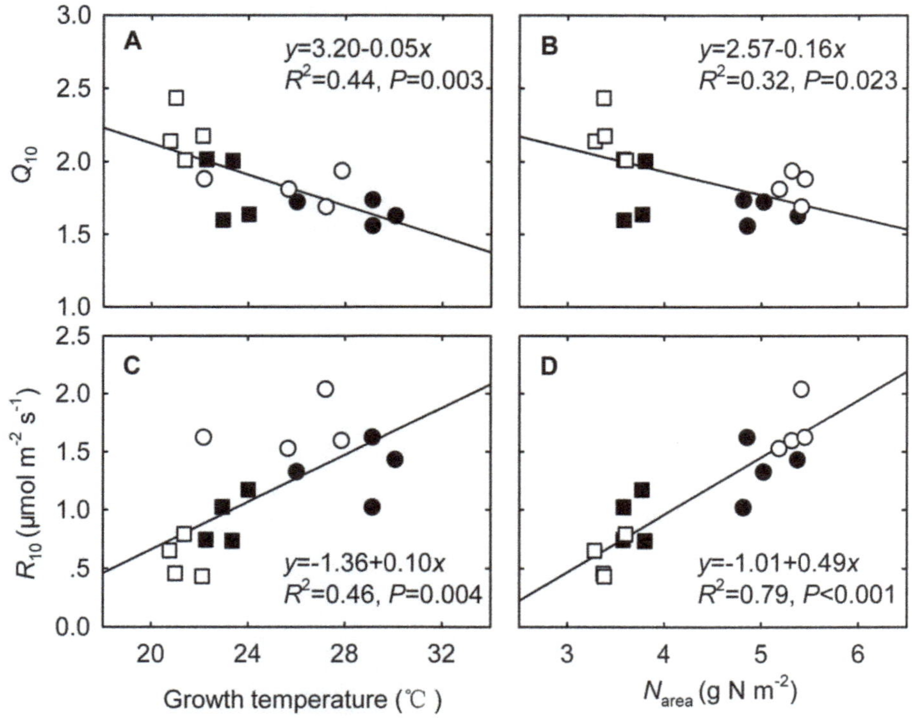

Figure 7. Responses of Q_{10} (the temperature sensitivity) (top panel) and R_{10} (the estimated basal respiration rate at the reference temperature of 10°C) (lower panel) in the control (open) and warming (filled) plots in 2010 (circles) and 2011 (squares) to T_{growth} (left panel) and foliar nitrogen concentrations (right panel), respectively. Values are means (n = 4, ± SE).

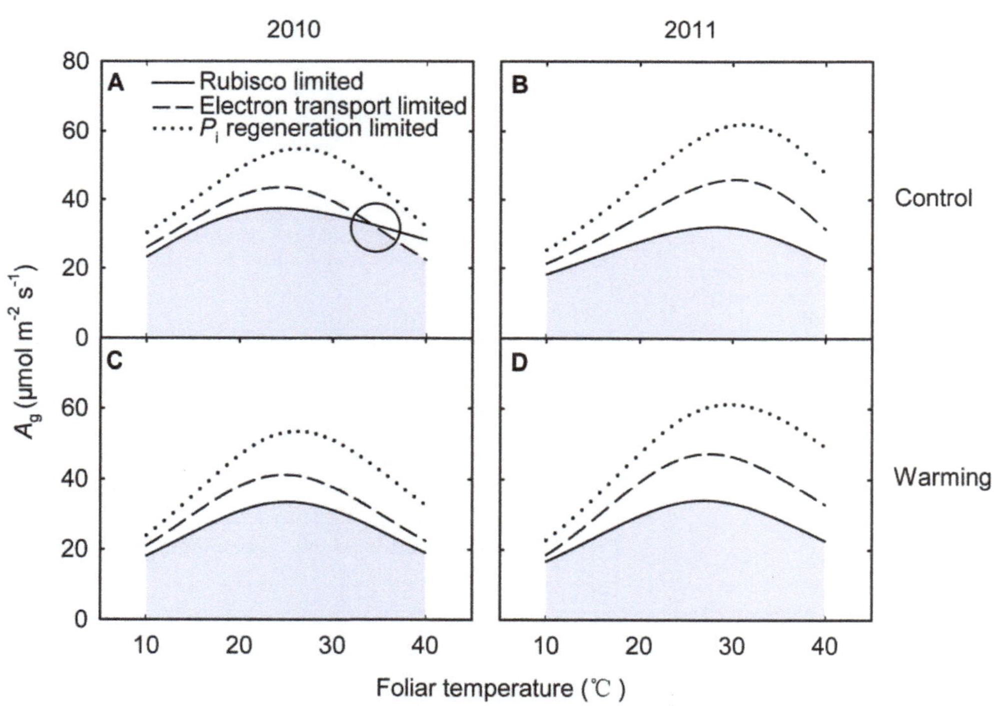

Figure 8. Warming effects on the responses of biochemical limitations in gross CO_2 assimilation (A_g) to foliar temperature (T_{leaf}) at chloroplast partial pressure of CO_2 (C_c) of 250.8 ppmv in 2010 (left panels) and 2011 (right panels). The top panels indicate the control plots and the lower panels for the warming plots. C_c was set at 250.8 ppmv considering that the mean C_c/C_a ratio was 0.66 at ambient CO_2 concentration (380 ppmv) for all the A_n/C_i curves measured. The response of A_g is delineated by the minimum value of either Rubisco-limited (solid curve), ribulose bisphosphate (RuBP) regeneration-limited (dashed curve) and P_i regeneration-limited (dotted curve). Circle indicates co-limited point, moving from the Rubisco-limited state to RuBP regeneration-limited state.

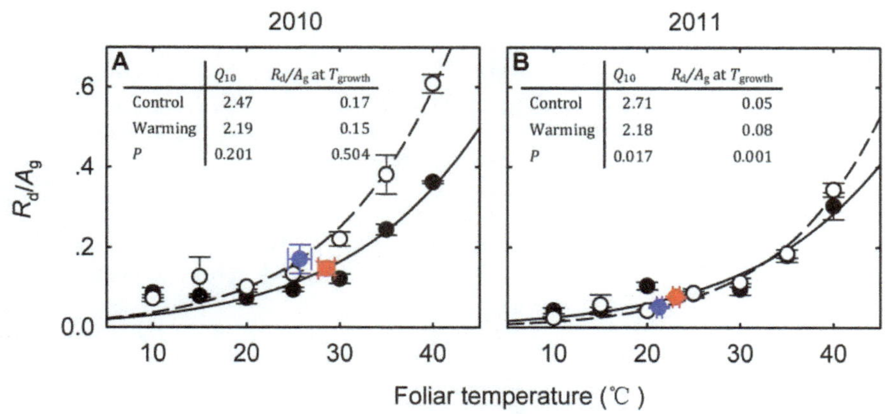

Figure 9. Warming effects on the response of R_d/A_g ratio (balance between dark respiration and gross CO_2 assimilation) to instantaneous change (10–40°C within a 5 h period) in T_{leaf} (foliar temperature) in the dry growing season (2010) (A) and the wet growing season (2011) (B). The filled circles indicate the warming plots and the open circles for the control plots. The blue and red circles indicate R_d/A_g ratio at growth temperature (T_{growth}), computed using the thermal dynamic properties (individual ΔH_a and c values for each plot) and the T_{growth}.

strongly correlated with mean daytime T_{air}. In addition, T_{growth} in the warming plots were approximatively calculated by adding warming effects (2.07°C) to the mean daytime T_{air} during the 5 d prior to gas exchange measurements in each plot.

Results

Microclimate and Experimental Warming

The meteorological data collected at the experimental site showed that the growing season of 2010 was dry while the growing season of 2011 was wet (Fig. 2). The daily mean T_{air} between 1 May, the onset of plant growth, and the time of the field measurements (27 July in 2010 and 2011) was 17.2°C in 2010 and 15.6°C in 2011 with the long-term average (1953–2011) of 15.5°C during the same period. Meanwhile, the precipitation during the same period was only 115 mm in 2010 and 183 mm in 2011 with the long-term average of 177 mm. The growing season precipitation in 2010 was only about 65% of that in a normal year, confirming 2010 was a dry year (Fig. 2).

The experimental warming significantly increased daytime T_{leaf} by 2.07°C ($P<0.001$), on average (Fig. 3). Warming increased daytime T_{growth} in the warming plots reaching 28.59 and 23.14°C in 2010 and 2011, respectively. Meanwhile, the daytime T_{growth} in the control plots was only 25.72 and 21.31°C in 2010 and 2011, respectively. The details of the warming effects on microclimate at the study site can be found in Wan *et al.* [44] and Xia *et al.* [45].

Respiration

Warming significantly decreased respiratory temperature sensitivity, Q_{10}, in both years (both $P<0.05$) (Fig. 4, Table 1). Q_{10} of R_d on a foliar area basis decreased from 1.83 in the control plots to 1.66 in the warming plots in 2010 ($P=0.049$) (Table 2) and from 2.19 to 1.81 in 2011 ($P=0.042$) (Table 3). Meanwhile, Q_{10} of R_d on a foliar N basis marginally decreased from 1.81 to 1.66 in 2010 ($P=0.094$) and significantly decreased from 2.25 to 1.76 in 2011 ($P=0.011$) (Table 2, 3). Warming marginally reduced base respiration rate at 10°C (R_{10}) on a foliar area basis from 1.70 to 1.35 μmol m^{-2} s^{-1} in 2010 ($P=0.090$) but increased that from 0.58 to 0.92 μmol m^{-2} s^{-1} in 2011 ($P=0.050$) (Table 2, 3). Warming effects on the R_{10} on a foliar N basis were similar to the area-based R_d (Fig. 4).

Photosynthesis

The A_n/T_{leaf} curves were typically bell-shaped in both warming and control plots (Fig. 4). Warming had little effect on T_{opt} of A_n in both years (both $P>0.05$) (Table 1). T_{opt} of A_n on a foliar area basis was 22.49 and 23.99°C for the control and the warming plots respectively in 2010, and 24.89 and 26.48°C respectively in 2011 (Fig. 4). T_{opt} of A_g on a foliar area basis was 22.53 and 24.30°C for the control and the warming plots respectively in 2010 ($P=0.328$), and 25.65 and 27.34°C respectively in 2011 ($P=0.637$) (Fig. 4). Warming also had little effects on T_{opt} of A_n and A_g on a foliar N basis in either 2010 or 2011 (all $P>0.05$) (Table 1).

Biochemical Limitations to Photosynthesis

The effects of warming on Q_{10} of V_{cmax} were not statistically significant between the warming and the control plots in both years (both $P>0.05$) (Table 1), but we found a general decreasing trend from the control to warming plots (Fig. 5). Q_{10} of V_{cmax} on a foliar area basis was 1.91 and 1.74 for the control and the warming plots respectively in 2010 ($P=0.062$), and 1.87 and 1.83 respectively in 2011 ($P=0.779$) (Fig. 5, Table 2, 3). Q_{10} of V_{cmax} on a foliar N basis was 1.87 and 1.72 for the control and the warming plots respectively in 2010 ($P=0.174$), and 1.90 and 1.78 respectively in 2011 ($P=0.668$) (Fig. 5, Table 2, 3). The warming effects on Q_{10} of J_{max} were not be detected in 2010 or 2011 (both $P>0.05$) (Table 1). In addition, the warming effects on the slope and y-intercept of the temperature-response curves for J_{max}/V_{cmax} ratio were not statistically significant (all $P>0.05$), though the ratio decreased linearly with the T_{leaf} (Fig. 5).

Diffusion Limitations to Photosynthesis

In 2010, a dry year, g_s in the warming plots was marginally greater than that in the control plots ($P=0.137$), and T_{opt} for g_s was about 17.42°C in the warming plots and less than 10°C in the control plots (Fig. 6). The g_m in the warming plots was significantly greater than that in the control plots ($P<0.001$), and T_{opt} for g_m appeared at 37.09°C in the warming plots and 27.86°C in the control plots (Fig. 6). C_c in the warming plots was approximately 35 ppmv greater than that in the control plots ($P<0.001$), but C_c was independent of T_{leaf} in both the warming and the control plots (both $P>0.05$) (Fig. 6). Similarly, C_c/C_a ratio was constant and independent of T_{leaf} in the warming and the control plots (both $P>0.05$) (Fig. 6). However, experimental warming significantly increased C_c/C_a ratio in 2010 ($P=0.001$) with an average value of 0.70 in the warming plots and 0.61 in the control plots (Fig. 6).

In 2011, a wet year, Warming had little effect on g_s and g_m (both $P>0.05$), which resulted in no difference in C_c between the warming and the control plots ($P=0.860$) (Fig. 6). Experimental warming also had little effect on C_c/C_a ratio in 2011 ($P=0.447$) with an average value of 0.67 in the warming plots and 0.65 in the control plots (Fig. 6).

Foliar Characteristics

Warming marginally decreased foliar N concentration in 2010 ($P=0.063$), but significantly increased that in 2011 ($P=0.002$) (Table 4). Warming had little effect on foliar carbon concentration in both years (both $P>0.05$). Foliar C/N ratio was significantly higher in the warming plots than in the control plots in 2010 ($P<0.001$) and the opposite was true in 2011 (Table 4).

Discussion

Acclimation of Respiration

R_d was sensitive to T_{leaf} with the R_d/T_{leaf} relationship following a typical exponential curve, but warming reduced the magnitude (Fig. 4, Table S1). Our results are consistent with previous studies [18,20,62] that the temperature sensitivity of R_d is negatively related to the T_{growth} (Fig. 7). According to the respiratory acclimation mechanisms proposed by Atkin and Tjoelker [11], the temperature-mediated change in Q_{10} is determined by the maximum enzyme activity and/or substrate availability [1,17,20]. Earlier results from the same warming experiment confirmed that day warming significantly reduced foliar starch concentrations (–6.1%, $P=0.009$), suggesting the reduction in Q_{10} in the current study might be attributed to the lower substrate concentrations.

Foliar N concentrations induced by experimental warming in our study may also affect the temperature sensitivity of R_d, Q_{10} (Fig. 7). To date, few studies have examined the role of N in the change in Q_{10}. Turnbull *et al.* [63] found that Q_{10} of R_d for the trees in a temperate rainforest increased with increasing N availability along a soil chronosequence in New Zealand. However, Ow *et al.* [64] have reported that N had little or no impact on Q_{10} of R_d when saplings grown at high and low N availabilities were transferred to a different T_{growth} regime. Here, we found a negative correlation between Q_{10} of R_d and foliar N concentrations (Fig. 7). The detailed mechanisms are not clear, but the confounding effect of foliar N concentrations with other factors, such as temperature and precipitation, may have played an important role in the "apparent" Q_{10} [11,65,66].

In the current study we found that experimental warming marginally reduced base respiration rate at 10°C (R_{10}) in 2010 but increased that in 2011 (Table 2, 3). This could have been attributed to the differential responses of foliar N concentration to warming in the two hydrologically contrasting growing seasons. Warming marginally decreased foliar N concentration in the dry growing season (2010), but increased that in the wet growing season (2011) (Table 4). A growing number of studies [8,14,17], including our current study, have found that foliar N concentration was strongly related to R_{10} (Fig. 7). Therefore, we believed that foliar N concentration played an important role in the diverging responses of R_{10} to warming in both years.

Acclimation of Photosynthesis

Photosynthesis has long been known to acclimate to prevailing T_{growth} by shifting the T_{opt} [67]. For example, Gunderson *et al.* [60] have reported that a 3-year warming of 2–4°C has resulted in a higher T_{opt} of A_n for five species of deciduous trees. In the current study we found that a 6-year warming of 2.07°C did not resulted in changes in T_{opt} of A_n (Fig. 4, Table S1). We also found that there were not statistically significant differences between the shift in T_{opt} of A_n and A_g in 2010 ($P=0.896$) or 2011 ($P=0.984$). This suggests that the instantaneous response of photosynthesis was independent of changes in R_d.

It has been proposed that the increase in the temperature sensitivity of V_{cmax}, indicated by ΔH_a of V_{cmax}, contributed to the thermal acclimation of photosynthesis to experimental warming [36,61,68]. However, in the current study we found that warming slightly decreased ΔH_a of V_{cmax} (Fig. 5, Table 1). Biochemically, the change in ΔH_a of V_{cmax} is closely related to the temperature dependence of Rubisco activity [69], Rubisco activation status [70,71], dimorphism of Rubisco [31], and the amount of Rubisco [72]. The lower ΔH_a of V_{cmax} obtained from the warming plots indicated that warming slightly decreased the temperature sensitivity of those processes.

Previous studies found that RuBP regeneration processes may play an important role in the thermal acclimation of photosynthesis [34,39,73]. The increase in the thermal stability of photosystem II, indicated by ΔH_a of J_{max}, has been shown to be related to the thermal acclimation of A_g to warming [34–36,74]. However, in the current study we found only minor response of ΔH_a of J_{max} to warming (Fig. 5, Table 1). This is also confirmed by our results that the RuBP regeneration seldom limited A_g (Fig. 8).

A number of studies have reported that the balance between the carboxylation and the regeneration of RuBP, indicated by J_{max}/V_{cmax} ratio, can also affect the thermal acclimation of photosynthesis [39,75]. In our study, the experimental warming had little effect on the linear trend of J_{max}/V_{cmax} ratio to T_{leaf} (Fig. 5). Nevertheless, in this study we found that J_{max}/V_{cmax} ratio declined sharply and linearly with the instantaneous increase in T_{leaf} (Fig. 5). Many ecosystem models, such as Biome-BGC [76], have set J_{max}/V_{cmax} ratio as a constant (2.1) which is independent of T_{leaf}. Wullschleger [77] analyzed 164 A_n/C_i curves for 109 C_3 plant species which were measured under T_{leaf} ranging from 13 to 35°C and found the average J_{max}/V_{cmax} ratio was 2.1. Others found that J_{max}/V_{cmax} ratio was not a constant instead varying with T_{leaf} through a linear [51,78–80] or nonlinear relationship [81]. Our current results show that the relationship (between J_{max} and V_{cmax}) itself is highly temperature dependent, suggesting that photosynthesis models have to consider the temperature dependence of J_{max}/V_{cmax} ratio.

In addition to biochemical limitations, the thermal acclimation of photosynthesis may also relate to CO_2 diffusion processes in leaves and chloroplasts, such as g_s and g_m, because changes in T_{growth} may affect CO_2 diffusivity, solubility, membrane permeability and stomatal movement [82–85]. Previous studies have found that increasing g_s and/or g_m can cause the increase of T_{opt} of A_n [36,40,41,67,86]. In the current study we found that warming increased g_m (Fig. 6) in 2010 which might contribute to the modest variation in T_{opt} of A_g in 2010. However, we found smaller increases in g_s and g_m (Fig. 6) in 2011, which may explain the weaker acclimation in 2011 (Fig. 4). The differential responses of CO_2 diffusion process to warming in the two hydrologically contrasting growing seasons could have been attributed to changes in soil moisture and N availability induced by warming [87]. It is noted that, so far, no consistent conclusions have been achieved on the warming effect on g_s and g_m. Some researchers found that warming increased g_s [39,88–90] and g_m [91], and others found warming decreased g_s [92] and g_m [61], or no effect on g_s [93] and g_m [40]. Those various studies suggest that other factors, such as warming-induced water depletion and change in N availability, may have interacting effects on responses of CO_2 diffusion process to warming. These results call for multi-factor experiments, such as the combination of warming with water manipulation and fertilization [21], for understanding the mechanisms of thermal acclimation of photosynthesis under future global change.

Balance between Respiration and Photosynthesis

The acclimation of foliar respiration and photosynthesis is also reflected in R/A ratio which indicates the balance between carbon gain, loss and accumulation [1,2]. Our results show that the instantaneous (<5 h) warming at foliage level has non-linearly increased R_d/A_g ratio, indicating proportionally more carbon loss through R_d as T_{leaf} goes up (Fig. 9). However, the 6-year experimental warming has resulted in thermal acclimation of the grasses as evidenced by the decrease of the curvature of the response curve of R_d/A_g ratio to T_{leaf} (Fig. 9). It is important to note that though the balance between R_d and A_g was re-established through the thermal acclimation [6,8,9,18], R_d/A_g ratio was still increasing with T_{growth} in a wet year (Fig. 9). This means that, at foliage level, acclimation can only partially compensate the negative impact from the global warming.

Supporting Information

Table S1 Results (*P*-values) of two-way ANOVA on the effects of warming, year, and both interactions on the responses of A_n (the net CO_2 assimilation rate), R_d (dark respiration), A_g (the gross CO_2 assimilation rate), V_{cmax} (the maximum rate of Rubisco carboxylation) and J_{max} (the maximum rate of photosynthetic electron transport) expressed per unit foliar area and nitrogen to instantaneous change (10–40°C within a 5 h period) in T_{leaf} (foliar temperature). *c* is a scaling constant, ΔH_a is the activation energy, ΔH_d is a term for deactivation, ΔS is an entropy term, T_{opt} is the thermal optimum, Q_{10} is the temperature sensitivity and ref_{10} is the estimated basal rate at the reference temperature of 10°C. Significant values (P<0.05) are shown bold.

Appendix S1 User's guide for the A/C_c curve fitting model with measured respiration, modified based on Sharkey *et al.*'s [50] Microsoft Excel spreadsheet-based software to reduce the number of fitting parameters (R_d is fixed in the model), version 1.2 (Last updated 25 July, 2012).

Acknowledgments

We thank Jianyang Xia and Diheng Zhong for their help with the field measurements.

Author Contributions

Conceived and designed the experiments: YC MX SW. Performed the experiments: YC RS QY. Analyzed the data: YC MX QY. Contributed reagents/materials/analysis tools: BH SW. Wrote the paper: YC MX RS QY BH SW.

References

1. Campbell C, Atkinson L, Zaragoza-Castells J, Lundmark M, Atkin OK, et al. (2007) Acclimation of photosynthesis and respiration is asynchronous in response to changes in temperature regardless of plant functional group. New Phytol 176: 375–389.
2. Ow LF, Whitehead D, Walcroft AS, Turnbull MH (2010) Seasonal variation in foliar carbon exchange in *Pinus radiata* and *Populus deltoides*: respiration acclimates fully to changes in temperature but photosynthesis does not. Global Change Biol 16: 288–302.
3. Way D, Sage R (2008) Thermal acclimation of photosynthesis in black spruce [*Picea mariana* (Mill.) B.S.P.]. Plant Cell Environ 31: 1250–1262.
4. Luo Y (2007) Terrestrial carbon-cycle feedback to climate warming. Annu Rev Ecol Evol Syst 38: 683–712.
5. Tjoelker MG, Oleksyn J, Reich PB (2001). Modelling respiration of vegetation: evidence for a general temperature-dependent Q_{10}. Global Change Biol 7: 223–230.
6. Gifford RM (1995) Whole plant respiration and photosynthesis of wheat under increased CO_2 concentration and temperature: long-term vs. short-term distinctions for modeling. Global Change Biol 1: 249–263.
7. Gunderson CA, Norby RJ, Wullschleger SD (2000) Acclimation of photosynthesis and respiration to simulated climatic warming in northern and southern populations of *Acer saccharum*: laboratory and field evidence. Tree Physiol 20: 87–96.
8. Loveys BR, Atkinson LJ, Sherlock DJ, Roberts RL, Fitter AH et al. (2003) Thermal acclimation of leaf and root respiration: an investigation comparing inherently fast-and slow-growing plant species. Global Change Biol 9: 895–910.
9. Dewar RC, Medlyn BE, McMurtrie RE (1999) Acclimation of the respiration photosynthesis ratio to temperature: insights from a model. Global Change Biol 5: 615–622.
10. Anderson JM, Chow WS, Park Y (1995) The grand design of photosynthesis: Acclimation of the photosynthetic apparatus to environmental cues. Photosynth Res 46: 129–139.
11. Atkin OK, Tjoelker MG (2003) Thermal acclimation and the dynamic response of plant respiration to temperature. Trends in Plant Sci 8: 343–351.
12. Rachmilevitch S, Lambers H, Huang B (2008) Short-term and long-term root respiratory acclimation to elevated temperatures associated with root thermotolerance for two *Agrostis* grass species. J Exp Bot 59: 3803–3809.
13. Atkin OK, Evans JR, Ball MC, Lambers H, Pons TL (2000) Leaf respiration of snow gum in the light and dark. Interactions between temperature and irradiance. Plant Physiol 122: 915–923.
14. Griffin KL, Turnbull M, Murthy R (2002) Canopy position affects the temperature response of leaf respiration in *Populus deltoides*. New Phytol 154: 609–619.
15. Lee TD, Reich PB, Bolstad PV (2005) Acclimation of leaf respiration to temperature is rapid and related to specific leaf area, soluble sugars and leaf nitrogen across three temperature deciduous tree species. Funct Ecol 19: 640–647.
16. Xu C, Griffin KL (2006) Seasonal variation in the temperature response of leaf respiration in *Quercus rubra*: foliage respiration and leaf properties. Funct Ecol 20: 778–789.
17. Tjoelker MG, Oleksyn J, Reich PB, Zytkowiak R (2008) Coupling of respiration, nitrogen, and sugars underlies convergent temperature acclimation in *Pinus banksiana* across wide-ranging sites and populations. Global Change Biol 14: 782–797.
18. Atkin OK, Bruhn D, Hurry VM, Tjoelker MG (2005) The hot and the cold: unraveling the variable response of plant respiration to temperature. Funct Plant Biol 32: 87–105.
19. Zhou X, Liu X, Wallace LL, Luo Y (2007) Photosynthetic and respiratory acclimation to experimental warming for four species in a tallgrass prairie ecosystem. J Integr Plant Biol 49: 270–281.
20. Crous KY, Zaragoza-Castells J, Löw M, Ellsworth DS, Tissue DT, et al. (2011) Seasonal acclimation of leaf respiration in *Eucalyptus saligna* trees: impacts of elevated atmospheric CO_2 conditions and summer drought. Global Change Biol 17: 1560–1576.
21. Luo Y, Weng E (2011) Dynamic disequilibrium of the terrestrial carbon cycle under global change. Trends Ecol Evol 26: 96–104.
22. Miroslavov EA, Kravkina IM (1991) Comparative analysis of chloroplasts and mitochondria in leaf chlorenchyma from mountain plants grown at different altitudes. Ann Bot-London 68: 195–200.
23. Armstrong AF, Logan DC, Atkin OK (2006) On the developmental dependence of leaf respiration: responses to short- and long-term changes in growth temperature. Am J Bot 93: 1633–1639.
24. Armstrong AF, Badger MR, Day DA, Barthet MM, Smith PMC, et al. (2008) Dynamic changes in the mitochondrial electron transport chain underpinning cold acclimation of leaf respiration. Plant Cell Environ 31: 1156–1169.
25. Rachmilevitch S, Lambers H, Huang B (2006). Root respiratory characteristics associated with plant adaptation to high soil temperature for geothermal and turf-type *Agrostis* species. J Exp Bot 57: 623–631.
26. Atkin OK, Macherel D (2009) The crucial role of plant mitochondria in orchestrating drought tolerance. Ann Bot-London 103: 581–597.

27. Lin YS, Medlyn BE, Ellsworth DS (2012). Temperature responses of leaf net photosynthesis: the role of component processes. Tree Physiol 32: 219–231.
28. Niu S, Luo Y, Fei S, Yuan W, Schimel D, et al. (2012) Thermal optimality of net ecosystem exchange of carbon dioxide and underlying mechanisms. New Phytol 194: 775–783.
29. Sage RF, Kubien DS (2007) The temperature response of C_3 and C_4 photosynthesis. Plant Cell Environ 30: 1086–1106.
30. Gutteridge S, Gatenby AA (1995) Rubisco synthesis, assembly, mechanism, and regulation. Plant Cell 7: 809–819.
31. Yamori W, Noguchi K, Terashima I (2005) Temperature acclimation of photosynthesis in spinach leaves: analyses of photosynthetic components and temperature dependencies of photosynthetic partial reactions. Plant Cell Environ 18: 536–547.
32. Weston DJ, Bauerle WL, Swire-Clark GA, Moore BD, Baird WV (2007) Molecular characterization of Rubisco activase from thermally contrasting genotypes of *Acer rubrum L.* (Aceraceae). Am J Bot 94: 926–934.
33. Yamori W, Noguchi K, Hikosaka K, Terashima I (2010) Phenotypic plasticity in photosynthetic temperature acclimation among crop species with different cold tolerances. Plant Physiol 152: 388–399.
34. Yamasaki T, Yamakawa T, Yamane Y, Koike H, Satoh K et al. (2002) Temperature acclimation of photosynthesis and related changes in photosystem II electron transport in winter wheat. Plant Physiol 128: 1087–1097.
35. Badger MR, Björkman O, Armond PA (1982) An analysis of photosynthetic response and adaptation to temperature in higher plants: temperature acclimation in the desert evergreen *Nerium oleander* L. Plant Cell Environ 5: 85–99.
36. Hikosaka K, Ishikawa K, Borjigidai A, Muller O, Onoda Y (2006) Temperature acclimation of photosynthesis: mechanisms involved in the changes in temperature dependence of photosynthetic rate. J Exp Bot 57: 291–302.
37. Sage RF, Way DA, Kubien DS (2008) Rubisco, Rubisco activase and global climate change. J Exp Bot 59: 1581–1595.
38. Hikosaka K (1997) Modelling optimal temperature acclimation of the photosynthetic apparatus in C_3 plants with respect to nitrogen use. Ann Bot-London 80: 721–730.
39. Hikosaka K, Murakami A, Hirose T (1999) Balancing carboxylation and regeneration of ribulose-1, 5-bisphosphate in leaf photosynthesis: temperature acclimation of an evergreen tree, *Quercus myrsinaefolia.* Plant Cell Environ 22: 841–849.
40. Yamori W, Noguchi K, Hanba YT, Terashima I (2006) Effects of internal conductance on the temperature dependence of the photosynthetic rate in spinach leaves from contrasting growth temperatures. Plant Cell Physiol 47: 1069–1080.
41. Kirschbaum MUF, Farquhar GD (1984) Temperature dependence of whole-leaf photosynthesis in *Eucalyptus pauciflora* Sieb Ex Spreng. Aust J Plant Physiol 11: 519–538.
42. Niu S, Li Z, Xia J, Han Y, Wu M et al. (2008) Climatic warming changes plant photosynthesis and its temperature dependence in a temperate steppe of northern China. Environ Exp Bot 63: 91–101.
43. Yan L, Chen S, Huang J, Lin G (2011) Water regulated effects of photosynthetic substrate supply on soil respiration in a semiarid steppe. Global Change Biol 17: 1990–2001.
44. Wan S, Xia J, Liu W, Niu S (2009) Photosynthetic overcompensation under nocturnal warming enhances grassland carbon sequestration. Ecology 90: 2700–2710.
45. Xia J, Niu S, Wan S (2009) Response of ecosystem carbon exchange to warming and nitrogen addition during two hydrologically contrasting growing seasons in a temperate steppe. Global Change Biol 15: 1544–1556.
46. von Caemmerer S, Farquhar GD (1981) Some relationships between the biochemistry of photosynthesis and the gas exchange of leaves. Planta 153: 376–387.
47. Galmés J, Medrano H, Flexas J (2007) Photosynthetic limitations in response to water stress and recovery in Mediterranean plants with different growth forms. New Phytol 175: 81–93.
48. Luo H, Ma L, Xi H, Duan W, Li S, et al. (2011) Photosynthetic responses to heat treatments at different temperatures and following recovery in grapevine (*Vitis amurensis* L.) leaves. PLoS one 6(8): e23033.
49. Wang X, Lewis JD, Tissue DT, Seemann JR, Griffin KL (2001) Effects of elevated atmospheric CO_2 concentration on leaf dark respiration of *Xanthium strumarium* in light and in darkness. P Natl Acad Sci USA 98: 2479–2484.
50. Sharkey TD, Bernacchi CJ, Farquhar GD, Singsaas EL (2007) Fitting photosynthetic carbon dioxide response curves for C_3 leaves. Plant Cell Environ 30: 1035–1040.
51. Ethier GJ, Livingston NJ (2004) On the need to incorporate sensitivity to CO_2 transfer conductance into the Farquhar-von Caemmerer-Berry leaf photosynthesis model. Plant Cell Environ 27: 137–153.
52. Miao Z, Xu M, Lathrop RG, Wang Y (2009) Comparison of the A–C_c curve fitting methods in determining maximum ribulose 1,5-bisphosphate carboxylase/oxygenase carboxylation rate, potential light saturated electron transport rate and leaf dark respiration. Plant Cell Environ 32: 109–122.
53. Harley PC, Thomas RB, Reynolds JF, Strain BR (1992) Modelling photosynthesis of cotton grown in elevated CO_2. Plant Cell Environ 15: 271–282.
54. Harley PC, Tenhunen JD (1991) Modeling the photosynthetic response of C_3 leaves to environmental factors. In: *Modeling Crop Photosynthesis: from Biochemistry to Canopy* (eds K.J. Boote & R.S. Loomis), 1–16. Crop Science Society of America, Madison, WI.
55. Sharkey TD (1985) O_2-insensitive photosynthesis in C_3 plants. Its occurrence and a possible explanation. Plant Physiol 78: 71–75.
56. Medlyn BE, Loustau D, Delzon S (2002) Temperature response of parameters of a biochemically based model of photosynthesis. I. Seasonal changes in mature maritime pine (*Pinus pinaster* Ait.). Plant Cell Environ 25: 1155–1165.
57. Teskey RO, Will RE (1999) Acclimation of loblolly pine (*Pinus taeda*) seedlings to high temperatures. Tree Physiol 19: 519–525.
58. Bolstad PV, Reich P, Lee T (2003) Rapid temperature acclimation of leaf respiration rates in *Quercus alba* and *Quercus rubra.* Tree Physiol 23: 969–976.
59. Hartley IP, Armstrong AF, Murthy R, Barron-Gafford G, Ineson P, et al. (2006) The dependence of respiration on photosynthetic substrate supply and temperature: integrating leaf, soil and ecosystem measurements. Global Change Biol 12: 1954–1968.
60. Gunderson CA, O'hara KH, Campion CM, Walker AV, Edwards NT (2010) Thermal plasticity of photosynthesis: the role of acclimation in forest responses to a warming climate. Global Change Biol 16: 2272–2286.
61. Dillaway DN, Kruger EL (2010) Thermal acclimation of photosynthesis: a comparison of boreal and temperate tree species along a latitudinal transect. Plant Cell Environ 33: 888–899.
62. Atkin OK, Scheurwater I, Pons TL (2006) High thermal acclimation potential of both photosynthesis and respiration in two lowland *Plantago* species in contrast to an alpine congeneric. Global Change Biol 12: 500–515.
63. Turnbull MH, Tissue DT, Griffin KL, Richardson SJ, Peltzer DA et al. (2005) Respiration characteristics in temperate rainforest tree species differ along a long-term soil-development chronosequence. Oecologia 143: 271–279.
64. Ow LF, Griffin KL, Whitehead D, Walcroft AS, Turnbull MH (2008) Thermal acclimation of leaf respiration but not photosynthesis in *Populus deltoids* × *nigra.* New Phytol 178: 123–134.
65. Xu M, Qi Y (2001) Spatial and seasonal variations of Q_{10} determined by soil respiration measurements at a Sierra Nevadan forest. Global Biogeochem Cy 15: 687–696.
66. Davidson EA, Janssens IA, Luo Y (2006) On the variability of respiration in terrestrial ecosystems: moving beyond Q_{10}. Global Change Biol 12: 154–164.
67. Berry J, Björkman O (1980) Photosynthetic response and adaptation to temperature in higher plants. Annu Rev Plant Physiol 31: 491–543.
68. Warren CR (2008) Does growth temperature affect the temperature responses of photosynthesis and internal conductance to CO_2 ? A test with *Eucalyptus regnans.* Tree Physiol 28: 11–19.
69. Galmés J, Ribas-Carbó M, Medrano H, Flexas J (2011) Rubisco activity in Mediterranean species is regulated by the chloroplastic CO_2 concentration under water stress. J Exp Bot 62: 653–665.
70. Crafts-Brandner SJ, Salvucci ME (2000) Rubisco activase constrains the photosynthetic potential of leaves at high temperature and CO_2. P Natl Acad Sci USA 97: 13430–43435.
71. Cen Y, Sage RF (2005) The regulation of Rubisco activity in response to variation in temperature and atmospheric CO_2 partial pressure in sweet potato. Plant Physiol 139: 979–990.
72. Delatorre J, Pinto M, Cardemil L (2008) Effects of water stress and high temperature on photosynthetic rates of two species of Prosopis. J Photoch Photobio B 92: 67–76.
73. Mitchell RAC, Barber J (1986) Adaptation of photosynthetic electron-transport rate to growth temperature in pea. Planta 169: 429–436.
74. Armond PA, Schreiber U, Björkman O (1978) Photosynthetic acclimation to temperature in the desert shrub, *Larrea divaricata.* II. Light-harvesting efficiency and electron transport. Plant Physiol 61: 411–415.
75. Onoda Y, Hikosaka K, Hirose T (2005) The balance between RuBP carboxylation and RuBP regeneration: a mechanism underlying the interspecific variation in acclimation of photosynthesis to seasonal change in temperature. Funct Plant Biol 32: 903–910.
76. Thornton PE, Law BE, Gholz HL, Clark KL, Falge E, et al. (2002) Modeling and measuring the effects of disturbance history and climate on carbon and water budgets in evergreen needleleaf forests. Agr Forest Meteorol 113: 185–222.
77. Wullschleger SD (1993) Biochemical limitations to carbon assimilation in C_3 plants-a retrospective analysis of the A/C_i curves from 109 species. J Exp Bot 44: 907–920.
78. Walcroft AS, Whitehead D, Silvester WB, Kelliher FM (1997) The response of photosynthetic model parameters to temperature and nitrogen concentration in *Pinus radiate* D. Don. Plant Cell Environ 20: 1328–1348.
79. Dreyer EW, Roux XL, Montpied P, Daudet FA, Masson F (2001) Temperature response of leaf photosynthetic capacity in seedlings from seven temperate tree species. Tree Physiol 21: 223–232.
80. Kattge J, Knorr W (2007) Temperature acclimation in a biochemical model of photosynthesis: a reanalysis of 36 species. Plant Cell Environ 30: 1176–1190.
81. Leuning R (2002) Temperature dependence of two parameters in a photosynthesis model. Plant Cell Environ 25: 1205–1210.
82. Flexas J, Bota J, Loreto F, Cornic G, Sharkey TD (2004) Diffusive and metabolic limitations to photosynthesis under drought and salinity in C_3 plants. Plant Biology 6: 269–279.
83. Bernacchi CJ, Portis AR, Nakano H, von Caemmerer S, Long SP (2002) Temperature response of mesophyll conductance. Implications for the

determination of Rubisco enzyme kinetics and for limitations to photosynthesis *in vivo*. Plant Physiol 130: 1992–1998.
84. Flexas J, Ortuño MF, Ribas-Carbó M, Díaz-Espejo A, Florez-Sarasa ID et al. (2007) Mesophyll conductance to CO_2 in *Arabidopsis thaliana*. New Phytol 175: 501–511.
85. Flexas J, Ribas-Carbó M, Díaz-Espejo A, Galmés J, Medrano H (2008) Mesophyll conductance to CO_2: current knowledge and future prospects. Plant Cell Environ 31: 602–621.
86. Farquhar GD, von Caemmerer S, Berry JA (1980) A biochemical model of photosynthetic CO_2 assimilation in leaves of C_3 species. Planta 149: 78–90.
87. Melillo JM, Steudler PA, Aber JD, Newkirk K, Lux H, et al. (2002) Soil warming and carbon-cycle feedbacks to the climate system. Science 298: 2173–2176.
88. Ferrar PJ, Slatyer RO, Vranjic JA (1989) Photosynthetic temperature acclimation in *Eucalyptus* species from diverse habitats, and a comparison with *Nerium oleander*. Aust J Plant Physiol 16: 199–217.
89. Williams DG, Black RA (1993) Phenotypic variation in contrasting temperature environments: growth and photosynthesis in *Pennisetum setaceum* from different altitudes on Hawaii. Funct Ecol 7: 623–633.
90. Hikosaka K (2005). Nitrogen partitioning in the photosynthetic apparatus of *Plantago asiatica* leaves grown at different temperature and light conditions: similarities and differences between temperature and light acclimation. Plant Cell Physiol 46: 1283–1290.
91. Makino A, Nakano H, Mae T (1994) Responses of ribulose-1,5-bisphosphate carboxylase, cytochrome *f*, and sucrose synthesis enzymes in rice leaves to leaf nitrogen and their relationships to photosynthesis. Plant Physiol 105: 173–179.
92. Mooney HA, Björkman O, Collatz GJ (1978) Photosynthetic acclimation to temperature in the desert shrub, Larrea divaricata. Plant physiol 61: 406–410.
93. Hendrickson L, Ball MC, Wood JT, Chow WS, Furbank RT (2004) Low temperature effects on photosynthesis and growth of grapevine. Plant Cell Environ 27: 795–809.

The Potential and Flux Landscape Theory of Ecology

Li Xu[1]◔, Feng Zhang[1,2]◔, Kun Zhang[1], Erkang Wang[1], Jin Wang[1,2,3]*

1 State Key Laboratory of Electroanalytical Chemistry, Changchun Institute of Applied Chemistry, Chinese Academy of Sciences, Changchun, Jilin, China, **2** College of Physics, Jilin University, Changchun, Jilin, China, **3** Department of Chemistry & Physics, State University of New York at Stony Brook, Stony Brook, New York, United States of America

Abstract

The species in ecosystems are mutually interacting and self sustainable stable for a certain period. Stability and dynamics are crucial for understanding the structure and the function of ecosystems. We developed a potential and flux landscape theory of ecosystems to address these issues. We show that the driving force of the ecological dynamics can be decomposed to the gradient of the potential landscape and the curl probability flux measuring the degree of the breaking down of the detailed balance (due to in or out flow of the energy to the ecosystems). We found that the underlying intrinsic potential landscape is a global Lyapunov function monotonically going down in time and the topology of the landscape provides a quantitative measure for the global stability of the ecosystems. We also quantified the intrinsic energy, the entropy, the free energy and constructed the non-equilibrium thermodynamics for the ecosystems. We studied several typical and important ecological systems: the predation, competition, mutualism and a realistic lynx-snowshoe hare model. Single attractor, multiple attractors and limit cycle attractors emerge from these studies. We studied the stability and robustness of the ecosystems against the perturbations in parameters and the environmental fluctuations. We also found that the kinetic paths between the multiple attractors do not follow the gradient paths of the underlying landscape and are irreversible because of the non-zero flux. This theory provides a novel way for exploring the global stability, function and the robustness of ecosystems.

Editor: Antoni Margalida, University of Lleida, Spain

Funding: This study was supported by NSFC grant no. 21190040, 91227114, and 11174105; 973 project 2010CB933600; and NSF. The funders had no role in study design, data collection and analysis, decision to publish, or preparation of the manuscript.

Competing Interests: The authors have declared that no competing interests exist.

* E-mail: jin.wang.1@stonybrook.edu

◔ These authors contributed equally to this work.

Introduction

Ecosystems are the ones in which their living and nonliving components interact with and depend on each other linking together the exchange of energy, material, information. The structure and the function of the ecosystems are determined by the interplay of both cooperation and competition [1,2]. Ecosystems are able to regulate themselves to maintain certain stability. Therefore, the stability is one of the most fundamental and essential features of the ecological systems. The study of stability is direct relevant to the existence of every species. The stability is influenced by many factors, such as the structure within the components and the features of the environment. The studies of the stability of ecosystems are significant for uncovering the underly ecological law of species and populations [1,2].

The ecosystems are complex and involve many kinds of interactions among the elements. The inherent interactions are often non-linear and intricate. These systems can be described by a set of nonlinear differential equations [3,4]. These nonlinear interactions lead to complex dynamics. There have been many investigations on the stability of ecosystems [3–11]. Most of the works have been focused on the local linear stability analysis. The global stability of the ecological systems is still challenging in general. Furthermore, the link between the global characterization of the ecological systems and the dynamics of the elements is still not clear.

The past researchers also explored the dynamical system with the approach of Lyapunov function which was developed to investigate the global stability. The analytical Lyapunov function for predation model was first proposed by Volterra in 1931 [4]. Since then, significant efforts have been devoted to find the analytical Lyapunov function [5–8,10,11] for ecological equations. However, it is still challenging to find the Lyapunov function of the ecological models with the general nonlinear response terms, even though a few highly simplified ones were worked out [5,8]. So far there is no general and universal approach for finding the Lypunov function. One has to work case by case. There is even no guarantee if a Lyapunov function exists for a more complex system. Furthermore the Lyapunov function and the stability of a predation model with a solution of limit cycle have hardly been discussed. Here we would like to suggest an universal and straightforward approach to explore the Lyapunov function and therefore the global stability of the general ecological systems.

In the earlier studies, people have aimed at macroscopic properties based on the deterministic models. However, both external and intrinsic fluctuations of mesoscopic and even macro scale systems are unavoidable. The environmental fluctuations can influence the ecological behaviors. The intrinsic fluctuations emerge when the size of system is finite. It is widely agreed that the analysis of the global stability is important for a full understanding of the robustness of ecological systems under fluctuations [5–7,11–13]. As mentioned, it is challenging to explore global stability with deterministic dynamics. The probability (P) description is necessary due to the presence of fluctuations of the real systems. The probability description has the advantage of quantifying the weights in the whole population state space and therefore is global. The potential landscape U linked with the probability P by $U \sim lnP$ can address global properties of ecosystems, such as the global stability, function and robustness.

Here we developed a potential and flux landscape theory to explore the global stability and dynamics for ecosystems [14–19]. We found that the underlying intrinsic potential landscape is a global universal Lyapunov function for the ecosystem dynamics and therefore topology of the landscape provides a quantitative measure for the global stability of the ecosystems. We also found the dynamics of the ecosystems is determined by the gradient of the potential landscape and additional curl flux force from breaking down of the detailed balance.

We applied our theory to three important ecosystems: predation, competition, mutualism and a realistic lynx-snowshoe hare model. Lotka-Volterra model for two species interactions is the famous ecological model proposed by Lotka [3]and Volterra [4] respectively. Over the years, this model has attracted attentions for exploring the dynamical process of the ecology. In the ecosystems, the relationship between species can be grouped into two categories: the negative antagonism interaction(−) and the positive mutualism interaction(+). We show the different modes in Figure 1. Predation shows the relationship (+/−) which one species S_A is disfavored, while the other species S_B benefits in Figure 1A; Competition shows the relationship(−/−) which each species S_A or S_B is influenced negatively by the other one in Figure 1B; Mutualism shows the relationship (+/+) which both species S_A and S_B benefit from interactions of the other in Figure 1C [2,6,11].

For the predation, predators sustain their lives by the consumption of preys. Without the presence of prey, predators can not survive. Preys can sustain their lives and grow without predators. The presence of the preys controls the predators' growth. This forms a predator-prey(predation) system [2–4]. The system can have one stable state or stable limit cycle. Competition between species often occurs when they are using the same resources. Competition can promote the ecological characteristics of species differentiation and produce certain biological structure of community. The system can have multi-stable states. Mutualism means that two different species biologically interact with each other and lead to benefit each individuals. The system can also have multi-stable states. These relations above demonstrate the complexity of biological communities,their stable structures, and the ecological balance [2,6,11]. These models are important for population biology because of its applications to the real biological world.

Limit cycle attractor shown in Figure 1D(Mexican hat shape)and multiple attractors shown in Figure 1E emerge from these three cases: predation, competition and mutualism. Figure 1D and Figure 1E show the potential landscapes closely associated with the probability distribution for the underlying ecosystem. The lower potential landscape means the higher probability and therefore more stable states. The closed ring valley is a continuous attractor for oscillation while the discrete basins represent the stable attractors on the potential landscapes. We found the quantitative criterion for stability in terms of barrier height between basins of attraction. The barrier height from the top of the Mexican hat to the ring valley of limit cycle attractor (the potential maximum along the limit cycle) in Figure 1D or between each two attractors(from the saddle point to the minimum of each basin) in Figure 1E quantitatively determine the global stability of these ecosystems. The curl flux is essential for driving the oscillation on the valley ring and maintaining the coherence. We studied stability and robustness of ecosystems against parameters and fluctuations. We also explored how the non-equilibrium free energy links with the different phases and phase transitions of the ecosystem with respect to the changes of parameters. These explained how the stability and robustness of ecosystem change under different conditions. Quantification of pathway is important for understanding the dynamics of ecosystem. We developed the path integral method to quantify the kinetic pathway of ecosystems. We found that the paths do not

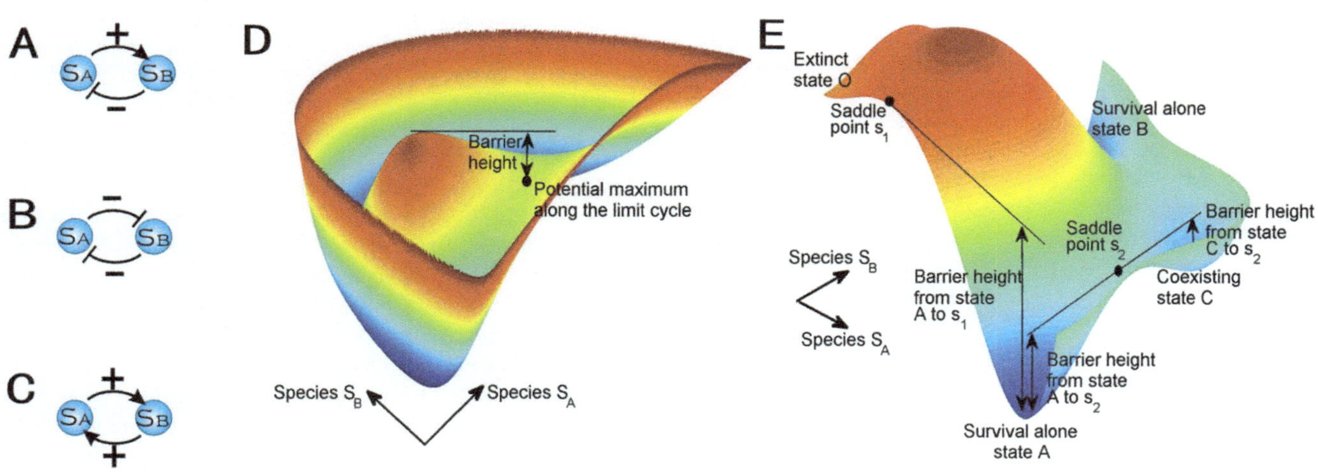

Figure 1. The schematic diagram for the ecological models. (A)Predation model. (B)Competition model. (C)Mutualism model. The potential landscape U is linked with the probability P by $U \sim lnP$ in species space. (D) Limit cycle attractor. The barrier height from the maximum inside the closed ring to the potential maximum along the ring can quantify the stability of the limit cycle attractor. (E) Multiple attractors. There are four stable states: survival alone state A of species S_A, survival alone state B of species S_B, coexisting state C, and both extinct state O. s_1 is the saddle points between the attractors A and O while s_2 is the saddle points between the attractors A and C. The barrier heights from the saddle points to the potential minimums of the basins can quantify the stability of each attractor.

follow gradient of the underlying potential landscape and are irreversible because of the non-zero flux.

Results and Discussion

The potential landscapes and fluxes of ecosystems: predation, competition and mutualism

We quantified the underlying potential landscape of the ecosystems through the exploration of the underlying dynamics of probability. We solved the Fokker-Planck diffusion equation describing the probability dynamics to obtain the population potential landscape U related to the steady state probability distribution P_{ss} through $-lnP_{ss}$ and steady state probability flux $\mathbf{J}$. We explored the Hamilton-Jacobi equation in zero noise limit to quantify the intrinsic potential landscape ϕ_0 with Lyapunov properties and the associated intrinsic flux velocity. The details are in the Materials and Methods section. We will discuss how to apply the landscape theory of intrinsic potential landscape ϕ_0, the population potential landscape U and the probability flux $\mathbf{J}$ to explore the global stability and dynamics of the three ecosystem models.

The population potential landscape U (the top row) and intrinsic potential landscape ϕ_0 (the bottom row)for predation, competition and mutualism model are shown respectively in Figure 2. The negative gradient of the population potential landscape $-\nabla U$ on the top row and the intrinsic potential landscape $-\nabla\phi_0$ on bottom row are represented by the black arrows while the steady state probability flux $\mathbf{J}_{ss}/P_{ss}$ on the top row and the intrinsic flux velocity on the bottom row are represented by the purple arrows. The arrows at the bottom of each sub-figures are the projection of the direction of the according arrows. The flux with purple arrows are almost orthogonal to the negative gradient of U with black arrows shown on the bottom plane of Figure 2A, Figure 2B and Figure 2C. The flux velocity with purple arrows are orthogonal to the negative gradient of ϕ_0 with black arrows shown on the bottom plane of Figure 2D, Figure 2E and Figure 2F.

Figure 2A and Figure 2D show the non-equilibrium population potential landscape U and intrinsic potential landscape ϕ_0 for predation model when the parameters are $a=1.5$, $b=0.1$, $d=0.2$, $D=3\times10^{-5}$. The mathematical model and the range of the parameters are discussed in Materials and Methods section. We can see when the fluctuations characterized by the diffusion coefficient are small, the underlying potential landscape has a distinct closed irregular and inhomogeneous closed ring valley or Mexican hat like shape shown in Figure 2A. We can clearly see the population potential landscape is not uniformly distributed along the limit cycle path or the closed ring. The intrinsic potential landscape ϕ_0 has a homogeneous closed ring valley along deterministic oscillation trajectories which has a constant value of ϕ_0 shown in Figure 2D. The non-equilibrium intrinsic potential landscape ϕ_0 with Lyapunov properties can characterize the global topography of the oscillation landscape of predation model. This figure shows the potential is lower along the oscillation path or on the ring. The potential landscape is higher with a mountain inside the oscillation ring and outside the oscillation ring. The system is attracted to the closed oscillation ring by the landscape's gradient-potential force ∇U or the $\nabla\phi_0$. The flux driving the system maintains the periodical continuous oscillation dynamics. Both landscape and flux are necessary to characterize the non-equilibrium predation ecosystems. This oscillation shows that when the number of predators increases, more preys will be consumed. Then due to the shortage of food, the number of the predator will go down. The reduction of the predators makes the preys multiply, then the number of predators increases again for the rich preys.

Figure 2B and Figure 2E show the non-equilibrium population potential landscape U and intrinsic potential landscape ϕ_0 for competitive model when the parameters are $a_1=a_2=0.1$, $L_1=L_2=0.3$, $\alpha=1.0$, $D=0.01$. We can see the underlying population potential landscape and intrinsic potential landscape both have four distinct basins which have been discussed in the Materials and Methods section. The basins are around the four stable states. These four stable states are: survival alone state A of species S_A, survival alone state B of species S_B, coexisting state C, and both extinct state O. These figures show the potential landscape is relatively higher (and probability is relative lower) on the extinct state (the state O) of these two species because the lower critical points L_1,L_2 for species are small. The potential landscapes of the survival alone state A and B are lower(more stable) than that of the coexisting state C. It shows that the coexisting state is less stable than the species survival alone states for they have competitive relation. The forces from negative gradient of the potential landscape are more significant away from the attractors and less significant near the basins. Therefore the system is attracted by the gradient of the landscape towards the four basins. The directions of the flux are curling among the basins.

Figure 2C and Figure 2F show the non-equilibrium population potential landscape U and intrinsic potential landscape ϕ_0 for mutualism model when the parameters are $a_1=a_2=0.1$, $L_1=L_2=0.5$, $\alpha=1.0$, $D=0.01$. The underlying population potential landscape and intrinsic potential landscape both have distinct four basins. The basins are around the four stable states. These figures show the potential landscape is the highest (and probability is lower) on the extinct state O of these two species. The potential landscape of mutualism coexisting state C is lowest than those of species survival alone state A and B, and the extinct state O. It shows that the coexisting state is more stable than the species alone state for the two species having the relationship of mutualism. The directions of the flux are curling among the four basins. And the system is also attracted by the gradient of landscape towards the four basins.

Quantitative measure of global stability for predation

We now study the stability and robustness of the ecosystems. The stability is related to the escape time from the basins. Since the system is characterized by the basins of attractor with large weights, the easier it is to escape, the less stable the system is. We will essentially explore the average time escaping from a basin of attraction.

We define the barrier height of U for predation model as: $Ba=U_{fix}-U_{max}$. U_{max} is the population potential maximum along the limit cycle attractor. U_{fix} is the population potential landscape at the local maximum point inside the limit cycle. And the barrier height of ϕ_0 for predation model is defined as: $\Delta\phi_0=\phi_{0fix}-\phi_{0max}$. ϕ_{0max} is the intrinsic potential maximum along the limit cycle attractor. ϕ_{0fix} is the intrinsic potential landscape at the local maximum point inside the limit cycle. The escape time τ can be solved by the formula [20]: $\mathbf{F}\cdot\nabla\tau+D\cdot\nabla^2\tau=-1$. It represents the average time the system spent from one position to another position [18].

We showed the change of population potential landscapes for increasing diffusion coefficient D (Figure S1 in File S1). In Figure 3A, the barrier heights associated with escape time from the limit cycle attractor Ba becomes higher when the diffusion coefficient characterizing the fluctuations decreases. In Figure 3B, we see a direct relationship between the escape time τ from the

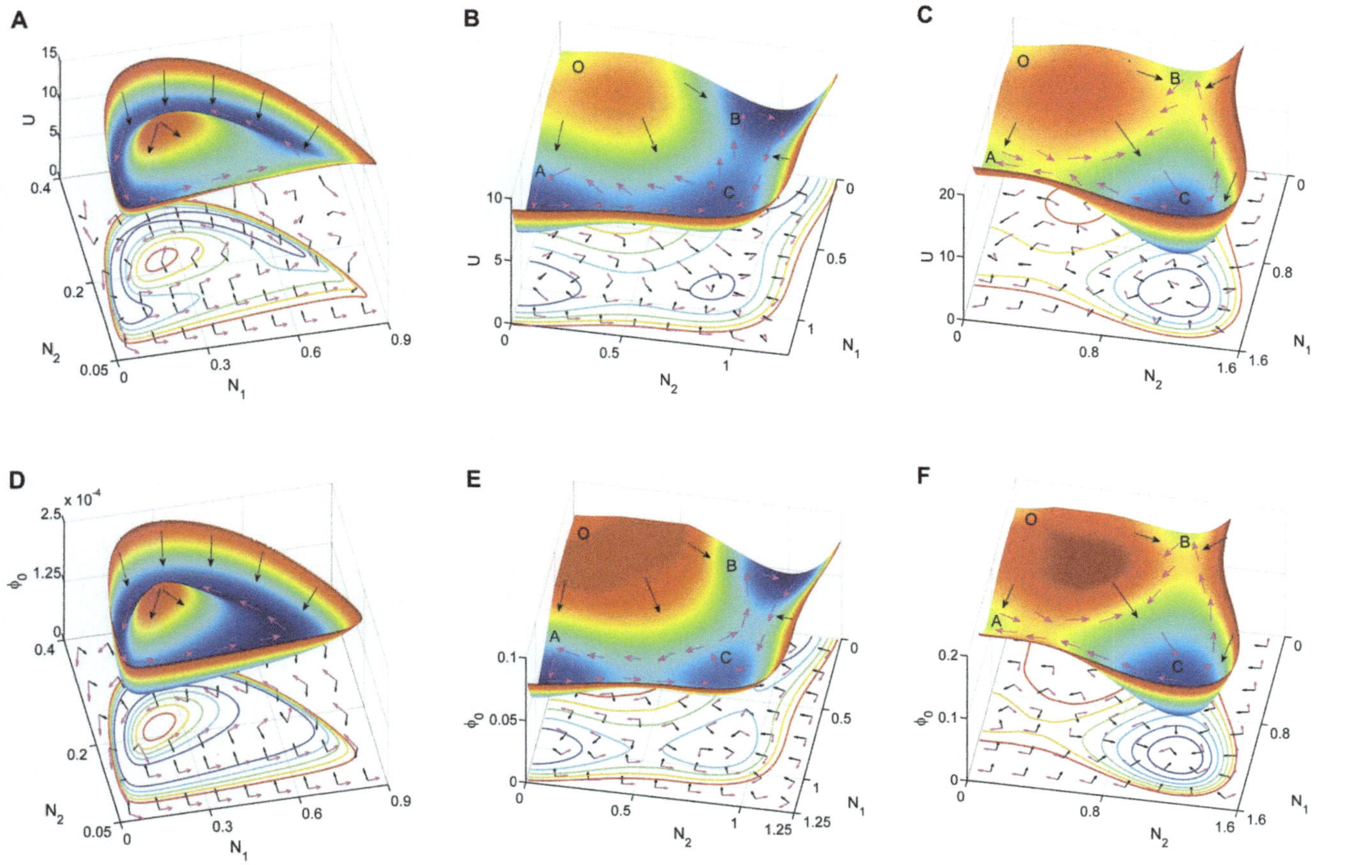

Figure 2. The potential landscapes for the predation, competition and mutualism models. Top row: The population potential landscape U ((A) predation model. (B) competition model. (C) mutualism model.) Purple arrows represent the flux velocity($\mathbf{J}_{ss}/P_{ss}$) while the black arrows represent the negative gradient of population potential($-\nabla U$). Bottom row: The potential intrinsic energy landscape ϕ_0. ((D) predation model. (E) competition model. (F) mutualism model.). Purple arrows represent the intrinsic flux velocity($\mathbf{V}=(\mathbf{J}_{ss}/P_{ss})_{D\to 0}$) while the black arrows represent the negative gradient of intrinsic potential($-\nabla\phi_0$).

limit cycle attractor [17,18] and barrier height for non-equilibrium ecosystems. As the barrier height for escape becomes higher, the escape time becomes longer. Therefore, the limit cycle attractor becomes more stable since it is harder to go from the valley ring to outside. The robustness and stability in the oscillatory predation system need small fluctuations and large barrier height. Figure 3C shows the heat dissipation rate $HDR=\int(\mathbf{J}\cdot(\mathbf{F}-\nabla\cdot\mathbf{D}))d\mathbf{N}=\int\mathbf{J}\cdot\mathbf{F}d\mathbf{N}$ (see the Materials and Methods section 1 for details)versus different diffusion coefficients. We can see the heat dissipation rate decreases when the diffusion coefficient decreases. This implies that when the fluctuations deduce, the associated heat dissipation rate becomes smaller. We have explored that less fluctuations lead to more robust and stable oscillation with higher barrier height. Therefore, the less dissipation can lead to less fluctuations and a more stable ecosystem with longer escape time.

We also explored the effects of the rate parameters representing interaction strengths between species on the robustness. We showed the change of population potential landscapes for increasing parameters a,b,d (Figure S2 in File S1, Figure S3 in File S1 and Figure S4 in File S1). We show the effects of rate parameters a,b,d on the robustness through barrier height in Figure 4A, Figure 4D and Figure 4G; the escape time τ in Figure 4B, Figure 4E and Figure 4H; the dissipation rate in Figure 4C, Figure 4F and Figure 4I with $D=3\times10^{-5}$. For the small parameter a, the relative death rate or the interaction strength for the prey is very small, the prey and the predator both are near their carrying capacities. The stable point is $N_1^*=N_2^*$, where the population of the predator is equal to the population of the prey. The system is attracted to its stable coexisting state. When the relative death rate or the interaction strength increases to a certain specific range, the system can reach a limit cycle state since it is a negative feedback system. When the parameter a continues to increase beyond the range of limit cycle, the death rate or the interaction strength of the preys increases, the preys decrease. The lack of food leads to the reduction of the predators. So the system reaches a stable state that both predator and preys have low population. The Ba versus a increases first, then decreases as shown in Figure 4A. Figure 5A shows barrier height $\Delta\phi_0$ of the intrinsic potential landscape(ϕ_0) has the same tendency with that of the population landscape. It implies the system first becomes more stable, and then less stable when the parameter a increases in the range of limit cycle. The system has an optimal stability in this range. This is due to the fact that both decreasing and increasing a from optimal stability of the limit cycle will promote the formation of mono-stability and make the limit cycle less stable. Figure 4B shows the average escape time versus the barrier height Ba. When the barrier height for escape becomes higher, the escape time becomes longer. Figure 4C shows the dissipation rate increases with the increasing parameter a. It implies that the system will consume more energy to keep order

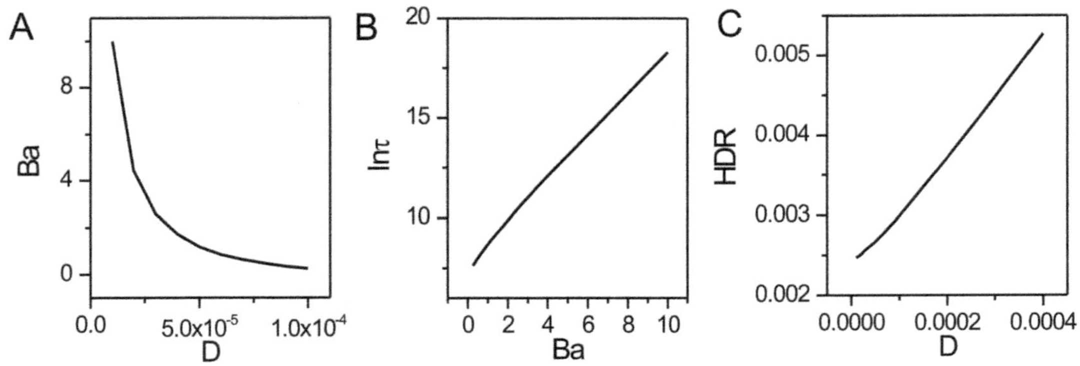

Figure 3. The barrier height of population landscape, escape time and dissipation rate versus the diffusion coefficient for predation model. (A) The barrier height of the population landscape U versus the diffusion coefficient D. (B) The escape time versus the barrier height of population landscape. (C) The dissipation rate versus the diffusion coefficient D.

when the system stays at limit cycle oscillation state than when it stays at one stable state with the transition point $a=1.109$. Figure 5D shows the intrinsic free energy(see the Materials and Methods section 2 for details) versus a. We can see clearly the non-equilibrium free energy is continuous versus parameter a, the first derivative of the free energy is discontinuous at the transition point $a=1.109$ from a stable state to a limit cycle oscillation state. It is a signal of non-equilibrium thermodynamic phase transition analogous to equilibrium statistical mechanics. The non-equilibrium free energy for ecosystem can measure and predict the global phases transitions and can be use to investigate the global natural stability and robustness of the ecosystem.

When the parameter b which represents the ratio of the linear growth rate of the predator to that of the prey is small, the system stays at the specific range of limit cycle state. When the parameter b continues to increase, the growth rate of the predator increases, so the prey decreases and therefore it will lead the predator to decrease. So the system will stay at a stable state where the predator and prey both keep relative lower population. The Ba versus b decreases shown in Figure 4D. Figure 5B shows barrier height of the intrinsic potential landscape(ϕ_0) has the same tendency with the population potential landscape. This implies that the limit cycle attractor becomes less stable when the ratio of the linear growth rate of the predator increases. Figure 4E shows the average escape time versus the increasing parameter b has the same tendency with the barrier height versus the increasing parameter b. Figure 4F shows that the dissipation rate decreases with the increasing parameter b. Figure 5E shows the free energy versus b. The first derivative of the non-equilibrium free energy is discontinuous at the transition point $b=0.1899$ from a limit cycle oscillation state to a stable state.

The predation term $\frac{aN_1N_2}{N_1+d}$ which is the response of the predator to the change in the prey density, shows the saturation effect. When the parameter d which represents the relative saturation effect rate of the prey is small, the system stays at the specific range of the limit cycle state as saturation point of predation is very low. When the parameter d continues to increase, the response term and the death rate of the preys decreases, so the preys and the predators are in a relative lower population stable state. The barrier height Ba decreases as d increasing shown in Figure 4G. Figure 5C shows barrier height of the intrinsic potential landscape(ϕ_0) has the same tendency with the population landscape. This implies that the limit cycle attractor of this system becomes less stable when the saturation point of predation increasing. Figure 4H shows the average escape time versus the increasing parameter d has the same tendency with the Ba versus the increasing parameter d. Figure 4I shows that the dissipation rate decreases with the increasing parameter d. The system becomes less stable. Figure 5F shows the free energy versus d. The transition point $d=0.2715$ shows the system transits from a limit oscillation state to a stable state.

We also can use the phase coherence to quantify the robustness and the stability of a cycle [17,18,21]. Coherence ζ represents the degree of the regularity in time sequence of cycle motion of species variables in the ecosystem. The vector $M(t)=n_1(t)e_1+n_2(t)e_2$ is set with $e_1=(0,1)$ and $e_2=(1,0)$ as the unit vectors, as well as n_1 and n_2 represent the population of each species at time t. $w(t)$ is the phase angle between $M(t)$ and $M(t+\tau)$. Then the ζ is defined as $\zeta=(2\sum_i\theta(w(t_i))w(t_i))/(\sum_i|w(t_i)|)-1$ [17,18,21]. ($\theta=1$ if $w>0$, and $\theta=0$ if $w\leq 0$) The larger value of coherence close to 1 means the more periodic while the smaller value of coherence close to 0 means the less periodic. In Figure 6A, ζ decreases when the diffusion coefficient increases. This means larger fluctuations will reduce the coherence and the robustness of the oscillations.Figure 6B, Figure 6C and Figure 6D show the ζ versus the parameter a,b,d. We found that they have nearly the same tendency with the barrier height versus the parameters a,b,d shown in Figure 4A, Figure 4D and Figure 4G. It implies that the higher barrier height lead the system to have more coherent periodic oscillation, and therefore the system is more stable.

Quantifying the global stability of competition ecosystems

For competition and mutualism models, we introduced the effective barrier heights for simplicity. The effective barrier heights are similar to the effective resistance of parallel circuits. They can measure the effect of average barrier on a single basin. We defined the effective barrier heights for state i in population potential U as: $Ba_i=\frac{(Us_{ij}-U_i)*(Us_{ik}-U_i)}{(Us_{ij}-U_i)+(Us_{ik}-U_i)}$ where $\{i,j,k\}\in\{\{A,O,C\},\{B,O,C\},\{C,A,B\}\}$. When the system has four stable states: survival alone state A of species S_A, survival alone state B of species S_B, coexisting state C, and both extinct state O. Us_{ij} is the population landscape value of saddle point between state i and state j. U_A, U_B, U_C and U_O are the values of population landscape at state A, B, C and O. When the coexisting state C vanishes, we defined the $Ba_i=\frac{(Us_{ij}-U_i)*(Us_{ik}-U_i)}{(Us_{ij}-U_i)+(Us_{ik}-U_i)}$ where $\{i,j,k\}\in$

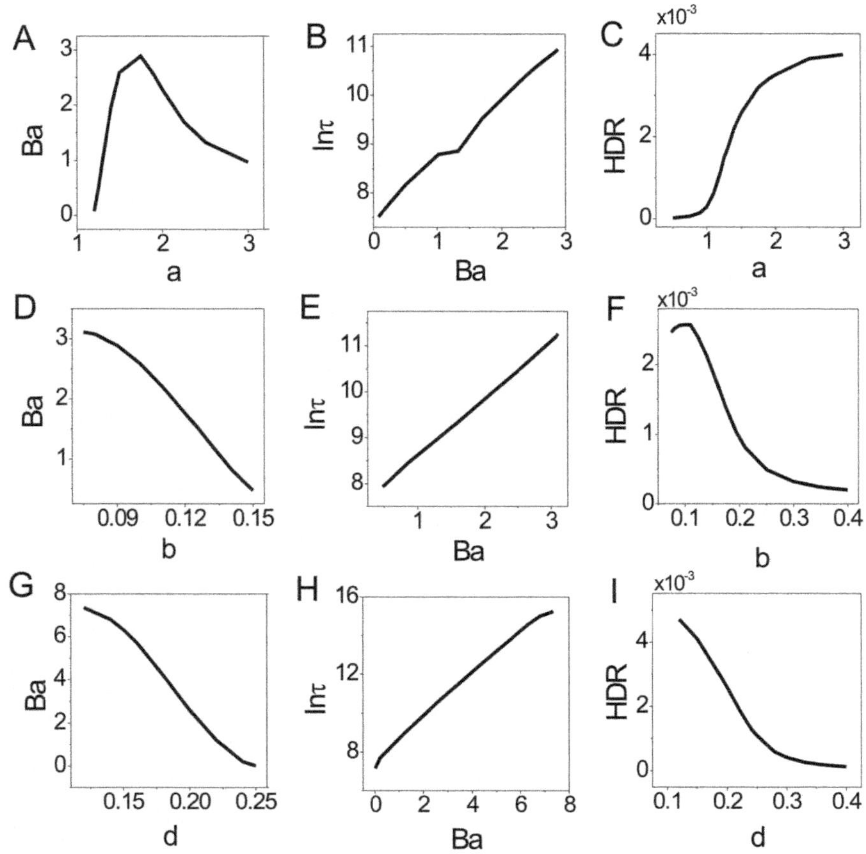

Figure 4. The barrier height of the population landscape, escape time and dissipation rate versus the rate parameters for predation model. (A) The barrier height of the population landscape versus *a*. (B) The escape time versus barrier height of the population landscape for changing *a*. (C) The dissipation rate versus *a*. (D) The barrier height of the population landscape versus *b*. (E) The escape time versus barrier height of the population landscape for changing *b*. (F) The dissipation rate versus *b*. (G) The barrier height of the population landscape versus *d*. (H) The escape time versus barrier height of the population landscape for changing *d*. (I) The dissipation rate versus *d*.

$\{\{A,O,B\},\{B,O,A\}\}$, We also defined the effective barrier heights in intrinsic potential as: $\Delta\phi_{0i}=\frac{(\phi_{0ij}-\phi_{0i})*(\phi_{0ik}-\phi_{0i})}{(\phi_{0ij}-\phi_{0i})+(\phi_{0ik}-\phi_{0i})}$ where $\{i,j,k\}\in\{\{A,O,C\},\{B,O,C\},\{C,A,B\}\}$ when the system has four stable states A,B,C,O. When the coexisting state C vanishes, we defined the $\Delta\phi_{0i}=\frac{(\phi_{0ij}-\phi_{0i})*(\phi_{0ik}-\phi_{0i})}{(\phi_{0ij}-\phi_{0i})+(\phi_{0ik}-\phi_{0i})}$ where $\{i,j,k\}\in\{\{A,O,B\},\{B,O,A\}\}$. ϕ_{0ij} is the intrinsic landscape value of saddle point between state i and state j. ϕ_{0A}, ϕ_{0B}, ϕ_{0C} and ϕ_{0O} are the intrinsic potential landscape values of state A, B, C and O.

We showed the change of population potential landscapes for increasing diffusion coefficient D (Figure S5 in File S1). Figure 7 shows diffusion effect on the competition model. In Figure 7A, as the diffusion coefficient characterizing the fluctuations decreases, the barrier heights $Ba_{A(B)}$ and Ba_C are higher when the parameters are $a_1=a_2=0.1, L_1=L_2=0.3, \alpha=1.0$. Figure 7B shows the escape time versus the barrier height Ba_A. We can see the system is harder to escape from the basins of attraction as the fluctuation decreases, and the barrier height also increases. Figure 7C shows the dissipation rate for different diffusion coefficients. We can see the heat dissipation rate decreases when the diffusion coefficient decreases and the fluctuations of the systems become smaller.

We showed the change of population potential landscapes for increasing parameters a_1, L_1, α (Figure S6 in File S1, Figure S7 in File S1 and Figure S8 in File S1). Figure 8A, Figure 8B and Figure 8C show the effects of rate parameter a_1 on this system at $D=0.001$. When a_1, the competitiveness of S_B on S_A decreases, the competitiveness of species S_B becomes weaker while that of species S_A becomes stronger, so the barrier height Ba_A increases while Ba_B decreases. With the parameter (a_1) further weakened the competitiveness of the species S_B, the two species can achieve their coexisting state C. The coexisting state C becomes more stable as the competitiveness of B on A decreases shown in Figure 8A. Figure 9A shows barrier height of the intrinsic potential landscape(ϕ_0) has the same tendency with that of the population landscape. This implies the two species can coexist stably as the mutual competitiveness is reduced. The logarithm of escape time $ln\tau_A$ from basin A has positive correlation with barrier height Ba_A shown in Figure 8B. The escape time increases when the associated barrier height becomes higher.

Figure 8D, Figure 8E and Figure 8F show the effects of rate parameter L_1 on competition system with $D=0.001$. When L_1 increases, the lower critical number or density of species S_A becomes larger, so the survival alone state A of species S_A becomes less stable and shallower while the other survival alone state B becomes more stable and deeper (The barrier height Ba_B increases,Ba_A and Ba_C decrease as L_1 increases shown in Figure 8D. The escape time $ln\tau_A$ from basin A has positive correlation with Ba_A shown in Figure 8E. Figure 9B shows barrier height of the intrinsic potential landscape(ϕ_0) has the same

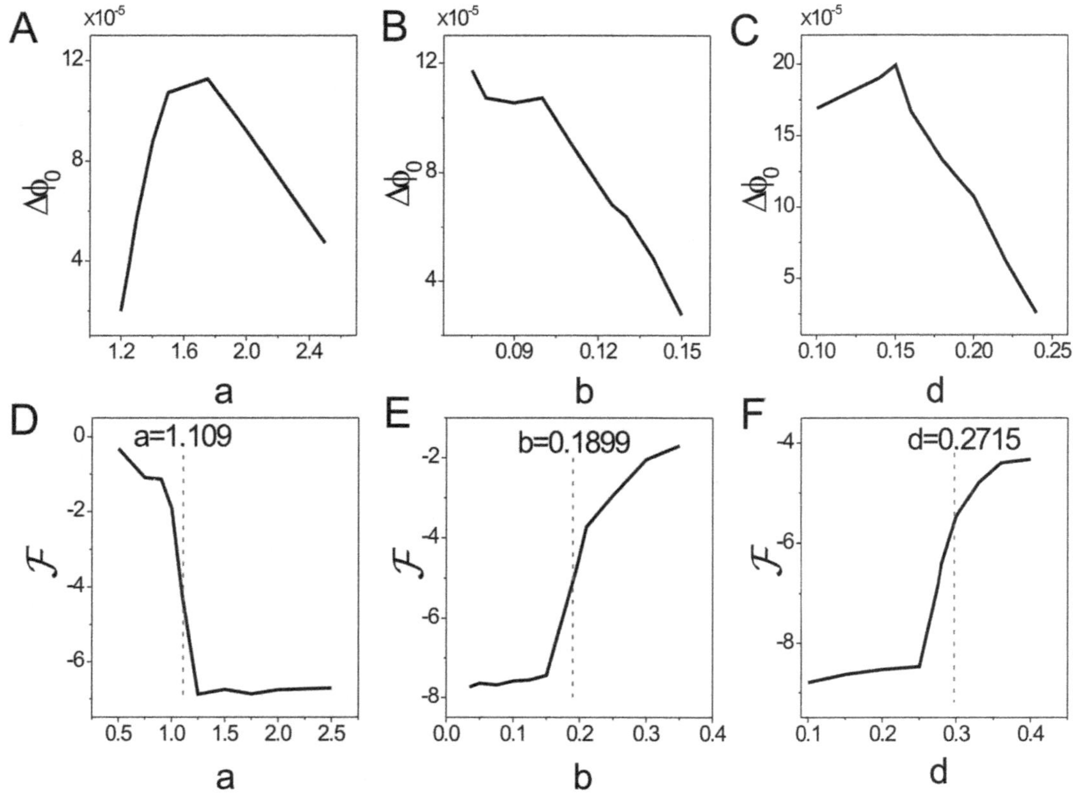

Figure 5. The barrier height of intrinsic potential landscape and free energy versus the rate parameters for predation model. The barrier heights of intrinsic potential landscape versus parameters a (A), b (B), d (C). The free energy versus a (D), b (E), d (F).

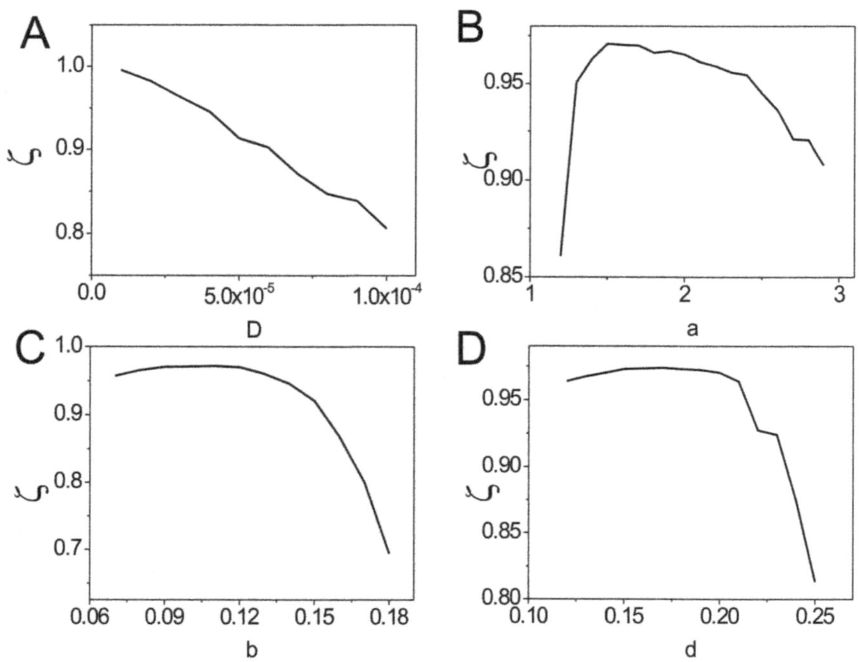

Figure 6. The coherence for predation model. (A)The coherence versus the diffusion coefficient. (B) The coherence versus the parameter a. (C) The coherence versus the the parameter b. (D) The coherence versus the the parameter d.

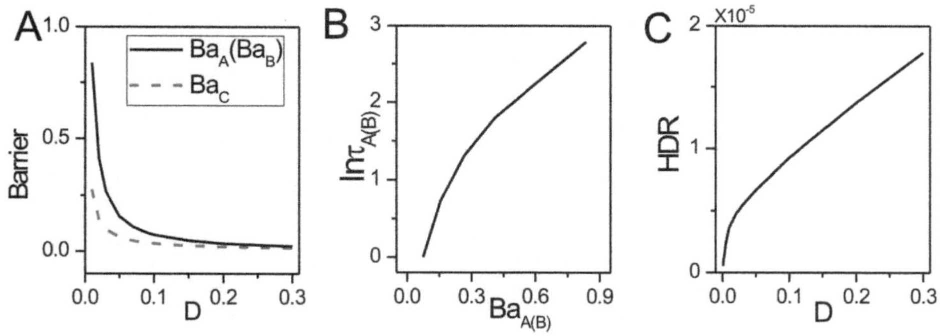

Figure 7. The barrier height of the population landscape, escape time and dissipation rate versus the diffusion coefficient for competition model. (A) The barrier height of the population landscape U versus the diffusion coefficient D. (B) The escape time versus the barrier height of the population landscape. (C) The dissipation rate versus the diffusion coefficient D.

tendency at those of the population landscape.), and the coexisting state C generally vanishes(The barrier height Ba_C decreases). It implies that when the lower critical number or density of species S_A becomes larger, the species S_A will extinct easily. The competitiveness for species S_B becomes larger, so it will live more stably.

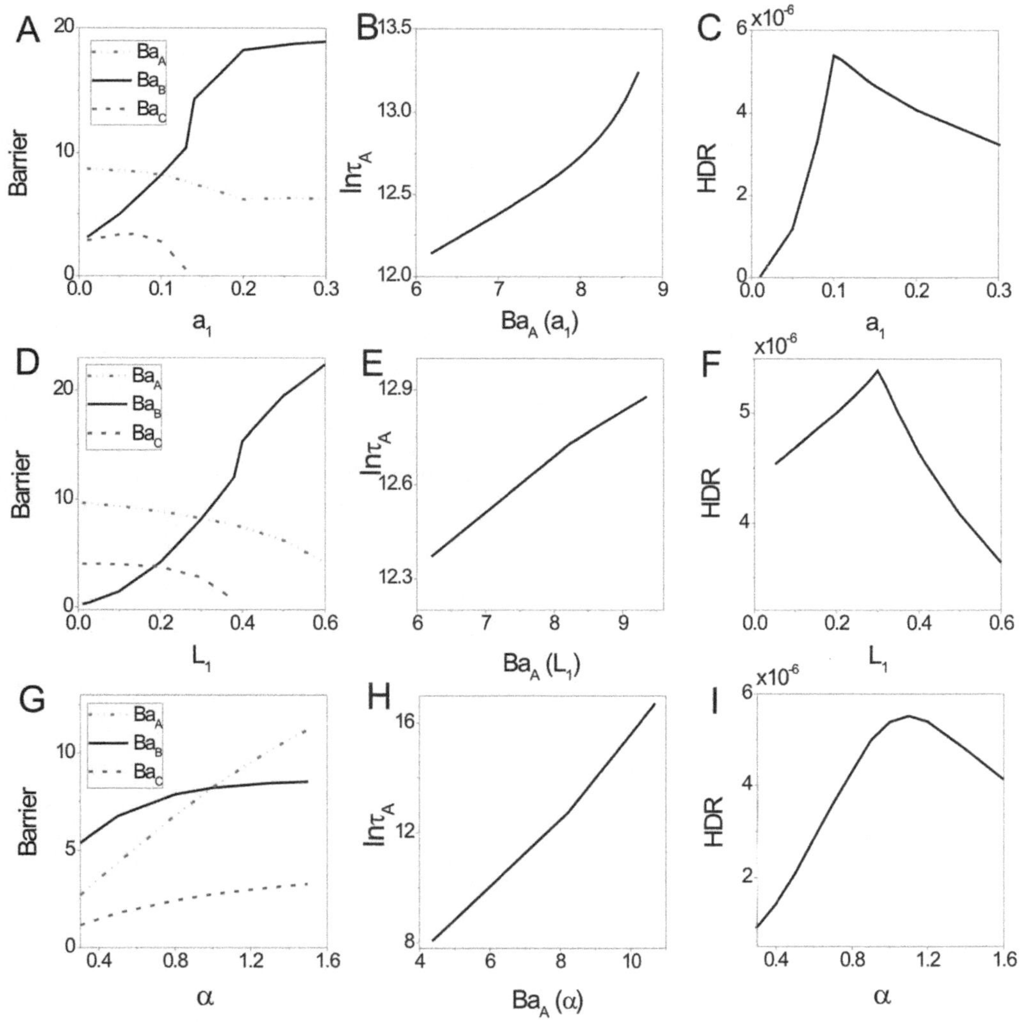

Figure 8. The barrier height of the population landscape, escape time and dissipation rate versus the rate parameters for competition model. (A) The barrier height of the population landscape versus a_1. (B) The escape time versus barrier height of the population potential landscape. (C) The dissipation rate versus a_1. (D) The barrier height of the population potential landscape versus L_1. (E) The escape time versus barrier height of the population potential landscape. (F) The dissipation rate versus L_1. (G) The barrier heights of the population landscape versus α. (H) The escape time versus barrier height of the population potential landscape. (I) The dissipation rate versus α.

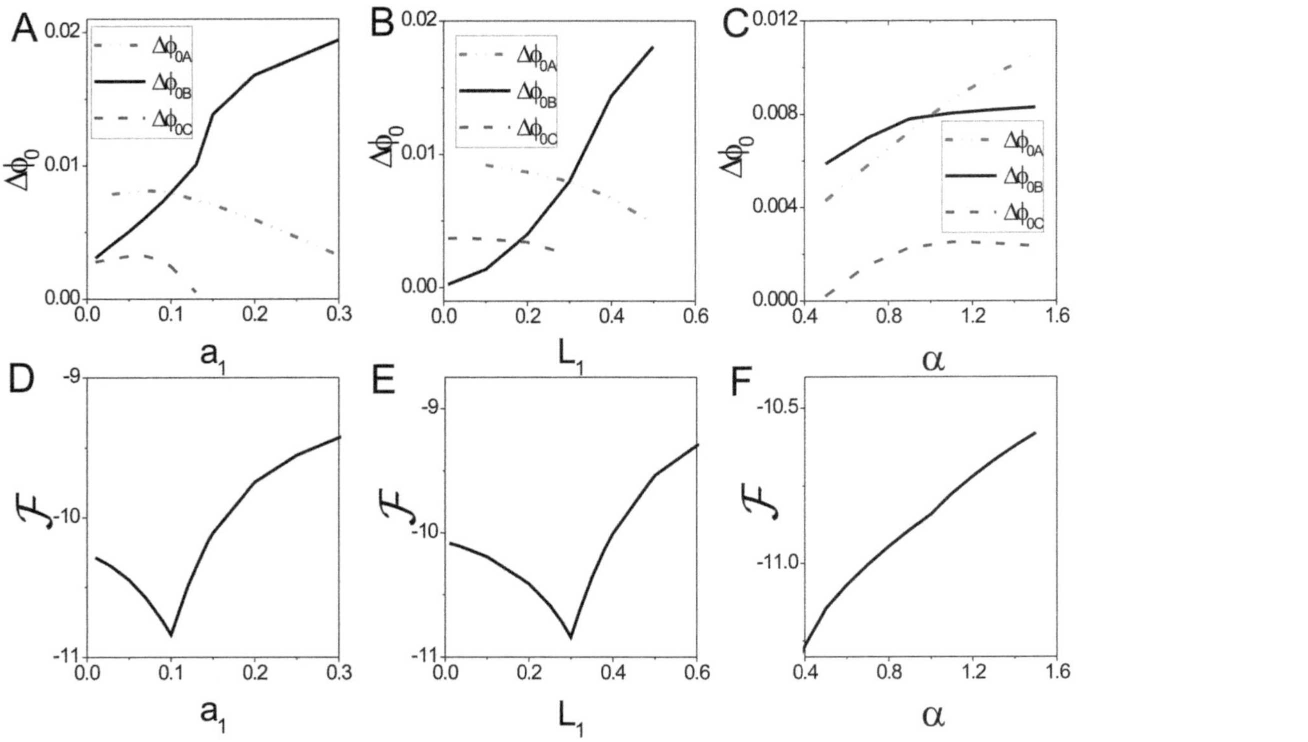

Figure 9. The barrier height of intrinsic potential landscape and free energy versus the rate parameters for competition model. The barrier heights of intrinsic potential landscape versus a_1 (A), L_1 (B), α (C). The free energy versus a_1 (D), L_1 (E), α (F).

Figure 8G, Figure 8H and Figure 8I show the effects of rate parameter α which means the relative birth rate of species S_B on this system with $D=0.001$. When α increases, the relative birth rate of species S_B becomes larger, so the survival alone basin B of species S_B becomes more stable and deeper. The other survival alone state A and the coexisting state C also become more stable and deeper (The barrier height Ba_A increases shown in Figure 8G and the escape time $ln\tau_A$ becomes longer shown in Figure 8H. They are positively correlated.Figure 9C shows the barrier height of the intrinsic potential landscape(ϕ_0) has the same tendency with that of the population landscape.) It implies that when the birth rate of B becomes larger, the species S_B will be more stable. This can lead the coexisting state C and the survival alone state A to become more stable.

Figure 8C, Figure 8F and Figure 8I show the heat dissipation rates increase then decrease when the parameter a_1,L_1,α increase. The flux makes a significant contribution to the heat dissipation rate $HDR=\int \mathbf{J}\cdot\mathbf{F}d\mathbf{N}=\int \mathbf{J}\cdot(\mathbf{J}/P+D\nabla U)d\mathbf{N}$. The contribution of the term $D\nabla U$ is numerically much smaller than that of term $\mathbf{J}/P$ because the values of $D\nabla U$ are small in less fluctuations and their directions are near orthogonality to the flux. When the parameters are given specific values: $a_1=a_2=0.1$, $L_1=L_2=0.3$, $\alpha=1.0$, the system stays at a symmetrical landscape topography, so the depths of the basin A and basin B have the same value. These two basins both have large areas of dominant flux. As the three parameters increase or decrease, the system becomes less symmetrical in landscape topography, one area of the basin A or B becomes more dominant. Since the area of the dominant flux becomes less, the heat dissipation rate becomes less and the system consume less energy. Figure 9D shows the phase transition from four stable states to three stable states nearby $a_1=0.1$. The first derivative of the non-equilibrium free energy is discontinuous at this point, which is a signal of thermodynamic phase transition. Figure 9E shows the phase transition from four stable states to three stable states nearby $L_1=0.3$. Figure 9F shows the free energy increases as α increases, and has no phase transition.

Due to competitive exclusion principle, the two species competing for the same resources are impossible to coexist in the same area. So we can see the landscape for larger $a_1(a_2)$ can not have the coexisting state, but it can have two stable states for living alone. The system will eventually select one stable state according to the initial condition (the slightest advantage for one species)and the fluctuations in the environment. If they coexist in the same area, there must have differences on the ecological factors, such as habitat, diet, activity time or other characteristics among the competitive species.

Quantifying the global stability of the mutualism ecosystems

We showed the change of population potential landscapes for increasing diffusion coefficient D (Figure S9 in File S1). Figure 10 shows diffusion effect on the mutualism model. In Figure 10A, as the diffusion coefficient characterizing the fluctuations decreases, the barrier heights $Ba_{A(B)}$ and Ba_C becomes higher when the parameters are $a_1=a_2=0.1$, $L_1=L_2=0.5$, $\alpha=1.0$. Figure 10B shows the escape time versus the barrier height Ba_A. We can see clearly that the higher the barrier height is, the longer the escape time is. The system is harder to escape from the basins of attraction as the barrier height increases. Figure 10C shows the heat dissipation rate for different diffusion coefficients. We can see the dissipation or the entropy production rate decreases when the diffusion coefficient decreases.

We showed the change of population potential landscapes for increasing parameters a_1,L_1,α (Figure S10 in File S1, Figure S11

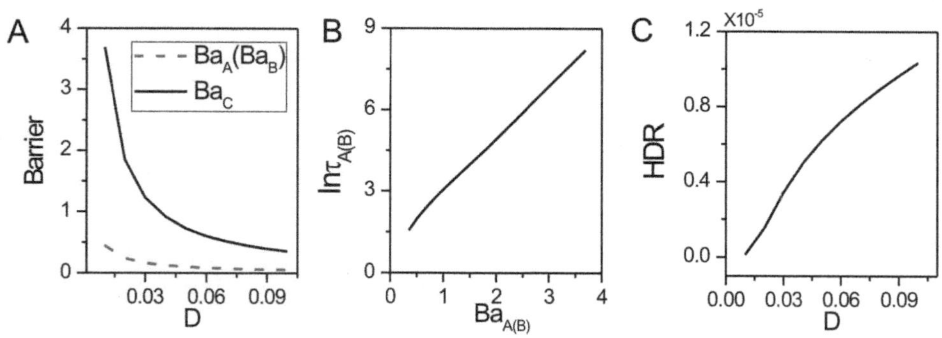

Figure 10. The barrier height of the population landscape, escape time and dissipation rate versus the diffusion coefficient for mutualism model. (A) The barrier height of the population landscape U versus the diffusion coefficient D. (B) The escape time versus the barrier height of the population landscape. (C) The dissipation rate versus the diffusion coefficient D.

in File S1 and Figure S12 in File S1). Figure 11A, Figure 11B and Figure 11C show the effects of rate parameter a_1 on mutualism system at $D=0.002$. We can see when the value of a_1, the mutualism ability is small, there are four stable states. Here we also included the lower critical number or density, so the system can go to their mutual extinction state. The coexisting state C is deeper than the other two states A and B which represent the species living alone shown in Figure 11A. Figure 12A shows that the barrier height of the intrinsic potential landscape(ϕ_0) has the same tendency with that of the population landscape. This implies an obvious rule, mutual benefit can help the species to live and reproduce. As a_1 increases, the cooperative effect on species S_A becomes stronger, the coexisting state C become more stable and the escape time $ln\tau_C$ becomes longer as the barrier height Ba_C increases shown in Figure 11B. It implies that when the cooperative effect increases, the system will go to the coexisting stable state. With the increasing of the cooperative effect, the species can coexist in larger population than their respective carrying capacity.

Figure 11D, Figure 11E and Figure 11F show the effects of rate parameter L_1 on mutualism system with $D=0.002$. When L_1 increases, the lower critical number or density of species S_A becomes larger, so the survival alone basin A of species S_A becomes less stable and shallower while the other survival alone state B becomes more stable and deeper (The barrier height Ba_B increases, Ba_A and Ba_C decrease as L_1 increases shown in Figure 11D. The escape time $ln\tau_C$ from basin C has positive correlation with Ba_C shown in Figure 11E.) When the lower critical number of species S_A becomes larger further, the species S_A will become extinct easily. Figure 12B shows barrier height of the intrinsic potential landscape(ϕ_0) has the same tendency with that of the population landscape.

Figure 11G, Figure 11H and Figure 11I show the effects of rate parameter α which means the relative birth rate of species S_B on mutualism system with $D=0.002$. When α increases, the relative birth rate of species S_B becomes larger, so the survival alone basin B of species S_B becomes more stable and deeper (The barrier height Ba_B increases slightly.) The other survival alone state A and the coexisting state C also becomes more stable and deeper shown in Figure 11G. The escape time $ln\tau_C$ becomes longer according to the tendency of Ba_C shown in Figure 11H. It implies a rule that the increase in population growth of species S_A can lead to the result of greater number of species S_B, and vice versa. Figure 12C shows the barrier height of the intrinsic potential landscape(ϕ_0) has the same tendency with that of the population landscape.

Figure 11C, Figure 11F and Figure 11I show the heat dissipation rates decrease then increase when the parameter a_1, L_1, α increase. When the parameters $a_1=a_2=0.1$, $L_1=L_2=0.5$, $\alpha=1.0$, the system stays at a symmetrical landscape topography that the depths of the basin A and basin B have the same value. In this case, these two basins both have less contribution to the HDR since the flux in coexisting basin C is much larger than those in basin A or B. As the three parameters increase or decrease, one area of the basin A or B becomes generally more dominant than that of the symmetric landscape topography in addition to the area of basin C. Since the area of dominant flux expands, the heat dissipation rate becomes larger and the system needs to consume more energy.

Figure 12D and Figure 12F shows that the free energy increases as a_1 and α increase.Figure 12E shows the free energy increases as L_1 decreases. We can see there is no discontinuous changes in the first derivative of the free energy since there is no phase transition phenomenon.

Quantifying the kinetic paths for the non-equilibrium ecosystems

We showed the kinetic paths for the ecosystems on the landscapes(see the Materials and Methods section 3 for details). Figure 13A and Figure 13B show the kinetic paths on the intrinsic potential landscape ϕ_0 for competition model (We use the parameters $a_1=0.25$, $a_2=0.1$, $L_1=L_2=0.3$, $\alpha=1.0$ which has no coexisting state C in order to see the paths clearly here.) and for mutualism ($a_1=a_2=0.1$, $L_1=L_2=0.5$, $\alpha=1.0$) model respectively. Figure 13C and Figure 13D show the kinetic paths on the population potential landscape U for competition model and mutualism model with $D=0.01$. We can see the paths by purple line from state A and state B are quite different from the paths by black line from state B to A in Figure 13A and Figure 13C. The paths from A to C (from B to C) by purple line and the paths from C to A (from C to B)by black line are shown in Figure 13B and Figure 13D. We can see these pathways do not follow the gradient paths on both the intrinsic potential landscape and the population potential landscape due to the non-zero flux. The pathways do not necessarily pass the saddle point which is not similar with the equilibrium system. The forward and backward kinetic paths are irreversible which provides a clear signature of the non-equilibrium system. We can explore the detailed dynamical mechanism of ecosystems by quantifying the kinetic paths.

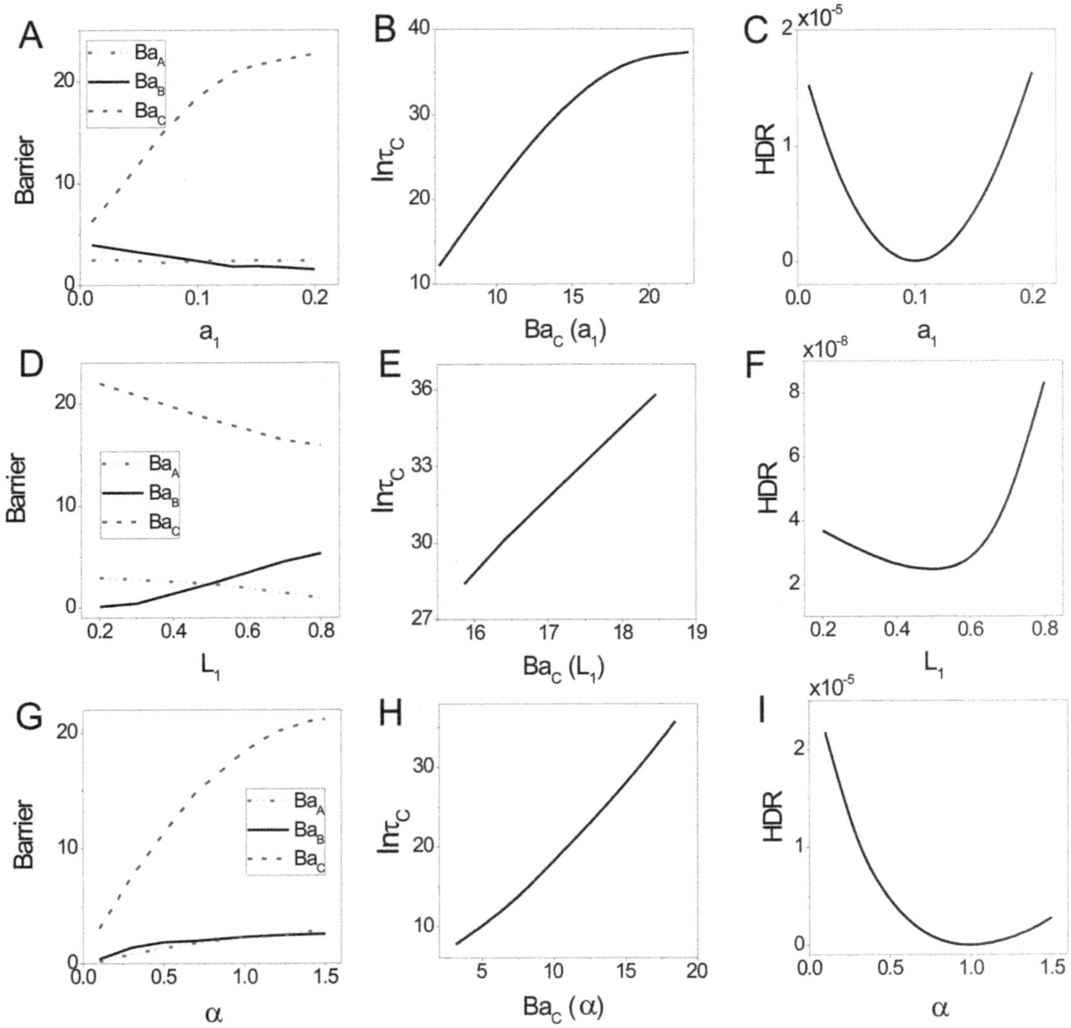

Figure 11. The barrier height of the population landscape, escape time and dissipation rate versus the rate parameters for mutualism model. (A) The barrier heights of the population landscape versus a_1. (B) The escape time versus barrier height of the population landscape for changing a_1. (C) The dissipation rate versus a_1. (D) The barrier height of the population landscape versus L_1. (E) The escape time versus barrier height of the population landscape for changing L_1. (F) The dissipation rate versus L_1. (G) The barrier height of the population landscape versus α. (H) The escape time versus barrier height of the population landscape for changing α. (I) The dissipation rate versus α.

Quantifying the global stability and dynamics for Canada lynx and snowshoe hare population cycle

The lynx-snowshoe hare cycles show the typical predator-prey behaviors which many researches have tried to do realistic modeling on [2,22,23]. The modeling equations describing the behavior of lynx-snowshoe hare cycles were proposed and there exists limit cycles under much wider scope of parameters with consideration of the Holling-type II functional responses for lynx and Holling-type III functional responses for general predators, such as coyote and great horned owl in the boreal forest [2,22–24]. Here we use a model of lynx-snowshoe hare cycles shown as [22–24]: $\frac{dH}{dt}=rH(1-H/k)-\frac{\gamma H^2}{H^2+\eta^2}-\frac{\alpha HL}{H+\mu}, \frac{dL}{dt}=sL(1-qL/H)$, where H and L are the population density of hares and lynx. r is the growth rate of hares, k is the carrying capacity, γ is the maximum killing rate of general predation, η is the half-saturation constant for general predation [22–24]. α is the maximum killing rate of lynx, μ is the half-saturation constant for lynx' predation [22–24].

We can not ensure which oscillation is more robust by only the analysis of the local stability of differential equations. Therefore, we have explored the stochastic dynamics of this model and quantified the global stability using our landscape and flux theory. Figure 14A shows the population potential landscape with the basic set of parameters [22]. Figure 14B shows the sensitivity of each parameter with respect to the landscape topography through the barrier heights which can be used to quantify the robustness and global stability of the lynx-snowshoe hare cycles. The parameters decrease by $-20\%, -10\%, -5\%$ and increase by 5%, 10%, 20% of their own values. The color bars which are not displayed for each parameter, such as the disappearances of black and red bars for parameter r, indicate the disappearance of the periodical oscillations for that value of certain parameters. We can see the changing of r have significant impact on the robustness of the oscillation while the changes of the parameter γ and η have less significant impacts on the robustness of the oscillation. We showed a more detailed analysis of the robustness for changing parameter r in Figure 14C. There exists an optimal value r in Figure 14C. The optimal value of r leads to a more robust and stable ecological

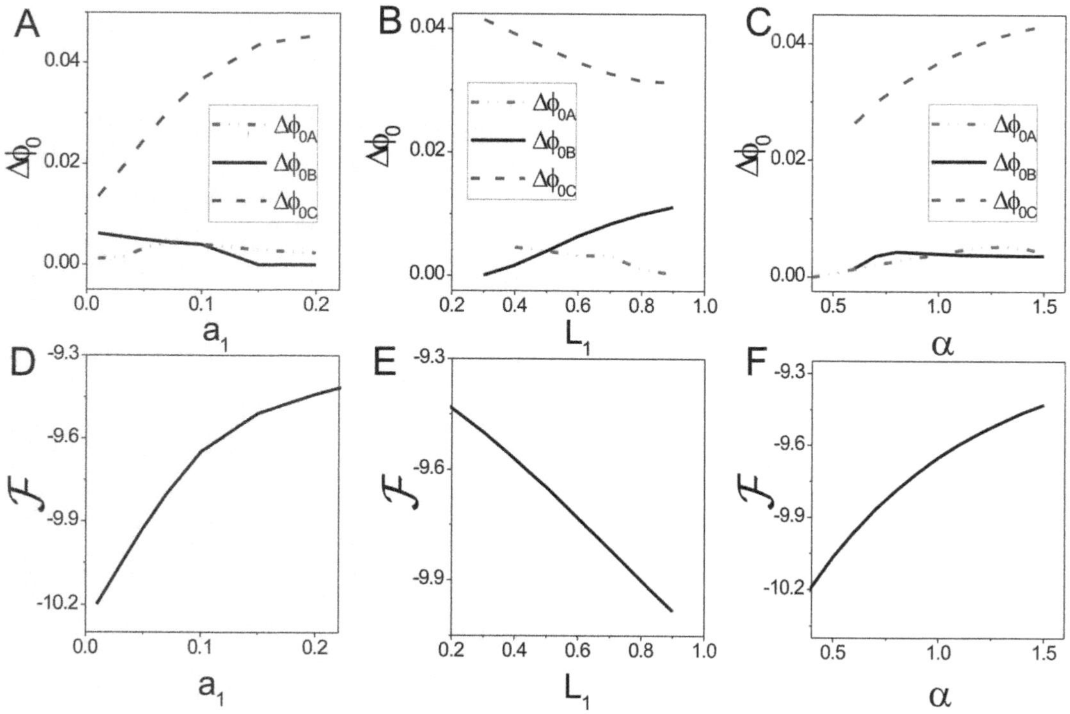

Figure 12. The barrier height of intrinsic potential landscape and free energy versus the rate parameters for mutualism model. The barrier heights of intrinsic potential landscape versus a_1 (A), L_1 (B), α (C). The free energy versus a_1 (D), L_1 (E), α (F).

oscillation of hares and lynx. The barrier heights from the potential landscape topography can be used to quantify the global stability and the robustness of the oscillations. We can explore which set of parameters will lead the ecosystem to have more robust oscillation with higher barrier heights. This will help to design strategy to preserve the ecosystems.

Conclusion

Stability and dynamics are crucial for understanding the structure and function for ecology. Ecological stability is commonly defined as Lyapunov stability to describe the global stable behavior of ecosystem upon perturbations.Unfortunately, in general, Lyapunov stability cannot be assessed because explicit Lyapunov function can hardly be found. In this study we have illustrated a general method to explore the Lyapunov global stability of the ecosystem through the quantification of the underlying intrinsic landscape. It can be used to explore more general complex ecosystem where the situation can only be studied case by case before. We found the dynamics of the ecosystems is not only determined by the gradient of the potential landscape but also by an additional curl flux force from breaking down the detailed balance. This provides a new way to explore the general dynamics in non-equilibrium regime for ecosystems.

We considered three important ecological systems with the interactions between two species: the predation model, the competition model, the mutualism model and a realistic lynx-snowshoe hare model. Multiple attractors and limit cycle attractor with a distinct Mexican hat shape emerge from these cases. We found the quantitative measure for global stability through barrier height. The non-equilibrium free energy can reflect the global phases of the underlying ecosystems and the transition regions between the global phases. We quantified the pathways of ecosystem which do not follow the gradient path on the landscape and are irreversible. We quantified the kinetic speed from one stable state basin to another of the ecosystems and linked with the underlying landscape topography through the barrier height between the basins of attractions.

For ecosystems, the stability is directly related to the survivals of every species. We showed the effects of parameters representing the interactions among species on the global natures such as landscape topography represented by barrier height, kinetics speed represented by escape time and the thermodynamic dissipation by the entropy production or heat dissipation rate in these ecosystems. Therefore we can quantify the change of the stability by increasing or decreasing the interaction parameters, respectively. These results can help us to design more stable ecosystems.

The ecosystem dynamics shares some common global features as the biochemical systems such as gene-gene regulations, in terms of the underlying landscapes, the global stability and dynamics, kinetic rate and pathways. There are also significant differences between these two types of systems. First of all, the ecosystems and gene-gene regulation systems are at completely different level, one is on the population species and the other is within the cell. Second, their components have different sensitivity against changes. On the one hand, it is relatively easier to mutate the genes and harder to mutate the species. On the other hand, it is relatively easier to change the link between the species due to the sensitivity of ecosystems to the environment rather than the gene-gene regulations. Third, the ecosystems depend more sensitively to the outside input than the gene-gene regulations.

It is worth mentioning here that the landscape ecology is an emerging subfield in ecology. The landscape ecology models concentrate on spatial heterogeneity with space probabilistic methods [25,26]. Although these methods and our theory all focus on the dynamical evolution in probability, we concentrate on the probability landscape and flux in the population space rather than in the geographical space as in the landscape ecology models.

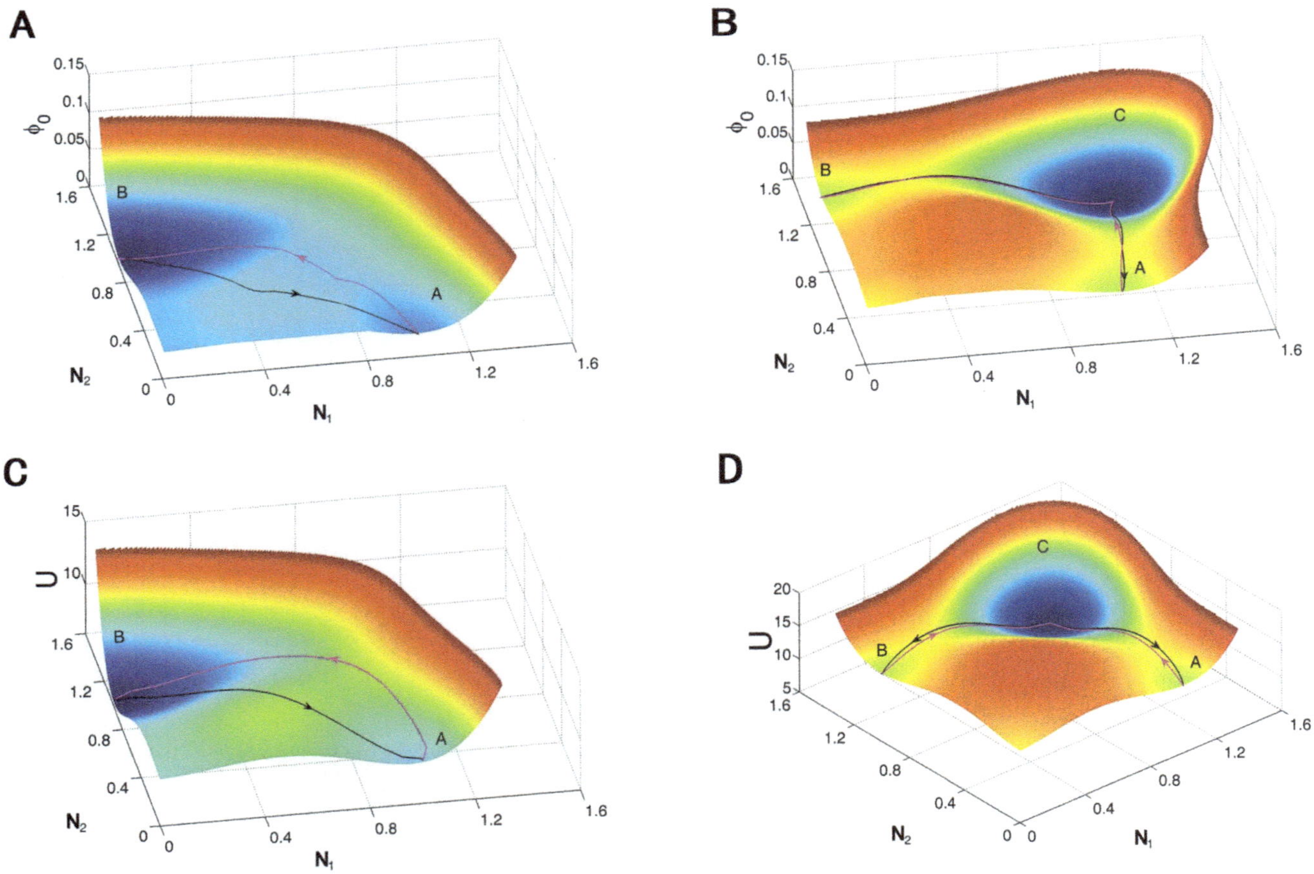

Figure 13. The paths for competition model and mutualism model. (A) The paths on the intrinsic potential landscape ϕ_0 for competition model. (B) The paths on the intrinsic potential landscape ϕ_0 for mutualism model. (C) The paths on the population potential landscape U for competition model. (D) The paths on the population potential landscape U for mutualism model.

The realistic models are usually more complex. The models often have more parameters controlling the dynamical behaviors of the ecosystem. We can apply our potential and flux landscape framework to quantify the stability of those systems irrespect to how complicated the underlying processes dictated by the differential equations may be. Furthermore, we can explore the set of parameters for more stable system by the underlying landscape topography quantified by the barrier heights. Thus the parameter-control can be realized to achieve the ecological stability and ecological decision makings.

For a mutualism model like honeybee and flowering plants, the local analysis of the differential equations can only give the existence of the stable states and local stability but not the global robustness of the stable states. The honeybee and flowering plants can coexist in a wide range of their dynamical parameters. Our potential and flux landscape framework can give in which conditions the honeybee and flowering plants can more robustly coexist. We can also quantify the changes of the robustness of ecosystem through the changes in controlling parameters for the underlying landscape topography.

We can also use our potential and flux landscape theory to explore the interactions of multi-species as a multi-node network [15,27]. The analysis of sensitivity from the potential landscape can explore the effects of specific parameters or wirings to robustness and stability of the ecosystem. We can identify the key species or wirings that are responsible for the global stability and function of the whole ecosystem through the potential landscape topology.

The stability of ecosystems is a challenging issue. We have focused on our discussion with low dimensional models in this study. In realistic ecosystems, the global stability can seldomly be quantified by just exploring the stability with a few differential equations. The stability must be viewed as a multi-dimensional problem. It is determined by many internal and external degrees of freedom including population density, temperature, space, age,other competitors and environmental factors. A population dynamics P system (PDP) model as a tool for modeling and simulating complex ecosystems was proposed with a high computational modularity, efficiency, and the parallelism [28,29]. We will need to go further to explore the even more complex ecosystems with the consideration of other factors.

The methods developed in this study can be applied to more complicated and realistic ecosystems to understand the global stability, function and robustness.

Materials and Methods

Quantifying the non-equilibrium population and intrinsic potential, flux, Lyapunov function and non-equilibrium thermodynamics for general ecological systems with finite fluctuations

The ecological systems are not in isolations since there are often intrinsic and extrinsic environmental noises around them [30]. So

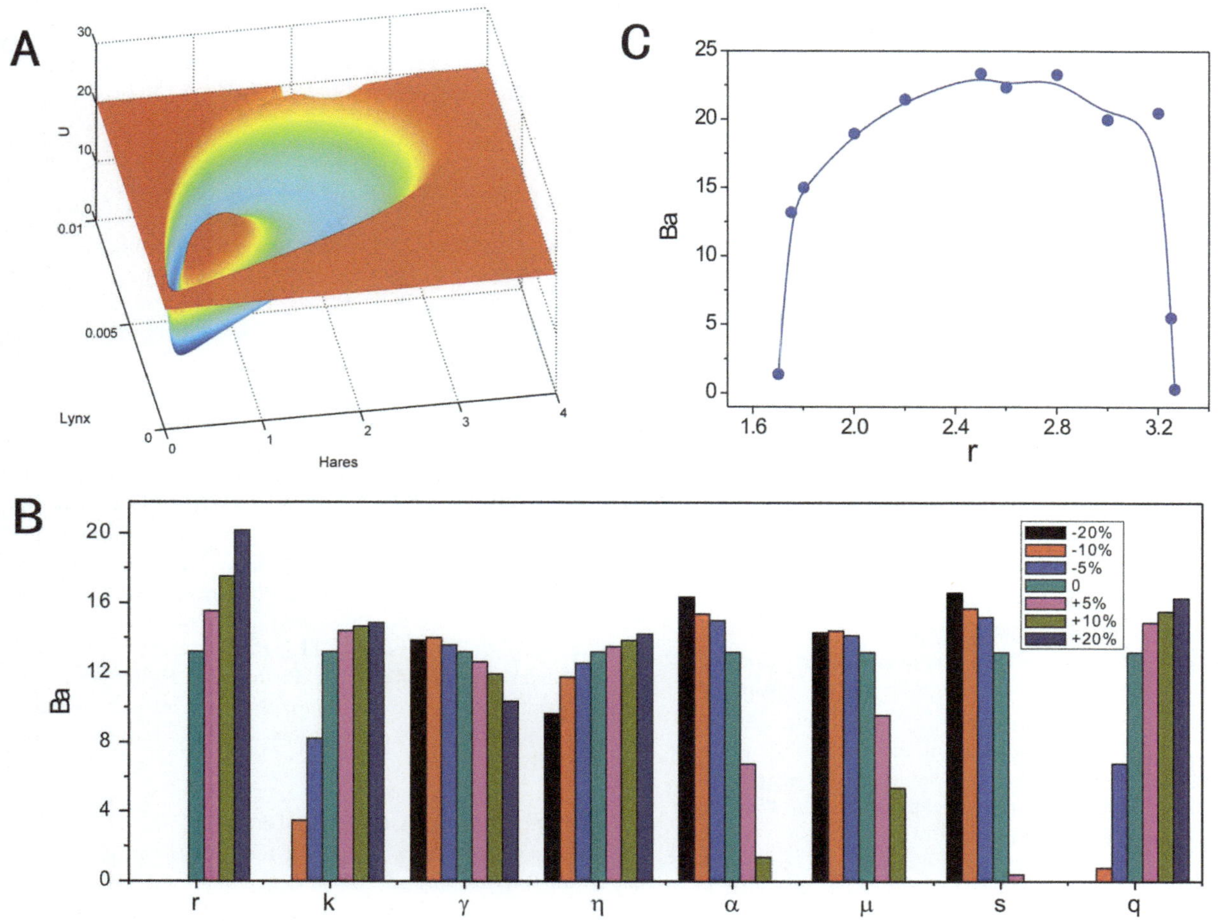

Figure 14. The potential landscape, barrier height of the population landscape and the sensitivity of parameters for lynx-snowshoe hare model. (A) The population potential landscape for lynx-snowshoe hare model. (B) The barrier heights versus changing parameters. The basic set of the parameters are: $r=1.75$, $k=8$, $\gamma=0.1$, $\eta=1.25$, $\alpha=505$, $\mu=0.3$, $s=0.85$, $q=212$. (C) The barrier heights versus the hares' rate of population growth.

the dynamics can be described as follows [20,31]: $d\mathbf{N}/dt=\mathbf{F}(\mathbf{N})+\mathbf{g}\cdot\mathbf{\Gamma}$. The vector $\mathbf{F}(\mathbf{N})$ is the deterministic force that drives the system, where $\mathbf{N}$ is the vector of dynamical variables, with each component of which representing different species in the ecosystems. $\mathbf{g}$ is the noise force from the fluctuations. The statistical nature of the noise can often be assumed as Gaussian white noise: $<\mathbf{\Gamma}_i(t)\mathbf{\Gamma}_j(t')>=2D\delta_{ij}(t-t')$ ($\delta_{ij}=1$, if $i=j$; $\delta_{ij}=0$ if $i\neq j$) and $<\mathbf{\Gamma}_i(t)>=0$. D is the diffusion coefficient representing the level of noise strength while $\mathbf{G}=\mathbf{g}\cdot\mathbf{g}^T/2$ is the scaled diffusion matrix described the anisotropy phenomenon. We set $\mathbf{D}=D\mathbf{G}$.

Under fluctuations, the individual trajectories are stochastic and do not have the predictive power, we should focus on the evolution of probability distributions rather than evolution of individual trajectories. We can explore the corresponding stochastic dynamics governed by the Fokker-Planck diffusion equation [20,31] for the temporal evolution of the probability distribution $P(\mathbf{N},t)$:

$$\partial P/\partial t=-\nabla\cdot\mathbf{J}(\mathbf{N},t)=-\nabla\cdot(\mathbf{F}P-D\nabla(\mathbf{G}P)) \tag{1}$$

This represents a conservation law of probability (local change is due to net flux in or out). The diffusion equation essentially describes the probability of finding the state $\mathbf{N}(t)$ in the state space which is driven by the driving force $\mathbf{F}$. The probability flux vector $\mathbf{J}$ of the system in population space $\mathbf{N}$ is defined as: $\mathbf{J}(\mathbf{N},t)=\mathbf{F}P-D\nabla(\mathbf{G}P)$.

In steady state, $\partial P/\partial t=0$, therefore $\nabla\cdot\mathbf{J}(\mathbf{N},t)=0$. $\mathbf{J}\neq 0$ means the detailed balance is broken and the ecosystem is in non-equilibrium state. The divergence-free property implies the flux $\mathbf{J}$ is rotational. From the definition of $\mathbf{J}(\mathbf{N},t)$, one can write the force decomposition: $\mathbf{F}=-D\mathbf{G}\nabla U+\mathbf{J}_{ss}/P_{ss}+D\nabla\cdot\mathbf{G}$. In this way, the driving force $\mathbf{F}$ can be decomposed into the gradient of the potential U, the divergence of the diffusion coefficient and the curl probability flux. Therefore the global nature can be determined by the non-equilibrium population potential U characterizing the probability landscape of the whole population space, the non-equilibrium local dynamics and their global properties can be determined by both the gradient of the potential landscape and the rotational curl flux [17]. This is in contrast with the zero flux case for detailed balance of equilibrium systems, where the global nature is determined by the equilibrium potential, and the dynamics is determined by the gradient of the equilibrium potential.

The steady state probability distribution under the fluctuation and the non-equilibrium population potential of ecosystems is quantitatively linked by: $P_{ss}(\mathbf{N})=\exp(-U)/Z$, where Z is the partition function for non-equilibrium ecosystems defined as $Z=\int\exp(-U)d\mathbf{N}$. Then the entropy of the non-equilibrium ecosystem is given as [14,18,32–34]: $S = -\int P(\mathbf{N},t)\ln P(\mathbf{N},t)d\mathbf{N}$, and the energy of the non-equilibrium ecosystem is given as: $E = D\int UP(\mathbf{N},t)d\mathbf{N} = -D\int \ln[ZP_{ss}]P(\mathbf{N},t)d\mathbf{N}$. Therefore, the free energy $\mathcal{F}$ of the non-equilibrium ecosystem can be defined as:

$$\mathcal{F} = E - DS = D(\int P\ln(P/P_{ss})d\mathbf{N} - \ln Z). \quad (2)$$

The free energy is a combination of non-equilibrium energy and entropy, suggesting the first law of non-equilibrium thermodynamics for ecosystems. The free energy decreases in time monotonically until it reaches its minimum value, $\mathcal{F} = -D\ln Z$ [14,18,32–34]. The free energy of these non-equilibrium ecosystems never increases, suggesting the second law of non-equilibrium thermodynamics for ecosystems. We can use the free energy to explore the global stability of non-equilibrium ecosystems with finite fluctuations from the environments or intrinsic sources.

We also investigated the derivatives of the entropy in the time evolution [14,18,32–34]. It can be decomposed as follows: $\dot{S} = -\int (dP/dt)\, lnPd\mathbf{N} = \int (\mathbf{J}\cdot\mathbf{D}^{-1}\cdot\mathbf{J})/Pd\mathbf{N} - \int (\mathbf{J}\cdot\mathbf{D}^{-1}\cdot(\mathbf{F} - \nabla\cdot\mathbf{D}))d\mathbf{N} = \dot{S}_t - \dot{S}_e$ The first term is entropy production rate $\dot{S}_t$, having the physical significance of the total entropy change of system and environment. It is always non-negative corresponding to thermodynamic second law. The last term is the heat dissipation term $\dot{S}_e$ from the environments, and it can be either positive or negative [14,18,32–35]. The entropy of the ecosystems is not always increasing, but the free energy of the ecosystems reduces itself to a minimum in time. It can be used as an optimal principles to explore the topologies and the design of ecosystems.

Quantifying the non-equilibrium intrinsic potential and flux, Lyapunov function and global stability for general ecosystems with zero fluctuations: deterministic ecosystems

The essence of the stability-problem is to analyze that how a ecosystem returns to the original state under an initial perturbation. Lyapunov function is traditional used to quantify the global stability of dynamical ecosystems. But the population potential U we have obtained does not necessarily have the property of the Lyapunov function. [14–18, 32–34, 36–40]. Some analytical solutions of the Lyapunov function can be found for special simplified ecosystem models [5,8] and they can only be analyzed case by case. For complex real ecosystems, it is still challenging to find a proper Lyapunov function. We show here a general approach to obtain the Lyapunov functions for dynamical ecosystems.

The underlying intrinsic potential for the dynamic system can be obtained under the weak noise limit $D \ll 1$. Therefore, we can expand the population potential $U(\mathbf{N})$ according to the fluctuation strength D, and the associate probability $P(\mathbf{N})$ is as the following form [18,35,36,38,41]:

$$P(\mathbf{N}) = \exp(-(\phi_0(\mathbf{N})/D + \phi_1(\mathbf{N}) + D\phi_2(\mathbf{N}) + \cdots))/Z \quad (3)$$

where $Z = \int \exp(-U(\mathbf{N}))d\mathbf{N}$. We can substitute the expanded equation to the Fokker-Plank diffusion equation.Therefore we have the D^{-1} order expansion of the Fokker-Plank diffusion equation: $H = \mathbf{F}\cdot\nabla\phi_0 + \nabla\phi_0\cdot\mathbf{G}\cdot\nabla\phi_0 = 0$. The equation ϕ_0 followed is called Hamilton - Jacobi equation(HJE). We can solve the Hamilton-Jacobi equation which is zero-fluctuation limit of the Fokker-Planck equation to get the intrinsic potential ϕ_0 by a numerical method - level set method using the Mitchell's level-set toolbox [42].

Then we can study the deterministic equation for $\phi_0(\mathbf{N})$ [18,41]. We can explore the time evolution of the $\phi_0(\mathbf{N})$ in the zero fluctuation (D) limit:

$$\dot{\phi}_0(\mathbf{N}) = \dot{\mathbf{N}}\cdot\nabla\phi_0 = \mathbf{F}\cdot\nabla\phi_0 = -\nabla\phi_0\cdot\mathbf{G}\cdot\nabla\phi_0 \leq 0 \quad (4)$$

Because $\mathbf{G}$ is positive definite, the value of $\phi_0(\mathbf{N})$ monotonously decreases under the deterministic evolution equation (zero fluctuation limit). Therefore ϕ_0 is a Lyapunov function, which can be used to quantify the global stability of the ecosystems. In addition, ϕ_0 is related to the population potential U under weak noise limit by $U = \phi_0/D$. So ϕ_0 is linked with the steady state probability which can quantify the global properties for ecosystem.

For the deterministic systems, we can take the zero-fluctuation limit and follow the procedures described above to obtain the non-equilibrium energy, entropy and free energy as well as the corresponding non-equilibrium thermodynamics. Furthermore, we can also recover the force decomposition for the deterministic ecosystems in zero fluctuation limit as gradient of non-equilibrium intrinsic potential $-\mathbf{G}\cdot\nabla\phi_0$ and steady state intrinsic divergent free curl flux $J_{ss}|_{D\to 0}$: $\mathbf{F} = -\mathbf{G}\cdot\nabla\phi_0 + (\mathbf{J}_{ss}/P_{ss})|_{D\to 0}$.. From the Hamilton-Jacobian equation above, we see that $(\mathbf{J}_{ss}/P_{ss})|_{D\to 0}\cdot\nabla\phi_0 = 0$. This means the gradient of the non-equilibrium intrinsic potential ϕ_0 is perpendicular to the intrinsic flux under the zero-fluctuation limit [18,36,38].

Quantifying the paths of ecosystems

Quantification of the pathways of the ecosystems gives us an opportunity to explore the ecological dynamical process. We assume the path probability starts from initial configuration $\mathbf{N}_i$ at $t = 0$, and end at the final configuration of $\mathbf{N}_f$ at time t. The path integral formula is [43,44]:

$$(\mathbf{N}_f,t|\mathbf{N}_i,0) = \int D\mathbf{N}exp[-\int L(\mathbf{N}(t))dt] \quad (5)$$

where L is the Lagrangian of the ecosystem [43,44]:

$$L = \frac{1}{4}D^{-1}(\dot{\mathbf{N}} - \mathbf{F})\cdot\mathbf{D}^{-1}\cdot(\dot{\mathbf{N}} - \mathbf{F}) + \frac{1}{2}(\mathbf{D}\cdot\nabla)\cdot(\mathbf{D}^{-1}\cdot\mathbf{F}) \quad (6)$$

The path integral over $D\mathbf{N}$ represents the sum over all possible paths connecting $\mathbf{N}_i$ at time $t = 0$ to $\mathbf{N}_f$ at time t. The term $(1/4)\ D^{-1}\ (\dot{\mathbf{N}} - \mathbf{F})\cdot\mathbf{D}^{-1}\cdot(\dot{\mathbf{N}} - \mathbf{F})$ gives the weight contribution from specific path from the underlying Gaussian noise. The term $(1/2)\ (\mathbf{D}\cdot\nabla)\cdot(\mathbf{D}^{-1}\cdot\mathbf{F})$ gives the contribution from the Jacobian variable transformation from the Gaussian noise to the path. The $L(\mathbf{N}(t))$ represents the weight for each path. The probability of ecosystem dynamics from initial configurations $\mathbf{N}_i$ to the final state $\mathbf{N}_f$ is equal to the sum of all possible paths with different weights. Not every dynamical path contributes to the same weight. We can identify the dominant paths which give the most contribution to the weight. This approximation is based on the fact the weight is exponentially weighted. The sub-leading contributions are exponentially small. Therefore dominant paths which give the most contribution to the weights can emerge. We explored the dominant kinetic paths from one state to another for ecosystems in this study.

The ecological dynamical models

The ecosystems can be described by a set of nonlinear ordinary differential equations for two species interactions. We will add some restrictions on the models to enable them to be more reasonable and closer to the real situation [2,45], such as avoidance of exponential growth and existence of lower critical bound for each species.

(1)Predation Model

The general Holling type II responses for the prey to account the nonlinearity in interactions can be added in a predation model proposed by Murray [2]:

$$\frac{dN_1}{dt} = N_1(1-N_1) - \frac{aN_1N_2}{N_1+d} = F_1(N_1,N_2)$$

$$\frac{dN_2}{dt} = bN_2(1-\frac{N_2}{N_1}) = F_2(N_1,N_2) \quad (7)$$

where N_1 is the normalized population of prey while N_2 is the normalized population of predator.The parameter a is the relative death rate or the interaction strength for the prey. The parameter b is the ratio of the linear birth rate of the predator to that of the prey. The parameter d is the relative saturation rate of the prey. The system has two general saddle points: one is (0,0) representing that none of the species exists. The second one (1,0) represents the prey at their carrying capacity in the absence of predators [2]. The second point is stable along the N_1 population axis and unstable along the N_2 population axis. There is also a critical point which is the unstable center of the limit cycle or the stable point in different parameter ranges. The system has a stable limit cycle oscillation when we set $a=1.5$, $b=0.1$, $d=0.2$.

(2) Competition Model

The realistic competitive model should have with a lower critical bound, which means the creatures would perish once the size of the population is below this threshold. The model is shown as follows [45]:

$$\frac{dN_1}{dt} = N_1(N_1-L_1)(1-N_1) - a_1N_1N_2 = F_1(N_1,N_2)$$

$$\frac{dN_2}{dt} = \alpha N_2((N_2-L_2)(1-N_2) - a_2N_1) = F_2(N_1,N_2) \quad (8)$$

where N_1 and N_2 are the normalized populations of the two competitive species S_A and S_B. L_1,L_2 are the lower critical bounds for species S_A,S_B, respectively. The ranges of L_1,L_2 are from 0 to 1. a_1,a_2 are the competitive ability for species S_A,S_B, respectively. α is the relative rate of natural increase for species S_B [45].

We have explored the phase analysis of the system. The two kinds of population which are both at zero mean the two species both are at extinct state (0,0) (Marked as O) of the system. This is because the two groups both have lower critical density. When there are no competitors for each species, the states: (1,0) (Marked as A, which means the species S_A exists alone.) and (0,1) (Marked as B, which means the species S_B exists alone.) that exist in isolation are locally stable. Besides the above three states, when the values of a_1 and a_2 meet certain conditions, the system can have another local stable state which corresponds to the coexistence of the two species (Marked as C). Here, we set $a_1=a_2=0.1$, $L_1=L_2=0.3$, $\alpha=1.0$ since it has these four states.

(3) Mutualism Model

We will consider the two mutualism species both having lower critical bound. This realistic mutualism model can be described as [45]:

$$\frac{dN_1}{dt} = N_1(N_1-L_1)(1-N_1) + a_1N_1N_2 = F_1(N_1,N_2)$$

$$\frac{dN_2}{dt} = \alpha N_2((N_2-L_2)(1-N_2) + a_2N_1) = F_2(N_1,N_2) \quad (9)$$

where N_1 and N_2 are the normalized populations of the two mutualism species S_A and S_B. L_1,L_2 are lower critical points for species S_A and S_B, respectively. The ranges of L_1,L_2 are from 0 to 1. a_1,a_2 are the mutualism ability for species S_A,S_B, respectively. α is the relative rate of natural increasing for species S_B [45].

We have explored the phase analysis of this system. The two kinds of population which are both at zero mean the trivial solution (0,0) (Marked as O) of the system. This is because the two groups having a lower critical density. When there is no mutual helper for each species, the states: (1,0) (Marked as A, which means the species S_A exists alone.) and (0,1) (Marked as B,which means the species S_B exists alone.) that exist in isolation are locally stable. Besides the above three points, the system has another local stable point which corresponds to the coexistence of the two species (Marked as C). Here, we set $a_1=a_2=0.1$, $L_1=L_2=0.5$, $\alpha=1.0$ for the system has these four states completely.

Supporting Information

File S1 Supporting figures. **Figure S1.** The population potential landscapes for increasing diffusion coefficient D of predation model. **Figure S2.** The population potential landscapes for increasing parameter a of predation model.**Figure S3.** The population potential landscapes for increasing parameter b of predation model. **Figure S4.** The population potential landscapes for increasing parameter d of predation model. **Figure S5.** The population potential landscapes for increasing diffusion coefficient D of competition model. **Figure S6.** The population potential landscapes for increasing parameter a_1 of competition model. **Figure S7.** The population potential landscapes for increasing parameter L_1 of competition model. **Figure S8.** The population potential landscapes for increasing parameter α of competition model. **Figure S9.** The population potential landscapes for increasing diffusion coefficient D of mutualism model. **Figure S10.** The population potential landscapes for increasing parameter a_1 of mutualism model. **Figure S11.** The population potential landscapes for increasing parameter L_1 of mutualism model. **Figure S12.** The population potential landscapes for increasing parameter α of mutualism model.

Author Contributions

Conceived and designed the experiments: JW EKW. Performed the experiments: LX FZ. Analyzed the data: LX FZ KZ JW. Contributed reagents/materials/analysis tools: JW EKW. Wrote the paper: LX FZ JW EKW.

References

1. Vandermeer J, Goldberg D (2003) Population ecology: First principles. Woodstock, Oxfordshire: Princeton University Press.
2. Murray J (1998) Mathematical Biology. New York: Springer-Verlag Berlin Heidelberg.
3. Lotka A (1925) Elements of Physical Biology. Baltimore: Williams and Wilkins.
4. Volterra V (1931) Lecons sur la Theorie Mathematique de la Lutte pour la Vie. Paris: Gauthier- Villars.
5. Harrison G (1979) Global stability of predator-prey interactions. J Math Biology 8: 159–171.
6. Goh B (1976) Global stability in two species interactions. J Math Biol 3: 313–318.
7. Goh B (1977) Global stability in many-species systems. Am Nat 111: 135–143.
8. Hsu S (1978) On global stability of a predator prey system. Math Biosci 39: 1–10.
9. Holling C (1965) The functional response of predators to prey density and its role in mimicry and population regulation. Mem Ent Soc Can 45: 1–60.
10. Murdoch W, Oaten A (1975) Predation and population stability. Advan EcoL ReS 9: 1–131.
11. Hastings A (1978) Global stability of two species systems. J Math Biol 5: 399–403.
12. Balzsi G, van Oudenaarden A, Collins J (2011) Cellular decision making and biological noise: from microbes to mammals. Cell 144: 910–925.
13. Karsai I, Balzsi G (2002) Organization of work via a natural substance: regulation of nest con- struction in social wasps. J Theor Biol 218: 549–565.
14. Wang J, Huang B, Xia XF, Sun ZR (2006) Funneled landscape leads to robustness of cell networks: Yeast cell cycle. PloS Comput Biol 2: 1385–1394.
15. Han B, Wang J (2007) Quantifying robustness of cell cycle network: Funneled energy landscape perspectives. Biophys J 92: 3755–3763.
16. Lapidus S, Han B, Wang J (2008) Intrinsic noise, dissipation cost, and robustness of cellular networks: The underlying energy landscape of mapk signal transduction. Proc Natl Acad Sci USA 105: 6039–6044.
17. Wang J, Xu L, Wang EK (2008) Potential landscape and flux framework of nonequilibrium net- works: Robustness, dissipation, and coherence of biochemical oscillations. Proc Natl Acad Sci USA 105: 12271–12276.
18. Zhang F, Xu L, Zhang K, Wang E, Wang J (2012) The potential and flux landscape theory of evolution. J Chem Phys 137: 065102.
19. Xu L, Zhang F, Wang E, Wang J (2013) The potential and flux landscape, lyapunov function and non-equilibrium thermodynamics for dynamic systems and networks with an application to signal-induced ca2+ oscillation. Nonlinearity 26: 69–84.
20. Van Kampen N (2007) Stochastic processes in physics and chemistry. Amsterdam: Elsevier.
21. Yoda M, Ushikubo T, Inoue W, Sasai M (2007) Roles of noise in single and coupled multiple genetic oscillators. J Chem Phys 126: 1–11.
22. Tyson R, Haines S, Hodges K (2010) Modelling the canada lynx and snowshoe hare population cycle: the role of specialist predators. Theor Ecol 3: 97–111.
23. Krebs C, Boutin S, Boonstra R (2001) Ecosystem dynamics of the Boreal forest: the Kluane project. Oxford: Oxford University Press.
24. ODonoghue M, Boutin S, Krebs C, Zultea G, Murray D, et al. (1998) Functional responses of coyotes and lynx to the snowshoe hare cycle. Ecol 79: 1193–1208.
25. Wu J (2008) Landscape ecology. In: S. E. Jorgensen (ed), Encyclopedia of Ecology. Oxford: Elsevier.
26. Wu J, Hobbs R (2007) Key Topics in Landscape Ecology. Cambridge: Cambridge University Press.
27. Wang J, Li C, Wang E (2010) Potential and flux landscapes quantify the stability and robustness of budding yeast cell cycle network. Proc Natl Acad Sci USA 107: 8195–8200.
28. Colomer M, Margalida A, Perez-Jimenez M (2013) Population dynamics p system (pdp) models: A standardized protocol for describing and applying novel bio-inspired computing tools. PLoS ONE 8: e60698.
29. Colomer M, Margalida A, Sanuyc D, Perez-Jimenez M (2011) A bio-inspired computing model as a new tool for modeling ecosystems: The avian scavengers as a case study. Ecol Model 222: 33–47.
30. Swain P, Elowitz M, Siggia E (2002) Intrinsic and extrinsic contributions to stochasticity in gene expression. Proc Natl Acad Sci USA 99: 12795–12800.
31. Gillespie D (1977) Exact stochastic simulation of coupled chemical reactions. J Phys Chem 81: 2340–2361.
32. Ao P (2005) Laws in darwinian evolutionary theory. Phys Life Rev 2: 117–156.
33. Qian H (2009) Entropy demystified: The "thermo"-dynamics of stochastically uctuating systems. Method Enzymol 467: 111–134.
34. Schnakenberg J (1976) Network theory of microscopic and macroscopic behavior of master equation systems. Rev Mod Phys 48: 571–585.
35. Ge H, Qian H (2010) Physical origins of entropy production, free energy dissipation, and their mathematical representations. Phys Rev E 81: 051133.
36. Graham R (1989) Macroscopic potentials, bifurcations and noise in dissipative systems. In: Moss F, McClintock P, editors, Noise in Nonlinear Dynamical Systems Vol. 1, Cambridge University Press. pp. 225–278.
37. Sasai M, Wolynes PG (2003) Stochastic gene expression as a many-body problem. Proc Natl Acad Sci USA 100: 2374–2379.
38. Haken H (1987) Advanced synergetics: instability hierarchies of self-organizing systems and devices. Berlin: Springer.
39. Qian H (2006) Open-system nonequilibrium steady-state: Statistical thermodynamics, uctuations and chemical oscillations. J Phys Chem B 110: 15063–15074.
40. Qian H (2001) Mesoscopic nonequilibrium thermodynamics of single macromolecules and dynamic entropy-energy compensation. Phy Rev E 65: 016102.
41. Hu G (1986) Lyapunov function and stationary probability distributions. Z Phys B: Condens Matter 65: 103–106.
42. Mitchell IM (2008) The exible, extensible and efficient toolbox of level set methods. J Sci Comput 35: 300–329.
43. Wang J, Zhang K, Xu L, Wang E (2011) Quantifying the waddington landscape and biological paths for development and differentiation. Proc Natl Acad Sci USA 108: 8257–8262.
44. Wang J, Zhang K, Wang E (2010) Kinetic paths, time scale, and underlying landscapes: A path integral framework to study global natures of nonequilibrium systems and networks. J Chem Phys 133: 125103.
45. Bazykin A (1985) Mathematical Biophysics of Interacting Populations(in Russian). Moscow: Nauka.

8

The EBM-DPSER Conceptual Model: Integrating Ecosystem Services into the DPSIR Framework

Christopher R. Kelble[1]*, Dave K. Loomis[2], Susan Lovelace[3], William K. Nuttle[4], Peter B. Ortner[5], Pamela Fletcher[6], Geoffrey S. Cook[1,5], Jerry J. Lorenz[7], Joseph N. Boyer[8]

1 Ocean Chemistry Division, Atlantic Oceanographic & Meteorological Laboratory, National Oceanic & Atmospheric Administration, Miami, Florida, United States of America, **2** Institute for Coastal Science and Policy, East Carolina University, Greenville, North Carolina, United States of America, **3** JHT Incorporated, Hollings Marine Laboratory, National Centers for Coastal Ocean Science, National Oceanic and Atmospheric Administration, Charleston, South Carolina, United States of America, **4** Eco-Hydrology, Ottawa, Canada, **5** Cooperative Institute for Marine and Atmospheric Studies, Rosenstiel School of Marine and Atmospheric Science, University of Miami, Miami, Florida, United States of America, **6** University of Florida Sea Grant, Gainesville, Florida, United States of America, **7** Tavernier Science Center, Audubon Florida, Tavernier, Florida, United States of America, **8** Center for the Environment, Plymouth State University, Plymouth, New Hampshire, United States of America

Abstract

There is a pressing need to integrate biophysical and human dimensions science to better inform holistic ecosystem management supporting the transition from single species or single-sector management to multi-sector ecosystem-based management. Ecosystem-based management should focus upon ecosystem services, since they reflect societal goals, values, desires, and benefits. The inclusion of ecosystem services into holistic management strategies improves management by better capturing the diversity of positive and negative human-natural interactions and making explicit the benefits to society. To facilitate this inclusion, we propose a conceptual model that merges the broadly applied Driver, Pressure, State, Impact, and Response (DPSIR) conceptual model with ecosystem services yielding a Driver, Pressure, State, Ecosystem service, and Response (EBM-DPSER) conceptual model. The impact module in traditional DPSIR models focuses attention upon negative anthropomorphic impacts on the ecosystem; by replacing impacts with ecosystem services the EBM-DPSER model incorporates not only negative, but also positive changes in the ecosystem. Responses occur as a result of changes in ecosystem services and include *inter alia* management actions directed at proactively altering human population or individual behavior and infrastructure to meet societal goals. The EBM-DPSER conceptual model was applied to the Florida Keys and Dry Tortugas marine ecosystem as a case study to illustrate how it can inform management decisions. This case study captures our system-level understanding and results in a more holistic representation of ecosystem and human society interactions, thus improving our ability to identify trade-offs. The EBM-DPSER model should be a useful operational tool for implementing EBM, in that it fully integrates our knowledge of all ecosystem components while focusing management attention upon those aspects of the ecosystem most important to human society and does so within a framework already familiar to resource managers.

Editor: Maura Geraldine Chapman, University of Sydney, Australia

Funding: This paper is a result of research under the Marine and Estuarine Goal Setting (MARES) for South Florida Project funded by the National Oceanic and Atmospheric Administration Center for Sponsored Coastal Ocean Research (Coastal Ocean Program), under award NA08OAR4320889 to the University of Miami, NA09NOS4780224 to Nova Southeastern University, NA09NOS4780225 to the University of Massachusetts Amherst, NA09NOS4780226 to the National Audubon Society, NA09NOS4780227 to Florida Gulf Coast University, NA09NOS4780228 to Florida International University, and to the NOAA Atlantic Oceanographic and Meteorological Laboratory. The funders had no role in study design, data collection and analysis, decision to publish, or preparation of the manuscript.

Competing Interests: Dr. William Nuttle works as a consultant for his firm Eco-Hydrology. In this capacity he functions similarly to a research professor by obtaining funding from competitive grants and writing research publications and technical reports. He received a portion of the MARES funding stated in the competing interests section to conduct the work related to this manuscript and thus has no additional competing interests. This does not alter the authors' adherence to all the PLOS ONE policies on sharing data and materials.

* E-mail: chris.kelble@noaa.gov

Introduction

Ecosystem Based Management

The concept of ecosystem based management (EBM) was developed to improve resource management efficacy by applying a holistic approach that accounts for ecosystem complexity and integration rather than managing for individual issues or sectors (including individual species) [1]. EBM recognizes that: 1) the biophysical and human components of an ecosystem interact in many complex ways, 2) society relies upon and benefits from the ecosystem through ecosystem services, and 3) ecosystem services are directly and indirectly affected by multiple human activities/uses [2]. The goal of EBM is to maximize and sustain the production of ecosystem services, thus shifting management's focus from short-term economic gains or purely environmental protection/restoration towards assuring the long-term ability of an ecosystem to yield a broad suite of services important to human well-being [3].

Most management of the marine ecosystem focuses on single species, single uses, or single sectors (e.g. toxins, nutrients, development). These single-issue management approaches focus upon avoiding major damage to the ecosystem due only to the pressures attributable to that sector, or to perhaps maximize short-term single-issue outcomes (e.g., economic profitability, historic preservation, etc.). Even as a collective whole they cannot provide sufficient understanding of the tradeoffs that unavoidably occur among sectors and the cumulative effect of the different pressures being placed on the ecosystem from all of the sectors. Although EBM has been widely hailed as an improvement upon the single-sector management paradigm [4], there remain few examples of EBM being successfully implemented [5]. This is partially due to the difficulty of bridging traditional management, disciplinary, and professional boundaries [1].

In the United States, EBM was mandated in the National Ocean Policy of 2010 [6]. To successfully fulfill this mandate, we must address several inherent challenges: 1) how to integrate diverse scientific disciplines with different methodologies and approaches [7], 2) how to quantify cumulative impacts on ecosystem services [2], 3) how to develop and articulate appropriate targets that provide sustainable ecosystem services to meet human society's needs, and 4) how to incorporate and communicate uncertainty in our scientific understanding [3]; not to mention developing the operational tools required to effectively implement EBM [8]. Some of these challenges are addressed by conducting Integrated Ecosystem Assessments (IEAs); an emerging approach to synthesize and analyze existing scientific information to guide EBM development [9]. IEAs include scoping, indicator selection, and risk analysis; all of which must be informed by a synthesis of our integrated scientific knowledge about the human and biophysical components of the ecosystem [10].

Conceptual Models to Inform Management

Integrating relevant existing knowledge is the foundation of EBM. Conceptual ecosystem models (CEMs) integrate and synthesize scientific knowledge in a manner familiar to managers and policymakers [11,12]. If formulated properly, a CEM can address some of the challenges listed above by 1) integrating across scientific disciplines, 2) qualitatively identifying cumulative pressures, and 3) representing the scientific consensus as to how an ecosystem functions, including how humans interact with all other ecosystem components. CEMs have also been used to identify potential ecosystem indicators that can assess management success [12–14].

Initial CEMs employed to synthesize science in support of decision-making relied upon a Pressure-State-Response model [15]. The Pressure-State-Response model implicitly incorporated human society into the pressures impacting the ecosystem state and the responses that feedback to pressures. This framework evolved into the Driver, Pressure, State, Impact, and Response (DPSIR) model (Figure 1) that more explicitly depicted how human society affects ecosystem state [10,14]. The initial impetus for the DPSIR model was to illustrate and communicate cause-and-effect relationships among indicators [16]. It has found broad application in environmental assessments of terrestrial and aquatic ecosystems due to its ability to improve communication between policymakers, stakeholders, and scientists facilitating collaborative model development [14,17–19]. Because DPSIR links scientific findings with "real world" issues, it has contributed to making resource management decisions science-based [18,20]. However, the current DPSIR model does not explicitly include ecosystem services or the values humans place on services from the ecosystem [21], and its focus upon drivers makes it difficult to fully capture the needs of local or regional human communities and less than ideal for EBM.

Drivers, which reflect underlying human needs and desires, and how they are manifest in pressures that impact the state of an ecosystem are highlighted in DPSIR CEMs. However, responses rarely directly affect drivers, but rather responses more typically

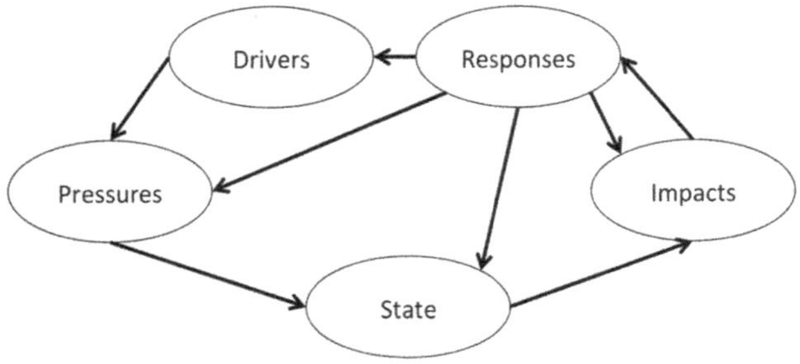

Figure 1. The DPSIR model. This is the DPSIR conceptual model that has conventionally been utilized for ecosystem management, assessment, indicator selection, and communication.

directly affect pressures or the consequences of pressures by altering the methods by which humans react to or express these drivers [22–24]. For example, one driver is the energy requirement of a growing human population. In many situations, this driver manifests itself by the extraction of oil and the burning of fossil fuels resulting in a suite of pressures on the ecosystem including air pollution, CO_2 emissions, ocean acidification, and anthropogenic climate change. Responses, such as changing behaviors and energy use through altering habits; carbon credits; investment in alternative energy sources; or the increased use of hybrid or electric vehicles, mitigate the burning of fossil fuels and therefore the pressures placed upon the ecosystem, but the responses do not significantly alter the driver; the energy requirement of the global human population.

The term impact in DPSIR unavoidably implies a negative environmental consequence of human activities [14]. As Svarstad et al. [19] observed traditional applications of DPSIR represented primarily the Preservationist "discourse" and the impacts module in particular did not capture the necessary information for alternative Traditionalist, Win-Win, or Promethean discourses. A partial exception to that generalization is if analyses of economic trade-offs are done in conjunction with DPSIR models both economic losses and gains are considered [25]. Nonetheless because impacts are negative environmental effects, DPSIR analyses more typically focus upon responses to these adverse environmental impacts [19,26] and do not facilitate proactive management to sustain and maximize ecosystem services. Proactive management implies that we actively work towards a particular standard or endpoint that we find optimal or acceptable at worst, when optimal is unobtainable. For example, we try to maintain services that meet our needs and values, such as protection from storms, recreational opportunities, clean air, clean water, carbon sequestration, etc. in the locations we desire. While many of these needs and desires are universal, there are regional differences in values and preferences. In nearly all cases, maintaining these services over extended periods involves tradeoffs and decisions. As such, there exists a need for regional management tools that move resource management from a reactive approach to a proactive approach; as an analogy, making EBM a form of preventative medicine for the planet.

Ecosystem Services and Human Well-Being

Ecosystem services are the benefits people receive from the ecosystem. As such, they reflect societal values, goals, desires, and benefits [9,27,28] and contribute to human well-being. "Well-being" is used by human dimensions scientists as a measure of quality of life in many contexts and is typically broken into components related to economics, environment, basic human needs, and the subjective well-being of people. On a global level, the Millennium Ecosystem Assessment describes the following components of well-being: basic material needs, freedom, health, good social relations, and personal security [27].

A distinction is often made between basic human needs and subjective well-being. Basic human needs are things required for survival such as food, water, and shelter. Subjective well-being, on the other hand, encompasses things that may not be absolutely necessary for immediate individual survival but are important to a positive emotional and psychological sense of life, such as culture and aesthetics, and may be important to long-term societal survival. Health is important to both. The absence of acute trauma and disease is a basic need, but chronic health issues contribute to subjective well-being. Developing countries focus upon meeting basic needs, while those in which those needs are being met, strive to achieve additional levels of well-being in search of a good life [27].

Aspects of well-being including environmental attributes have been addressed in the scientific literature [29] and have become the focus of assessments such as the Canadian Index of Well-being [30] and the Organization for Economic Cooperation and Development's Better Life Index [31]. Food, recreation, and storm protection are ecosystem services that benefit people directly. Not only do they provide life's basic needs, but changes in them affect economic conditions, movement of people, regulation of climate and disease, recreation and cultural opportunities, and security. As a result, changes in these ecosystem services have a wide-ranging impact upon personal well-being [27,32]. Effective EBM assumes that regardless of an individual's recognition of ecosystem services in their lives, these services are nonetheless reflected in well-being attributes of their communities and can be measured through indicators such as health, safety, economic security, effective governance, education, food/water, housing, access to critical services, social cohesion, social conflict and environmental use [33]. Indicators of these attributes are therefore included in EBM models to provide managers with information about social and economic conditions and how these interact with natural resources.

Because ecosystem services describe the benefits that society derives from the ecosystem, both directly and indirectly [34], they are a natural bridge between the biophysical and human dimensions sciences [35]. Bridging human dimensions and biophysical sciences provides the holistic, integrated perspective of interrelated ecosystem components necessary for applying EBM [36]. Extensive scientific effort has been devoted to develop methodologies that identify, locate, and quantify the services we value [c.f. 34,37,38,39]. Despite these efforts, there are few examples employing ecosystem services to improve decision-making indicating they have yet to truly penetrate into the realm of resource management [40]. Ecosystem services will become a staple of resource management only when practical methodologies and metrics are developed that make consideration of ecosystem services tractable for decision-makers [35].

EBM must focus upon the production of and trade-offs amongst ecosystem services when evaluating the relative merit of potential management strategies or responses. Given how important ecosystem services are to EBM, it is not enough that ecosystem services be incorporated into a CEM; they need to be the focal point and explicit in the CEM. While ecosystem services can be implicitly incorporated into the DPSIR framework [17] and alternative integrated frameworks such as the Press-Pulse Dynamics model for organizing long-term research [21], we propose herein a CEM framework that explicitly merges ecosystem services directly with DPSIR to form an EBM-DPSER conceptual model. The goal of this model is to depict how the ecosystem functions and produces the ecosystem services that benefit human well-being. By including ecosystem services instead of impacts, the EBM-DPSER model captures a greater diversity of discourses providing more comprehensive information to decision–makers than a traditional DPSIR model. Although such analyses are outside the scope of this study, the EBM-DPSER model should facilitate other ecosystem service analyses, such as economic valuation, required to conduct quantitative scenario and trade-off analyses. We apply the EBM-DPSER model to the Florida Keys and Dry Tortugas marine ecosystem as a case study demonstrating the benefits and utility of this approach.

Methods

EBM-DPSER Framework

The EBM-DPSER model was developed within the Marine and Estuarine Goal Setting for south Florida project (MARES) [41]. The goal of MARES was to reach a science-based consensus about the defining characteristics and fundamental regulating processes of a south Florida coastal marine ecosystem that is both sustainable and capable of providing the diverse ecosystem services upon which our society depends. To achieve this goal, it was necessary to consider regional, social, political, cultural, economic, and public health factors, in both a research and management context, along with ecological variables [7,42–45].

Multiple state, federal, and regional agencies share responsibility for managing the south Florida coastal marine ecosystem, but they operate with inconsistent mandates and answer to a diverse group of citizens, organizations, agencies, and businesses with diverse concerns. The need to adopt a more integrated approach to coastal ocean management has been widely recognized for many years [45]. MARES was based on the premise that through a systematic consensus-building process, science could contribute more directly and effectively to decisions being made by various management agencies. To build scientific consensus and develop a common framework for management agencies in south Florida, the MARES project required fully integrated CEMs that synthesized our knowledge about both the human dimensions and biophysical components of it's constituent ecosystem to inform integrated, holistic management, i.e. EBM, of the overall ecosystem.

The first step in the MARES process was to convene the relevant scientific experts (both biophysical and human dimensions), stakeholders, and agency representatives in a facilitated workshop and charge them with developing an integrated conceptual ecosystem model. An initial attempt was made employing the DPSIR model (Fig. 1). However, we found the DPSIR model was not a good fit for this process, in particular, or EBM in general, because of DPSIR's inability to adequately capture the full range of interactions between humans and the environment due to its reliance upon the restrictive term "impacts" which emphasizes the negative consequences of human activities upon an ecosystem [11,14,18,20,46].

A small, but increasing number of researchers applying DPSIR have also noted the omission of ecosystem services. A Web of Science search returned 127 papers on a search for "DPSIR" and only 11 papers on a search for "DPSIR" and "Ecosystem Services." Many of these 11 papers discussed the lack of ecosystem services within DPSIR [17,47,48] and two actually suggested that impacts be modified to reflect changes in ecosystem services [47,49]. Of these two, one defined impacts as the effects of environmental degradation on ecosystem services [49]. In fact only one paper did not define impacts as the strictly negative results of environmental change [47]. At least one group has attempted to link the ecosystem services cascade proposed by Haines-Young and Potschin [50] with DPSIR by associating indicators of ecosystem services and human well-being with the impacts module [51]. It is our basic contention that merely changing the definition of the impacts module is insufficient to avoid its negative connotations given its application to problems such as water pollution, poor drinking water, eutrophication, etc. [46,52].

To overcome this challenge and provide a more explicit definition of the relationships involved in EBM we replaced the impacts module with an ecosystem services module (Fig. 2). The exact definition of ecosystem services is still actively evolving as our scientific knowledge about ecosystem services advances [53,54] (Table 1). Fisher et al. [54] state the need for an ecosystem services classification scheme to be determined based on the decision context. There is good reason for the definition of ecosystem services to be malleable based not just on the decision context, but also on the intended audience for your analyses and products. Our goal with EBM-DPSER was to build consensus among a broad range of scientists, managers, and stakeholders. To achieve this goal most effectively, we opted to employ the most commonly accepted definition of ecosystem services; the Millennium Ecosystem Assessment definition of ecosystem services as the benefits people obtain from ecosystems [27]. The decision context for applying the EBM-DPSER framework in MARES was to build consensus and educate decision-makers about the variety of benefits that are produced by the coastal ecosystem. Thus, ecosystem services were categorized into cultural, regulating, provisioning, and supporting services following Farber et al. [55]. In the EBM-DPSER model, state sub-models are linked to one another and as such capture supporting ecosystem services with state-to-state interactions. For example, the seagrass sub-model connects to nutrient concentrations in the water column sub-model, thus capturing the supporting service of nutrient cycling by seagrass (Fig. 3). The cultural, regulating, and provisioning ecosystem services defined by Farber et al. [55] were used as the basis for identifying ecosystem services in the EBM-DPSER conceptual model. Because the state sub-models captured supporting services, they did not need to be included explicitly within the ecosystem services module.

Sub-models were developed for each state component linking the myriad pressures to state variables that we measure to endpoints, hereafter termed *ecosystem attributes people care about.* The quantifiable state variables included in the sub-model are themselves a parsimonious subset of the descriptive characteristics that represent the overall condition of that state component [56]. The *ecosystem attributes people care about* are ecological components of the state module utilized to produce human well-being. This is the same definition of ecosystem services suggested by Fisher et al. [54]. Thus, these attributes directly correspond with ecosystem services as defined by the MEA and this project [27]. An example of these sub-models is illustrated in Figure 3, which depicts how pressures influence the attributes of the water column we measure via monitoring programs and thus affect the *ecosystem attributes people care about.*

Under the EBM guidelines, management responses should aim to provide the sustainable level of ecosystem services desired by society, making a natural link from ecosystem services to responses. Because ecosystem services play this central role within EBM, they are the key module of the EBM-DPSER model. Not only have ecosystem services been substituted into the DPSIR model, the visual representation of the EBM-DPSER model itself is inverted so ecosystem services are the pinnacle and drivers the base (Fig. 2). This de-emphasizes drivers, which are without question important; however, it is ecosystem services that relate most directly to the goals and values that motivate society to respond to changes in environmental condition.

As with more recent DPSIR applications, it is intended that the EBM-DPSER model be a causal network in the sense used by Niemeijer and de Groot [57] rather than a unidirectional chain, in order to better incorporate the multitude of complex interactions observed within an ecosystem. Figures 2 and 3 display the application of this network approach to the EBM-DPSER model and the water column element of the state module in MARES.

Florida Keys And Dry Tortugas Case Study

The EBM-DPSER model for the Florida Keys and Dry Tortugas (Fig. 4) was developed with input from over 60 scientists, agency resource managers, and representatives from environmental non-governmental organizations (NGOs). This group of experts worked to identify the key characteristics within each module of the EBM-DPSER framework needed to effectively synthesize our scientific knowledge and our understanding of society's relationship to the environment within the Florida Keys and Dry Tortugas study area (Fig. 5). The full report detailing the Florida Keys and Dry Tortugas EBM-DPSER model and its development is available as a technical report [41].

Results

Florida Keys and Dry Tortugas EBM-DPSER Case Study

The cultural, regulating, and provisioning ecosystem services put forth by Farber et al. [55] were modified and the list reduced based on input from the regional group of experts to include only those services produced by the Florida Keys and Dry Tortugas (Table 2). The cultural ecosystem services furnished by the Florida Keys and Dry Tortugas marine ecosystem are aesthetics and existence, recreation, science and education, and cultural amenity. The provisioning ecosystem services furnished by the Florida Keys and Dry Tortugas marine ecosystem are food/fisheries, ornamental resources, and medicinal and biotechnology resources. The regulating ecosystem services furnished by the Florida Keys and Dry Tortugas marine ecosystem are hazard moderation, waste treatment, climate regulation, atmospheric regulation, and biological interactions. All of these ecosystem services result in benefits to society that can be evaluated and therefore used to determine the efficacy of specific responses.

In the Florida Keys and Dry Tortugas marine ecosystem EBM-DPSER model, the ecosystem state was delineated into 6 sub-models. These were the water column, fish and shellfish, and marine birds and the three dominant habitat types: coral and hardbottom, seagrass, and mangroves. Each of these sub-models, includes the linkages from pressures to *ecosystem attributes people care about* (c.f. Fig. 3). For example, in the water column ten pressures (ocean acidification, air and water temperature, dredging, groundings, changes in freshwater runoff, altered rainfall and evaporation, changes in tropical cyclone intensity and/or frequency, contaminant releases, shoreline alteration, and sea-level rise) affect the key attributes that we measure. These key attributes are grouped by the four water column indicators: eutrophication, human health, abiotic, and land-based sources of pollution. In addition, the other sub-models that influence the water column (seagrass, fish and shellfish, mangroves, and coral and hardbottom) are shown with their connections to water column attributes. The state of the water column includes nine *ecosystem attributes people care about*: water clarity, carbon sequestration, abundant and diverse wildlife, protected species, seafood safety, human health, quality of beaches and shorelines, aesthetics, and nutrient/pollution filter. In turn, these attributes directly affect

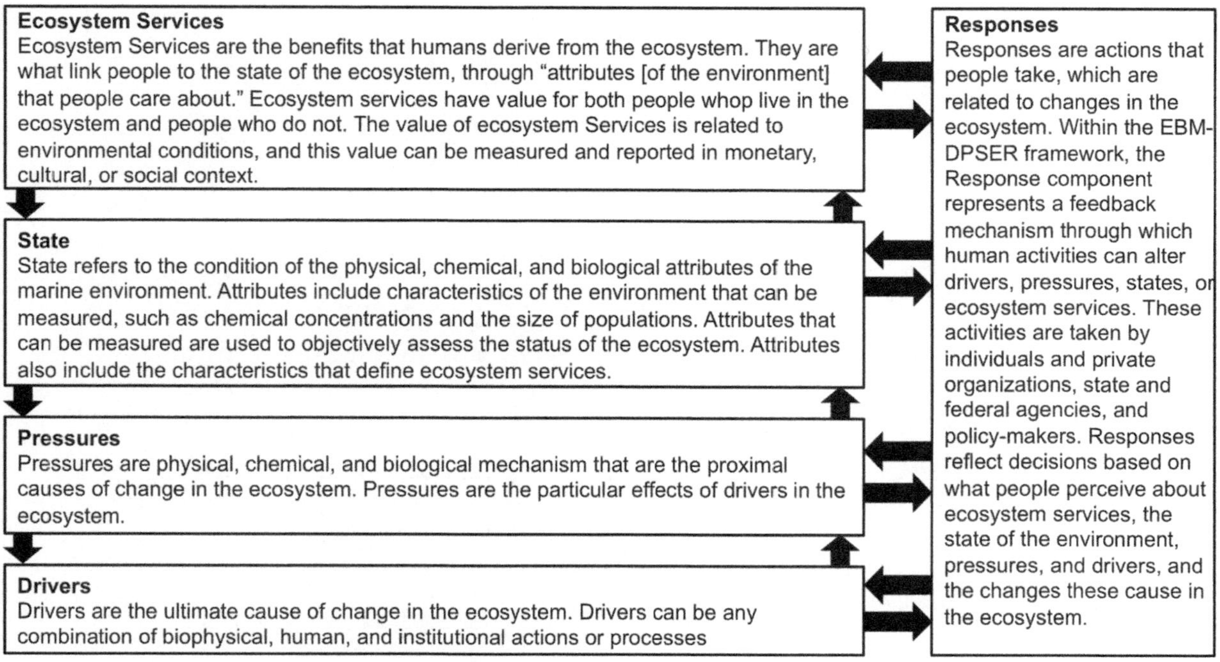

Figure 2. The EBM-DPSER model. The DPSIR model was modified by replacing the impacts module with ecosystem services facilitating a more complete representation of ecosystem interactions including those with human society and the associated feedbacks. Ecosystem services are at the top of the model, instead of drivers to focus attention upon the module that should be the focus of EBM decision-making.

the ecosystem services: aesthetics and existence, recreation, science and education, food/fisheries, waste treatment, and biological interactions.

There were 3 primary drivers linked to 14 primary pressures (Fig. 5). Drivers and pressures were organized to remain consistent with the mandates of major resource management agencies in the Florida Keys and Dry Tortugas, which have primary responsibility for either terrestrial or marine issues. The three primary drivers are climate change, land-based human activities and water-based human activities. Climate change is exclusively a far-field driver; whereas, land-based and water-based human activities produce both near-field and far-field pressures on the ecosystem. Far-field drivers and pressures are defined as those originating outside of the study area, while near-field drivers and pressures originate within the study area. Thus, the near-field drivers and pressures can be manipulated by responses that occur within the study site; whereas, far-field drivers and pressures can only be effectively managed with responses spanning a broader spatial scale, including areas outside of the study site. This organization of drivers and pressures was done to help responsible management agencies easily identify within-system responses they could employ to move the ecosystem towards a desired state.

Climate change produces five far-field pressures on the Florida Keys and Dry Tortugas ecosystem (ocean acidification, sea-level rise, increasing water and air temperatures, altered regional rainfall & evaporation, and changes in storm intensity, duration, and/or frequency). Water-based activities, in both the near- and far-field, result in fishing, marine debris, and contaminant releases. In addition, near-field water-based activities result in groundings, dredging, noise, and invasive species. Land-based activities, in both the near- and far-field, cause contaminant releases and changes in freshwater inflow. In addition, near-field land-based activities alter the adjacent shoreline.

The response module was defined in the EBM-DPSER model as actions taken as a result of the ecosystem condition. Responses occur to move the ecosystem toward a desired state or as a result of the current state of the ecosystem. For example, management regulations may be implemented that are intended to move the ecosystem to a desired state or scuba diving tourists may visit the Keys less frequently if they perceive the ecosystem to be in a degraded less desirable state. Responses occur across a range of human scales from the individual via behavioral changes to the global community via international treaties. Within the Florida Keys and Dry Tortugas these responses come from individuals, NGOs, and local, state, and federal agencies. Some identified responses include behavioral changes, technological advancements, education, outreach, social marketing, research, monitoring, infrastructure improvements, and possible regulations by responsible agencies.

Three especially significant responses were taken over the past two decades in an attempt to move the Florida Keys and Dry Tortugas marine ecosystem towards a more desirable condition. These responses were: 1) the creation of the Florida Keys National Marine Sanctuary, 2) the implementation of a rate of growth ordinance in the Florida Keys, and 3) passage of state laws requiring improvements to wastewater treatment throughout the Keys. These responses spanned various governmental levels from the Monroe County rate of growth ordinance that restricted building permits to federal government creation of the Florida Keys National Marine Sanctuary.

Application of EBM-DPSER

To demonstrate the utility and versatility of the EBM-DPSER model, the Florida Keys and Dry Tortugas model was applied to two distinct issues relevant to EBM of the Florida Keys. First to examine the holistic ecosystem effects of a specific pressure and second to attempt to improve a specific ecosystem service in a hypothetical situation.

Growth of the human population in the Florida Keys has increased the quantity of wastewater that must be treated.

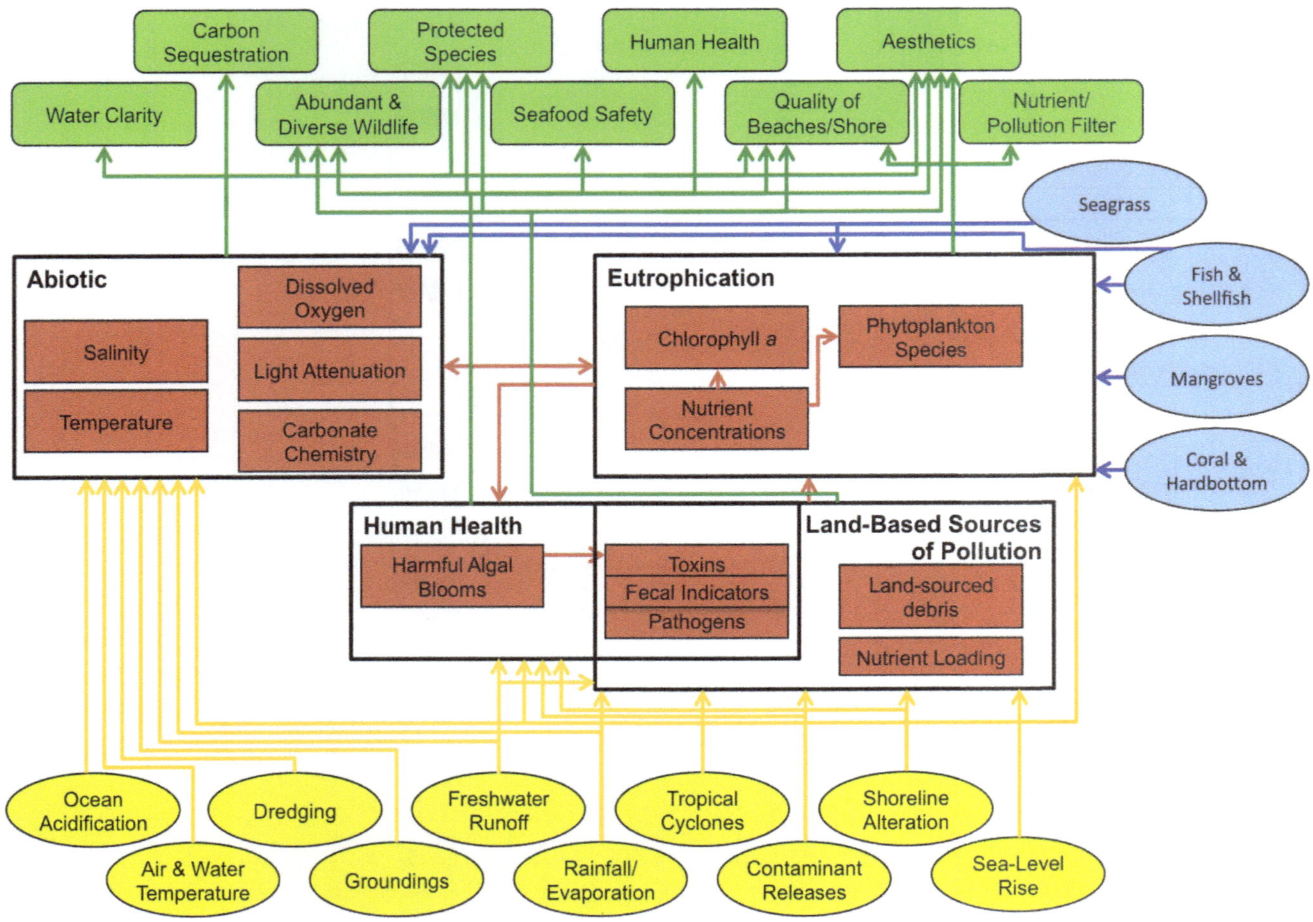

Figure 3. The south Florida water column sub-model. The sub-model for the water column of south Florida depicts the linkage from pressures (yellow ovals) to the state attributes that we measure (red boxes) with yellow arrows. These state attributes that we measure are organized into indicators for the water column (black outlined boxes and black text). The other states that influence the water column are depicted in the blue boxes and arrows to the right. The state attributes that we measure produce *ecosystem attributes people care about* (green boxes and arrows), which can be directly translated to ecosystem services.

Historically, septic tanks or cesspools were used, especially in individual residences. However, the presence of porous limestone throughout the Florida Keys permits seepage from these pits resulting in contaminant releases into canals and inshore waters.

Contaminant releases from wastewater carry nutrients and microbes that can have detrimental impacts on the state of the nearshore environment [58,59]. Nutrients cause phytoplankton blooms that decrease water clarity and decay causing hypoxia in

Table 1. Three common definitions of ecosystem services show significant disparity.

This Study	MEA 2005	Boyd and Banzhaf 2007	Fisher et al. 2009
the benefits people obtain from ecosystems	the benefits people obtain from ecosystems	the ecological components directly consumed or enjoyed to produce human well-being	the aspects of ecosystems utilized (actively or passively) to produce human well-being
Ecosystem Services	Ecosystem Services	Benefits and Ecosystem Services	Benefits, Intermediate and Final Ecosystem Services
Ecosystem attributes people care about		Ecosystem Services	Intermediate and Final Ecosystem Services
Ecosystem attributes people care about		Ecosystem Services	Final Ecosystem Services

The first row shows that the Millenium Ecosystem Assessment applies the broadest definition of ecosystem services with more detailed definitions provided by Boyd and Banzhaf [53] and Fisher et al [54]. In our initial EBM-DPSER model development we employed the MEA (2005) definition, because this was the most familiar definiton to the majority of participants and our goal was to build consensus. However, when the EBM-DPSER model is applied to conduct trade-off analyses of management options the other definitions for ecosystem services may be more appropriate. To help facilitate the application of other ecosystem service definitions, the table shows the linkages between the definitions of *ecosystem attributes people care about*, ecosystem services, and benefits used in this study, the Millenium Ecosystem Assessment, Boyd and Banzhaf 2007, and Fisher et al. 2009.

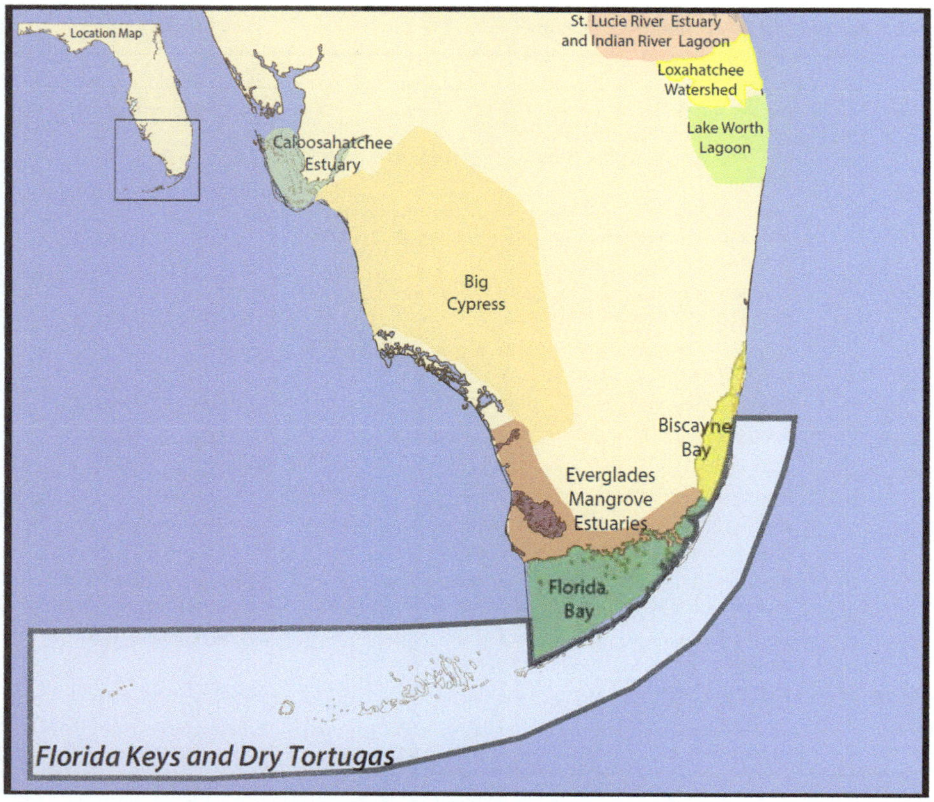

Figure 4. The Florida Keys and Dry Tortugas study site. The area shaded in white is the study site for the development of the Florida Keys and Dry Tortugas marine ecosystem EBM-DPSER model.

nutrient-enriched canals [59]. Nutrients can also cause macroalgal overgrowth of seagrasses and corals leading to less desirable habitats [58,60]. The microbes released may create health problems for both humans and marine organisms, such as corals.

These impacts on the state of the nearshore environment decrease the quantity and quality of some of the ecosystem services it provides. Phytoplankton blooms decrease water clarity impacting the quality of recreation, such as snorkeling and sight fishing.

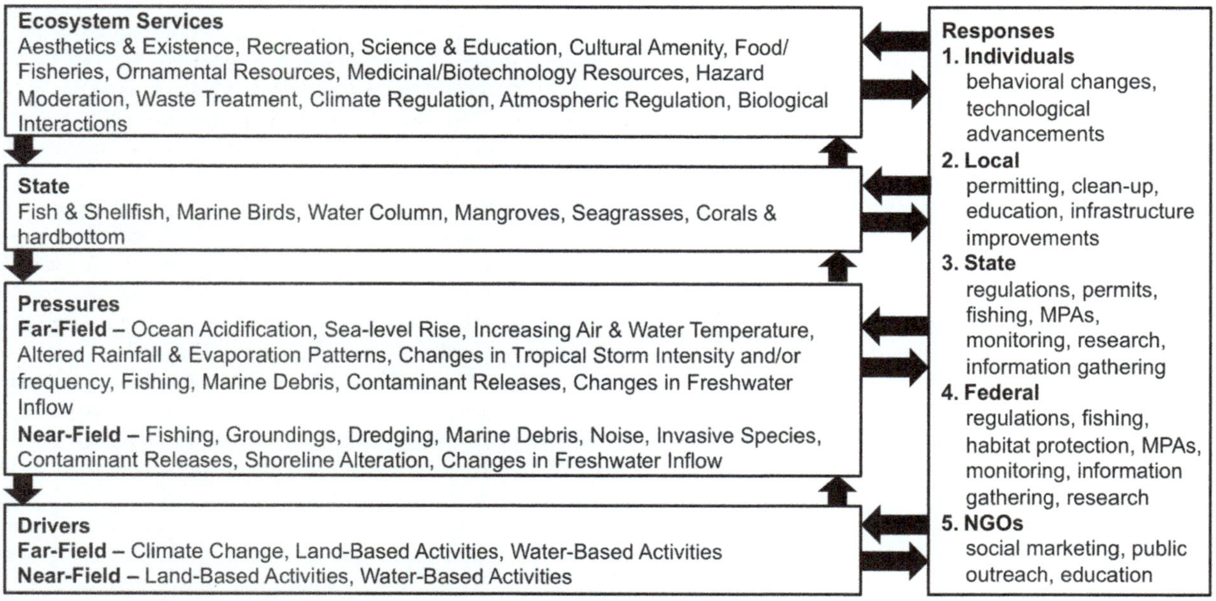

Figure 5. The Florida Keys and Dry Tortugas EBM-DPSER model. The EBM-DPSER model for the Florida Keys and Dry Tortugas marine ecosystem identifies the key components of each module within the CEM.

Table 2. Ecosystem services provided by the Florida Keys marine ecosystem.

Cultural	*Aesthetics & Existence*	Provide aesthetic quality of aquatic and terrestrial environments (visual, olfactory, and auditory), therapeutic benefits, and pristine wilderness for future generations	*Final Ecosystem Service*
	Recreation	Provide a suitable environment or setting for beach activities and other marine activities such as fishing, diving, snorkeling, motor, an non-motor boating	*Benefit*
	Science & Education	Provide a living laboratory for formal and informal education, and scientific research	*Benefit*
	Cultural Amenity	Support a maritime way of life, sense of maritime tradition, spiritual experience	*Benefit*
Provisioning	*Food/Fisheries*	Provide safe to eat seafood	*Final Ecosystem Service*
	Ornamental Resources	Provide materials for jewelry, fashion, aquaria, etc.	*Final Ecosystem Service*
	Medicinal & Biotechnology Resources	Provide natural materials and substances for inventions and cures	*Final Ecosystem Service*
Regulating	*Hazard Moderation*	Moderate extreme environmental events (e.g. mitigation of waves and stormsurge during hurricanes)	*Final Ecosystem Service*
	Waste Treatment	Retain storm water; remove nutrients, contaminants, and sediment from water; and dampen noise	*Final Ecosystem Service*
	Climate Regulation	Moderate temperature and influence or control other processes such as wind, precipitation, and evaportation	*Final Ecosystem Service*
	Atmospheric Regulation	Exchange CO_2, O_2, mercury, etc. with the atmosphere	*Final Ecosystem Service*
	Biological Interactions	Regulate species interactions to maintain beneficial functions such as seed dispersal, pest/invasive control, herbivory, etc.	*Intermediate Ecosystem Service*

The last column identifies these services as benefits, final or intermediate ecosystem services according to Fisher et al. 2009 [54].

Hypoxia can enrich the sediments and bottom-water of stratified canals with sulfur. When wind events overturn the canal waters, this results in an unpleasant odor decreasing aesthetics and recreational potential. Replacement of seagrass and coral with macroalgae significantly degrades the quality of marine recreation for divers and snorkelers, reduces habitat quality for fish and other wildlife, and affects pollution treatment by altering ecosystem nutrient cycling rates. The release of the wastewater-associated microbial community can cause health impacts in humans making some areas of the marine ecosystem unusable. Thus, the use of septic tanks and cesspits, decreased the ecosystem services recreation, aesthetics and existence, food and fisheries, waste treatment, and biological interactions.

Threatened and observed degradation in ecosystem services resulted in a response by the EPA, which encouraged Monroe County to reduce reliance on cesspits and septic fields by providing municipal wastewater treatment. A state law was then passed requiring advanced wastewater treatment and disposal in the Florida Keys. Analyzing this response in the EBM-DPSER context one can see it minimized the pressures placed on the ecosystem by cesspits and septic tanks, but doing so involved a trade-off; the cost of upgrading all septic tanks and cesspits to comply with the new regulations and a marked slowdown in new residential construction. The removal of septic tanks and cesspits does not require additional trade-offs with respect to ecosystem sustainability or services, because septic tanks and cesspits provide "no net benefits" to the ecosystem that are not also realized by the new advanced wastewater systems beyond the lower monetary cost to the human population.

The second application considers how one might improve the ecosystem service of recreation in the Florida Keys, specifically scuba diver recreation. This is a hypothetical scenario that would occur if it was determined that scuba diver satisfaction in the Florida Keys was less than desirable. There is some data suggesting that this is becoming an issue [61]. Ideally as is the case with this data, there will be information on the *ecosystem attributes people care about* to pinpoint the ecosystem attributes to target in response. With or without this information, the resource manager or decision-maker could apply the EBM-DPSER model to determine responses that would improve scuba diver satisfaction in the Florida Keys. Without the additional information on the *ecosystem attributes people care about*, you have to examine all the *ecosystem attributes people care about* which contribute to scuba diver recreation making the response less targeted and likely less effective. As identified through the MARES process, these attributes are water clarity, abundant and diverse marine wildlife, protected species, abundant and diverse fish, and abundant healthy coral.

For simplicity, we assume water clarity was the primary cause of scuba diver dissatisfaction similar to what was found in a satisfaction survey for the Florida Keys from 1995–1996 to 2000–2001 [61]. First, it is necessary to understand whether the conditions have actually changed over time. Leeworthy et al. [61] found that scuba divers perceived a decrease in water clarity, but systematic monitoring data did not support this perception; showing no decrease in water clarity [61]. In our fictive example we assume there was a measured decrease in water clarity.

The MARES water column sub-model has water clarity as an end-point associated with a measurable attribute (i.e. light attenuation), which itself is a function of suspended sediments, chromophoric dissolved organic matter (CDOM), and phytoplankton (Fig. 3) [62,63]. Suspended sediment concentrations are a product of land-based sources and sediment re-suspension, a function of bottom habitat state and human activities (e.g. dredging, groundings, etc.). Land-based sources are unlikely to have a significant effect on offshore scuba diver experience, since the reefs are approximately 10km from shore. However, dredging of channels and loss of benthic habitat, especially seagrass, increases suspended sediments, negatively impacting water clarity, and indirectly impacting the abundance and diversity of fish, wildlife, and protected species. CDOM is largely land-based and thus unlikely to affect water clarity on offshore coral reefs. Algal blooms over the coral reefs would impact water clarity for scuba diving. Typically, algal blooms in the Florida Keys are caused by increased nutrient loading or a loss of grazers. Given the multiplicity of potential causes for degraded water clarity, it is necessary to determine the most significant cause(s). The need to first determine if an actual change has occurred and second determine the cause of that change to make a directed, efficient response; highlights the benefits gained from regularly collected monitoring data.

Assuming monitoring data was analyzed and it was determined that increased suspended sediment caused the degraded water clarity, the EBM-DPSER model would then be consulted to identify the pressures that affect suspended sediment concentrations. These pressures are dredging, groundings, and shoreline alteration (Fig. 3), as well as the pressures that degrade seagrass beds (groundings, dredging, contaminants releases, fishing, and changes to freshwater inflow). Responses to improve water clarity could include establishing marine protected areas that minimize damage by groundings, stabilizing sediments and promoting seagrass growth; restricting contaminant releases from vessels near the coral reefs; and ceasing dredging operations for some or all channels.

When evaluating which response or suite of responses should be implemented to improve water clarity, the EBM-DPSER model provides a useful scenario analysis tool that encapsulates the holistic effects of these responses and illuminates tradeoffs. If the management response were to ban dredging throughout the Florida Keys, this would reduce the concentration of suspended sediments in waters near channels during times dredging operations would have occurred; however, it would also restrict access for some vessels and possibly decrease the availability of deep-water refugia for some organisms, particularly during extreme temperature events. By restricting access of some vessels, there could be decreased recreational and commercial fishing opportunities and decreased tourism from cruise ships. Thus, this potential response might improve scuba diver recreation, but it would negatively impact the ecosystem services of food/fisheries, cultural amenity, ornamental resources, and possibly science & education.

Discussion

In practice, we manage natural resources, our natural capital, for people. People are an integral participant in the ecosystem; their actions produce or reduce pressures upon the environment, but they are also the benefactors of ecosystem services. In some instances, such as wildlife refuge management or through restoration initiatives, human society chooses to enhance the environment and increase the benefits that it provides. Goals may compete, but highlighting the complex roles people play within the ecosystem should assist managers in balancing competing goals by making trade-offs explicit facilitating their efforts to balance acceptable levels of ecosystem change with protection of needed services, and provision of desired services.

By explicitly incorporating ecosystem services, use of the EBM-DPSER model facilitates and encourages ecosystem service analyses. Ecosystem services have values that can potentially be measured by human dimensions scientists. Quantitative and qualitative analytical methods can be used to produce data and tools for decision-making and to estimate the relative importance of different natural resources to particular human populations. Knowing the values that people place upon ecosystem services informs decisions that involve tradeoffs between environmental and other societal objectives and between competing objectives. Assessing the value of ecosystem services can occur within either economic or social contexts. While there is great utility in monetization for cost-benefit analyses [38,64,65], this must not be considered the complete valuation of the ecosystem [66]. Recreational services, as an example, are valued by society in ways that are not economic, but are still possible to quantify and interpret using other common methods [67]. Identifying the most appropriate approach whether quantitative or qualitative for the circumstances is important in order to evaluate the full range of benefits and make the most well-informed decisions [68].

Other considerations including distributive justice (the fairness associated with allocating scarce resources), sustainability, ecological stewardship, human well-being, and cultural and ethical values are important to consider in the decision-making process [69,70]. Equity analysis (one approach to allocating scarce resources) requires estimation of the differences between groups who receive benefits and those who lose benefits under different management alternatives. There are other allocation norms associated with who does or does not receive ecosystem service benefits in the amount they want or feel they deserve, such as equality or need based allocations [71]. Sustainability and stewardship analyses focusing on the past, present, and future distributions of those services consider additional layers of complexity. Cultural and ethical considerations may place further constraints on the acceptability of different management decisions [55]. Human societies are complex with diverse perspectives on the use of ecosystem services depending on circumstances at the global-regional-local level of political or societal organization.

When conducting ecosystem service analyses it is important that ecosystem services be selected and refined based on the specific decision context [54]. Thus, if the goal is to use EBM-DPSER to examine changes in ecosystem services under different management scenarios leading to economic valuation and trade-off analyses, the definitions and classifications of ecosystem services have to be appropriate for that objective. Where those include economic valuation, it is helpful to follow either Boyd and Banzhaf [53] or Fisher et al. [54] (Table 1) and classify services as either intermediate or final. This is necessary to ensure only final services are utilized in economic valuation which avoids double-counting, a common problem in ecosystem service valuation [72].

The EBM-DPSER model is flexible and can accommodate alternative definitions and classifications of ecosystem services. The ecosystem services currently identified within the Florida Keys and Dry Tortugas case study are a mixture of intermediate and final ecosystem services, as well as benefits based upon the Fisher et al. [54] scheme (Table 2). As mentioned previously, the *ecosystem attributes people care about* which link the state to ecosystem services in the current EBM-DPSER model are defined as ecosystem services, themselves, by Boyd and Banzhaf [53] and Fisher et al. [54]. Thus, the *ecosystem attributes people care about* can and should become the ecosystem services in the EBM-DPSER model when warranted by the decision-context, such as to undertake cost-benefit analysis or to calculate green GDP.

Within the EBM-DPSER model, responses encompass human actions motivated by changes in the condition in the environment (state) or in the ecosystem services provided. Responses can affect drivers, pressures, states, or ecosystem services and represent a mechanism for anthropogenic feedback, and therefore the possibility of anthropogenic alteration. Included in responses are activities of gathering information, decision-making and program implementation that are conducted by agencies charged with making policies and implementing management actions that affect the ecosystem. The value and location of services both regulated and unregulated can have a large effect on the drivers and pressures acting on the ecosystem.

The EBM-DPSER model allows for the inclusion of the complete suite of complex positive and negative interactions between humans and their environment, thus presenting a more holistic picture of how the ecosystem, including humans, functions. This is an improvement over the antecedent DPSIR approach which primarily defines impacts as results from a degraded system and highlights societies role in causing negative impacts on the environment [49]. Capturing positive human responses that increase the production of ecosystem services or restore ecosystem health and the many positive effects of the ecosystem on human society is essential if EBM is ever to be widely implemented.

Applying scenario analysis to the DPSIR model has been recognized as a useful approach to incorporate DPSIR into management decisions [20]. This is equally true for EBM-DPSER. By applying scenario analysis to the EBM-DPSER model, the holistic effect of impending management decisions on the production of ecosystem services and ecosystem condition may be evaluated and tradeoffs among these services made explicit.

The case study and example applications in the Florida Keys and Dry Tortugas confirm the utility of applying the EBM-DPSER model to inform resource management decision-making. The EBM-DPSER model reflected a consensus as to the underlying mechanisms that create the current status of ecosystem services within the Florida Keys and Dry Tortugas; i.e., an understanding of how the *ecosystem attributes people care about* are altered by components of the ecosystem state (Figs. 3 and 5). Describing how the ecosystem states relate to ecosystem services can be useful to managers attempting to implement EBM. Identifying which ecosystem states produce a given ecosystem service and which pressures impact these states highlights potential underlying causes of ecosystem service production levels. This enables ready recognition of those responses that improve the delivery and sustainability of an ecosystem service. Placing ecosystem services into a model (DPSIR) that has been widely accepted by the management community should provide a politically and socially realistic path towards the incorporation of ecosystem service analysis into decision-making [18,20].

As in the example of scuba diver satisfaction, once a problem is identified it will often be necessary to undertake a more detailed study that characterizes the ecosystem attributes, which are below a desired level. The information from that study will facilitate the most effective targeted response. The EBM-DPSER model may also be applied to conduct scenario analyses of alternative responses to determine their holistic effect and identify potential trade-offs. The EBM-DPSER model illustrated that a hypothetical response to ban dredging does not just decrease suspended sediment, but also has negative effects including access for vessels and providing refugia. These positive benefits provided by dredged channels increase the production of ecosystem services. By illuminating these positive effects, in addition to the negative impacts, EBM-DPSER provides a more complete identification of the tradeoffs inherent in any response. Identifying these tradeoffs between ecosystem services with respect to management goals is recognized as a key informational need for EBM [73]. Managers must consider tradeoffs. If they determine that detrimental use of the area is necessary for society and that negative impacts are unavoidable then it will be important for managers to identify alternative areas and direct recreational users and commercial fishers to them to obtain the ecosystem services that they demand.

Identifying the trade-offs qualitatively is a first step towards quantification of these trade-offs. Quantification is difficult in many cases, because of the frequent asymmetry of effects of management actions on ecosystem services. The effect of management responses are often direct and localized, while the benefits from ecosystem services are more diffuse and indirect [74]. Moreover, quantification is confounded by a lack of consistent units. One approach to overcome the lack of consistent units would be to use expert opinion to scale or ordinate these linkages following a method similar that employed by Altman et al. [2] to quantify the linkage between pressures and ecosystem services.

Progress towards quantification of interactions among pressures, states, and ecosystem services would make scenario analyses and the inherent trade-offs among scenarios quantitative rather than qualitative and better inform resource management decisions. It would also allow for holistic risk assessment following a modification of the Altman et al. approach [2] to incorporate the effect of multiple pressures upon ecosystem states and in turn upon ecosystem services. This holistic risk analysis would determine the pressures causing the largest loss of ecosystem services and the ecosystem services subject to the largest stress from the cumulative effect of all pressures.

Both scenario analyses and holistic risk assessment account for the impact of multiple human uses on ecosystem services simultaneously. Although this is an underlying assumption in EBM, there are few practical tools that account for the cumulative effect of multiple human uses on ecosystem services [2]. If we fail to account for the cumulative effect of pressures upon ecosystem services and ecosystem state, the science used to inform decision-making will be incomplete. At the same time we need to account for human societal uses to calculate tradeoffs. By coupling risk assessment and scenario analysis, the relative reduction or increase in risk to ecosystem services could be evaluated for each potential response.

The EBM-DPSER model is an important step towards effectively informing ecosystem management for the benefit of humans; whereas, the DPSIR model is designed to inform ecosystem management to protect the ecosystem from human impacts. The EBM-DPSER model directly addresses several of the inherent challenges to EBM, while maintaining a focus on EBM's fundamental objectives. It integrates diverse scientific disciplines and captures more completely the multitude of complex interactions between the biophysical and human components of the ecosystem. By highlighting ecosystem services it emphasizes the extent to which society relies upon and benefits from the ecosystem. This inclines EBM toward proactive intervention rather than strictly reactive management. The linking of pressures to states to ecosystem services permits at a minimum the qualitative assessment of the cumulative impacts of pressures upon ecosystem services and captures the direct and indirect effect of multiple human uses on ecosystem services, as well as the loss of ecosystem services to human society. This is an important first step towards quantifying these complex interactions. As shown in the example applications, the EBM-DPSER model can assist in bridging the communication gap between human dimensions scientists, biophysical scientists, and resource managers thereby providing EBM with a useful operational tool.

Acknowledgments

We would like to thank all of the participants in MARES for their invaluable contribution to this research.

Author Contributions

Conceived and designed the experiments: PO CK WN DL JL JB. Performed the experiments: CK DL SL WN PO PF GC JL JB. Analyzed the data: CK PO DL SL. Wrote the paper: CK DL SL WN PO PF GC JL JB.

References

1. Slocombe DS (1993) Implementing Ecosystem-Based Management. Bioscience 43: 612–622.
2. Altman I, Blakeslee AMH, Osio GC, Rillahan CB, Teck SJ, et al. (2011) A practical approach to implementation of ecosystem-based management: a case study using the Gulf of Maine marine ecosystem. Frontiers in Ecology and the Environment 9: 183–189.
3. Rosenberg AA, McLeod KL (2005) Implementing ecosystem-based approaches to management for the conservation of ecosystem services. Marine Ecology-Progress Series 300: 270–274.
4. McLeod KL, Lubchenco JL, Palumbi SR, Rosenberg AA (2005) Scientific Consensus Statement on Marine Ecosystem-Based Management. In: Sea CPfSat, editor. Communication Partnership for Science and the Sea (COMPASS). Signed by 219 academic scientists and policy experts with relevant expertise.
5. Curtice C, Dunn DC, Roberts JJ, Carr SD, Halpin PN (2012) Why Ecosystem-Based Management May Fail without Changes to Tool Development and Financing. Bioscience 62: 508–515.
6. Lubchenco J, Sutley N (2010) Proposed US Policy for Ocean, Coast, and Great Lakes Stewardship. Science 328: 1485–1486.
7. Lubchenco J (1998) Entering the century of the environment: A new social contract for science. Science 279: 491–497.
8. Arkema KK, Abramson SC, Dewsbury BM (2006) Marine ecosystem-based management: from characterization to implementation. Frontiers in Ecology and the Environment 4: 525–532.
9. Levin PS, Fogarty MJ, Murawski SA, Fluharty D (2009) Integrated ecosystem assessments: Developing the scientific basis for ecosystem-based management of the ocean. Plos Biology 7: 23–28.
10. Levin PS, Fogarty MJ, Matlock GC, Ernst M (2008) Integrated ecosystem assessments. NOAA Tech. Memo.: US Department of Commerce. 20 p.
11. Sekovski I, Newton A, Dennison WC (2012) Megacities in the coastal zone: Using a driver-pressure-state-impact-response framework to address complex environmental problems. Estuarine Coastal and Shelf Science 96: 48–59.
12. Ogden JC, Davis SM, Jacobs KJ, Barnes T, Fling HE (2005) The use of conceptual ecological models to guide ecosystem restoration in South Florida. Wetlands 25: 795–809.
13. Doren RF, Trexler JC, Gottlieb AD, Harwell MC (2009) Ecological indicators for system-wide assessment of the greater everglades ecosystem restoration program. Ecological Indicators 9: S2–S16.
14. Bowen RE, Riley C (2003) Socio-economic indicators and integrated coastal management. Ocean & Coastal Management 46: 299–312.
15. Gentile JH, Harwell MA, Cropper W, Harwell CC, DeAngelis D, et al. (2001) Ecological conceptual models: a framework and case study on ecosystem management for South Florida sustainability. Science of the Total Environment 274: 231–253.
16. Smeets E, Weterings R (1999) Environmental Indicators: Typology and Overview. 19 p.
17. Atkins JP, Burdon D, Elliott M, Gregory AJ (2011) Management of the marine environment: Integrating ecosystem services and societal benefits with the DPSIR framework in a systems approach. Marine Pollution Bulletin 62: 215–226.
18. Mangi SC, Roberts CM, Rodwell LD (2007) Reef fisheries management in Kenya: Preliminary approach using the driver-pressure-state-impacts-response (DPSIR) scheme of indicators. Ocean & Coastal Management 50: 463–480.
19. Svarstad H, Petersen LK, Rothman D, Siepel H, Watzold F (2008) Discursive biases of the environmental research framework DPSIR. Land Use Policy 25: 116–125.
20. Tscherning K, Helming K, Krippner B, Sieber S, Paloma SGy (2012) Does research applying the DPSIR framework support decision making? Land Use Policy 29: 102–110.
21. Collins SL, Carpenter SR, Swinton SM, Orenstein DE, Childers DL, et al. (2010) An integrated conceptual framework for long-term social'Äiecological research. Frontiers in Ecology and the Environment 9: 351–357.
22. Gray JS, Elliot M (2009) Ecology of Marine Sediments – From Science to Management, second edition. Oxford: Oxford University Press. 256 p.
23. Elliott M (2011) Marine science and management means tackling exogenic unmanaged pressures and endogenic managed pressures – A numbered guide. Marine Pollution Bulletin 62: 651–655.
24. Mee LD, Jefferson RL, Laffoley DD, Elliott M (2008) How good is good? Human values and Europe's proposed Marine Strategy Directive. Marine Pollution Bulletin 56: 187–204.
25. Turner RK, Lorenzoni I, Beaumont N, Bateman IJ, Langford IH, et al. (1998) Coastal Management for Sustainable Development: Analysing Environmental and Socio-Economic Changes on the UK Coast. The Geographical Journal 164: 269–281.
26. Carr ER, Wingard PM, Yorty SC, Thompson MC, Jensen NK, et al. (2007) Applying DPSIR to sustainable development. International Journal of Sustainable Development & World Ecology 14: 543–555.
27. MEA (2005) Ecosystems and human well-being: synthesis; Assessment ME, editor. Washington, DC USA: Island Press.
28. TEEB (2010) The Economics of Ecosystems and Biodiversity: Mainstreaming the Economics of Nature: A synthesis of the approach, conclusions and recommendations of TEEB. 36 p.
29. PCAST (2011) Sustaining Environmental Capital: Protecting Society and the Economy. Washington, DC USA: Executive Office of the President of the United States. 125.
30. Wellbeing CIo (2012) How are Canadians Really Doing? The 2012 CIW report. WaterlooON: Canadian Index of Wellbeing and University of Waterloo. 72 p.
31. Kerenyi A (2011) The Better Life Index of the Organization for Economic Co-operation and Development. Public Finance Quarterly 56: 518–538.
32. Lubchenco J (2009) NOAA Annual Guidance Memorandum. In: Commerce Do, editor. 14.
33. Lovelace S, Goedeke TL, Dillard M (2012) Prioritizing County-Level Well-Being: Moving Toward Assessment of Gulf Coast Counties Impacted by the Deepwater Horizon Industrial Disaster. 48 p.
34. Costanza R, dArge R, deGroot R, Farber S, Grasso M, et al. (1997) The value of the world's ecosystem services and natural capital. Nature 387: 253–260.
35. Burkhard B, Petrosillo I, Costanza R (2010) Ecosystem services ' Bridging ecology, economy and social sciences. Ecological Complexity 7: 257–259.
36. de Jonge VN, Pinto R, Turner RK (2012) Integrating ecological, economic and social aspects to generate useful management information under the EU Directives' 'ecosystem approach'. Ocean & Coastal Management 68: 169–188.
37. Feld CK, Sousa JP, da Silva PM, Dawson TP (2010) Indicators for biodiversity and ecosystem services: towards an improved framework for ecosystems assessment. Biodiversity and Conservation 19: 2895–2919.
38. Turner RK, Paavola J, Cooper P, Farber S, Jessamy V, et al. (2003) Valuing nature: lessons learned and future research directions. Ecological Economics 46: 493–510.
39. Sherrouse BC, Clement JM, Semmens DJ (2011) A GIS application for assessing, mapping, and quantifying the social values of ecosystem services. Applied Geography 31: 748–760.
40. Cowling RM, Egoh B, Knight AT, O'Farrell PJ, Reyers B, et al. (2008) An operational model for mainstreaming ecosystem services for implementation. Proceedings of the National Academy of Sciences of the United States of America 105: 9483–9488.
41. Nuttle WK, Fletcher PJ (2013) Integrated Conceptual Ecosystem Model Develoment for the Florida Keys/Dry Tortugas Coastal Marine Ecosystem. Silver Springs, Md. USA: National Oceanic & Atmospheric Administration. 28 p.
42. Weinstein MP (2009) The Road Ahead: The Sustainability Transition and Coastal Research. Estuaries and Coasts 32: 1044–1053.
43. Turner RK (2000) Integrating natural and socio-economic science in coastal management. Journal of Marine Systems 25: 447–460.
44. Visser L (1999) Coastal zone management from the social scientific perspective. Journal of Coastal Conservation 5: 145–148.
45. Cheong S-M (2008) A new direction in coastal management. Marine Policy 32: 1090–1093.
46. Borja A, Galparsoro I, Solaun O, Muxika I, Tello EM, et al. (2006) The European Water Framework Directive and the DPSIR, a methodological approach to assess the risk of failing to achieve good ecological status. Estuarine Coastal and Shelf Science 66: 84–96.
47. Rounsevell MDA, Dawson TP, Harrison PA (2010) A conceptual framework to assess the effects of environmental change on ecosystem services. Biodiversity and Conservation 19: 2823–2842.
48. Kohsaka R (2010) Developing biodiversity indicators for cities: applying the DPSIR model to Nagoya and integrating social and ecological aspects. Ecological Research 25: 925–936.
49. Rehr A, Small M, Bradley P, Fisher W, Vega A, et al. (2012) A Decision Support Framework for Science-Based, Multi-Stakeholder Deliberation: A Coral Reef Example. Environmental Management 50: 1204–1218.
50. Haines-Young RH, Potschin MP (2010) The links between biodiversity, ecosystem services and human well-being. In: Reaffaelli D, Frid C, editors. Ecosystem Ecology: A New Synthesis. Cambridge, UK: Cambridge University Press. 110–139.
51. Müller F, Burkhard B (2012) The indicator side of ecosystem services. Ecosystem Services 1: 26–30.
52. Fassio A, Giupponi C, Hiederer R, Simota C (2005) A decision support tool for simulating the effects of alternative policies affecting water resources: an application at the European scale. Journal of Hydrology 304: 462–476.

53. Boyd J, Banzhaf S (2007) What are ecosystem services? The need for standardized environmental accounting units. Ecological Economics 63: 616–626.
54. Fisher B, Turner RK, Morling P (2009) Defining and classifying ecosystem services for decision making. Ecological Economics 68: 643–653.
55. Farber S, Costanza R, Childers DL, Erickson J, Gross K, et al. (2006) Linking ecology and economics for ecosystem management. Bioscience 56: 121–133.
56. Ogden JC (2005) Everglades Ridge and Slough Conceptual Ecological Model. Wetlands 25: 810–820.
57. Niemeijer D, de Groot R (2008) Framing environmental indicators: moving from causal chains to causal networks. Environment, Development and Sustainability 10: 89–106.
58. Lapointe BE, Barile PJ, Matzie WR (2004) Anthropogenic nutrient enrichment of seagrass and coral reef communities in the Lower Florida Keys: discrimination of local versus regional nitrogen sources. Journal of Experimental Marine Biology and Ecology 308: 23–58.
59. Lapointe BE, Clark MW (1992) Nutrient inputs from the watershed and coastal eutrophication in the Florida keys. Estuaries 15: 465–476.
60. Collado-Vides L, Caccia VG, Boyer JN, Fourqurean JW (2007) Tropical seagrass-associated macroalgae distributions and trends relative to water quality. Estuarine Coastal and Shelf Science 73: 680–694.
61. Leeworthy VR, Wiley PC, Hospital JD (2004) Importance-Satisfaction Ratings Five-year Comparison, SPA & ER Use, and Socioeconomic and Ecological Monitoring Comparison of Results 1995–96 to 2000–01. Silver Spring, MD USA. 59 p.
62. Kelble CR, Ortner PB, Hitchcock GL, Boyer JN (2005) Attenuation of photosynthetically available radiation (PAR) in Florida Bay: Potential for light limitation of primary producers. Estuaries 28: 560–571.
63. Kirk JTO (1994) Light and Photosynthesis in Aquatic Ecosystems. CambridgeUK: Cambridge University Press. 530 p.
64. Loomis J, Kent P, Strange L, Fausch K, Covich A (2000) Measuring the total economic value of restoring ecosystem services in an impaired river basin: results from a contingent valuation survey. Ecological Economics 33: 103–117.
65. Hein L, van Koppen K, de Groot RS, van Ierland EC (2006) Spatial scales, stakeholders and the valuation of ecosystem services. Ecological Economics 57: 209–228.
66. Pimm SL (1997) The value of everything. Nature 387: 231–232.
67. Anderson LE, Loomis DK (2012) Normative Standards for Coral Reef Conditions: A Comparison of SCUBA Divers by Specialization Level.
68. Wegner G, Pascual U (2011) Cost-benefit analysis in the context of ecosystem services for human well-being: A multidisciplinary critique. Global Environmental Change-Human and Policy Dimensions 21: 492–504.
69. Costanza R, Folke C (1997) Valuing ecosystem services with efficiency, fairness, and sustainability as goals. In: Daily GC, editor. Nature's Services: Societal Dependence on Natural Ecosystems. Washington, DC USA: Island Press. 47–70.
70. Deutsch M (1975) Equity, Equality, and Need: What Determines Which Value Will Be Used as the Basis of Distributive Justice? Journal of Social Issues 31: 137–149.
71. Loomis DK, Ditton RB (1993) Distributive Justice in Fisheries Management. Fisheries 18: 14–18.
72. Fisher B, Turner K, Zylstra M, Brouwer R, Groot Rd, et al. (2008) Ecosystem services and economic theory: integration for policy-relevant research. Ecological Applications 18: 2050–2067.
73. Steffan-Dewenter I, Kessler M, Barkmann J, Bos MM, Buchori D, et al. (2007) Tradeoffs between income, biodiversity, and ecosystem functioning during tropical rainforest conversion and agroforestry intensification. Proceedings of the National Academy of Sciences of the United States of America 104: 4973–4978.
74. Cook GS, Heinen JT (2005) On the uncertain cost and tenuous benefits of marine reserves: a case study of the Tortugas Ecological Reserve, South Florida, USA. Natural Areas Journal 25: 390–396.

9

Representation of Dormant and Active Microbial Dynamics for Ecosystem Modeling

Gangsheng Wang[1,2]*, Melanie A. Mayes[1,2], Lianhong Gu[1,2], Christopher W. Schadt[1,3]

1 Climate Change Science Institute, Oak Ridge National Laboratory, Oak Ridge, Tennessee, United States of America, **2** Environmental Sciences Division, Oak Ridge National Laboratory, Oak Ridge, Tennessee, United States of America, **3** Biosciences Division, Oak Ridge National Laboratory, Oak Ridge, Tennessee, United States of America

Abstract

Dormancy is an essential strategy for microorganisms to cope with environmental stress. However, global ecosystem models typically ignore microbial dormancy, resulting in notable model uncertainties. To facilitate the consideration of dormancy in these large-scale models, we propose a new microbial physiology component that works for a wide range of substrate availabilities. This new model is based on microbial physiological states and the major parameters are the maximum specific growth and maintenance rates of active microbes and the ratio of dormant to active maintenance rates. A major improvement of our model over extant models is that it can explain the low active microbial fractions commonly observed in undisturbed soils. Our new model shows that the exponentially-increasing respiration from substrate-induced respiration experiments can only be used to determine the maximum specific growth rate and initial active microbial biomass, while the respiration data representing both exponentially-increasing and non-exponentially-increasing phases can robustly determine a range of key parameters including the initial total live biomass, initial active fraction, the maximum specific growth and maintenance rates, and the half-saturation constant. Our new model can be incorporated into existing ecosystem models to account for dormancy in microbially-driven processes and to provide improved estimates of microbial activities.

Editor: Jonathan H. Badger, J. Craig Venter Institute, United States of America

Funding: This research was funded by the Laboratory Directed Research and Development (LDRD) Program of the Oak Ridge National Laboratory (ORNL) and by the U.S. Department of Energy Biological and Environmental Research (BER) program. ORNL is managed by UT-Battelle, LLC, for the U.S. Department of Energy under contract DE-AC05-00OR22725. The funders had no role in study design, data collection and analysis, decision to publish, or preparation of the manuscript.

Competing Interests: The authors have declared that no competing interests exist.

* E-mail: wangg@ornl.gov

Introduction

Ecologically-important processes such as soil organic carbon and nutrient cycling largely depend on the active fraction of microbial communities [1]. At any given time in a given environment, microorganisms can be in active, dormant, or dead states [2]. Dormancy is considered an evolutionary strategy designed to maintain the genetic code until conditions improve to allow replication [3]. When environmental conditions are unfavorable for growth, e.g., resource limitation, microbes may enter a reversible state of low to zero metabolic activity to alleviate the loss of biomass and metabolic functions [4,5]. The maintenance coefficient (i.e., maintenance cost of C per unit microbial biomass C per unit time) can be two to three orders of magnitude lower in dormant microbes than in metabolically active microbes [6,7]. Many soils have slow organic matter turnover rates with seasonal changes in substrate supply, temperature, and moisture. The complexity of soils in space and time may result in uneven distributions of multiple potentially limiting resources, leading to significant rates of dormancy even when some resources are abundant. When spatial and temporal complexity is combined with differential resource partitioning among species in a community, high rates of dormancy could be a prominent feature in soil systems. Thus it is essential to understand dormancy in order to more accurately predict how active microorganisms contribute to ecosystem processes such as decomposition and nutrient turnover [1].

A complicating factor in studying microbial dormancy is that no single approach can be easily employe to simultaneously measure individual microbial states (active, dormant or dead), and a combination of different techniques is required. Differential staining is often used to segregate physiological states with direct microscopic counting of bacteria and fungi. 'Life-indicating' stains that require the presence of 'standard' physiological abilities, such as the esterase activity needed for fluorescein diacetate cleavage, may distinguish active from dormant+dead cells [8]. When combined with general-purpose stains, these strains can distinguish dormant cells by difference [9]. Combining membrane-permeant with membrane-impermeant nucleophilic stains (e.g., SYTO-9 and propidium iodide respectively) may distinguish live from dead, but not active from dormant [10,11]. Active microbes may or may not be 'viable' with common culture-based techniques, which complicates classification and measurement of dormancy phenomena [5]. Methods such as direct plating, serial dilution and most probable number (MPN) techniques will not distinguish between active and dormant organisms [12]. Substrate Induced Respiration (SIR) or Substrate Induced Growth Response (SIGR) method [13,14] can distinguish active and dormant communities if growth respiration curves are modeled (using initial exponentially-increasing respiration); however, the technique often needs to be combined with microscopy or chloroform fumigation/extraction in order to obtain total live microbial biomass [15,16].

Despite limitations in distinguishing active, dormant and dead microbial biomass, abundant evidence indicates that the majority of environmental microorganisms in a given community may be dormant under natural conditions [1,17]. Alvarez *et al.* [18] reported that only 3.8–9.7% of the total biomass is active in a Typic Argiudoll soil from the Argentinean Pampa. Khomutova *et al.* [19] showed that the fraction of active microbial biomass ranged from 0.02% to 19.1% in the subkurgan paleosoils of different age and 9.2–24.2% in modern background soils. Microbial biomass measured through SIR or SIGR is thought to reflect only the active portion because the maintenance respiration of dormancy biomass is negligible in the initial exponentially-increasing phase [13,16,20]. Through a mathematical analysis of respiration curves, Van de Werf & Verstraete [21] examined 16 soils and found that 4–49% of the total biomass was in an active state; and the active component in undisturbed natural ecosystems (18.8±8.8%, mean±standard deviation) was about 70% of that in arable agricultural soils (25.7±14.8%). Stenström *et al.* [22] showed that the fraction of active biomass typically varied from 5% to 20% in soils without recent addition of substrates. Lennon & Jones [5] found much lower active fractions in soils (18±15%) than in marine (65±19%) and fresh (54±11%) water environments. From the above studies it seems conservative to extrapolate that the active fraction is very likely below 50% of live microbes under most natural soil conditions.

Microbially-mediated processes have been incorporated into ecosystem models [23–28] although continued development is still required to bring microbial processes into global climate models [29–31]. However, these recent models do not consider physiological state changes and assume that measures of microbial biomass constitute the active biomass. The exclusion of dormancy from the microbially-driven ecosystem processes could result in incorrect estimates of total live microbial biomass, which further leads to deficiencies in model parameterization and predictions of soil organic carbon and nutrients.

Generally, there are two strategies to represent physiological states in microbial-ecology models: one is to explicitly separate the total live biomass into two pools, i.e., active and dormant [4,32]; the other is to directly regard the active fraction (i.e., ratio of active biomass to total live biomass) as a state variable [33,34]. Both of these two approaches predict the total live biomass, active and dormant biomass, and the flux between the active and dormant components. Apparently the introduction of the 'active fraction' as a state variable in the latter approach simplifies the model structure since the adaptive variation of microbial composition might be represented by one single variable (active fraction) [34,35]. However, another state variable indicating the microbial biomass pool size (e.g., total live biomass, active biomass or dormant biomass) is still essential for ecosystem modeling since the carbon and nutrient fluxes are pool-size dependent. For example, if we define active fraction and total microbial biomass as state variables, the active and dormant biomass could be determined by them, and the net flux between active and dormant fractions and other related fluxes could also be computed according to the active and dormant biomass constrained by mass balance. The above-mentioned modeling efforts have shown that adequate representation of dormancy and the transitions between the dormant and active states is crucial for modeling important microbially-mediated ecosystem processes.

Here, we review state-of-the-art microbial dormancy modeling approaches and discuss the rationales of these models with a focus on transformation processes between active and dormant states. We propose an improved synthetic microbial physiology model based on accepted assumptions and examine the model behavior with theoretical and experimental analyses. In this paper, the 'total microbial biomass' refers to the 'total live microbial biomass' unless otherwise stated. Our objective is to clarify the applicability of existing microbial dormancy models and provide a new theoretical basis for representing microbial activity and dormancy in ecosystem models.

Dormancy In Microbial Models

Transformation between active and dormant states

Although Buerger *et al.* [36] argued that dormant microbial cells could reactivate stochastically and might be independent of environmental cues, environmental factors such as substrate availability are often thought to control the transformation between active and dormant states [5]. Most models (see Appendix S1) distinguish the active biomass pool from the dormant pool and define them as two state variables (B_a and B_d) (Fig. 1). Only active microbes (B_a) can uptake substrate and produce new cells. The connection between the active and dormant states is a reversible process including two directional sub-processes, i.e., dormancy (from active to dormant) and reactivation (or resuscitation, from dormant to active). Losses from active biomass include growth respiration and maintenance (maintenance respiration, mortality, enzyme synthesis, etc.) [23]. Dormant microbes still require energy for maintenance and survival although at a lower metabolic rate [5].

The net transformation rate ($B^N_{a\to d}$) from active to dormant state is the difference between the flux from active to dormant ($B_{a\to d}$) and the flux from dormant to active state ($B_{d\to a}$), i.e., $B^N_{a\to d} = B_{a\to d} - B_{d\to a}$. The models of Hunt [37] (Equation S1-1 in Appendix S1) and Gignoux *et al.* [38] (Equation S1-2 in Appendix S1) directly formulate the net flux ($B^N_{a\to d}$) without explicit components for $B_{a\to d}$ and $B_{d\to a}$. The direction of the net flux depends on the maintenance requirement relative to the substrate availability. If the available substrate is less than the maintenance requirement, there is a positive net flux from active to dormant pool, and vice versa. In addition, Hunt [37] assumed a 'buffer zone' for the change of states: when the maintenance requirement surpasses the substrate supply but the deficit is within a small fraction (1% d^{-1}) of B_a, there is no flux between the two states.

Some models define rates for both dormancy and reactivation. In the model of Ayati [39] (Equation S1-3 in Appendix S1), the

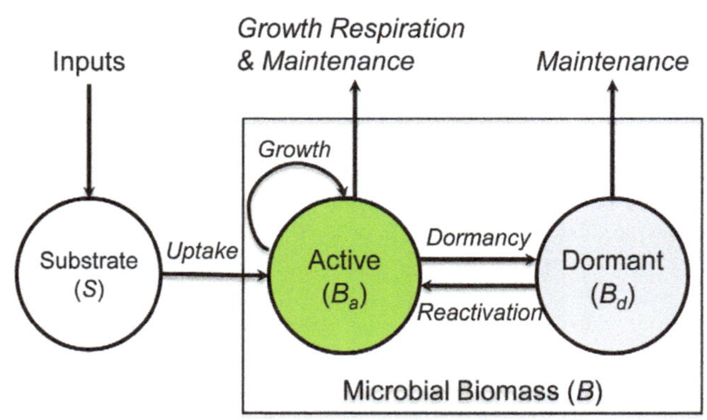

Figure 1. Active and dormant microbial biomass pools in microbial physiology models (modified from Fig. 2 in Lennon & Jones, 2011).

dormant rate ($\gamma_{a\rightarrow d}$) increases with declining substrate concentration, and the reactivation ($\gamma_{d\rightarrow a}$) only occurs when substrate concentration is higher than the half-saturation constant (K_s). Konopka [32] modified the potential rates for dormancy and reactivation by the relative growth rate ($\mu/\mu_{\max}$, ratio of true specific growth rate to maximum specific growth rate), i.e, the two rates are multiplied by ($1-\mu/\mu_{\max}$) and $\mu/\mu_{\max}$, respectively (Equation S1-4 in Appendix S1). Similarly, Jones & Lennon [40] postulated two complementary rates (1-*R* and *R*) for dormancy and resuscitation (Equation S1-5 in Appendix S1).

Two other models also explicitly formulate the two conversion rates between states but do so using concepts of probability. Bär *et al.* [41] used two complementary factors (1-$\mathcal{J}$ and $\mathcal{J}$) to represent the probability for the transition between active and dormant state in addition to an identical potential rate constant for the two processes (Equation S1-6 in Appendix S1). The conceptual model of Locey [42] applies a deterministic dormant rate and a stochastic resuscitation rate (Equation S1-7 in Appendix S1). The potential resuscitation rate is modified by (1-*p*), where *p* is the probability that a disturbance in the active pool will result in the immigration of one individual from the metacommunity. The probability ($\mathcal{J}$) in Bär *et al.* [41] is explicitly calculated from the environmental cues (e.g., soil moisture), while the cause of the probability (*p*) in Locey [42] is not elucidated.

Switch function model

In addition to the dormancy and reactivation processes, a key concept in the model developed by Stolpovsky *et al.* [4] is 'switch function' (Equation S1-8 in Appendix S1). The switch function (θ) determines the fraction of active cells taking up dissolved organic carbon (DOC). This function refers to the growth fraction in active biomass (B_a) that consumes substrate and thus is not the same as the active fraction (r) in total biomass (B). Furthermore, the dormancy and reactivation fluxes are set to be proportional to ($1-\theta$) and θ, respectively. θ is formulated by the Fermi-Dirac statistics [4]. Another feature of this model is the consideration of 'depth' of dormancy in reactivation, where the reactivation rate is negatively dependent on the duration of dormancy. The switch function model includes at least 15 model parameters and it is difficult to compute the Gibbs energy change of the oxidation of DOC [4].

The switch function (θ) sets it apart from the conventional Michaelis-Menten (M-M) or Monod kinetics because of its new perspective of thermodynamics. According to the M-M kinetics [43], the substrate saturation level represents the fraction of enzyme-substrate complex (*ES*) in active enzyme (E_0), where the substrate saturation level is formulated by $S/(K_s + S)$ with S and K_s being the substrate concentration and the half-saturation constant [43]. When the M-M kinetics is applied to describe microbial uptake of substrate, the substrate (or combined with TEA) saturation level (i.e., $\mu(S, TEA)$ in Equation S1-8a of Appendix S1) is a measure of the actively growing fraction in the active microbial community. The switch function is also determined by the saturation levels of substrate and terminal electron acceptor (TEA), i.e., $\mu(S, TEA)$ (Equation S1-8e in Appendix S1). Mathematically the inclusion of both the switch function (θ) and the M-M kinetics (i.e., $\mu(S, TEA)$) might result in double counting of the impact of substrate and TEA. We would recommend using the switch function (θ) to modify the microbial uptake rate if the Gibbs energy change of the oxidation of substrate (ΔG) is tractable and the thermodynamic threshold (G_0) and the steepness of the step function (*st*) are identifiable.

Physiological state index models

As an alternative to models with two microbial biomass pools (i.e., active and dormant), a further state variable indicating the dormant or active fraction in total biomass has been proposed. Wirtz [44] developed a simple index ($r_d = 0.5–1.0$) representing the dormant microbial biomass as a fraction of the steady-state total biomass (B_{stat}) under the condition of $B_d << B_a$. In case of a net loss of total biomass ($dB/dt<0$), the dormant biomass $B_d = B_{stat} \cdot r_d$; otherwise ($dB/dt>0$), $B_d = B_{stat} \cdot (1-r_d)$. This model has a sudden change of dormant biomass at the transition point (i.e., $dB/dt = 0$) since $r_d>0.5$.

Different from the dormant index of Wirtz [44], the concept of an active index (i.e., index of physiological state) of soil microbial community has been employed in soil carbon and nutrient cycling models [33,34]. The index of physiological state (r), referring to the activity state, is often defined as the ratio of metabolically active microbial biomass to the total soil microbial biomass [22,33,34].

In the Synthetic Chemostat Model (SCM), the rate of change of the state variable r is described as follows [33,45]:

$$\frac{dr}{dt} = \frac{1}{B}\frac{dB}{dt} \cdot (\phi - r) = \mu \cdot (\phi - r) \tag{1}$$

with

$$\phi = \phi(S) = S^n/(K_r + S^n), \text{or} \phi = S/(K_r + S) \tag{2}$$

where $r = B_a/B$, representing the fraction (hereinafter referred to as 'active fraction') of active biomass in total biomass; μ is the specific growth rate of total biomass; ϕ denotes the saturation level of substrate (S); the simple power ($n = 1$) has been widely used [35] and, in this case ($n = 1$), K_r is called the half-saturation constant.

Blagodatsky & Richter [34] used the expression $\mu(S) = \mu_{\max} \cdot \phi(S)$ in their model development. This expression was not derived in the original definition of the specific growth rate (see Equation 3) by Panikov [45] and because its validity cannot be inferred, the concepts will not be addressed here.

According to Panikov's derivation [45], the specific growth rate (μ) follows the general definition [46,47]:

$$\mu = \frac{1}{B}\frac{dB}{dt} \tag{3}$$

Based on Equations 1 and 3, we can derive (see Equation S2-1 in Appendix S2):

$$dB_a/dt = \phi \cdot (dB/dt) \tag{4}$$

$$dB_d/dt = (1-\phi) \cdot (dB/dt) \tag{5}$$

We find that the model described by Equation 1 is not applicable under low substrate availability, as described below. Generally, the rates of change in biomass pools (B, B_a, and B_d) can be expressed as

$$dB/dt = g^{\pm}(S,B_a) - f^{+}(S,B_d) \tag{6}$$

$$dB_a/dt = g^{\pm}(S,B_a) - B^N_{a\to d} \tag{7}$$

$$dB_d/dt = -f^{+}(S,B_d) + B^N_{a\to d} \tag{8}$$

where $B^N_{a\to d}$ denotes the net dormancy flux; $g^{\pm}(S, B_a)$ is a function that represents the difference between the substrate uptake and the maintenance requirements of B_a, i.e., the net growth of B_a; and $f^{+}(S, B_a)$ is a function denoting the maintenance and survival energy costs of B_d. The superscript '±' in $g^{\pm}$ indicates the function value of g could be positive at high S or non-positive when the substrate uptake cannot satisfy the maintenance requirements of B_a at low S. The superscript '+' in f^{+} implies $f \geq 0$. Note that the function $f^{+}(S, B_a)$ is not necessarily dependent on S [4].

From Equations 4, 6 and 7, we can obtain

$$B^N_{a\to d} = (1-\phi)\cdot g^{\pm}(S,B_a) - \phi\cdot f^{+}(S,B_d) \tag{9}$$

The two terms in the right side of Equation 9 may be regarded as the conversion of B_a to B_d (i.e., $B_{a\to d}$) and the transformation of B_d to B_a (i.e., $B_{d\to a}$), respectively. At high S resulting in $g \geq 0$, Equation 9 may be one of the possible expressions for $B_{a\to d}$ and $B_{d\to a}$. However, at low S leading to $g<0$ and $B_{a\to d}<0$, i.e., no active cells become dormant under insufficient substrate, which is inconsistent with the strategy of dormancy for microorganisms when faced with unfavorable environmental conditions [5].

Based on the above analysis, we conclude that the physiological state index model (Equation 1) needs to be improved. In other words, the empirical assumption that the steady state active fraction (r^{ss}) approaches the substrate saturation level (ϕ^{ss}) may not be necessary because this assumption could lead to impractical flux (Equation 9) between dormant and active states under low substrate availability.

A Synthetic Microbial Physiology Model

Based on the aforementioned review and analysis, we have developed a synthetic microbial physiology model component relating to substrate availability. As indicated by Fig. 1, the growth and maintenance functions of active microbes (B_a) are characterized by the maximum specific growth rate (μ_G) and maintenance rate (m_R); whereas the dormant microbes (B_d) cost energy to maintain their basic cellular functions at a much lower specific maintenance rate (denoted by $\beta\cdot m_R$, where $\beta<1$) [48].

General assumptions

First we define the substrate saturation level (ϕ) as

$$\phi = S/(K_S + S) \tag{10}$$

where the parameter K_s is the half saturation constant for substrate uptake as indicated by the M-M kinetics [43].

Based on the above review of existing dormancy models, the following assumptions are accepted in our new model: (1) the dormancy rate is proportional to the active biomass and the reactivation rate is proportional to the dormant biomass, i.e., $B_{a\to d} \propto B_a$ and $B_{d\to a} \propto B_d$; (2) under very high substrate concentration ($S >> K_s$), $\phi \to 1$, $B_{a\to d} \to 0$ and $B_{d\to a} \geq 0$; (3) under very low substrate ($S << K_s$), $\phi \to 0$, $B_{a\to d} \geq 0$ and $B_{d\to a} \to 0$; (4) based on the assumptions (1–3), we derive that $B_{a\to d} \propto (1-\phi)\cdot B_a$ and $B_{d\to a} \propto \phi\cdot B_a$; (5) further we assume that the maximum specific maintenance rate for active microbes (m_R with units of h^{-1}) controls both transformation processes since the maintenance energy cost is the key factor regulating the dormancy strategy [5,37,38]. As a result we postulate that

$$B_{a\to d} = (1-\phi)\cdot m_R\cdot B_a \tag{11a}$$

$$B_{d\to a} = \phi\cdot m_R\cdot B_d \tag{11b}$$

Model Description

Equations 11a and 11b only describe the transformation between the active and dormant states. They need to be linked to a microbial growth and maintenance model for depicting microbial dynamics. Our recent work to develop the Microbial-ENzyme-mediated Decomposition (MEND) model [23] suggested that it might be adapted to serve this purpose due to its focus on microbial processes for which we have developed a firm theoretical basis [47,49]. Combining Equations 11a and 11b with the MEND model [23,47], we express the microbial physiology component (see Fig. 1) as a group of differential equations

$$dS/dt = I_s - \frac{1}{Y_G}\cdot\frac{\phi}{\alpha}m_R\cdot B_a \tag{12a}$$

$$dB/dt = d(B_a + B_d)/dt = (\phi/\alpha - 1)\cdot m_R\cdot B_a - (\beta\cdot m_R)\cdot B_d \tag{12b}$$

$$dB_a/dt = (\phi/\alpha - 1)\cdot m_R\cdot B_a - (1-\phi)\cdot m_R\cdot B_a + \phi\cdot m_R\cdot B_d \tag{12c}$$

$$dB_d/dt = -(\beta\cdot m_R)\cdot B_d + (1-\phi)\cdot m_R\cdot B_a - \phi\cdot m_R\cdot B_d \tag{12d}$$

where t is the time scale; ϕ is defined by Equation 10; I_s is the input to substrate pool; Y_G is the true growth yield; m_R denotes the specific maintenance rate at active state (h^{-1}); $\alpha = m_R/(\mu_G + m_R)$ is the ratio of m_R to the sum of maximum specific growth rate (μ_G) and m_R, $\alpha \in (0,0.5)$ since usually $m_R \leq \mu_G$; and β (0–1) is the ratio of dormant maintenance rate to active maintenance rate, i.e., ($\beta\ m_R$) denotes the maximum specific maintenance rate at dormant state.

In summary, there are five parameters (α, β, m_R or μ_G, Y_G, K_s) in the proposed model (hereinafter referred to as the MEND model). From Equations 12b and 12c, we can derive the change rate of active fraction (r) (see Equation S2-2 in Appendix S2)

$$dr/dt = m_R\cdot[(\phi - r) + (\phi/\alpha + \beta - 1)\cdot r\cdot(1-r)] \tag{12e}$$

This equation for r is more complicated than Equation 1 but still practical, given currently available data. Additionally, it implies that r needs not approach ϕ at steady state in our model, whereas $r \equiv \phi$ at steady state is required by the model of Panikov [45].

Steady state analysis

Assuming the input (I_s) is time-invariant, we can obtain the steady state solution to the above new MEND model (see Equations S2-3(a–e) in Appendix S2). Fig. 2 shows the steady

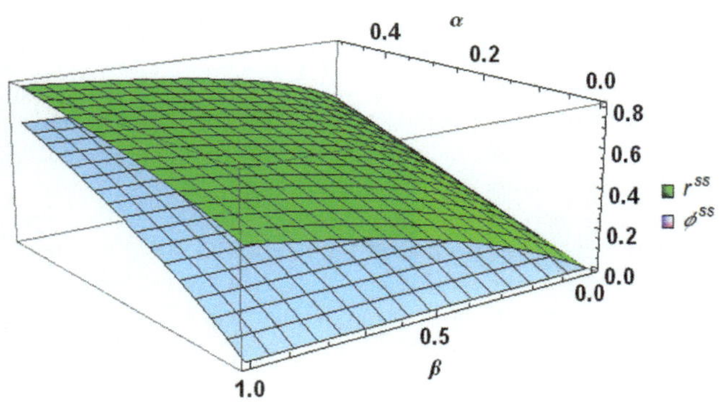

Figure 2. Steady state active fraction (r^{ss}) and substrate saturation level (ϕ^{ss}) as a function of α and β; $\alpha = m_R/(\mu_G+m_R)$, μ_G and m_R (h^{-1}) are maximum specific growth rate and specific maintenance rate for active microbial biomass, respectivly; β denotes the ratio of dormant specific maintenance rate to m_R.

state active fraction (r^{ss}) and substrate saturation level (ϕ^{ss}) as a function of the two physiological indices, i.e., α (0–0.5) and β (0–1). Both r^{ss} and ϕ^{ss} positively depend on α and β and $r^{ss} \geq \phi^{ss}$ for any combinations of α and β. If we consider two extreme values of $\beta \to 0$ or $\beta \to 1$, the r^{ss} and ϕ^{ss} (see Equations S2-4 and S2-5 in Appendix S2) can be simplified to

$$r^{ss}_{\beta\to 0} = \phi^{ss}_{\beta\to 0} = \alpha \tag{13a}$$

$$\begin{cases} r^{ss}_{\beta\to 1} = \frac{1+\sqrt{1+8\alpha}}{4} \\ \phi^{ss}_{\beta\to 1} = \frac{4\alpha-1+\sqrt{1+8\alpha}}{3+\sqrt{1+8\alpha}} \end{cases} \tag{13b}$$

Equation 13 and Fig. 2 indicate that: (1) the steady state active fraction (r^{ss}) is equal to ϕ^{ss} and they are identical to $\alpha = m_R/(\mu_G+m_R)$ only under the condition of $\beta \to 0$; (2) the upper bound of r^{ss} is approximately 0.8 at $\alpha \to 0.5$ and $\beta \to 1$; and (3) with $\alpha \leq 0.5$, the maximum r^{ss} is ca. 0.5 if the magnitude of β is around 0.001–0.01 [6]. This threshold value (0.5) of r^{ss} is a reasonable estimate that can explain how the measured active fraction of microbes in undisturbed soils is usually considerably less than the total biomass [5,21,22].

Model simplification under sufficient substrate condition

As mentioned in the Introduction, SIR or SIGR method can distinguish active from dormant composition and the data from these experiments have been widely used to estimate the active microbial biomass and the maximum specific growth rate [13,14]. The simplification of the microbial model under excess substrate has also been employed to estimate maximum specific growth rate (μ_G), active microbial biomass (B_a), and/or total microbial biomass (B) using the SIR or SIGR data [1,13,35]. Here we show the simplification of our model (Equation 12) for conditions appropriate to SIGR or SIR experiments, e.g., the short-term period of exponentially-increasing respiration of active biomass following substrate addition. We will test our reduced and full model with the SIGR data of Colores *et al.* [13] in the next section.

Under sufficient substrate (i.e., $S >> K_s$ in Equation 10 thus $\phi \to 1$), Equations 12(a–e) can be simplified and integrated for initial conditions, i.e., $S = S_0$, $B = B_0$ and $r = r_0$ at $t = 0$ (see Equations S2-6 and S2-7 in Appendix S2):

$$S(t) = S_0 - \frac{B(t) - B_0}{Y_G(1-\alpha)} \tag{14a}$$

$$B(t) = B_0 r_0 \cdot e^{(1/\alpha - 1)m_R t} + B_0(1-r_0)\cdot[\alpha \cdot e^{(1/\alpha-1)m_R t} + (1-\alpha)\cdot e^{-m_R t}] \tag{14b}$$

$$r(t) = \frac{[r_0 + \alpha(1-r_0)]\cdot e^{(m_R/\alpha)\cdot t} - \alpha(1-r_0)}{[r_0 + \alpha(1-r_0)]\cdot e^{(m_R/\alpha)\cdot t} + (1-\alpha)(1-r_0)} \tag{14c}$$

The CO_2 production rate, $v(t)$, during the exponential growth stage is derived as an explicit function of t (see Equation S2-7d in Appendix S2):

$$v(t) = \frac{dCO_2}{dt} = \frac{B_0(1-Y_G)}{Y_G}\left\{[(m_R/\alpha)\cdot r_0 + m_R\cdot(1-r_0)]\cdot e^{(1/\alpha-1)m_R t} - [m_R\cdot(1-r_0)]\cdot e^{-m_R t}\right\} \tag{14d}$$

The respiration rate, $v(t)$, is associated with two exponential items, i.e., $e^{(1/\alpha-1)m_R t} = e^{\mu_G t}$ and $e^{-m_R t}$. Considering an extreme case that $m_R << \mu_G$ (i.e., $\alpha \to 0$), Equations 14(b–d) can be further simplified to Equations S2-8(b–d) (see Appendix S2). Equations S2-8b and S2-8c (denoting $B(t)$ and $r(t)$, respectively) are similar to Equations 11 and 10 in Panikov & Sizova [35], respectively. However, Equation S2-8d (for $v(t)$) is different from Equation 13 of Panikov & Sizova [35], where a constant '*A*' was added to the exponential. Equation S2-8d is identical to Equation 7 derived for SIGR experiments in Colores *et al.* [13].

Panikov & Sizova [35] used their Equation 13 to fit respiration rates during the exponentially-increasing (i.e., no substrate limitation) phase (see Fig. 2 in Panikov & Sizova [35] for data and curve fittings). However, these data are based on glucose-induced respiration that includes both basal respiration of native SOC and respiration due to the addition of glucose [13]. The basal respiration rate may be regarded as a constant in certain cases (see Colores *et al.* [13] and data in Fig. 1 of Blagodatsky *et al.* [50]). The constant '*A*' representing the basal respiration rate was included in Equation 13 of Panikov & Sizova [35] in order to fit the combined respiration from the addition of glucose and basal respiration. However, this constant '*A*' cannot be derived from such governing equations as Equations S2-6(a–c) (see Appendix S2) that assume respiration is the sole result of substrate addition. In other words, the equations do not include basal respiration. Certainly, our predicted respiration could include basal respiration as long as (i) a basal respiration rate is added to Eq. 14d *ad hoc* or (ii) Equations S2-6(a–c), or more commonly Equations 12(a–e), are linked to a soil organic matter (SOM) decomposition model, which can produce decomposed native soil C in addition to the respiration of substrate addition. Because Equation 13 of Panikov & Sizova [35] is not linked to a native C decomposition model, fitting the model to combined native C and substrate respiration data is not appropriate.

Model test I: substrate-induced respiration

In this section, we used the respiration data from ^{14}C-labeled glucose SIGR experiments by Colores et al. [13] to calibrate our

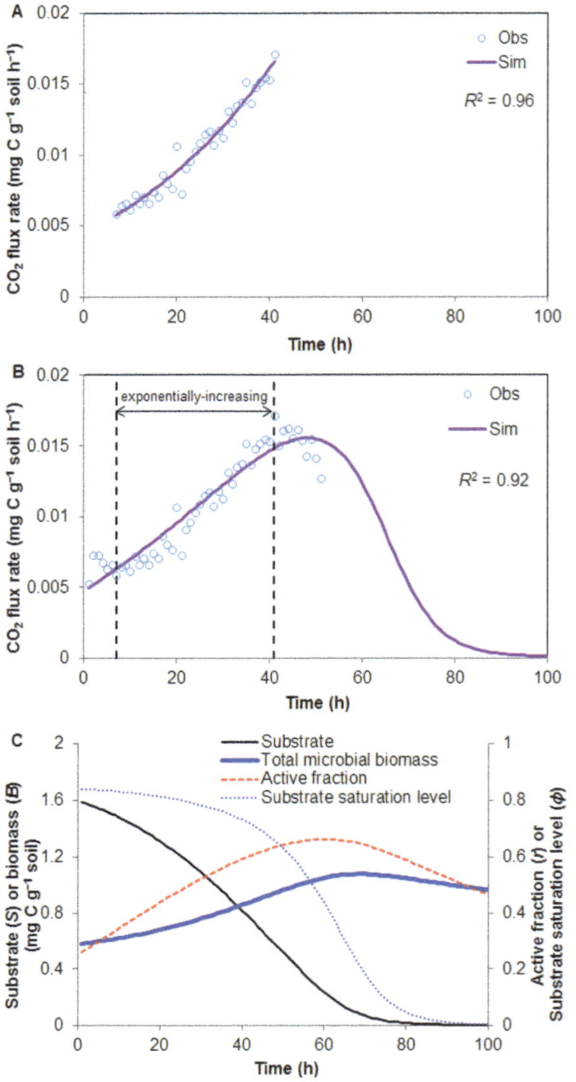

Figure 3. MEND model simulations against the respiration rates due to added ^{14}C-labeled glucose in Colores et al. [13]. (a) Fitting of the respiration rates in the exponentially-increasing phase using Equation 14, 'Obs' and 'Sim' denote observed and simulated data, respectively. (b) Fitting of the respiration rates in both exponentially-increasing and non-exponentially-increasing phases using Equation 12. (c) Simulated substrate (S), total live microbial biomass (B), active fraction (r) and substrate saturation level (ϑ) based on Equation 12.

MEND model. The respiration data only represented the CO_2 production from the added substrate and did not include basal respiration from the native C.

First we employed Equation 14d to fit the respiration rates during the exponentially-increasing stage and the result is shown in Fig. 3a (see original data in Fig. 3 of Colores et al. [13]). The true growth yield (Y_G) was set to 0.5 according to Colores *et al.* [13]. There are four undetermined parameters (B_0, r_0, μ_G, α) in Equation 14d (with $m_R = \mu_G \cdot \alpha/(1-\alpha)$). We found that only the maximum specific growth rate (μ_G) could be determined with high confidence (coefficient of variation (CV) = 5%) from the exponentially-increasing respiration rates. The CVs of the other three optimized parameters (B_0, r_0, α) were as high as 55–77% (Table 1). However, the initial active microbial biomass ($B_{a0} = B_0 \times r_0$) had a lower uncertainty (CV = 20%) compared to B_0 and r_0. The above results indicate that the exponentially-increasing respiration rates can only be used to obtain μ_G and B_{a0}.

We then conducted numerical simulations in terms of all data including both exponentially-increasing and non-exponentially-increasing respiration rates (Fig. 3b). The non-exponentially-increasing respiration rates include the lag period before the exponentially-increasing phase and the respiration at longer times after the rates cease to increase exponentially [13]. The latter phase is likely because of the substrate saturation levels (ϕ) become limiting to respiration. We used Equations 12a, 12b, 12e and the corresponding expression for CO_2 flux rate, to allow the substrate saturation level (ϕ) to change with time. Additionally, we used the ranges of μ_G determined above. We used the SCEUA (Shuffled Complex Evolution at University of Arizona) algorithm [51,52] to determine model parameters. The SCEUA is a widely used stochastic optimization algorithm for calibrating hydrological and environmental models [51].

When exponentially-increasing and non-exponentially-increasing data are included together, the CVs of all parameters (B_0, r_0, μ_G, α, K_s, β) are within 25% except β with a high CV of 76% (Table 1). The optimized μ_G values (0.030 ± 0.001 h^{-1}) are almost the same as obtained by Colores *et al.* [13]. Model estimates of $\alpha = 0.228 \pm 0.031$ indicate that the maximum specific maintenance rate of active microbes (m_R) is about 30% of μ_G and thus cannot be ignored. The initial active biomass (B_{a0}) is 0.145 ± 0.004 mg C g^{-1} soil (Table 1), which is lower than the values (0.194 ± 0.004 mg C g^{-1} soil) using the SIGR method [13]. This is likely due to the inclusion of maintenance respiration (characterized by m_R, see Equation 14d) in our model even for the exponentially-increasing stage; thus a lower B_{a0} could produce similar CO_2 flux to the case with higher B_{a0} that does not include the contributions from maintenance respiration. Our results also show that the initial active fraction (r_0) is 28.5 ± 6.4% and β is 0.025 ± 0.019. The magnitude of β is comparable to the estimation by Anderson & Domsch [6,7]. In addition, the half-saturation constant (K_s) was estimated as 0.275 ± 0.038 mg C g^{-1} soil, which is very close to the values derived from 16 soils by Van de Werf & Verstraete [21]. This K_s value indicates the substrate saturation level (ϕ) is higher than 0.7 before the transition from exponentially-increasing to non-exponentially-increasing phase (Fig. 3c). The changes of substrate (S), total microbial biomass (B) and active fraction (r) with time are also shown in Fig. 3c. In conclusion, the five parameters (B_0, r_0, μ_G, α, K_s) can be effectively determined using both exponentially-increasing and non-exponentially-increasing respiration rates, whereas β may also be determined but with a relatively high uncertainty (CV = 76%) than the other parameters.

Through this experimental analysis, we identified the need for isotopic data to discriminate between basal and substrate-induced respiration. We also discovered that the exponentially-increasing period due to substrate addition can be used to identify only a select set of model parameters (i.e., μ_G and B_{a0}) as also demonstrated by the method of Colores *et al.* [13]. These parameters, however, can be further applied to longer-term respiration experiments to enable fitting to obtain the remainder of model parameters by using our MEND model. Thus, we have found a new and unique solution to identify different parameters as a function of time, and to effectively use isotopic labeling to yield a specific set of model parameters.

Model test II: intermittent substrate supply

In order to further validate this additional physiological component in the MEND model, we also tested it against a laboratory experimental dataset with intermittent substrate supply [4]. In addition to the substrate, another limiting factor (i.e.,

Table 1. MEND model parameters values used for simulation of respiration rates due to added ^{14}C-labeled glucose in Colores et al. [13].

Parameter	Exponentially-increasing respiration*			All data†			Description
	Mean	SD‡	CV§	Mean	SD	CV	
B_0	0.504	0.279	55%	0.525	0.080	15%	Initial microbial biomass, (mg C g^{-1} soil)
r_0	0.394	0.263	67%	0.285	0.064	23%	Initial active fraction
μ_G	0.027	0.001	5%	0.030	0.001	3%	Maximum specific growth rate (h^{-1})
α	0.185	0.142	77%	0.228	0.031	13%	$m_R/(\mu_G+m_R)$, m_R is maximum specific maintenance rate for active microbes (h^{-1})
K_s	—	—	—	0.275	0.038	14%	Half-saturation constant for substrate (mg C g^{-1} soil)
β	—	—	—	0.025	0.019	76%	Ratio of dormant maintenance rate to m_R
Y_G	0.5	—	—	0.5	—	—	True growth yield, constant
B_{a0}	0.135	0.027	20%	0.145	0.004	3%	Initial active biomass (mg C g^{-1} soil), calculated by $B_0 \times r_0$

*Only the respiration rates during exponentially-increasing phase are used.
†All data including both exponentially-increasing and non-exponentially-increasing respiration.
‡SD: standard deviation.
§CV: Coefficient of variation.

oxygen, O_2) was included in this study. For this reason, we also introduced one more parameter (K_o: half saturation constant for O_2) to represent the limitation of O_2 on the microbial processes sketched in Fig. 1. Similar to substrates, the saturation level of O_2 is computed as $O_2/(O_2+K_o)$, where O_2 denotes the concentration of oxygen. The simulated oxygen concentrations by Stolpovsky et al. [4] were used as an input to our model. We used the SCEUA algorithm to determine the six model parameters in addition to the initial value for active fraction (r_0).

A summary of the seven parameters (one of them is r_0) and their fitted values is presented in Table 2. The initial active fraction (r_0) has a median of 0.925 with the 95% confidence interval (CI) of [0.628–1.000]. It means that a high r_0 is required for this experiment, but not necessary to be 1.0 set by Stolpovsky et al. [4]. The model and data are not sensitive to β since its 95% CI covers a wide range from 0.001 to 1. The reason is that the experiment only lasts for a very short time (33 h) so the influence of low metabolic rate at dormant state is insignificant.

Fig. 4 shows that the simulated total biomass (B) and substrate (S) concentrations agree very well with the observations (the coefficients of determination are 0.98 and 0.78 for biomass in Fig. 4a and substrate in Fig. 4b, respectively). Our simulation results indicate that, under limited O_2 between 12h and 24 h of the experiment, the active biomass decreases and the dormant biomass increases. As a result, the active fraction (r) declines from ca. 0.9 to 0.7 (Fig. 4a). For the same period Stolpovsky et al. [4] predicted a decrease of r from 1.0 to ca. 0, which means that all active biomass becomes dormant. Although there were not adequate measurements to confirm either prediction, our predicted changes in the active fraction (r) appear to be more reasonable during such a short experimental time period. This demonstration also shows that our model is capable of producing reasonable change in total, active, and dormant microbial biomass in response to substrate supply as well as an important forcing function (O_2).

Conclusions

We show that the physiological state index model (Equation 1) of Panikov [33] can be improved by eliminating the assumption that the steady state active fraction (r^{ss}) approaches the substrate saturation level (ϕ^{ss}). In particular, the model of Panikov [33] indicates that no active cells become dormant under insufficient substrate, which disregards the general nature of the strategy of dormancy in microorganisms when faced with unfavorable environmental conditions [5]. Our analysis also implies that the estimate of respiration rates under sufficient substrate by the

Table 2. MEND model parameter values used for simulation of the experiment described in Fig. 3 of Stolpovsky et al. (2011).

Parameter	Fitted Value*	Initial Range	Description
m_R	0.032, [0.011–0.048]	0.001–0.1	Specific maintenance rate for active biomass (h^{-1})
α	0.099, [0.045–0.181]	0.001–0.50	$m_R/(\mu_G+m_R)$, μ_G is specific growth rate (h^{-1})
K_s	3.110, [1.387–5.652]	0.1–9.0	Half-saturation constant for substrate (mg L^{-1})
Y_G	0.573, [0.463–0.600]	0.2–0.6	Growth yield factor (–)
K_o	0.0008, [0.0007–0.001]	0.005–0.1	Half-saturation constant for dissolved oxygen (mM)
β	0.351, [0.001–1.000]	0.001–1	Ratio of dormant maintenance rate to m_R
r_0	0.925, [0.628–1.000]	0–1	Initial fraction of active biomass (–)

*Medians and 95% confidence intervals of the fitted values from 100 optimization runs, i.e., 100 different random seeds are used for the stochastic optimization algorithm.

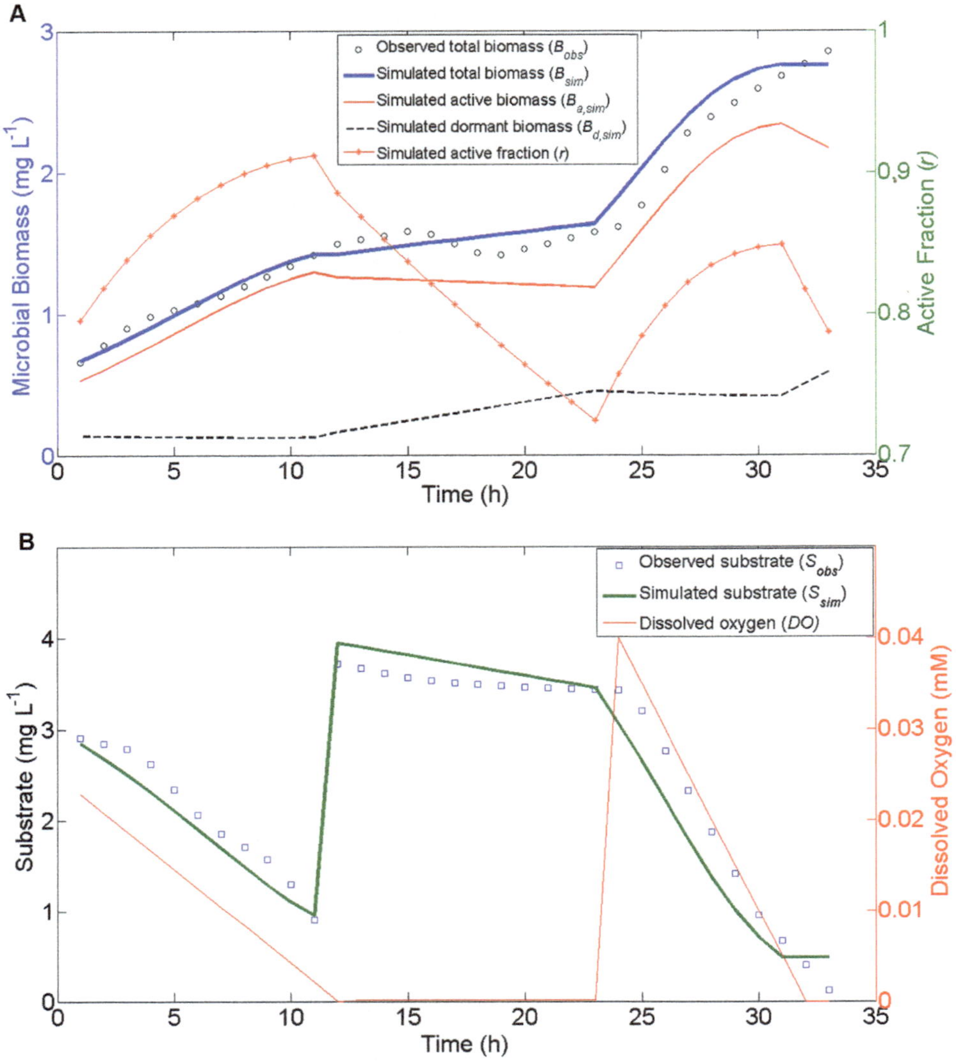

Figure 4. MEND model simulations against the experimental dataset used by Stolpovsky et al. (2011). (a) total live biomass, active and dormant biomass, and active fraction; (b) observed and simulated substrate concentration and prescribed O_2 concentration. There are three manipulations on the substrate and oxygen: (1) at time 0, the substrate (3 mg/L) and O_2 (0.025 mM) are added to the system; (2) after 12 h, the same amount of substrate is injected; (3) at 24 h, additional O_2 (0.04 mM) is injected to the system. The observed concentrations of substrate and total biomass are hourly data interpolated from the original observations in Stolpovsky et al. (2011). We scaled the substrate concentrations (with units of mM in original data) to match the magnitude of biomass concentration in units of mg/L.

physiological state index model is deficient. Pertaining to the switch function model, we argue that the switch function (θ) is also determined by the substrate (or combined with other impact factors) saturation level thus we would recommend using the switch function to modify the microbial uptake rate if the Gibbs energy change of the oxidation of substrate (ΔG) is tractable and the thermodynamic threshold (G_0) and the steepness of the step function (st) are identifiable. Based on the generally accepted assumptions summarized from existing dormancy models, we postulate a synthetic microbial physiology component to account for dormancy. Both the steady state active fraction (r^{ss}) and substrate saturation level (ϕ^{ss}) can be expressed as functions of two physiological indices: α and β. The index $\alpha = m_R/(\mu_G+m_R)$ is composed of μ_G and m_R denoting the maximum specific growth and maintenance rates, respectively, for active microbes. The index β represents the ratio of dormant to active maintenance rates. The value of r^{ss} is no less than ϕ^{ss}, and is equal only under the condition of $\beta \rightarrow 0$, where they are both identical to α. The upper bound of r^{ss} is ca. 0.8 at $\alpha \rightarrow 0.5$ and $\beta \rightarrow 1$. The maximum r^{ss} is ca. 0.5 if β (≤ 0.01) following the estimation of Anderson & Domsch [6]. It is evident that r^{ss} could be attenuated further by other limiting factors. The application of the MEND microbial physiology model to an experimental dataset with intermittent substrate supply shows satisfactory model performance (the determination coefficients are 0.98 and 0.78 for microbial biomass and substrate, respectively). The case study on the SIGR dataset indicate that the exponentially-increasing respiration rates can only be used to determine μ_G and B_{a0} (initial active biomass), while the major parameters (B_0, r_0, μ_G, α, K_s) can be effectively determined using both exponentially-increasing and non-exponentially-increasing respiration rates.

In conclusion, the microbial physiology model presented here can be incorporated into existing ecosystem models to account for dormancy in microbially-mediated processes. We have illustrated the impacts of substrate and oxygen availabilities on the physiological states through this study. Other environmental factors, such as soil temperature and soil water potential, could also be introduced into this framework to affect the transformation processes between the two microbial compositions. The changes in the physiological states of microbes could further alter the microbially-driven carbon and nutrient dynamics in ecosystems. Traditional measures of microbial biomass include the entire microbial population, even though dormancy is an important evolutionary strategy for preservation of microbial genetics and function until conditions for growth and replication improve. Parameterizing microbial ecosystem models assuming the entire population is active could therefore lead to significant errors. The approach described here provides a tractable and testable method to include dormancy as a response to external forcing.

Acknowledgments

The authors thank Dr. Sindhu Jagadamma for her helpful comments. Thanks also go to the two anonymous reviewers for their constructive comments.

Author Contributions

Conceived and designed the experiments: GW MAM. Analyzed the data: GW MAM CWS. Wrote the paper: GW MAM LG CWS. Mathematical analysis: GW LG.

References

1. Blagodatsky SA, Heinemeyer O, Richter J (2000) Estimating the active and total soil microbial biomass by kinetic respiration analysis. Biology and Fertility of Soils 32: 73–81.
2. Mason C, Hamer G, Bryers J (1986) The death and lysis of microorganisms in environmental processes. FEMS microbiology letters 39: 373–401.
3. Price PB, Sowers T (2004) Temperature dependence of metabolic rates for microbial growth, maintenance, and survival. Proceedings of the National Academy of Sciences of the United States of America 101: 4631–4636.
4. Stolpovsky K, Martinez-Lavanchy P, Heipieper HJ, Van Cappellen P, Thullner M (2011) Incorporating dormancy in dynamic microbial community models. Ecological Modelling 222: 3092–3102.
5. Lennon JT, Jones SE (2011) Microbial seed banks: the ecological and evolutionary implications of dormancy. Nature Reviews Microbiology 9: 119–130.
6. Anderson TH, Domsch KH (1985) Determination of ecophysiological maintenance carbon requirements of soil microorganisms in a dormant state. Biology and Fertility of Soils 1: 81–89.
7. Anderson TH, Domsch KH (1985) Maintenance carbon requirements of actively-metabolizing microbial populations under *in situ* conditions. Soil Biology and Biochemistry 17: 197–203.
8. Adam G, Duncan H (2001) Development of a sensitive and rapid method for the measurement of total microbial activity using fluorescein diacetate (FDA) in a range of soils. Soil Biology and Biochemistry 33: 943–951.
9. Jones KH, Senft JA (1985) An improved method to determine cell viability by simultaneous staining with fluorescein diacetate-propidium iodide. Journal of Histochemistry & Cytochemistry 33: 77–79.
10. Boulos L, Prévost M, Barbeau B, Coallier J, Desjardins R (1999) LIVE/DEAD® BacLight™: application of a new rapid staining method for direct enumeration of viable and total bacteria in drinking water. Journal of Microbiological Methods 37: 77–86.
11. Stocks S (2004) Mechanism and use of the commercially available viability stain, BacLight. Cytometry Part A 61: 189–195.
12. Schulz S, Peréz-de-Mora A, Engel M, Munch JC, Schloter M (2010) A comparative study of most probable number (MPN)-PCR vs. real-time-PCR for the measurement of abundance and assessment of diversity of alkB homologous genes in soil. Journal of Microbiological Methods 80: 295–298.
13. Colores GM, Schmidt SK, Fisk MC (1996) Estimating the biomass of microbial functional groups using rates of growth-related soil respiration. Soil Biology and Biochemistry 28: 1569–1577.
14. Anderson JPE, Domsch KH (1978) A physiological method for quantitative measurement of microbial biomass in soils. Soil Biology & Biochemistry 10: 215–221.
15. Jenkinson DS, Powlson DS (1976) The effects of biocidal treatments on metabolism in soil-V: A method for measuring soil biomass. Soil Biology and Biochemistry 8: 209–213.
16. Lodge DJ (1993) Nutrient cycling by fungi in wet tropical forests. In: Isaac S, editor. Aspects of Tropical Mycology: Symposium of the British Mycological Society Cambridge, UK: Cambridge University Press. pp. 37–57.
17. Yarwood S, Brewer E, Yarwood R, Lajtha K, Myrold D (2013) Soil microbe active community composition and capability of responding to litter addition after 12 years of no inputs. Applied and Environmental Microbiology 79: 1385–1392.
18. Alvarez C, Alvarez R, Grigera M, Lavado R (1998) Associations between organic matter fractions and the active soil microbial biomass. Soil Biology and Biochemistry 30: 767–773.
19. Khomutova T, Demkina T, Demkin V (2004) Estimation of the total and active microbial biomasses in buried subkurgan paleosoils of different age. Microbiology 73: 196–201.
20. Orwin KH, Wardle DA, Greenfield LG (2006) Context-dependent changes in the resistance and resilience of soil microbes to an experimental disturbance for three primary plant chronosequences. Oikos 112: 196–208.
21. Van de Werf H, Verstraete W (1987) Estimation of active soil microbial biomass by mathematical analysis of respiration curves: calibration of the test procedure. Soil Biology and Biochemistry 19: 261–265.
22. Stenström J, Svensson K, Johansson M (2001) Reversible transition between active and dormant microbial states in soil. FEMS Microbiology Ecology 36: 93–104.
23. Wang G, Post WM, Mayes MA (2013) Development of microbial-enzyme-mediated decomposition model parameters through steady-state and dynamic analyses. Ecological Applications 23: 255–272.
24. Moorhead DL, Lashermes G, Sinsabaugh RL (2012) A theoretical model of C- and N-acquiring exoenzyme activities, which balances microbial demands during decomposition. Soil Biology and Biochemistry 53: 133–141.
25. Sinsabaugh RL, Manzoni S, Moorhead DL, Richter A (2013) Carbon use efficiency of microbial communities: stoichiometry, methodology and modelling. Ecology Letters 16: 930–939.
26. Schimel JP, Weintraub MN (2003) The implications of exoenzyme activity on microbial carbon and nitrogen limitation in soil: a theoretical model. Soil Biology & Biochemistry 35: 549–563.
27. Lawrence CR, Neff JC, Schimel JP (2009) Does adding microbial mechanisms of decomposition improve soil organic matter models? A comparison of four models using data from a pulsed rewetting experiment. Soil Biology & Biochemistry 41: 1923–1934.
28. Wang G, Post WM, Mayes MA, Frerichs JT, Jagadamma S (2012) Parameter estimation for models of ligninolytic and cellulolytic enzyme kinetics. Soil Biology and Biochemistry 48: 28–38.
29. Todd-Brown KE, Hopkins FM, Kivlin SN, Talbot JM, Allison SD (2012) A framework for representing microbial decomposition in coupled climate models. Biogeochemistry 109: 19–33.
30. Treseder KK, Balser TC, Bradford MA, Brodie EL, Dubinsky EA, et al. (2012) Integrating microbial ecology into ecosystem models: challenges and priorities. Biogeochemistry 109: 7–18.
31. Wieder WR, Bonan GB, Allison SD (2013) Global soil carbon projections are improved by modelling microbial processes. Nature Clim Change 3: 909–912.
32. Konopka A (1999) Theoretical analysis of the starvation response under substrate pulses. Microbial Ecology 38: 321–329.
33. Panikov NS (1996) Mechanistic mathematical models of microbial growth in bioreactors and in natural soils: explanation of complex phenomena. Mathematics and Computers in Simulation 42: 179–186.
34. Blagodatsky S, Richter O (1998) Microbial growth in soil and nitrogen turnover: a theoretical model considering the activity state of microorganisms. Soil Biology and Biochemistry 30: 1743–1755.
35. Panikov NS, Sizova MV (1996) A kinetic method for estimating the biomass of microbial functional groups in soil. Journal of Microbiological Methods 24: 219–230.
36. Buerger S, Spoering A, Gavrish E, Leslin C, Ling L, et al. (2012) Microbial scout hypothesis, stochastic exit from dormancy, and the nature of slow growers. Applied and Environmental Microbiology 78: 3221–3228.
37. Hunt HW (1977) A simulation model for decomposition in grasslands. Ecology 58: 469–484.
38. Gignoux J, House J, Hall D, Masse D, Nacro HB, et al. (2001) Design and test of a generic cohort model of soil organic matter decomposition: the SOMKO model. Global Ecology and Biogeography 10: 639–660.
39. Ayati BP (2012) Microbial dormancy in batch cultures as a function of substrate-dependent mortality. Journal of Theoretical Biology 293: 34–40.

40. Jones SE, Lennon JT (2010) Dormancy contributes to the maintenance of microbial diversity. Proceedings of the National Academy of Sciences 107: 5881–5886.
41. Bär M, Hardenberg J, Meron E, Provenzale A (2002) Modelling the survival of bacteria in drylands: the advantage of being dormant. Proceedings of the Royal Society of London Series B: Biological Sciences 269: 937–942.
42. Locey KJ (2010) Synthesizing traditional biogeography with microbial ecology: the importance of dormancy. Journal of Biogeography 37: 1835–1841.
43. Wang G, Post WM (2013) A note on the reverse Michaelis–Menten kinetics. Soil Biology and Biochemistry 57: 946–949.
44. Wirtz KW (2003) Control of biogeochemical cycling by mobility and metabolic strategies of microbes in the sediments: an integrated model study. FEMS Microbiology Ecology 46: 295–306.
45. Panikov NS (1995) Microbial growth kinetics. London, UK: Chapman & Hall. 378 p.
46. Pirt SJ (1965) Maintenance energy of bateria in growing cultures. Proceedings of the Royal Society of London Series B-Biological Sciences 163: 224–231.
47. Wang G, Post WM (2012) A theoretical reassessment of microbial maintenance and implications for microbial ecology modeling. FEMS Microbiology Ecology 81: 610–617.
48. Hoehler TM, Jorgensen BB (2013) Microbial life under extreme energy limitation. Nat Rev Micro 11: 83–94.
49. Beeftink HH, Van der Heijden RTJM, Heijnen JJ (1990) Maintenance requirements: energy supply from simultaneous endogenous respiration and substrate consumption. FEMS microbiology letters 73: 203–209.
50. Blagodatsky S, Yevdokimov I, Larionova A, Richter J (1998) Microbial growth in soil and nitrogen turnover: model calibration with laboratory data. Soil Biology and Biochemistry 30: 1757–1764.
51. Duan QY, Sorooshian S, Gupta V (1992) Effective and efficient global optimization for conceptual rainfall-runoff models. Water Resources Research 28: 1015–1031.
52. Wang G, Xia J, Chen J (2009) Quantification of effects of climate variations and human activities on runoff by a monthly water balance model: A case study of the Chaobai River basin in northern China. Water Resources Research 45: W00A11.

Indirect Effects of Conservation Policies on the Coupled Human-Natural Ecosystem of the Upper Gulf of California

Hem Nalini Morzaria-Luna[1]*, Cameron H. Ainsworth[1¤], Isaac C. Kaplan[2], Phillip S. Levin[2], Elizabeth A. Fulton[3]

1 Marine Resources Assessment Group Americas Incorporated, Seattle, Washington, United States of America, **2** Conservation Biology Division, Northwest Fisheries Science Center, National Marine Fisheries Service, National Oceanic and Atmospheric Administration, Seattle, Washington, United States of America, **3** The Commonwealth Scientific and Industrial Research Organisation Wealth from Oceans Flagship, Division of Marine and Atmospheric Research, Hobart, Tasmania, Australia

Abstract

High bycatch of non-target species and species of conservation concern often drives the implementation of fisheries policies. However, species- or fishery-specific policies may lead to indirect consequences, positive or negative, for other species or fisheries. We use an Atlantis ecosystem model of the Northern Gulf of California to evaluate the effects of fisheries policies directed at reducing bycatch of vaquita (*Phocoena sinus*) on other species of conservation concern, priority target species, and metrics of ecosystem function and structure. Vaquita, a Critically Endangered porpoise endemic to the Upper Gulf of California, are frequently entangled by finfish gillnets and shrimp driftnets. We tested five fishery management scenarios, projected over 30 years (2008 to 2038), directed at vaquita conservation. The scenarios consider progressively larger spatial restrictions for finfish gillnets and shrimp driftnets. The most restrictive scenario resulted in the highest biomass of species of conservation concern; the scenario without any conservation measures in place resulted in the lowest. Vaquita experienced the largest population increase of any functional group; their biomass increased 2.7 times relative to initial (2008) levels under the most restrictive spatial closure scenario. Bycatch of sea lions, sea turtles, and totoaba decreased > 80% in shrimp driftnets and at least 20% in finfish gillnet fleets under spatial management. We found indirect effects on species and ecosystem function and structure as a result of vaquita management actions. Biomass and catch of forage fish declined, which could affect lower-trophic level fisheries, while other species such as skates, rays, and sharks increased in both biomass and catch. When comparing across performance metrics, we found that scenarios that increased ecosystem function and structure resulted in lower economic performance indicators, underscoring the need for management actions that consider ecological and economic tradeoffs as part of the integrated management of the Upper Gulf of California.

Editor: Myron Peck, University of Hamburg, Germany

Funding: This project was supported by the David and Lucile Packard Foundation. Atlantis code development was supported by the Gordon and Betty Moore Foundation. The funders had no role in study design, data collection and analysis, decision to publish, or preparation of the manuscript.

Competing Interests: The authors have read the journal's policy and have the following conflicts. Dr. Morzaria-Luna and Dr. Ainsworth were previously government contractors employed through a commercial company, Marine Resources Assessment Group (MRAG) Americas Inc. This does not alter the authors' adherence to all the PLOS ONE policies on sharing data and materials.

* E-mail: HemNalini.MorzariaLuna@noaa.gov

¤ Current address: College of Marine Science, University of South Florida, Saint Petersburg, Florida, United States of America

Introduction

Management of natural resources includes setting limits on exploitation or setting aside areas as reserves [1]. In marine systems, fisheries management aims to ameliorate the negative effects of fishing (i.e. population collapse, bycatch of non-target species, reduced habitat complexity, altered predator-prey relationships) through actions such as closed seasons and areas, limited entry, and gear restrictions [2–5]. Fisheries policies often regulate particular fleets or gears in an effort to protect species of conservation concern, such as marine mammals, sea turtles, and birds (i.e. [6–8]). However, species- or fishery- specific policies may lead to indirect consequences, both positive and negative, for other species, fisheries or whole ecosystems [9]. For example, closures could shift fishers into areas with higher bycatch of more vulnerable species or size classes [10] and length limits could actually expose target species and the ecosystem to increased negative effects [11].

Indirect consequences of bycatch reduction measures are evident in policies directed at vaquita (*Phocoena sinus*), a Critically Endangered [12] porpoise endemic to the Upper Gulf of California that is frequently entangled in finfish gillnets and shrimp driftnets. The vaquita population has declined rapidly from an estimated 567 individuals in 1997 (95% CI 177-1073; [13] to 245 in 2008 (95% CI 68-884; [14]. In 2009, the instantaneous annual bycatch mortality rate was high, 0.07 year^{-1} (7%), despite the implementation of bycatch reduction measures [14]. To protect vaquita, the Mexican government initially established the Upper Gulf of California and Colorado River Delta Biosphere Reserve [15] (See Methods for more details). However, a subsequent survey of vaquita distribution [13] found that sightings were concentrated outside of the Reserve's boundaries, so a marine refuge was established in the area where vaquita sightings were concentrated [16]. The refuge excludes finfish gillnets and shrimp driftnets, which entangle vaquita, and industrial trawling, which may disrupt vaquita behavior [17,18]. Currently, direct and indirect economic incentives (Table S1) are coupled with spatial restrictions with the goal of eliminating nets from the entire vaquita distribution area by 2012 (~8432 km^2; Figure 1), as specified in the vaquita conservation program [19]. The economic incentives are designed to limit the economic impact of area closures on local fishers [20,21]. The evolution of fisheries management aimed at reducing vaquita mortality is described in more detail in Rojas-Bracho et al. [22], Bobadilla et al. [23], and Avila-Forcada et al. [21].

The fisheries policies directed at reducing bycatch of vaquita have been designed as single-species management concerned with preventing extinction of the vaquita and reducing the socioeconomic impact of vaquita conservation on the region's fishers [19]. There is a high probability that eliminating nets from the vaquita distribution area would result in an increase in vaquita abundance after 10 yrs, as determined by a single-species population model [24]. Vaquita population trajectories under alternative fisheries policies, obtained using an ecosystem model that incorporated vaquita age structure and diets, were consistent with results of the single-species model [25]. However, calls for fishery management to address broader ecological and conservation goals in the Gulf [26–28] and globally [29–31] suggest the need to weigh the impacts of fisheries policies on the broader ecosystem.

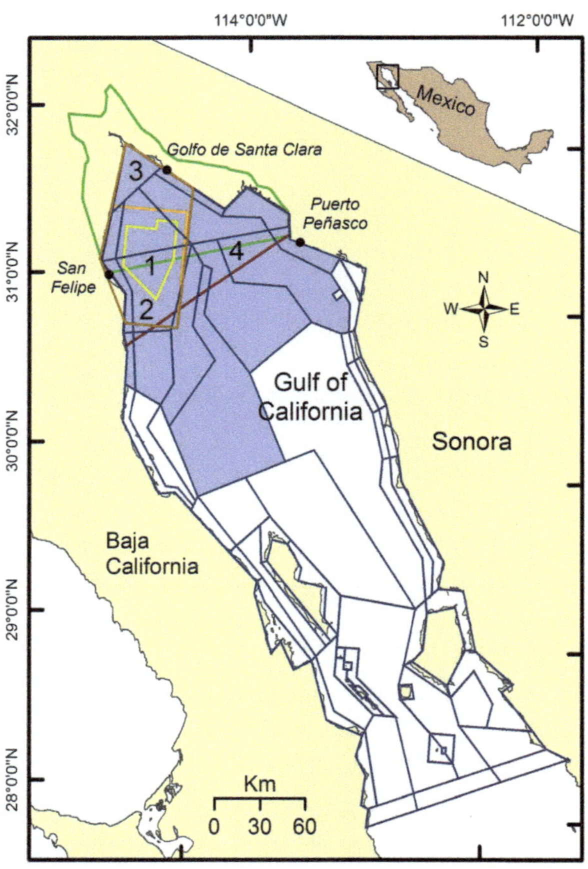

Figure 1. Northern Gulf of California Atlantis model extent, including Atlantis polygon geometry and polygons affected by management scenarios (blue shading). The main fishing communities in the Upper Gulf are indicated. Simulated spatial management restrictions in the Upper Gulf (yellow to red lines), including the Upper Gulf Biosphere Reserve (green line). Numbers correspond to scenarios: 1) Vaquita refuge (1264 km^2), 2) Extended refuge (3579 km^2), 3) Primary area (5339 km^2) and 4) Distribution area (8432 km^2).

Fisheries policies directed at reducing vaquita bycatch could have significant ecosystem-level effects for the Upper Gulf of California, because of their spatial extent and their focus on fisheries with high ecological impacts [32,33]. The Upper Gulf is characterized by high productivity driven by seasonal upwelling and constant tidal mixing [34]. This productivity drives highly profitable artisanal gillnet fisheries, mainly for curvina (*Cynoscion othonopterus*), sharks, skates and rays; driftnet fisheries for blue shrimp (*Litopenaeus stylirostris*); and industrial benthic trawl fisheries for blue and brown shrimp (*Farfantepenaeus aztecus* [32,35]). Both artisanal gillnets and industrial trawl fisheries have high bycatch rates [36,37], while trawling has significant physical effects on the seafloor and reduces the diversity of benthic and demersal communities [38].

'End-to-end' ecosystem models have proven useful for exploring the implications of fisheries policies on management objectives [39]. For example, Kaplan et al. [40] used an end-to-end model, built in the Atlantis software, to explore the consequences of various gear switching and spatial management scenarios in the California Current. Atlantis models include a coupled dynamic representation of biophysical, ecological, economic and social components of the system, enabling users to explore alternate fisheries management strategies in the context of diverse, interactive ecosystem processes [41].

We used an Atlantis model of the Northern Gulf of California [42,43] to examine the effects of existing and proposed fisheries policies directed at reducing vaquita bycatch. We compared a reference scenario that did not include any actions for vaquita protection with four scenarios that simulate the current vaquita refuge [17–19,44], planned expansions of the refuge within the species recovery plan [19], and the most recent expert-recommended spatial closure [45]. The scenarios tested combine spatial closures for industrial shrimp trawls, finfish gillnets, and shrimp driftnets while allowing operation of a new artisanal light shrimp trawl that eliminates vaquita bycatch [46]. We examined how the simulated scenarios affected biomass, catch, and diet composition of species of conservation concern and target species. We also analyzed the effects of fisheries policies on performance metrics of ecosystem function and structure; these metrics are intended to reflect changes in ecosystem attributes and can be linked to specific management objectives [47]. The ecosystem metrics included were biodiversity, trophic level of the system, trophic level of catch, system organization, and habitat integrity.

Our overall aim was to illustrate how the indirect consequences of fisheries policies can support conservation objectives, reveal potential tradeoffs, and strengthen long-term management plans. We found indirect effects on species and ecosystem function and structure as a result of vaquita management actions. In general, scenarios that increased ecosystem function and structure resulted in lower economic performance indicators, pointing to the need to consider ecological and economic tradeoffs as part of integrated management.

Methods

Atlantis Ecosystem Model

The technical specifications of the Atlantis code base and a review of existing applications are detailed in Fulton et al. [41,48]. Atlantis is a spatially explicit modeling framework that incorporates multiple submodels that represent oceanography (flux of water, heat and salt), biogeochemistry (primarily N cycling), food web interactions and fisheries. Atlantis is a deterministic model, tracking flows of limiting nutrients through the main biological groups using a system of differential equations solved on a 12-hour time step. Atlantis appears to capture the dynamics observed in real ecosystems and produces spatial zonation and long-term cycles characteristic of natural systems [49]. Aspects of structural uncertainty and parameterization of the Atlantis model are considered by Fulton [50] and Fulton et al. [48,51], while Link et al. [52] summarize both challenges and future directions for handling uncertainty in ecosystem models. However model limitations are not obstacles to using Atlantis as a strategic model for illustrating broad-scale tradeoffs [49].

Atlantis can be used as a policy exploration tool to predict management policy efficacy for target populations and to understand associated effects on ecosystem components. Importantly, ecosystem models such as Atlantis are meant for strategic evaluations (i.e. ranking policy options) and not for tactical management decisions (i.e. setting management quotas) [41,53]. Currently, there are 13 Atlantis models being used to support ecosystem-based management, with several others under development [41].

Biotic ecosystem components are typically represented in functional groups: groups of species aggregated according to life history, feeding, or niche similarities. The main model dynamics and processes in Atlantis include two-way trophodynamic coupling, meaning that predators influence prey abundance and vice versa; dynamic weights-at-age; multiple options for describing predator-prey relationships; density dependence arising from both stock-recruit relationships and explicitly modeled resource limitation; and directed movements (i.e. seasonal migrations and foraging) [40]. In existing Atlantis models, target species are represented with sufficient detail to evaluate direct effects of fishing, while other species are aggregated into functional groups with enough resolution to capture human, trophic, and climate impacts on the ecosystem [40,41,48,54]. The model includes a three-dimensional representation of the spatial extent; irregular boxes or polygons represent important bioregional features. Exchange of biomass occurs between polygons based on seasonal migration and foraging behavior, while fluxes of water, heat and salinity across polygon boundaries can be represented by a coupled hydrodynamic model. Subroutines represent biological processes between functional groups, including consumption, production, waste production, recruitment, habitat dependency and mortality, including predation, senescence, and fishery removals; the equations for these processes are described in [48].

Northern Gulf of California Atlantis model

The Northern Gulf of California model has been applied to test the future ecosystem-level impacts of current fisheries policy and a range of potential policies [25,43,55]. Initial model conditions are described in Ainsworth et al. [42]; they represent the ecosystem structure and function for 2008 and provide a detailed representation of the Northern Gulf's oceanography, historical fishing patterns, migration and movement of key species, and variability in diet compositions. The model has been calibrated to fit historical catch series per functional group and tuned through the analysis of catch and biomass equilibria under a range of fishing pressures. The calibration process is iterative since the slow-run time in Atlantis prevents automated estimation of model parameters. Instead, state and rate parameters (i.e. recruitment variables, prey availabilities, predator consumption, mortality, and growth) are adjusted in order to generate realistic system behavior and fit predictions to observations. This overall strategy has been used in all Atlantis models built to date [48,56–58]. A summary of the calibration and tuning process is provided in Text S1.

The model domain extends over 57800 km^2, from the Colorado River Delta to the northern tip of Baja California Sur (Figure 1). The model area is divided into 66 boxes or polygons. The design of the polygons considered four major factors: the locations of the marine reserves in the region; bathymetry at the 25, 80, 150, 500, and 1,000 m isobaths; the location of fishing ports; and fishery-use areas indicated by Cudney-Bueno and Turk-Boyer [35]. Each polygon includes one sediment layer and up to six water depth layers. The irregular polygons allow the model to capture the critical dynamics of the system while being computationally efficient in homogeneous space. The model is driven by biological, chemical and physical processes that are replicated within each polygon and layer. Fluxes of water, heat, and salt are forced by a Regional Oceanographic Model System (ROMS) that reflects oceanographic conditions in the region from 1985–2008 [34]. Water flux drives the advection of plankton, nutrients and waste cycling; heat affects growth, consumption and primary production rates.

The biological components of the model include 63 functional groups, including 27 fishes, one seabird, 2 sea turtles, 5 mammals, 5 plankton, 18 invertebrates, algae, seagrass, and 2 forms of bacteria (pelagic and benthic), as well as 3 detritus groups: carrion (dead matter, large particles), refractory (cohesive, small particles), and labile (easily disassociated, small particles). The spatial distribution and abundance of each functional group are defined per model polygon and depth layer. The vertebrate groups are age-structured, but invertebrates are modeled as biomass pools. Atlantis tracks abundance and biomass for each pool and age class as mg $N \cdot m^{-2}$. Calculations of predation rates use a Holling Type II functional response; this allows diet composition to vary through time, considering density-dependent effects related to varying abundance of prey items. Feeding rates also vary dynamically according to gape limitation and the state of any prey refuges (for habitat dependent groups). The predation rate is also affected by the spatiotemporal segregation of predator and prey; maximum feeding rates will only occur when the prey and predator coincide in the same polygons and depth layers. Thus, rates of feeding respond to seasonal and diel movement patterns.

Scenarios

We simulated the impact of five management scenarios. The scenarios began with the same parameterization of ecology and oceanography, such that the differences between scenarios result from the dynamics of fishing. Fishing is simulated on a per-fleet basis; 32 fleets (Table S2) represent our best understanding of the current fishing patterns in the Northern Gulf of California. The fleets are defined based on gear used, targets, bycatch, base ports and fishery utilization areas (see [43]). Each of these fleets has specific fishing areas; we specify the proportion of each model polygon that is open or closed to individual fleets. Fishing mortality is imposed by the fishing fleets onto all relevant functional groups. The scenarios ran over 30 years, from 2008 to 2038. This time period allows the model to reach stable long-term biomass dynamics (quasi-equilibrium), capturing the effect of management scenarios on functional groups with varying life spans.

Initially, we simulated a reference scenario that did not include any management actions for vaquita protection ('No vaquita management' scenario). We began simulations for the No vaquita management scenario at 2008 biomass levels. We based initial catches on the average of 2000–2007 catches (Table S3), assembled from official fishery statistics, port-level surveys, and fisher log books [42]. We used data on catch and bycatch composition to assign a proportion of the catch of each functional group to each of the 32 fleets in the model (Table S4; [43]). Vaquita abundance was set at 245, the most recent estimate [14] and vaquita mortality rate was 0.15 $year^{-1}$ (15%), the median estimate for 2007 [24]. This scenario utilizes an estimate of the current degree of compliance with existing fisheries restrictions made by Ainsworth et al. [43] and incorporates a 30% reduction in trawl effort within the Upper Gulf Biosphere Reserve (Figure 1) implemented in 2008 [17], part of a voluntary program to reduce shrimp trawl effort nationwide [59].

The No vaquita management scenario was then compared to four scenarios that simulate management actions directed at eliminating vaquita bycatch [19,60]. These scenarios (Figure 1) each include a 1264-km^2 spatial closure to industrial shrimp trawls within the current vaquita refuge [17,18]. The scenarios then simulate progressively larger spatial closures for shrimp driftnets and finfish gillnets in the area where vaquita sightings are concentrated [13]; they also allow the shrimp driftnet fleet to switch to a light trawl with no vaquita bycatch [61], instead of being excluded. The four scenarios are as follows, in order of increasing restrictions on fisheries:

1) Vaquita refuge scenario: includes a 1264-km^2 spatial closure [17,19], representing the 2010 status quo.
2) Extended refuge scenario: includes a 3579-km^2 spatial closure representing an option from the species recovery plan [19].
3) Primary area scenario: includes a 5339-km^2 closure that excludes nets from the main vaquita distribution area [14]. This corresponds to a recent recommendation of the International Committee for the Recovery of the Vaquita, an ad-hoc scientific committee charged with making management recommendations to the Mexican government [45].
4) Distribution area: closes off the entire known vaquita range (8432 km^2). This is equivalent to the 2012 target in the species recovery plan [19].

The closures were simulated as partial or complete spatial closures to the shrimp driftnet and finfish gillnet fleets in the model polygons affected, with fishing mortality reduced proportionally to area closed. The conservation program being implemented to eliminate vaquita bycatch is designed to minimize redistribution of fishing effort. Fishers either receive a payment for conservation (rent-out), replace their gears for vaquita-safe gears (switchout), or are paid to leave the fishery entirely (buyout) (Table S1); thus, we do not consider possible increases in illegal fishing.

In each of these four spatial closure scenarios a new light shrimp trawl fleet that eliminates vaquita bycatch [46] was allowed to operate within the area closed to shrimp driftnets and finfish gillnets. We assumed that as the area closed to these gears increases, adoption of the light trawl will increase, as many fishers want to continue fishing [32] and shrimp are profitable [62]. This is consistent with data showing that fishers that enrolled in the buyout were those close to retirement and that no fishers have opted to leave the fishery since 2010 [21]; we do not consider participation in the rent-out option. To simulate the new light shrimp trawl, we reduced shrimp catch by 13% per unit fishing effort relative to the shrimp driftnet fleet [63] and increased bycatch of species other than vaquita by 11% [63,64]. Bycatch composition of the light trawl fleet also varied relative to the shrimp driftnet fleet. Dominant groups in the shrimp driftnet include crabs and lobsters, drums and croakers, flatfish, and small demersal fish [64]. In the shrimp light trawl fleet, drums and croakers, flatfish, and small demersal fish also represent a large proportion of bycatch in addition to small reef fish and large pelagics ([63,64]; Table S5). Fishing mortality of vaquita was set to 0 for this fleet.

Analysis

We present functional group biomass and catch results only for the fifteen polygons directly affected by management scenarios, rather than the model extent (Figure 1), to better illustrate the effect of fisheries policies directed at reducing vaquita bycatch in the Upper Gulf of California. We focus on species of conservation concern and priority target species as indicated in the Upper Gulf Biosphere Reserve management plan, excluding blue crabs (*Callinectes* spp.) because of unstable behavior in this high-productivity model group (Table S6; [60]). We also examined additional target groups for unexpected responses. As performance metrics for individual groups we include biomass, catch, exploitation rate (catch/biomass), and diet composition. Unless specified, in the Results we present biomass for the end of the simulation (2038) to show long-term trends and we present catch for the end of the first year (2009), to reflect the immediate effect of management scenarios not confounded by long-term biomass trends.

We then determined performance metrics of ecosystem function and structure: biodiversity, trophic level of the system, trophic level of catch, system organization, and habitat integrity. Biodiversity was calculated using the Q-90 statistic [65], which represents the slope of the cumulative species abundance curve and reflects both species evenness and richness. We used the 51 major vertebrate and invertebrate functional groups for the calculations.

Trophic level of the system and trophic level of catch were determined as:

$$TL = \frac{\sum_i (TL_i * B_{i,2038})}{\sum_i B_{i,2038}}$$

where TL is trophic level for each functional group i (Table S7), B is biomass in 2038 for that functional group.

To characterize which scenario had the largest impact on ecosystem organization, we used a reorganization index, modified from Samhouri et al. (2009). This index is calculated as the sum, across all functional groups, of the absolute difference in the relative biomass (B_i/B_{Total}) of each functional group (i) at the end of the first year (2009) and at end the simulation (2038).

$$BI = \sum \left| \left(\frac{B_i}{B_{Total}}\right) 2009 - \left(\frac{B_i}{B_{Total}}\right) 2038 \right|$$

.

This index is highest in the scenario where groups exhibited the largest differences in biomass between 2009 and 2038.

Though bottom gears are known to damage habitat [66,67], quantifying this impact for the dynamic ecosystem model is difficult with available data from the Gulf of California. Instead, for each scenario we calculated a simple index of spatial overlap between gear impacts and habitat [40]. This habitat index estimates the amount of habitat left undisturbed by fishing; for each scenario, the index was calculated based on the relative impacts of particular gear types on substrate [68], substrate per polygon, and fishing effort per gear type and polygon. We assumed that each gear type acted independently on a polygon; therefore the proportion of intact habitat, taking into account the effects of all gears, was the product of the proportion of remaining intact habitat from each gear:

$$Pp = \prod_{g=1}^{num\ gears} \left(1 - E_{g,p} * A_{g,p} * \sum_{s=1}^{num\ substrates} (I_{g,s} * H_{s,p}) \right)$$

where Pp is the proportion of habitat in polygon p that remains intact; $A_{g,p}$ is the proportion of polygon p open to fishing by gear g, $E_{g,p}$ is the effort by that gear in that polygon, relative to initial levels; $I_{g,s}$ is the impact factor per gear and substrate [68], and $H_{s,p}$ is the proportion of the habitat that is substrate s. The habitat integrity metric is then:

$$Hm_i = \frac{\sum_{p=1}^{num\ polygons} P_{p,i} * a_p}{\sum_{p=1}^{num\ polygons} P_{p,No\ vaquita\ management} * a_p}$$

where the habitat integrity metric (Hm) is the undisturbed habitat in scenario i relative to the No vaquita management scenario, and a_p is area of each polygon (km^2). The shrimp light trawl is designed to have less impact than the industrial bottom trawlers [46], but we could not obtain quantitative data on its expected benthic impact; thus we assigned it 50% of the impact factor for bottom trawlers.

We also analyzed economic benefit for artisanal net fleets and other artisanal fleets, since they are directly affected by vaquita management actions. For each scenario, Net benefit per fleet was calculated as the average Net benefit for the last five years of the simulation. For any given year, Net benefit is the sum of the Net benefits (NB) derived from the harvest of all functional groups caught by the fleet:

$$NB = \sum_{ij} GB_t[1 - C]$$

where GB is gross benefit (i.e. value of catch) for year t and C is cost rate for fishing. We assumed a cost rate of fishing of 32% for artisanal fleets [32]. Value of catch is dollars $tonne^{-1}$ by functional group (Table S7) for 2010 or the most recent year for which data was available. Catch per fleet are tonnes over the whole area where the fleet can operate.

Results

Our goal is to evaluate the impacts of the four management strategies in terms of four types of metrics: species of conservation concern, priority target species, ecosystem function and structure, and economic benefit. Responses of vaquita and economics are detailed in Morzaria-Luna et al. [25]. Here, we present a combined discussion of all these axes of management performance, to allow evaluation of the tradeoffs inherent in these policy choices.

Species of conservation concern

Overall, populations of species of conservation concern increased as the spatial area of closures increased. Generally, the most restrictive spatial closures (Distribution area scenario) resulted in the highest biomass of species of conservation concern in year 2038 relative to the No vaquita management scenario (Figure 2). Biomass of sea lions (*Zalophus californianus californianus*; pinniped functional group) in the Distribution area scenario was 99% higher than in the No vaquita management scenario. Increases in biomass for other species of conservation concern were less than 10%. Here we focus on responses (Figure 2) in the area affected by management actions (blue polygons in Figure 1). Nonetheless, these responses were consistent with results calculated at the scale of the entire model, once spatial distribution of the species is taken into account (Text S2).

Incidental catch of groups of conservation concern decreased under highly restrictive management scenarios; these reductions were evident by the first year of the simulations. For example, 2009 bycatch of sea lions, whales and dolphins, and totoaba (*Totoaba macdonaldi*), a threatened endemic sciaenid fish, decreased over 60% in the Distribution area scenario relative to the No vaquita management scenario (Figure 2). Exploitation rate of sea lions, sea turtles, and totoaba decreased by>80% in shrimp driftnets and by at least 20% in finfish gillnet fleets in the Distribution area scenario relative to the No vaquita management scenario (Figure 3). The amount of prey consumed by each group varied; in general, species consumed more prey in scenarios where reduced fishing effort led to subsequent increases in biomass (Figure S1).

Priority target species

In the case of priority target species, species in higher trophic levels, mackerel (trophic level 3.84) and drums and croakers (TL 3.95), experienced lower catch and subsequently higher biomass in spatial management scenarios relative to the No vaquita management scenario (Figure 4). Catch of mackerel in 2009 declined 84% and catch of drum and croaker declined 62% in the Distribution area scenario relative to the No vaquita management scenario (Figure 4). Both these species groups are primary targets in the finfish gillnet fisheries, and so their catch decreased as the fleet was excluded from the Upper Gulf. Similarly, catch of shrimp increased 16% in the Distribution area scenario relative to the No vaquita management scenario, resulting in a 10% decrease in biomass (Figure 4). Catch of several non-priority target species targeted by finfish gillnets also decreased > 50% in the Distribution area scenario relative to the No vaquita management scenario, including Amarillo snapper (*Lutjanus argentiventris*) (−79%), guitarfish (−57%), large pelagics (−54%), Pacific angel shark (*Squatina californica*) (−83%), and small migratory sharks (−83%), leading to biomass increases by year 30.

The exploitation rate of mackerel and drums and croakers in shrimp driftnets decreased 87% and 85% respectively (Figure 3) in the Distribution area scenario relative to the No vaquita management scenario; at the same time the exploitation rate increased in the shrimp light trawl as the gear was allowed to operate in a larger area. However, the exploitation rate of drums and croakers by the shrimp light trawl in the Distribution area scenario was still 20% lower than in the shrimp driftnet under the No vaquita management scenario. The amount of prey consumed by each group varied; in general, species consumed more prey in the Distribution area scenario (Figure S1; Table S8).

Ecosystem structure and function metrics

Biodiversity (Kempton's Q-90) showed small increases in management scenarios relative to the No vaquita management scenario (Table 1). The largest value in 2038 (4.26) was achieved in the Distribution area scenario, and the lowest in No vaquita management (4.052). A small decrease in Q-90 represents a large change in the ecosystem because the metric represents the cumulative species abundance curve [65]. Trophic level of the system varied 0.07 units between management scenarios (Table 1).

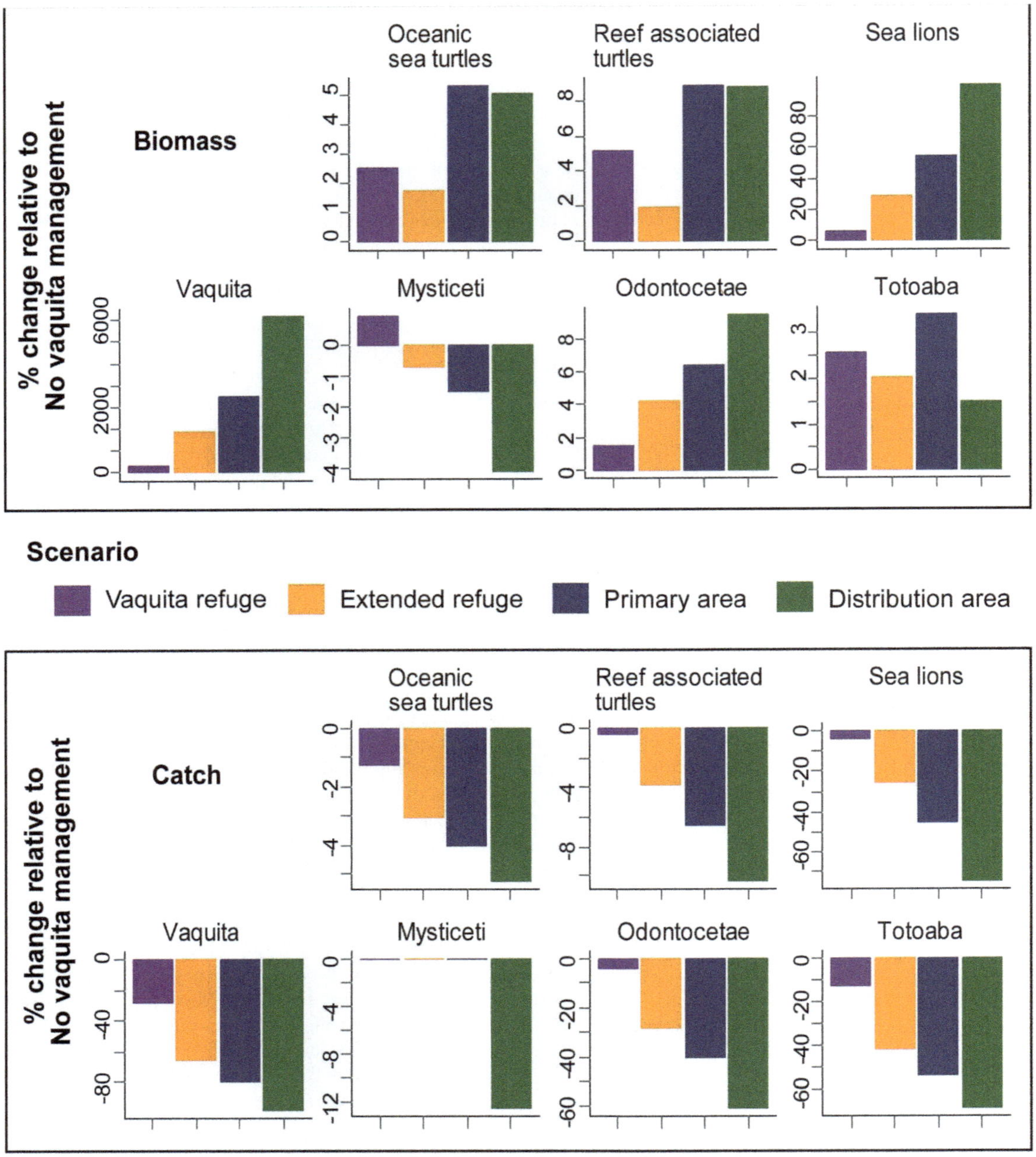

Figure 2. Biomass (top panel) and incidental catch (bottom panel) for species of conservation concern under various management scenarios. Bars show percent change relative to the No vaquita management scenario, relative to 2038 for biomass and to 2009 for catch.

The lowest value in 2038 (3.24) occurred under the No vaquita management scenario and the highest in the Vaquita refuge scenario. Trophic level of catch varied 0.05 units between management scenarios; the lowest value in 2038 (3.58) occurred under the Distribution area scenario and highest (3.63) in the No vaquita management scenario (Table 1); trophic level of catch decreased as total catch decreased. When finfish gillnets and shrimp driftnets were excluded from the Upper Gulf in the Distribution area scenario, catch of all target groups (except shrimp) decreased. From a bioenergetics perspective, the difference in trophic level represented a 3% difference in the primary production necessary to sustain a given amount of catch [69]. The reorganization index was highest in the Extended refuge scenario (1.52), where groups showed the largest response to management restrictions (Table 1). The index was lowest under the No vaquita management scenario (1.38), where biomass of individual functional groups showed the smallest changes throughout the simulation compared to other scenarios. The most restrictive spatial management scenario (Distribution area) resulted in the largest increase in habitat integrity (value of 1.5x No vaquita management scenario). The improvement in habitat integrity was less in other scenarios as the area subject to spatial restrictions was smaller.

Food web effects

We found that the fisheries policies implemented in the management scenarios led to cascading effects throughout the food web by 2038. These complex effects were examined using a

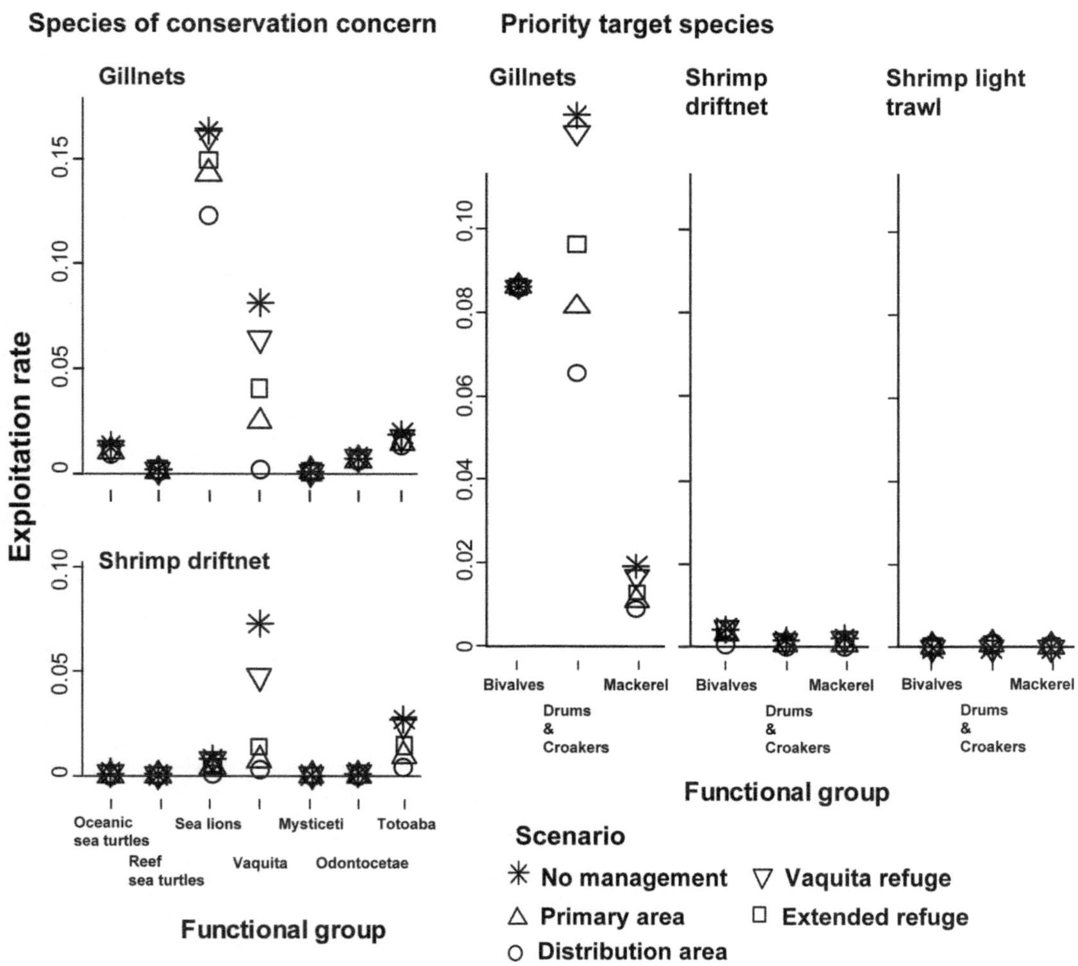

Figure 3. Exploitation rate (catch/biomass) of species of conservation concern and priority target species in the shrimp industrial trawl and shrimp driftnet fleets of the Upper Gulf and the gillnet fleet for the complete model extent. The exploitation rate across scenarios was calculated using catch relative to biomass in 2009.

combination of catch and biomass ratios relative to the No vaquita management scenario for individual scenarios (Figure 5) and ratios of prey mortality and predator consumption (Table S8). The spatially-restrictive Distribution area and Primary area scenarios resulted in large trophic effects. Most groups in trophic levels 3 and 4 increased in biomass and decreased in catch as fishing mortality decreased; almost all species are either target or bycatch of the shrimp driftnet and finfish gillnet fleets. Groups such as scorpionfish (TL 3.7), skates, rays and sharks (TL 3.3), and Gulf coney (TL 3.4) experienced a release from predation in combination with a reduction in fishing mortality, leading to increases in both catch and biomass relative to the No vaquita management scenario. In response to these biomass increases, prey groups such as bivalves (TL 2), small pelagics (TL 3.1), and small demersal fish (TL 3.8) declined in biomass and catch.

Economics

The economic cost of vaquita management actions was unequally divided between fishing fleets; the loss of value from finfish gillnet fisheries resulting from spatial restrictions was never recovered. The average annual net benefit (value of catch minus costs) of artisanal net fleets, which include finfish gillnets, shrimp driftnets, and shrimp light trawl, was US\$49±0.5 million in the Distribution area scenario by 2038; this represents a US\$23 million decrease relative to the No vaquita management scenario. Fishery gains from spatial management are modest because net benefit is driven by abundant finfish (over 65% of net benefit across scenarios) rather than harvest of sedentary species and overfished species more likely to benefit in reality from spatial closures. Overall, functional groups contributed unequally to net benefit dependent on spatial closures. For example, the catch value of Gulf grouper, large pelagics, and drums and croakers decrease relative to the No vaquita management scenario. A greater proportion of catch value under spatial management comes from herbivorous fish and sharks. Other artisanal fleets, including longline, handline, traps, and dive fisheries benefited from spatial management, and average net benefit for these fleets outperformed the No vaquita management scenario by US\$ 2–8 million by the end of the simulation. These gains in net benefit result from higher catch of groups that experience decreased fishing pressure in scenarios with spatial closures, including herbivorous fish, sharks, Amarillo snapper, and drums and croakers. A detailed analysis on the effects on vaquita management policies on economic benefit of fisheries catch is found in Morzaria-Luna et al. [25], including consideration of discounting (the current capital value of future income, reflecting uncertainty and lost opportunity costs).

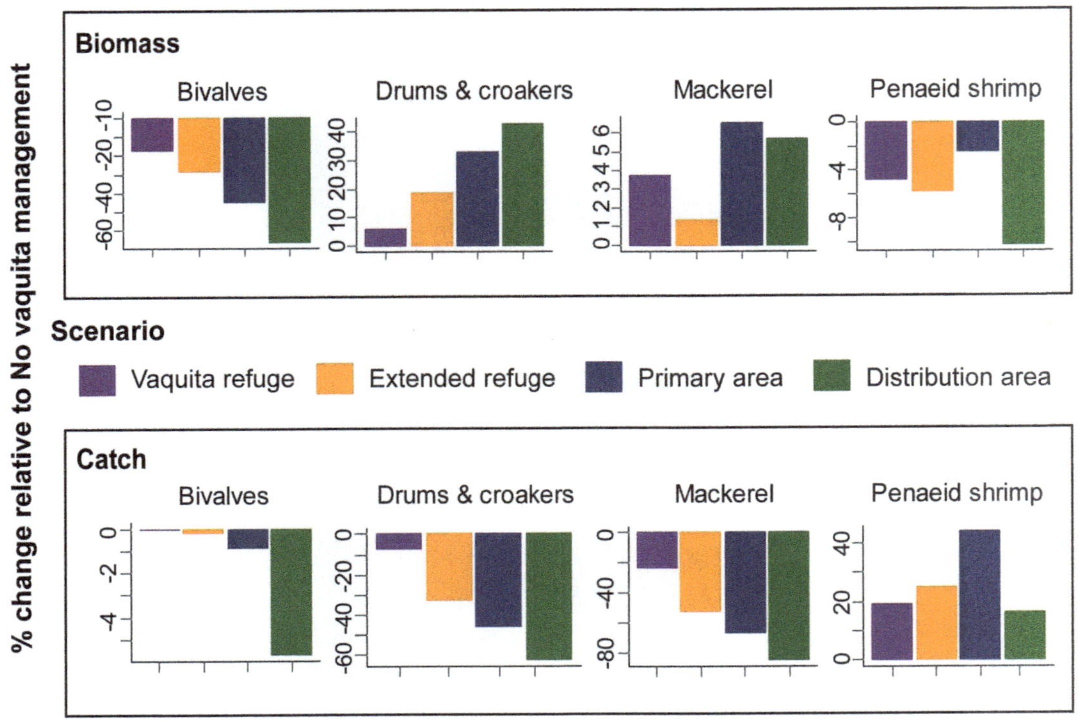

Figure 4. Biomass (top panel) and incidental catch (bottom panel) for priority target species under various management scenarios. Bars show percent change relative to the No vaquita management scenario, relative to 2038 for biomass and to 2009 for catch.

Discussion

In recent years, there has been a shift toward the implementation of an ecosystem-based approach to fisheries management that recognizes and addresses the indirect effects of fishing [70]; ecosystem-based management can help ameliorate indirect effects by taking into account a variety of ecosystem components (i.e. non-target species, trophic interactions, protected species) [71]. Nonetheless, it is still common for fisheries policies to focus on a single species as if it was an autonomous system rather than being embedded in wider ecological, socioeconomic, and institutional structures and processes [72]. In the Gulf of California, one of the most biodiverse seas in the world and Mexico's chief source of fishery resources for national and international markets [73], single-species management has led to cases of fishery collapse (and subsequent recovery) related to overfishing of stocks including totoaba [74], shrimp [75], sardine [76], and bigeye croaker (*Micropogonias megalops*) [77]; as well as declines in species of conservation concern including vaquita [22], sea lions [78], sea turtles [79], and whales [80].

We found indirect effects on species and ecosystem function and structure for the Upper Gulf of California as a result of vaquita management actions. Our results exemplify the potential for both positive and negative indirect effects of single-species management and point to complex interactions and tradeoffs. For example, vaquita management actions directly benefited other species of conservation concern, sea lions, sea turtles, whales and dolphins, and totoaba. Most importantly, sea lions showed increasing biomass and lower incidental catch as a result of the exclusion of finfish gillnets and shrimp driftnets. This is a meaningful finding because total abundance in the Gulf of this protected species [81] has declined > 20% between 1994 and 2004 [78]. The smaller increases in totoaba and sea turtle biomass are still important, as lower abundance of these groups has been attributed not only to bycatch but also to other processes and stresses unrelated to fishing that operate at distinct temporal and spatial scales. For example, reductions in the flow of the Colorado River, associated loss in

Table 1. Results for performance metrics of ecosystem function and structure for the management scenarios tested.

Scenario	Biodiversity	Trophic level of system	Trophic level of catch	Reorganization index	Habitat integrity
No vaquita management	4.046	3.24	3.635	1.384	1
Vaquita refuge	4.096	3.31	3.620	1.484	1.090
Extended refuge	4.172	3.31	3.604	1.517	1.285
Primary area	4.218	3.30	3.585	1.445	1.315
Distribution area	4.260	3.30	3.584	1.500	1.516

The habitat integrity metric is scaled relative to the No vaquita management scenario.

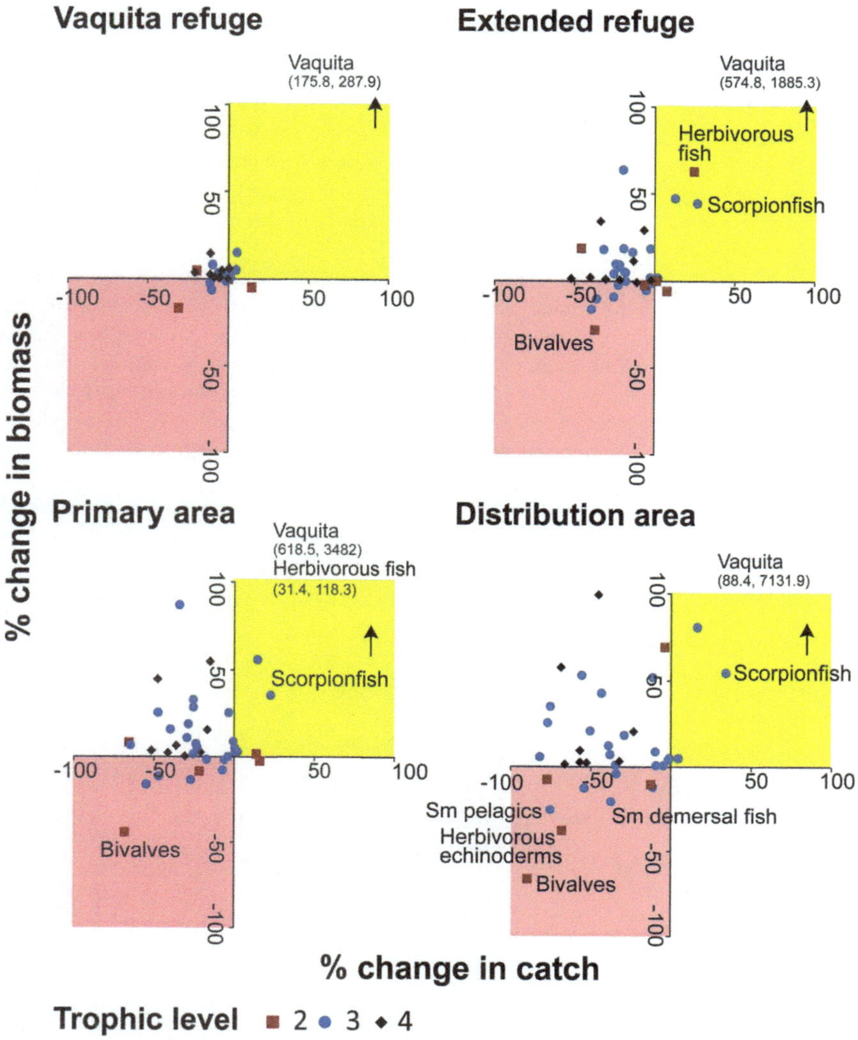

Figure 5. Percent change in catch and biomass for each management scenario relative to the No vaquita management scenario in 2038. Species are binned by trophic level as indicated by the markers. We indicate groups with>±20% change in both catch and biomass. The pink shading indicates species with aa decrease in both catch and biomass. The yellow shading indicates species withan increase in both catch and biomass. The arrows indicate the groups that are off the scale.

spawning and nursery habitats, and environmental variability interacted in the decline of the totoaba population [74,82]. In the case of sea turtles, habitat loss and egg poaching are major concerns [79].

Some fish groups targeted by commercial fisheries also benefited from vaquita management actions. Particularly, both catch and biomass of skates, rays, and sharks and Gulf coney increased due to a combination of lower predation and reduced fishing mortality in spatially restrictive scenarios. These increases could benefit the multispecies artisanal fisheries in the Northern Gulf of California, since skates, rays, and sharks are an overexploited but important component in the fisheries of the region [83] and Gulf coney is a species with high market value [84].

We also found negative indirect effects of vaquita management actions. The increase in biomass of higher-trophic level groups (TL 3 and 4), including species of conservation concern, resulted in higher predation pressure on lower trophic levels in scenarios with reduced fishing effort. We previously found that increased predation on small pelagics negatively affected the small pelagic industrial purse seine fishery; lower biomass in spatially restrictive scenarios led to a decrease in net benefit (purse seine vessel profits) relative to the No vaquita management scenario [25]. The sardine fishery, a Marine Stewardship Council-certified sustainable fishery, is an important economic driver in the Northern Gulf [85]. The fishery is characterized by extreme variability in landings due to environmental factors and/or food web feedbacks; thus, the indirect effects of vaquita management could further complicate management of the fishery [86].

The improvements in the performance metrics for individual species' and ecosystem function and structure metrics were limited, since management restrictions only exclude finfish gillnets and shrimp driftnets, while allowing other gears to operate. Most significantly, vaquita management policies only exclude industrial shrimp trawlers from the 1264-km^2 current vaquita refuge [17,18]. This fleet has high environmental impacts, including high bycatch of juveniles and threatened species [67], changes in the community structure of the benthos [87], physical changes in the sea floor and water column (caused by sediment suspension), and changes in organic and inorganic matter loading [88]. In contrast, a range of studies within existing marine reserves (those that prohibit fishing) have found reserves can maintain a diverse age-structure, and higher stock abundance and reproductive output in a variety of taxa [89].

To evaluate tradeoffs between scenarios, the performance metrics for species of conservation concern and priority target species and the metrics for ecosystem function and structure can be combined with the vaquita population response and the economic effects for fisheries in the Upper Gulf under alternative fishing policies. Previously [25], we found only the most extensive spatial management scenarios recovered the vaquita population above the threshold necessary to delist the species from Critically Endangered; in the Distribution area scenario, vaquita biomass increased 2.7 times relative to 2008 levels. When all performance metrics are evaluated simultaneously (Figure 6), we find that scenarios that increase ecosystem function and structure result in lower economic indicators. The No vaquita management scenario and Vaquita refuge scenario have high catch of priority target species, shrimp catch, and trophic level of catch but lower performance on other ecosystem function and structure metrics, biomass of species of conservation concern, and vaquita biomass. The rank order of the results was consistent; the Distribution area scenario resulted in the highest ecosystem function and structure metrics, except for trophic level and reorganization index. The Primary area scenario could be a more tenable management goal than eliminating shrimp driftnets and finfish gillnets from the complete vaquita distribution area. This option could provide ecological benefits while representing a compromise between vaquita conservation and fisheries, where vaquita bycatch is low, higher ecosystem function and structure metrics relative to the No vaquita management scenario, and there is a modest decrease in net benefit of fisheries [25].

Thus, preventing extinction of vaquita would require eliminating fishing nets from its distribution area or known range; this would be a process with high economic costs for the fishing community [21,25]. Economic incentives within the vaquita conservation plan [19] are designed to eliminate net fisheries through payments for conservation, subsidies to accelerate adoption of vaquita-safe technologies, or compensations for fishers to permanently exit the fishery [21]. These incentives are thus designed to reduce fishing effort overall, which could carry high cultural and social costs as fishers might not want or be able to shift to alternate economic activities [21,32]; limiting the effectiveness of vaquita management actions.

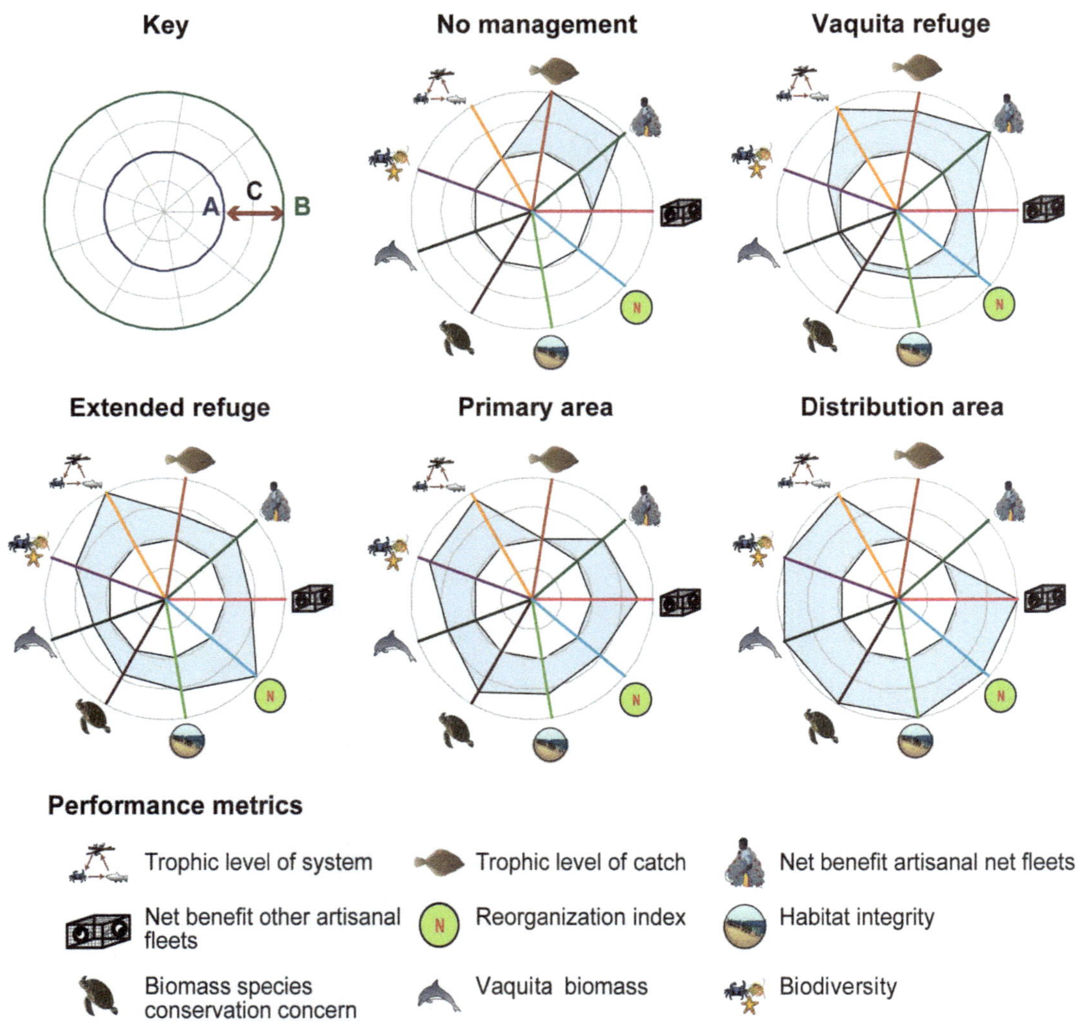

Figure 6. Performance of selected metrics across management scenarios. Since the metrics are not directly comparable (in absolute or relative change), we have scaled the performance metrics between the worst result observed (A/blue circle) and the best result observed (B/green circle); the range in between (C) shows the scope of possible outcomes. Symbols courtesy of IAN/UMCES Symbol and Image Libraries.

The cumulative impacts of vaquita management actions and other stressors, including those derived from climate and global change, could lead to surprising outcomes not considered here. Climate change could have direct effects on marine food webs, for example causing increases in species mortality, or indirect effects through predator–prey interactions [90]. The effects of climate change on commercially important species could be comparable to the ones produced by fishing [91]. For example, in the Gulf of California environmental variability is dominated by the interaction of decadal and multidecadal events derived from the El Niño–Southern Oscillation (ENSO) and Pacific Decadal Oscillation (PDO); changes in the frequency and severity of these events could lead to ecosystem reorganization [34].

End-to-end models, such as Atlantis, are best suited for strategic analyses such as the one presented here, where the questions involve the interaction of multiple species, biophysical processes, fleets, and management options [41]. Since no model can fully represent the dynamics and behavior of a natural ecosystem, there will always be factors that are not addressed [92]. In our case, these include the effects of simulated scenarios on larval dispersal and connectivity, which could have important management implications as ocean currents transport fish and invertebrate larvae from the northern to the southern Gulf during winter [93]. We used aggregated functional groups rather than species for some important commercial target species (i.e. Gulf corvina, *Cynoscion othonopterus*) which could confound the effects of particular gear restrictions. Importantly, we did not consider displacement of fishing effort nor economically-driven changes in effort, which could oversimplify fishers' response to management actions [94].

This is the first analysis of the effects of alternate vaquita policies on other species of conservation concern, target species, and ecosystem-level effects. Previous analyses of vaquita management actions have focused on vaquita population dynamics [24,95] and the socioeconomic impact on fishers and fisheries in the Upper Gulf of California [21,32,96,97]. Our findings illustrate the need for integrated management that reduces conflicts and simultaneously achieves conservation, ecological, and socioeconomic objectives. Extensive work would be needed to implement ecosystem-based management in the Upper Gulf of California, most importantly defining a desired ecosystem state that takes into account the needs and concerns of all stakeholders and aspects of the ecosystem and considers uncertainty derived from stochastic factors including climate change [98,99]. There is a clear need for an integrated perspective that regulates other activities in the Upper Gulf in addition to commercial fishing (i.e. conservation, aquaculture, sport fishing, tourism) [100–102]. Already, Mexican legislation specifies tools (i.e. Marine Protected Areas, 'ecological ordinance plans') that are science-based and can coordinate environmental conservation and fisheries management [73,103]. Given the potential ecological and economic tradeoffs resulting from conservation and management of vaquita, consideration of all possible fisheries and environmental impacts is urgently needed.

Supporting Information

Figure S1 Prey consumed by species of conservation concern and priority target species, at the end of the 30 yr simulation. Heat map of normalized values for each functional group (across rows), such that the color gradient from yellow to red represents a linear increase between the minimum and the maximum amount of prey consumed across management scenarios.

Table S1 Direct economic incentives that support phasing-out finfish gillnets and shrimp driftnets from the vaquita distribution area. Incentives are for small-scale fishers and fishing cooperatives [1] in the Upper Gulf of California, Mexico. Payments and guidelines for fiscal year 2011 [2].

Table S2 Fishery fleets for Atlantis model. Modified from Ainsworth et al. [1]. Includes number of functional groups targeted (out of 63) in the model, including target and bycatch groups. See Ainsworth et al. [1] for more information on fleets and functional groups used in the Atlantis model.

Table S3 Catch by functional group used as baseline in the No management scenario. Catches are the average of the 2000–2007 model catch series from Ainsworth et al. [1] summed for all fleets. Vaquita mortality rate was set at 0.15 year-1(15%), the median estimate prior to 2007 [2]. Since the publication of Ainsworth et al. [1], the model has been simplified to only include catch for a generic Penaeid shrimp group rather than for separate shrimp groups. Ainsworth et al. [3] provide species composition for each functional group.

Table S4 Catch per fleet used as a baseline. Asterisk indicates group is bycatch; fleet name is in bold. Vaquita catch was updated considering a mortality rate of 0.15 year-1 (15%). Bycatch composition for the Upper Gulf shrimp driftnet fleet was updated using recent monitoring data [12]. Otherwise catch per fleet values are unmodified from Ainsworth et al. [3,10].

Table S5 Catch for the shrimp light trawl fleet. Bycatch composition was based on INAPESCA & NMFS [1] and considering a reduction in shrimp catch of 10% [1] and an increase in the ratio of shrimp to bycatch (other than vaquita) of 11% [1,2]. Vaquita bycatch was set to 0 for this fleet.

Table S6 Species of conservation concern and priority target species in the Upper Gulf of California and Colorado River Delta Biosphere Reserve management plan [1]. Atlantis functional groups that contain the species are indicated. Note that the functional group Drums and croakers includes both priority target species and species of conservation concern.

Table S7 Trophic levels and value for each functional group. Trophic level are from Lozano [1] and Froese and Pauly [2]. Values are dollars tonne-1 for 2010 or the most recent year for which data was available. For Penaeid shrimp, prices were set by fleet, weighted by the amount of blue, brown and Japanese shrimp caught. Value information from National statistics for Sonora and Baja California, (Anuarios Estadísticos www.inegi.org.mx), state statistics for Sonora (www.oeidrus-sonora.gob.mx/), and port-level data for both states (unpublished data, A. Cinti, The University of Arizona, acinti@email.arizona.edu); data in Mexican pesos was converted to dollars using the exchange rate from 2005–2010 (www.x-rates.com). † Grouped as cabrilla in statistics. ‡ Grouped as sharks in statistics.

Table S8 Ratios of prey mortality (prey) and predator consumption (pred) for each functional group across management scenarios relative to the No-management scenario.

Text S1 Northern Gulf of California model tuning and diagnostics. Excerpted from Ainsworth et al. [3].

Text S2 Model-wide responses of biomass for species of conservation concern.

Acknowledgments

We acknowledge researchers from the PANGAS research initiative (http://www.pangas.arizona.edu) who participated in development of the Atlantis model for the Northern Gulf of California. We thank S. Perez-Valencia whose input helped with scenario development.

Author Contributions

Conceived and designed the experiments: HML CHA ICK PSL. Performed the experiments: HML CHA ICK. Analyzed the data: HML CHA ICK. Contributed reagents/materials/analysis tools: EAF CHA HML ICK. Wrote the paper: HML CHA ICK PSL EAF. Built the Ecosystem model: CHA ICK EAF.

References

1. Sumaila UR, Guénette S, Alder J, Chuenpagdee R (2000) Addressing ecosystem effects of fishing using marine protected areas. ICES Journal of Marine Science: Journal du Conseil 57: 752–760. doi:10.1006/jmsc.2000.0732.
2. Jackson JBC, Kirby MX, Berger WH, Bjorndal KA, Botsford LW, et al. (2001) Historical overfishing and the recent collapse of coastal ecosystems. Science 293: 629–638.
3. Kaiser MJ, Collie JS, Hall SJ, Jennings S, Poiner IR (2002) Modification of marine habitats by trawling activities: prognosis and solutions. Fish and Fisheries 3: 114–136. doi:10.1046/j.1467-2979.2002.00079.x.
4. Lewison RL, Crowder LB, Read AJ, Freeman SA (2004) Understanding impacts of fisheries bycatch on marine megafauna. Trends Ecol Evol 19: 598–604. doi:10.1016/j.tree.2004.09.004.
5. Salomon AK, Gaichas SK, Shears NT, Smith JE, Madin EMP, et al. (2010) Key features and context-dependence of fishery-induced trophic cascades. Conserv Biol 24: 382–394. doi:10.1111/j.1523-1739.2009.01436.x.
6. Bachem S (2003) Bycatch mitigation tools: selecting fisheries, setting limits, and modifying gear. Ocean Coast Manage 46: 103–125.
7. Bull LS (2007) Reducing seabird bycatch in longline, trawl and gillnet fisheries. Fish Fisheries 8: 31–56. doi:10.1111/j.1467-2979.2007.00234.x.
8. Gilman E, Kobayashi D, Swenarton T, Brothers N, Dalzell P, et al. (2007) Reducing sea turtle interactions in the Hawaii-based longline swordfish fishery. Biol Conserv 139: 19–28. doi:16/j.biocon.2007.06.002.
9. Moore JE, Wallace BP, Lewison RL, Zydelis R, Cox TM, et al. (2009) A review of marine mammal, sea turtle and seabird bycatch in USA fisheries and the role of policy in shaping management. Marine Policy 33: 435–451. doi:16/j.marpol.2008.09.003.
10. Martell SJ, Essington TE, Lessard B, Kitchell JF, Walters CJ, et al. (2005) Interactions of productivity, predation risk, and fishing effort in the efficacy of marine protected areas for the central Pacific. Canadian Journal of Fisheries and Aquatic Sciences 62: 1320–1336.
11. Coggins Jr LG, Catalano MJ, Allen MS, Pine III WE, Walters CJ (2007) Effects of cryptic mortality and the hidden costs of using length limits in fishery management. Fish and Fisheries 8: 196–210. doi:10.1111/j.1467-2679.2007.00247.x.
12. IUCN (2010) IUCN Red List of Threatened Species. Version 2010.3. Available:www.iucnredlist.org. Accessed 2010 September 2.
13. Jaramillo-Legorreta A, Rojas-Bracho L, Gerrodette T (1999) A new abundance estimate for vaquitas: first step for recovery. Mar Mammal Sci 15: 957–973.
14. Gerrodette T, Taylor BL, Swift R, Rankin S, Jaramillo-Legorreta AM, et al. (2011) A combined visual and acoustic estimate of 2008 abundance, and change in abundance since 1997, for the vaquita, *Phocoena sinus*). Mar Mammal Sci 27: E97–E100. doi:10.1111/j.1748-7692.2010.00438.x.
15. DOF (1993) Decreto por el que se declara área natural protegida con el carácter de Reserva de la Biosfera, la región conocida como Alto Golfo de California y Delta del Rio Colorado, ubicada en aguas del Golfo de California y los municipios de Mexicali, B.C., del Puerto Peñasco y San Luis Rio Colorado, Sonora. Diario Oficial de la Federación. 10 de Junio de 1993.
16. DOF (2005) ACUERDO mediante el cual se establece el área de refugio para la protección de la vaquita (*Phocoena sinus*). Secretaría del Medio Ambiente y Recursos Naturales. Diario Oficial de la Federación. 8 de Septiembre de 2005.
17. SGPA (2009) Resolutivo a la Manifestación de Impacto Modalidad regional. Pesca de camarón en el Alto Golfo de California. 26SO2009P0006. Promovente Armadores Unidos de Puerto Peñasco, S.A. de C.V. Subsecretaria de Gestión para la Protección Ambiental. Dirección General de Impacto y Riesgo Ambiental. S.G.P.A. – DGIRA - DG. 6690.09. 89 p.
18. SGPA (2010) Resolutivo a la Manifestación de Impacto Modalidad regional. Pesca sustentable del camarón azul y café y reconversión productiva, en el Alto Golfo de California. 26SO2010PD053. Promovente Federacion de Armadores del Piedron del Alto Golfo S.C. de R.L.de C.V. Subsecretaria de Gestión para la Protección Ambiental. Dirección General de Impacto y Riesgo Ambiental. S.G.P.A. – DGIRA - DG. 9150.10. 58 p.
19. SEMARNAT (2008) Programa de Acción para la Conservación de la Especie: Vaquita (*Phocoena sinus*). Estrategia integral para el manejo sustentable de los recursos marinos y costeros del Alto Golfo de California. Mexico, D.F.: Secretaría del Medio Ambiente y Recursos Naturales. 106 p. Available: http://www.conanp.gob.mx/pdf_especies/PACEvaquita.pdf.
20. Ruiz-López DM (2009) Diagnóstico socioéconomico y evaluación de una estrategia de compensación al sector pesquero de El Golfo de Santa Clara, Sonora [M.S. Thesis. Recursos Naturales y Medio Ambiente]. Guasave, Sinaloa: Centro Interdisciplinario de Investigación para el Desarrollo Integral Regional. Instituto Politécnico Nacional. Available: http://itzamna.bnct.ipn.mx:8080/dspace/handle/123456789/8211.
21. Avila-Forcada S, Martínez-Cruz AL, Muñoz-Piña C (2012) Conservation of vaquita marina in the Northern Gulf of California. Mar Policy 36: 613–622. doi:10.1016/j.marpol.2011.10.012.
22. Rojas-Bracho L, Reeves RR, Jaramillo-Legorreta A (2006) Conservation of the vaquita *Phocoena sinus*. Mammal Rev 36: 179–216.
23. Bobadilla M, Alvarez-Borrego S, Avila-Foucat S, Lara-Valencia F, Espejel I (2011) Evolution of environmental policy instruments implemented for the protection of totoaba and the vaquita porpoise in the Upper Gulf of California. Environ Sci Policy 14: 935–1220. doi:16/j.envsci.2011.06.003.
24. Gerrodette T, Rojas-Bracho L (2011) Estimating the success of protected areas for the vaquita, *Phocoena sinus*. Mar Mammal Sci 27: E101–E125. doi:10.1111/j.1748-7692.2010.00449.x.
25. Morzaria-Luna HN, Ainsworth CH, Kaplan IC, Levin PS, Fulton EA (2012) Exploring trade-offs between fisheries and conservation of the vaquita porpoise (*Phocoena sinus*) Using an Atlantis ecosystem model. PLoS ONE 7: e42917. doi:10.1371/journal.pone.0042917.
26. Underwood JG, Hernandez Camacho CJ, Aurioles-Gamboa D, Gerber LR (2008) Estimating sustainable bycatch rates for California Sea Lion populations in the Gulf of California. Conservation Biology 22: 701–710. doi:10.1111/j.1523-1739.2008.00919.x.
27. Cisneros-Mata MA (2010) The importance of fisheries in the Gulf of California and ecosystem-based sustainable co-management for conservation. In: Brusca R, editor. The Gulf of California: biodiversity and conservation. Tucson Ariz.: University of Arizona Press; Arizona-Sonora Desert Museum. pp. 119–134.
28. Erisman BE, Paredes GA, Plomozo-Lugo T, Cota-Nieto JJ, Hastings PA, et al. (2011) Spatial structure of commercial marine fisheries in Northwest Mexico. ICES J Mar Sci 68: 564–571. doi:10.1093/icesjms/fsq179.
29. Link JS (2002) Ecological considerations in fisheries management: when does it matter? Fisheries 27: 10–17. doi:10.1577/1548-8446(2002)027<0010:ECIFM>2.0.CO;2.
30. Leslie HM, McLeod KL (2007) Confronting the challenges of implementing marine ecosystem-based management. Front Ecol Environ 5: 540–548. doi:10.1890/060093.
31. Kellner JB, Sanchirico JN, Hastings A, Mumby PJ (2011) Optimizing for multiple species and multiple values: tradeoffs inherent in ecosystem-based fisheries management. Cons Letters 4: 21–30. doi:10.1111/j.1755-263X.2010.00132.x.
32. Aragón-Noriega EA, Rodríguez-Quiroz G, Cisneros-Mata MA, Ortega-Rubio A (2010) Managing a protected marine area for the conservation of critically endangered vaquita (*Phocoena sinus* Norris, 1958) in the Upper Gulf of California. Int J Sust Dev World 17: 410. doi:10.1080/13504509.2010.500823.
33. Moreno-Báez M, Orr BJ, Cudney-Bueno R, Shaw WW (2010) Using fishers' local knowledge to aid management at regional scales: spatial distribution of small-scale fisheries in the northern Gulf of California, Mexico. B Mar Sci 86: 339–353.
34. Lluch-Cota SE, Parés-Sierra A, Magaña-Rueda VO, Arreguín-Sánchez F, Bazzino G, et al. (2010) Changing climate in the Gulf of California. Prog Oceanogr 87: 114–126. doi:10.1016/j.pocean.2010.09.007.
35. Cudney-Bueno R, Turk Boyer PJ (1998) Pescando entre mareas del Alto Golfo de California: Una guía sobre pesca artesanal, su gente y sus propuestas de manejo. Puerto Peñasco, Sonora. México: CEDO Intercultural, A.C. 166 p.
36. Garcia-Caudillo JM, Cisneros-Mata MA, Balmori-Ramirez A (2000) Performance of a bycatch reduction device in the shrimp fishery of the Gulf of California, Mexico. Biological Conservation 92: 199–205.
37. Rodríguez-Valencia JA, Cisneros-Mata MA (2006) Captura Incidental de las Flotas Pesqueras Ribereñas del Pacífico Mexicano. Programa Golfo de California de WWF-México. 127 p. Available: http://www.wwf.org.mx/wwfmex/publicaciones.php?tipo = reps.
38. Nava-Romo JM (1994) Impactos, A corto y largo plazo, en la diversidad y otras características ecológicas de la comunidad béntico-demersal capturada por la pesquería de camarón en el norte del Alto Golfo de California. [M.S. . Thesis],

México: Instituto Technológico y de Estudios Superiores de Monterrey, División de Graduados e Investigación. 84 p.
39. Fulton EA (2010) Approaches to end-to-end ecosystem models. J Marine Syst 81: 171–183. doi:10.1016/j.jmarsys.2009.12.012.
40. Kaplan IC, Horne PJ, Levin PS (2012) Screening California Current fishery management scenarios using the Atlantis end-to-end ecosystem model. Prog Oceanogr 102: 5–18.
41. Fulton EA, Link JS, Kaplan IC, Savina-Rolland M, Johnson P, et al. (2011) Lessons in modelling and management of marine ecosystems: The Atlantis experience. Fish Fisheries 12: 171–188.
42. Ainsworth C, Kaplan IC, Levin PS, Cudney-Bueno R, Fulton EA, et al. (2011) Atlantis model development for the Northern Gulf of California. Seattle, WA: U.S. Department of Commerce. National Oceanic and Atmospheric Administration. National Marine Fisheries Service. NOAA Technical Memorandum NMFS-NWFSC-110. 293 p.
43. Ainsworth CH, Morzaria-Luna H, Kaplan IC, Levin PS, Fulton EA (2012) Full compliance with harvest regulations yields ecological benefits: Northern Gulf of California case study. J Appl Ecol 49: 63–72. doi:10.1111/j.1365-2664.2011.02064.x.
44. SGPA (2009) Resolutivo a la Manifestación de Impacto Modalidad regional. Pesca ribereña multiespecífica. 02BC2009P0005. Promovente. Federación de Sociedades Cooperativas de Producción pesquera Ribereña del Puerto de San Felipe, S. de R.L. de C.V. Subsecretaria de Gestión para la Protección Ambiental. Dirección General de Impacto y Riesgo Ambiental. S.G.P.A. - DGIRA - DG. 6766.09. 86 p. Available: http://sinat.semarnat.gob.mx/dgiraDocs/documentos/son/resolutivos/2009/26SO2009P0006.pdf.
45. CIRVA (2012) Report of the fourth meeting of the International Committee for the Recovery of the Vaquita (CIRVA), Ensenada, Baja California, México, 21–24 February 2012. 36 p.
46. INAPESCA (2011) Evaluación biotecnológica de la red de arrastre prototipo "RS-INP-MEX" para captura de camarón en el Alto Golfo de California. México: SAGARPA. INAPESCA. 22 p. Available:http://inapesca.gob.mx/portal/images/pdf/evaluacion_biotecnologica_de_la_red_de_arrastre_prototipo_rs-inp-mex_para_captura_de_camaron_en_el_alto_golfo_de_california_informe_tecnico_final.pdf.
47. Samhouri JF, Levin PS, Harvey CJ (2009) Quantitative evaluation of marine ecosystem indicator performance using food web models. Ecosystems 12: 1283–1298. doi:10.1007/s10021-009-9286-9.
48. Fulton EA, Parslow JS, Smith ADM, Johnson CR (2004) Biogeochemical marine ecosystem models. 2. The effect of physiological data on model performance. Ecol Model 173: 371–406.
49. Fulton EA, Smith ADM, Punt AE (2005) Which ecological indicators can robustly detect effects of fishing? ICES Journal of Marine Science: Journal du Conseil 62: 540–551. doi:10.1016/j.icesjms.2004.12.012.
50. Fulton EA (2001) The effects of model structure and complexity on the behaviour and performance of marine ecosystem models [PhD dissertation] University of Tasmania, School of Zoology. 427 p.
51. Fulton EA, Smith ADM, Johnson CR (2004) Biogeochemical marine ecosystem models I: IGBEM - a model of marine bay ecosystems. Ecological Modelling 174: 267–307.
52. Link JS, Ihde TF, Harvey CJ, Gaichas SK, Field JC, et al. (2012) Dealing with uncertainty in ecosystem models: The paradox of use for living marine resource management. Prog Oceanogr 102: 102–114. doi:10.1016/j.pocean.2012.03.008.
53. Rose KA, Allen JI, Artioli Y, Barange M, Blackford J, et al. (2010) End-to-end models for the analysis of marine ecosystems: challenges, issues, and next steps. Mar Coast Fish Dyn Mgmt Ecosys Sci 2: 115. doi:10.1577/C09-059.1.
54. Fulton EA, Smith ADM, Smith DC (2007) Alternative management strategies for Southeast Australian Commonwealth Fisheries: Stage 2: Quantitative Management Strategy Evaluation. Canberra, AU: Australian Fisheries Management Authority. Fisheries Research and Development Corporation. 398 p.
55. Ainsworth CH, Morzaria-Luna H, Kaplan IC, Levin PS, Fulton EA, et al. (2012) Effective ecosystem-based management must encourage regulatory compliance: A Gulf of California case study. Mar Policy 36: 1275–1283. doi:10.1016/j.marpol.2012.03.016.
56. Brand EJ, Kaplan IC, Harvey CJ, Levin PS, Fulton EA, et al. (2007) A spatially explicit ecosystem model of the California Current's food web and oceanography. Seattle, WA: U.S. Department of Commerce. National Oceanic and Atmospheric Administration. National Marine Fisheries Service. NOAA Technical Memorandum NOAA TMNMFSNWFSC84. 163 p.
57. Horne PJ, Kaplan IC, Marshall KN, Levin PS, Harvey CJ, et al. (2010) Design and parameterization of a spatially explicit ecosystem model of the central California Current. U.S. Department of Commerce. National Oceanic and Atmospheric Administration. National Marine Fisheries Service. NOAA Technical Memorandum NMFS-NWFSC-104. 140 p.
58. Link JS, Fulton EA, Gamble RJ (2010) The northeast US application of ATLANTIS: A full system model exploring marine ecosystem dynamics in a living marine resource management context. Prog Oceanogr 87: 214–234. doi:10.1016/j.pocean.2010.09.020.
59. CONAPESCA (2011) Proyecto estratégico de retiro voluntario de embarcaciones camaroneras. Available:http://pervec.conapesca.gob.mx/RetiroEmbarcacion/.Accessed 2011 November 1.
60. CONANP (2007) Programa de conservación y manejo de la Reserva de la Biosfera Alto Golfo de California y Delta del Río Colorado. México, D.F.: Comisión Nacional de Áreas Naturales Protegidas. 319 p.
61. INAPESCA WWF (2010) Tecnologías para reducir la captura incidental en las pesquerías de camarón del Golfo de California. Instituto Nacional de Pesca de México. World Wildlife Fund. 50 p. Available: http://www.wwf.org.mx/wwfmex/descargas/rep-tec-RS-INP-INAPESCA-WWF-1030.pdf.
62. Lluch-Cota SE, Aragón-Noriega EA, Arreguín-Sánchez F, Aurioles-Gamboa D, Baustista-Romero JJ, et al. (2007) The Gulf of California: review of ecosystem status and sustainability challenges. Prog Oceanogr 73: 1–26.
63. INAPESCA, NMFS (2012) Fishing technology research collaboration: Efficiency and selectivity of two trawl nets design to fish blue shrimp (*Litopenaeus stylirostris*) in the artisanal shrimp fishery of the Upper Gulf of California. 9 p.
64. Pérez-Valencia S, Gorostieta-Monjaraz M, Castañeda-Fernández de Lara V, Loaiza-Villanueva R, Turk-Boyer P, et al. (2011) Manifestación de Impacto Ambiental para la pesca ribereña responsable en la Reserva de la Biosfera Alto Golfo de California y Delta del Río Colorado: Costa Este. Puerto Peñasco, Sonora: Centro Intercultural de Estudios de Desiertos y Océanos, A.C. 221 p.
65. Ainsworth CH, Pitcher TJ (2006) Modifying Kempton's species diversity index for use with ecosystem simulation models. Ecol Indic 6: 623–630. doi:10.1016/j.ecolind.2005.08.024.
66. Kaiser MJ, Ramsay K, Richardson CA, Spence FE, Brand AR (2000) Chronic fishing disturbance has changed shelf sea benthic community structure. Journal of Animal Ecology 69: 494–503. doi:10.1046/j.1365-2656.2000.00412.x.
67. Meltzer L, Blinick NS, Fleishman AB (2012) Management implications of the biodiversity and socio-economic impacts of shrimp trawler by-catch in Bahía de Kino, Sonora, México. PLoS ONE 7: e35609. doi:10.1371/journal.pone.0035609#pone.0035609-Gillett1.
68. NMFS (2005) Essential fish habitat designation and minimization of adverse impacts final environmental impact statement. See Table 2, Page 137, National Marine Fisheries Service.
69. Pauly D, Christensen V (1995) Primary production required to sustain global fisheries. Nature 374: 255–257. doi:10.1038/374255a0.
70. Aswani S, Christie P, Muthiga NA, Mahon R, Primavera JH, et al. (2012) The way forward with ecosystem-based management in tropical contexts: Reconciling with existing management systems. Marine Policy 36: 1–10. doi:10.1016/j.marpol.2011.02.014.
71. Pikitch EK, Santora C, Babcock EA, Bakun A, Bonfil R, et al. (2004) Ecosystem-based fishery management. Science 305: 346–347. doi:10.1126/science.1098222.
72. Van Ginkel R (1999) Contextualizing marine resource use: a case from The Netherlands. Journal of Environmental Policy & Planning 1: 223–233. doi:10.1080/714038537.
73. Cinti A, Shaw W, Cudney-Bueno R, Rojo M (2010) The unintended consequences of formal fisheries policies: Social disparities and resource overuse in a major fishing community in the Gulf of California, Mexico. Mar Policy 34: 328–339. doi:10.1016/j.marpol.2009.08.002.
74. Lercari D, Chávez EA (2007) Possible causes related to historic stock depletion of the totoaba, *Totoaba macdonaldi* (Perciformes: Sciaenidae), endemic to the Gulf of California. Fish Res 86: 136–142. doi:10.1016/j.fishres.2007.05.010.
75. Meltzer L, Chang JO (2006) Export market influence on the development of the Pacific shrimp fishery of Sonora, Mexico. Ocean Coast Manage 49: 222–235. doi:10.1016/j.ocecoaman.2006.02.009.
76. Velarde E, Ezcurra E, Cisneros-Mata MA, Lavin MF (2004) Seabird ecology, El Niño anomalies, and prediction of sardine fisheries in the Gulf of California. Ecol Appl 14: 607–615.
77. Alberto Aragón-Noriega E, Valenzuela-Quiñones W, Esparza-Leal H, Ortega-Rubio A, Rodríguez-Quiroz G (2009) Analysis of management options for artisanal fishing of the Bigeye Croaker *Micropogonias megalops* (Gilbert, 1890) in the Upper Gulf of California. Int J Biodivers Sci Ecosyst Serv Manage 5: 208–214. doi:10.1080/17451591003709371.
78. Szteren D, Aurioles-Gamboa D, Gerber LR (2006) Population status and trends of the California Sea Lion (*Zalophus californianus californianus*) in the Gulf of California, Mexico. Sea Lions of the World. Alaska Sea Grant College Program. AK-SG-06-0. 17 p.
79. Felger R, Nichols WJ, Seminoff JA (2005) Sea turtles in Northwestern Mexico: Conservation, ethnobiology, and desperation. Biodiversity, ecosystems, and conservation in Northern Mexico. New York, NY: Oxford University Press. 405–424.
80. Sáenz-Arroyo A, Roberts CM, Torre J, Carino-Olvera M, Hawkins JP (2006) The value of evidence about past abundance: marine fauna of the Gulf of California through the eyes of 16th to 19th century travellers. Fish Fisheries 7: 128–146.
81. DOF (2010) NORMA Oficial Mexicana NOM-059-ECOL-2010, Protección ambiental-Especies nativas de México de flora y fauna silvestres-Categorías de riesgo y especificaciones para su inclusión, exclusión o cambio-Lista de especies en riesgo. Secretaría de Medio Ambiente y Recursos Naturales. Diario Oficial de la Federación. Jueves 30 de diciembre de 2010. Available: http://www.profepa.gob.mx/innovaportal/file/435/1/NOM_059_SEMARNAT_2010.pdf.
82. Rowell K, Flessa KW, Dettman DL, Román M, Gerber LR, et al. (2008) Diverting the Colorado River leads to a dramatic life history shift in an endangered marine fish. Biol Conserv 141: 1138–1148.

83. Bizzarro JJ, Smith WD, Márquez-Farías JF, Tyminski J, Hueter RE (2009) Temporal variation in the artisanal elasmobranch fishery of Sonora, Mexico. Fisheries Research 97: 103–117. doi:10.1016/j.fishres.2009.01.009.
84. Aburto-Oropeza O, Erisman B, C V-O, Danemann G (2008) Serránidos de importancia comercial del Golfo de California: Ecología, pesquerías y conservación. Ciencia y Conservación 2008: 1–23.
85. Pérez-Ramírez M, Lluch-Cota SE (2010) Fisheries certification in Latin America: recent issues and perspectives. Interciencia 35: 855–861.
86. Bakun A, Babcock EA, Lluch-Cota SE, Santora C, Salvadeo CJ (2009) Issues of ecosystem-based management of forage fisheries in "open" non-stationary ecosystems: the example of the sardine fishery in the Gulf of California. Rev Fish Biol Fisheries 20: 9–29. doi:10.1007/s11160-009-9118-1.
87. Brusca RC, Findley LT, Hastings PA, Hendrickx ME, Torre Cosio J, et al. (2005) Macrofaunal biodiversity in the Gulf of California. In: Cartron J-LE, Ceballos G, Felger RS, editors. Biodiversity, Ecosystems, and Conservation in Northern Mexico. New York, NY: Oxford University Press. 179–202.
88. Heredia-Quevedo JA (2001) Shrimp trawl design improvements suggested for Mexican fisheries [M.S. Thesis]. Reykjavik, Iceland,: United Nations University. Available:http://www.unuftp.is/static/fellows/document/joseprf.pdf.
89. Roberts CM, Hawkins JP, Gell FR (2005) The role of marine reserves in achieving sustainable fisheries. Phil Trans R Soc B 360: 123–132. doi:10.1098/rstb.2004.1578.
90. Micheli F (1999) Eutrophication, fisheries, and consumer-resource dynamics in marine pelagic ecosystems. Science 285: 1396–1398. doi:10.1126/science.285.5432.1396.
91. Jurado-Molina J, Livingston P (2002) Climate-forcing effects on trophically linked groundfish populations: implications for fisheries management. Can J Fish Aquat Sci 59: 1941–1951. doi:10.1139/f02-160.
92. Christensen V, Walters CJ (2004) Trade-offs in ecosystem-scale optimization of fisheries management policies. Bull Mar Sci 74: 549–562.
93. Peguero-Icaza M, Sánchez-Velasco L, Lavín MF, Marinone SG, Beier E (2011) Seasonal changes in connectivity routes among larval fish assemblages in a semi-enclosed sea (Gulf of California). J Plankton Res 33: 517–533. doi:10.1093/plankt/fbq107.
94. Abernethy KE, Allison EH, Molloy PP, Côté IM (2007) Why do fishers fish where they fish? Using the ideal free distribution to understand the behaviour of artisanal reef fishers. Can J Fish Aquat Sci 64: 1595–1604.
95. Ortiz I (2002) Impacts of fishing and habitat alteration on the population dynamics of the vaquita, (*Phocoena sinus*) [M.S. Thesis]. Seattle, WA: University of Washington, School of Aquatic and Fishery Sciences. 72 p.
96. INE WWF (2006) Conservación de la Vaquita y su hábitat en el Alto Golfo de California. Fondo de compensación económica para la pesca con redes de enmalle y el fomento de alternativas a la pesca en el Alto Golfo de California. Instituto Nacional de Ecología. Dirección General de Investigación en Política y Economía Ambiental. WWF programa México. 46 p. Available: http://www.ine.gob.mx/descargas/dgipea/pea-ar-2006-021.pdf.
97. Rodríguez-Quiroz G (2008) Sociedad, pesca y conservación en la Reserva de la Biosfera del Alto Golfo de California y Delta del Río Colorado [Ph.D.]. La Paz, B.C.S.: Centro de Investigaciones Biologicas del Noroeste, S.C. 240 p.
98. Hall SJ, Mainprize BM (2005) Managing by-catch and discards: how much progress are we making and how can we do better? Fish Fisheries 6: 134–155. doi:10.1111/j.1467-2979.2005.00183.x.
99. Ruckelshaus M, Doney SC, Galindo HM, Barry JP, Chan F, et al. (2013) Securing ocean benefits for society in the face of climate change. Mar Policy 40: 154–159. doi:10.1016/j.marpol.2013.01.009.
100. Sala E, Aburto-Oropeza O, Paredes G, Parra I, Barrera JC, et al. (2002) A general model for designing networks of marine reserves. Science 298: 1991–1993.
101. Leslie HM, Schlüter M, Cudney-Bueno R, Levin SA (2009) Modeling responses of coupled social–ecological systems of the Gulf of California to anthropogenic and natural perturbations. Ecol Res 24: 505–519. doi:10.1007/s11284-009-0603-8.
102. Cinti A, Shaw W, Torre J (2010) Insights from the users to improve fisheries performance: Fishers' knowledge and attitudes on fisheries policies in Bahía de Kino, Gulf of California, Mexico. Mar Policy 34: 1322–1334. doi:10.1016/j.marpol.2010.06.005.
103. Hernandez A, Kempton W (2003) Changes in fisheries management in Mexico: Effects of increasing scientific input and public participation. Ocean Coast Manage 46: 507–526. doi:10.1016/S0964-5691(03)00032-2.

11

Marine Mammal Impacts in Exploited Ecosystems: Would Large Scale Culling Benefit Fisheries?

Lyne Morissette[1,2]*, Villy Christensen[2], Daniel Pauly[2]

1 Institut des Sciences de la Mer de Rimouski, Université du Québec à Rimouski, Rimouski, Québec, Canada, **2** Fisheries Centre, University of British Columbia, Vancouver, British Columbia, Canada

Abstract

Competition between marine mammals and fisheries for marine resources—whether real or perceived—has become a major issue for several countries and in international fora. We examined trophic interactions between marine mammals and fisheries based on a resource overlap index, using seven *Ecopath* models including marine mammal groups. On a global scale, most food consumed by marine mammals consisted of prey types that were not the main target of fisheries. For each ecosystem, the primary production required (PPR) to sustain marine mammals was less than half the PPR to sustain fisheries catches. We also developed an index representing the mean trophic level of marine mammal's consumption (TL_Q) and compared it with the mean trophic level of fisheries' catches (TL_C). Our results showed that overall TL_Q was lower than TL_C (2.88 *versus* 3.42). As fisheries increasingly exploit lower-trophic level species, the competition with marine mammals may become more important. We used mixed trophic impact analysis to evaluate indirect trophic effects of marine mammals, and in some cases found beneficial effects on some prey. Finally, we assessed the change in the trophic structure of an ecosystem after a simulated extirpation of marine mammal populations. We found that this lead to alterations in the structure of the ecosystems, and that there was no clear and direct relationship between marine mammals' predation and the potential catch by fisheries. Indeed, total biomass, with no marine mammals in the ecosystem, generally remained surprisingly similar, or even decreased for some species.

Editor: Steven J. Bograd, National Oceanic and Atmospheric Administration/National Marine Fisheries Service/Southwest Fisheries Science Center, United States of America

Funding: This research was carried out as a contribution to a doctoral thesis at the Fisheries Centre of the University of British Columbia, with financial support from the Lenfest Ocean Program. The funders had no role in study design, data collection and analysis, decision to publish, or preparation of the manuscript.

Competing Interests: The authors have declared that no competing interests exist.

* E-mail: lyne.morissette@globetrotter.net

Introduction

Interactions between marine mammals and fisheries have received growing attention during the last decades [1–4]. Many studies have examined how fisheries may impact marine mammal populations [5–7], but the degree to which marine mammals compete for food with fisheries is still poorly known [5,8–9]. Nevertheless, competition for fish resources may be a primary source of current and future conflicts [10–14]. Even though many authors now document a growing concern about the widespread decline of many marine mammal populations [8,15–19], there is a serious need to address their competition with fishery for the same food resources.

Understanding how, where, and when marine mammals and fisheries compete is not an easy task [20–21]. First of all, detailed information on predation rates and how these relate to fluctuations in fish availability or marine mammal population size is lacking [22–23]. Furthermore, it is usually very difficult to observe marine mammal feeding and/or interacting with fisheries [24–25]. Quantifying their diets with estimation models (scats, stomach contents, fatty acids, etc.) is also problematic as diets can vary substantially over time and space [26–29]. Finally, even on the fisheries side, exhaustive data on yield and precise estimates on catches, bycatch (especially the commercially less important species), or discards are relatively hard to obtain [30–31].

Ecosystem models can be used address the trophic role of marine mammals in ecosystems, and their potential competition with fisheries. *Ecopath with Ecosim* (*EwE*) models are generally constructed to address fisheries questions, and merely consider the commercially important species, but in some cases marine mammal are included, providing a better representation of the trophic interactions in the upper trophic levels of ecosystems [32–33]. They then become useful to address questions about the competition between marine mammals and fisheries.

The present analysis has the following objectives: 1) to calculate the target species (prey or catch) overlap between marine mammals and fisheries; 2) to examine the global trophic impacts of marine mammals and fisheries on the commercially important species of each ecosystem; and 3) to simulate the extirpation of marine mammal populations and analyze the resulting changes on the structure of the food web. Also, we examine the impact of whales versus pinnipeds in ecosystems, in order to investigate if whaling and seal culling would have the similar impacts on ecosystems.

Materials and Methods

Modeling approach and ecosystem representation

EwE is a widespread software package widely used for the analysis of exploited aquatic ecosystems ([34], freely available through http://www.ecopath.org). This modeling approach creates a simple static model to describe the average interactions of the populations within an ecosystem during a certain period. The model assumes mass-balance, i.e., that we account for all energy flows in a food web. Such an approach is different and much easier to implement than attempts to model multispecies interactions such as multispecies virtual population analysis (MSVPA; [35]) for which an enormous quantity of catch-at-age data and stomach contents analyses is required [36].

While there is spectrum of models ranging from single species assessments through so-called 'minimum realistic ecosystem models' to very complex end-to-end models, *EwE* models generally represent a less-data intensive approach that allows evaluation of ecosystem-level questions with focus on the higher trophic levels [37]. In *EwE* models, input values (mainly biomass, production, consumption, diet composition and harvest) often are available for several species or groups in the ecosystem, and it is an approach allowing the construction and rapid evaluation of balanced ecosystem models [38]. *EwE* has the advantage of including all/most of the important species groups in contrast to minimum realistic models (MRM), and allows evaluation of unexpected indirect interactions that an MRM could miss.

For this study we used a series of ecosystem models. Each *Ecopath* model was based on mass balance principles, assuming that production of a given prey group (i) was equal to the biomass lost to fishing or export, predation, and natural mortality other than predation (other mortality). This mass balance can be expressed as:

$$\text{Consumption} = \text{production} + \text{respiration} + \text{unassimilated food} \qquad (1)$$

and

$$\text{Production} = \text{predation} + \text{fishing mortality} + \text{other mortality} \qquad (2)$$

where consumption is composed of consumption within the system and consumption of imports (i.e., consumption "outside the system"), and production may be consumed by predators, be exported from the system or contribute to the detritus. The terms of these equations may be replaced by:

$$\text{Production by } i = B_i \cdot P_i / B_i \qquad (3)$$

$$\text{Predatory losses of } i = \sum_j (B_j * Q_j / B_j * DC_{ij}) \qquad (4)$$

and

$$\text{Other losses of } i = (1 - EE_i) \cdot B_i \cdot P_i / B_i \qquad (5)$$

For any species or group of species of the system, this leads to the linear equation:

$$B_i \cdot P_i / B_i \cdot EE_i - \sum_j (B_j \cdot Q_j / B_j \cdot DC_{ij}) - Ex_i = 0 \qquad (6)$$

where:

i indicates a component (stock, species, group of species) of the model;

j indicates any of the predators of *I*;

B_i indicates the biomass of *I*;

P_i/B_i indicates the production/biomass ratio, which is equivalent to total mortality; (Z) under the most circumstances [39];

Q_i/B_i indicates the food consumption per unit biomass of *i*;

DC_{ij} indicates the contribution of i to the diet of j (in terms of mass);

EE_i indicates the ecotrophic efficiency of *i*, or the fraction of production that is consumed or caught within the system;

Ex_i indicates the export of i from the system (by emigration or fisheries catches).

Algorithms in the model also allow for the estimation of one missing parameter (B_i, Q_i/B_i, P_i/B_i, or EE_i) in each group [38].

In *Ecopath*, several system indices (see below) are computed to describe the food web, its complexity, and the way trophic groups interact with one another. The software also allows dynamic simulations through the *Ecosim* module, a dynamic modelling approach for exploring past and future impacts of fishing and environmental disturbances [40].

Ecosim provides temporal simulations using the initial parameters of the *Ecopath* master equation (Eq. 7). It works with a couple of differential equations to estimate biomass fluxes as follows:

$$\frac{dB_i}{dt} = g_i \sum_j Q_{ij} - \sum_j Q_{ij} + I_i - (M_i + F_i + e_i) B_i \qquad (7)$$

where dB_i/dt is the biomass growth rate of group (*i*) during the interval *dt*, g_i is the net growth efficiency (production/consumption ratio), I_i is the immigration rate, M_i and F_i are natural and fishing mortality rates of group (*i*), e_i is emigration rate [41]. *Ecosim* describes the interactions between predators and prey by attributing a density-dependent term ('vulnerability') for each of these interactions. This vulnerability parameter sets the maximum increase in predation mortality a given predator can cause on a given prey [40].

Because there is good coverage based on the same modelling methodology available from throughout the world's oceans, we chose *EwE* models as our sample units to quantify and analyze the impact of marine mammals in marine food webs. This approach is important because it also represents a rational way of quantifying the trade-offs between sustainable exploitation of natural marine resources and conservation of charismatic fauna [42]. The models also have the advantageous possibility of being validated to conventional stock assessment data or surveyed biomass estimates [33].

Seven models are part of this analysis, selected in terms of their location and the quality of their documentation (Table 1). Particular effort was made to cover both northern and southern hemispheres, in an attempt to have a wide coverage of the world's oceans for the global extrapolation. Kaschner and Pauly [43] have shown that the prominent hotspots of overlap and potential competition between marine mammals and fisheries include the Bering Sea where the potential negative impacts of the US groundfish fisheries on the endangered western population of Steller sea lions (*Eumetopias jubatus*) have been of great concern [44–45], the Benguela system off southwest Africa with the potential impacts of the increasing population of South African fur seals on the hake stocks [46] and the east coast of North America where the largest annual marine mammal cull worldwide is in part being justified based on the perception that the growing harp seal (*Phoca groenlandica*) population impedes the recovery of the northwest Atlantic cod (*Gadus morhua*) stocks (see review in [47]). To these three ecosystems, we added other ecosystems models that didn't necessarily featured a marine mammal-fisheries issue, but that were characterized by important and well-documented fisheries and marine mammals: the eastern tropical Pacific Ocean; the North Sea; the Gulf of Thailand and the Strait of Georgia (Figure 1). The seven models show different levels of aggregation (how the species are grouped to represent the whole foodweb) and cover from 26 to 40 trophic groups. They were selected because they included marine mammals and used high-pedigree data to describe their diet of these groups. The 'pedigree' of an Ecopath input is here understood as a coded statement categorizing the origin a given input (i.e., the type of data on which it is based), and specifying the likely uncertainty associated with the input, for diet data, it varies between 0 (general knowledge) to 1 (qualitative, detailed, diet composition) for diet composition [48] (Table 1). They were also selected based on the fact that they have been shown to reproduce well the past patterns of change in relative biomass of major species given historical disturbance patterns (fishing mortality rates and/or effort over time, and in some cases changes in oceanographic or nutrient-loading indices of relative primary productivity). For these seven ecosystems, an *EwE* model was obtained from the scientists who created it, and verified for its ability to reproduce observed time series of biomass changes (Table 1).

Ecosystem indicators

For each model, a comparison of the *Ecopath* outputs for food consumption by marine mammals *versus* the catch by fisheries was performed. The estimated annual catches (i.e., 'food consumption') of fisheries and marine mammals was calculated for each ecosystem. We also compared the estimated catch composition of the fisheries to the diet of marine mammals. Finally, the primary production required (PPR) to sustain fisheries was compared with the PPR to sustain marine mammals groups.

The mean trophic level of marine mammals' consumption (TL_Q) and of fisheries catches (TL_C) were derived from *Ecopath* outputs. TL_C is an indicator of the ecosystem health and the state of the fisheries [49], based on Lindeman's [50] concept of trophic levels, and calculated as:

$$TL_C = \sum_i \left(TL_i \left(\frac{Y_i}{\sum Y} \right) \right) \tag{8}$$

where Y_i is the total landings of species i (in tonnes), ΣY is the sum of landings for all species, and TL_i is the trophic level for species i, which can be fractional as suggested by Odum and Heald [51].

Similarly, the trophic level of consumption (TL_Q) by marine mammals was computed:

$$TL_Q = \sum_i \left(TL_i \cdot \left(\frac{\sum_{j=1}^{n} Q_{ij}}{\sum_{j=1}^{n} Q_j} \right) \right) \tag{9}$$

where Q_{ij} is the consumption of prey i (in tonnes) by marine mammal j, Q_j is the total consumption of all species by marine mammal j, and TL_i is the trophic level for species i. This equation represents the average trophic level on which marine mammals feed, i.e., the average TL of each species, multiplied by their proportion in the consumption matrix (tonnes·km^{-2}·year^{-1} that marine mammals consume).

Using estimates of fisheries catches and marine mammal consumption, the assessment of overlap between marine mammal and fisheries for each ecosystem was performed using an equation derived from Kaschner and Pauly [43]:

$$\alpha_{f,m} = \left(\frac{2 * \sum_k (p_{m,k} * p_{f,k})}{\sum_k p_{m,k}^2 + \sum_k p_{f,k}^2} \right) * \left(\frac{Q_m}{(Q_m + C_f)} * \frac{C_f}{(Q_m + C_f)} \right) \tag{10}$$

where $\alpha_{f,m}$ is the quantitative overlap between a fishery f and a marine mammal group m in the ecosystem, and the first term expresses the qualitative similarity in diet/catch composition between the marine mammal group m and fisheries f sharing the resource or food type k, with $p_{m,k}$ and $p_{f,k}$ representing the proportions of group k in the diet of marine mammals m or the catch by fishery f. This term is multiplied by the product of the proportion of total food consumption by marine mammals Q_m and the proportion of total fisheries catches C_f in the ecosystem. This index scales from 0 (no overlap) to 0.250 (identical resource). When resource use is identical between two groups, the first term of equation 10 is equal to 1, and each proportion of the second term is 0.5 (or 0.25 for the product). In order to make ecosystems comparable despite their different trophic structures, overlap by food types was also calculated, based on food categories that were first described in Pauly et al. [52]: benthic invertebrates (all crustaceans except krill, seasquirts, bivalves, gastropods, octopus, etc.), large zooplankton (mainly krill), small squid (mantle length <50 cm), large squid (mantle length >= 50 cm), miscellaneous fishes (Fishbase [53] habitat attributes: demersal, benthic, benthopelagic, bathydemersal, reef-associated, pelagic >80 cm), mesopelagic fishes, small pelagic fishes (fishbase attributes: pelagic <80 cm). The details on how the trophic groups of each model were classified into these different food categories are given in Table S1.

The mixed trophic impact (*TI*) analysis of *Ecopath* was used to compare the 'with/without' impact of predation by marine mammals on the whole ecosystem [54]. This quantifies all the direct and indirect trophic impacts of all groups in the system:

$$MTI_{ij} = DC_{ij} - FC_{j,i} \tag{11}$$

where DC_{ij} is the diet composition term expressing how much j contributes to the diet of i, and $FC_{j,i}$ is a host composition term giving the proportion of the predation on j that is due to i as a predator. When calculating the host compositions, the fishing fleets are included as 'predators'.

Beneficial predation refers to a situation where a predator may have a direct negative impact on its prey, which is counterbalanced by indirect positive effects through the consumption of other predators and competitors of the prey [14]. It was calculated as the percentage of the overall trophic impact by marine mammals that is positive for any prey group of this predator.

Dynamic simulations

In *Ecosim*, the predator-prey relationships are based on the foraging arena theory, dividing the prey biomass into vulnerable and invulnerable pools [55]. The vulnerability parameter (v) represents the transfer rate between these two pools, and has implications for how a given predator will impact predation mortality for a given prey, and can range from 1 to ∞. Low

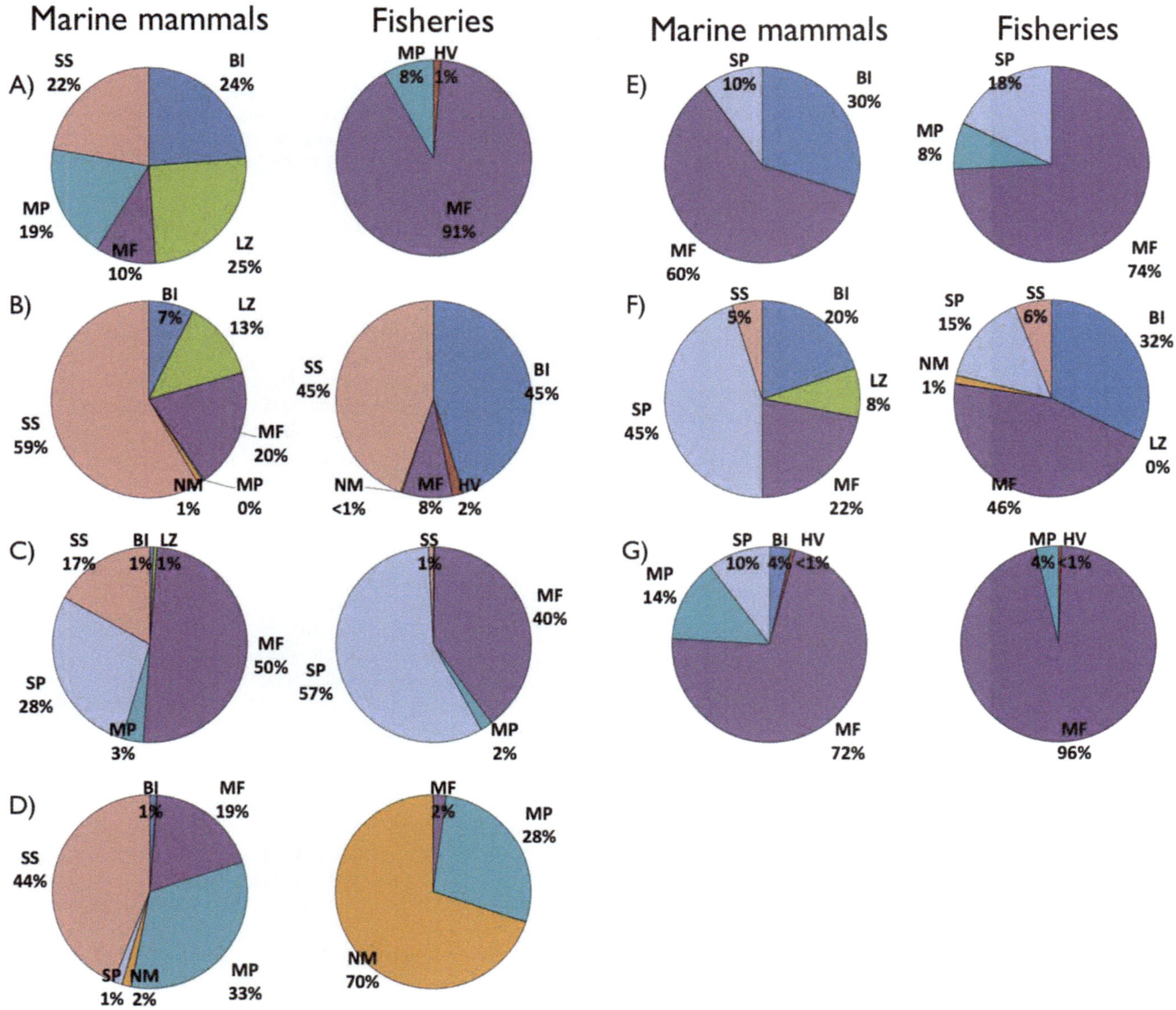

Figure 1. Location of the ecosystem modeled with *Ecopath* and used for this analysis.

vulnerability factors (e.g. = close to 1) imply that an increase in predator biomass will not cause any noticeable increase in the predation mortality the predator will cause on the given prey. High vulnerability factors (e.g. = 100) contrarily indicate that if the predator biomass is for instance doubled, it will cause close to a doubling in the predation mortality rate on a given prey. This then relates directly to assumptions about the carrying capacity for the predator in question [40], and *Ecosim* predictions are very sensitive to this parameter [4]. The default vulnerability (2.0) assumes that each predator group can at most increase the predation mortality they impose on their prey by a factor of 2.0, while a lower value implies a donor-driven density-dependent interaction, and a much higher value involves a predator-driven density-independent interaction, in which predation mortality is proportional to the product of prey and predator abundance (i.e., Lotka-Volterra).

Here, we used a set of models whose vulnerabilities were derived by fitting to historical data following Walters et al [56]. In order to quantify the potential impact of marine mammal predation on the ecosystem, and to examine if there really is strong competition with the fisheries, *Ecosim* simulations were run for 22 to 89 years, depending on the time series available. The seven ecosystem models were first analyzed with *Ecosim* using time series of fishing mortality (F) to see which groups' biomass decline or increase over time. The models covered different sources of fishing mortality or fishing fleets (trawls, long lines, coastal, deep-water, whaling, etc), which were combined in every model to see the overall effect of fisheries on the entire ecosystem. A first simulation was done with the original ecosystem structure (and original catches of fish and marine mammals), while a second was performed with a very high catch of marine mammals, with the purpose of driving them extinct. Vasconcellos et al. [57] showed that for fish species, a 5-fold increase in fishing mortality leads to serious depletion. Also, such an extreme scenario is routinely applied to many fish populations and often associated with stock collapse [58]. Consequently, an F value of 1.0 year-1 (representing an average five-fold increase for marine mammal species that were already hunted) was applied to each marine mammal group in the models. The higher values of F were kept constant for the first 20% of the

Table 1. *Ecopath* models used for analyses of marine mammals consumption.

Ecosystems	Area covered (km^2)	No of trophic groups	Marine mammals groups	Average pedigree or marine mammals' diets	Reference
Eastern Bering	484,500	26	1) Baleen whales	0.7	National Research
Sea			2) Toothed whales		Council [30]
			3) Sperm whales		
			4) Beaked whales		
			5) Walrus and bearded seals		
			6) Other seals		
			7) Steller sea lions		
Northern Gulf	103,812	32	1) Cetaceans	0.6	Morissette et al. 2003
of St.			2) Harp seals		[109]
Lawrence			3) Hooded seals		
			4) Grey seals		
			5) Harbour seals		
Benguela	179,000	32	1) Seals	0.7	Shannon et al. 2004 [78]
system			2) Cetaceans		
Eastern	32,800,000	39	1) Baleen whales	0.8	Olson and Watters [93]
tropical Pacific			2) Toothed whales		
			3) Spotted dolphins		
			4) Meso. dolphins		
North Sea	570,000	32	1) Seals	0.7	Christensen et al. 2002 [77]
Gulf of Thailand	101,384	40	1) Marine mammals	0.5	FAO/FISHCODE 2001 [110]
Strait of	6,900	27	1) Transient orcas	0.6	Martell et al. 2002 [111]
Georgia			2) Dolphins & Resident orcas		
			3) Seals		
			4) Sea lions		

time series, and then returned to the baseline, with the model running for the remaining 80% of the time.

Results

Total food consumption by marine mammals *vs* fisheries catches

At the global scale, when considering all ecosystems at once, the major source of overlap between marine mammals and fisheries for all models combined was for 'miscellaneous fish' (i.e., demersal; benthic; benthopelagic; bathydemersal; reef-associated habitat & common length <150 cm; or pelagic habitat & common length >60 cm and <150 cm) and 'small pelagic' (pelagic habitat & common length <60 cm). Marine mammals' consumption was diversified and represented a great array of marine organisms (36% of miscellaneous fish, 21% of small squids, and approximately equal proportions [10–16%] of benthic invertebrates, large zooplankton, meso-pelagic, and small pelagic fish), while fisheries catches were concentrated at 51% on 'miscellaneous fish', 13% of small pelagic fish, and 11% of benthic invertebrates. While marine mammals could consume different prey groups (mainly large zooplankton, cephalopods, small pelagic fish and macrobenthos), fisheries in the seven studied ecosystems combined were mainly targeting small crustaceans such as shrimp, pelagic fish (redeye, *Etrumeus whiteheadi*; redfish, *Sebastes* spp.; anchovy, *Eugraulis capensis*; sprat, *Sprattus sprattus*), and demersal species such as hake species, lingcod (*Ophiodon elongates*), and sandeel (*Ammodytes tobianus*). The detail of each ecosystem is given below.

In the Eastern Bering Sea system (Figure 2A), all fisheries catches fell into three types: mostly miscellaneous fishes (91%), but also mesopelagic fishes and higher vertebrates. In contrast, these food types accounted for less than a third of marine mammal consumption, which was more diverse and principally composed of large zooplankton (25%), benthic invertebrates (24%), and miscellaneous fish. In the Northern Gulf of St. Lawrence (Figure 2B), miscellaneous fish were the main target, accounting for 32% and 74% for marine mammal consumption and fisheries catches, respectively. However, the remaining marine mammal consumption was shared between three important groups (small pelagics, benthic invertebrates, and large zooplankton), while the fishery mainly caught miscellaneous fish (cod, redfish, and large Greenland halibut), benthic invertebrates (shrimp, crab, and molluscs) and small pelagics (herring). Marine mammal harvest (mainly seal hunt) also occurred in the Gulf of St. Lawrence, accounting for about 1% of the total catch. In the Benguela system (Figure 2C), more than 95% of all fisheries catches fell into three food types: small pelagic (57%), miscellaneous,(40%) and mesopelagic (2%) fishes. These food types were also the most important for marine mammals of this ecosystem (50% miscellaneous, 28% small pelagic, and 3% mesopelagic fish), whose diets also included an important proportion (17%) of small squids. In the eastern tropical Pacific model (Figure 2D) most fisheries catches were of two types: non-mammal food and mesopelagic, accounting for 70 and 28%, respectively. There, marine mammals fed on a variety of food types, mainly small squids (44%), mesopelagic (33%), and miscellaneous fish (19%). The North Sea model (Figure 2E) showed that about 75% of resources taken by marine mammals or by fisheries was composed of miscellaneous fish. However, the difference between marine mammals and fisheries was in what kind of miscellaneous fish they exploited. The main fishes eaten by marine mammals were dab and cod, while fisheries (at the time) mostly targeted Norway pout (*Trisopterus esmarkii*), sprat and sandeel. In the Gulf of Thailand (Figure 2F), marine mammals fed on a great variety of groups, while fisheries mainly caught miscellaneous fish (46%) and benthic invertebrates (32%). These two food types represented about a third of consumption by marine mammals, which was mainly composed of small pelagic fish (45%). 'Trash fish' (bycatch catches that are used in the production of fishmeal) was one of the most important miscellaneous fish to be taken by fisheries and marine mammals, but then the two competing groups differed as marine mammals consumed more small pelagic and benthos, and fisheries caught more shellfish and shrimp. In the Strait of Georgia ecosystem, almost all fish caught by the fisheries were miscellaneous fish (96%; Figure 2G), which also represented 72% of the consumption by marine mammal. The marine mammal prey was dominated by adult hake and demersal fishes, which was different from the miscellaneous fish (adult herring) caught by the fisheries.

Overall, the degree of overlap depended largely on the resolution of the marine mammals' prey and fisheries catches. Instances of direct overlap at the level of a trophic group (species of group of similar species) occurred in the Eastern Bering Sea (mainly for small flatfish, large flatfish, adult pollock [*Pollachius virens*], and other demersal fish), Gulf of St. Lawrence (planktivorous small pelagics, piscivorous small pelagics, shrimps and large crustaceans), Benguela (anchovy, redeye, and sardine [*Sardinops sagax*]), eastern tropical Pacific (small yellowfin tuna [*Thunnus albacares*], skipjack [*Katsuwonus pelamis*], and *Auxis* sp.), Gulf of Thailand ('trashfish', *Rastrelliger* spp., cephalopods, small demersals, and small pelagics), and Strait of Georgia (resident coho salmon [*Oncorhynchus kisutch*], resident chinook salmon, and lingcod). On the other hand, little overlap occurred between marine mammals prey and fisheries catches in the North Sea, even at the species level.

Trophic interactions between marine mammals and fisheries

In most ecosystems, marine mammals fed on lower trophic level species than what was caught by fisheries (Table 2). Except for the northern Gulf of St. Lawrence and the Strait of Georgia, TL_Q had lower values than TL_C. The largest discrepancy between these two values was observed in the eastern tropical Pacific, where TL_C is one trophic level higher than the TL_Q (4.70 *versus* 3.76).

The primary production required (PPR) to sustain marine mammal consumption was always lower than PPR to sustain the fisheries (Table 2). Globally, PPR for fisheries was twice as high as PPR for marine mammals' consumption (20% *vs.* 10%). Marine mammals of the Benguela system had the lowest PPR, requiring only 2.2% of total primary production of the system, compared to marine mammals in the Eastern Bering Sea, where the PPR was 31.8% of total primary production. For fisheries, the lowest value was in the Benguela system (3.2%), while the highest PPR for fisheries catch was in the Eastern Bering Sea (53.9%), closely followed by the North Sea (50.1%).

When marine mammals were considered all together, their resource overlap with the fisheries varied considerably within the seven ecosystems (Table 2). Ecosystems with higher resources overlap had a lower diversity of food groups caught/eaten by fisheries/marine mammals. Models with a very high proportion of miscellaneous fish had a higher resource overlap than systems where other food types were more important. When analyzed per trophic group instead than per food type, the overlap was always lower. Highest overlap value was seen in the Benguela system, while lowest overlap was in the eastern tropical Pacific. Interestingly, the North Sea ecosystem, which had the highest overlap per food type, had the third lowest value when overlap was calculated by trophic group. The number of trophic links in the ecosystem also had an effect on the resource overlap between marine mammal and fisheries. Indeed, food webs with lower connectance (less trophic links) generally showed higher overlap values (Table 2).

Marine mammals, fisheries and their impact on the trophic structure

The mixed trophic impact evaluated direct and indirect trophic effects. In the Eastern Bering Sea both marine mammals and fishery had an overall negative impact on the entire ecosystem (MM = −2.98; fishery = −3.04). The groups that were mostly impacted by marine mammal consumption were deep-water fish, large flatfish and other demersal fish. Conversely, small flatfish, deep pelagics and flatfish trawl seemed to benefit from the presence of marine mammals while marine mammals and flatfish were the most impacted by fisheries (Table S1).

In the Northern Gulf of St. Lawrence, fisheries had an overall negative impact (−4.93) that was much higher than that of marine mammals (−2.93). The groups that were the most negatively impacted by marine mammals were large demersals, large pelagics, and Greenland halibut (*Reinhardtius hippoglossoides*; large and small). In contrast, skates, small demersals, shrimp and most benthic invertebrates seemed to benefit from marine mammals. All marine mammals and seabirds, large cod, shrimp, small demersals and most benthic invertebrates were negatively impacted by fisheries in the Gulf of St. Lawrence. Small Greenland halibut, large demersals, and small cod seemed to benefit from fisheries (Table S1).

In the Benguela model, fisheries' negative impact on the groups' biomass was larger than that of marine mammals by an order of magnitude (−0.105 vs −0.011, respectively). The groups that were most negatively impacted by marine mammal consumption were cape hake (*Merluccius capensis*), horse mackerel (*Trachurus capensis*), and cephalopods. Conversely, apex chondrichthyans, mesopelagics, and redeye seemed to benefit from marine mammals. The fleet "other fisheries" would also benefit from an increase in marine mammal biomass in terms of mixed trophic impacts (Table S1). The main groups negatively impacted by fisheries in the Benguela ecosystem were snoek (*Thyrsites atun*), sardine, other large pelagics, shallow-water cape hake, and deep-water cape hake (*Merluccius paradoxus*) (Table S1).

In the eastern tropical Ocean, the *TI* of marine mammals and fisheries showed that fishery (−1.869) and marine mammals (−2.334) had an overall negative impact on the whole ecosystem. Groups that were the most negatively impacted by marine mammal consumption were small bigeye (*Priacanthus arenatus*), small wahoo (*Acanthocybium solandri*), albacore (*Thunnus alalunga*), and skipjack. On the other hand, large bigeye, large dorado (*Coryphaena hippurus*), large wahoo, large sharks, rays, and flyingfish (*Cypselurus naresii*) all benefited from the presence of marine mammals, while fisheries negatively affected them. Some fisheries also benefit from the presence of marine mammals in the system: the horse mackerel fisheries (midwater and demersal fleets) and

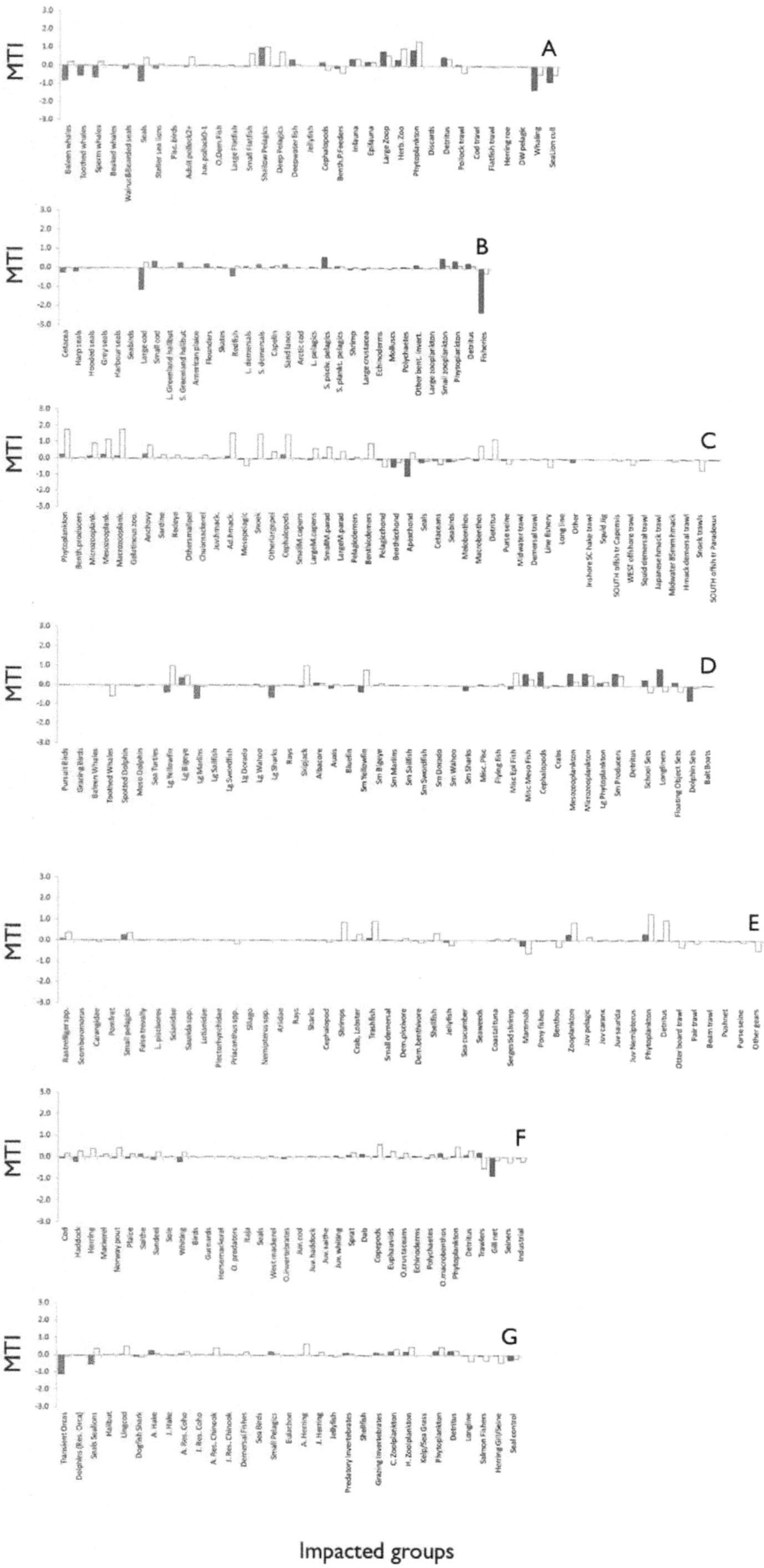
MTI
A
B
C
D
E
F
G
Impacted groups

Figure 2. Estimated mean annual catch and food consumption by food types expressed as proportions of total amounts taken ($t{\cdot}km^{-2}$) in the Eastern Bering Sea (A), Gulf of St. Lawrence (B), Benguela (C), Eastern tropical Pacific (D), Gulf of Thailand (E), North Sea (F), and Strait of Georgia (G) ecosystems. Food types categories defined by Pauly et al. 1998*a*): Non-marine mammal food (NM), miscellaneous fishes (MF), small pelagic fishes (SP), benthic invertebrates (BI), small squids (SS), large squids (LS), mesopelagic fishes (MP) large zooplankton (LZ), higher vertebrates (HV).

snoek trawls (Table S1) are examples. The *TI* also showed that groups that were the most positively impacted by the combined effect of all fisheries were mainly the juveniles of important commercial fish (wahoo, dorado, swordfish [*Xiphias gladius*], sailfish [*Istiophorus platypterus*], marlins, bigeye). When all the fisheries were grouped, their combined effect on the individual fishing fleets was mostly negative, suggesting a high level of competition between the fleets.

In the Gulf of Thailand, the *TI* of the fisheries was overall negative (−7.735), and much higher than that of marine mammals (−1.899). Marine mammals had larger negative impact on small pelagics, jacks (Carangidae), *Rastrelliger* spp., and on the purse seine fishery (Table S1).

In the North Sea model, the negative impact of fisheries (−7.79) was three times larger than the impact of seals (−2.22). Many of the groups that were seriously impacted by fisheries showed positive impact by seals; for instance, for gurnards (*Lepidotrigla* spp.), horse mackerel (*Trachurus trachurus*), sole (*Solea solea*), juvenile saithe (*Pollachius virens*), rays, and herring. Some groups were positively impacted by fisheries, such as dab (*Limanda limanda*), sandeel, juvenile haddock (*Melanogrammus aeglefinus*), juvenile cod, and birds. All of the groups (except for dab) were also positively impacted by seals. Negative impacts from seals were mainly observed for cod, saithe, plaice and dab (Table S1). Here again, the overall impact of all fisheries grouped together was damaging for each single fishing fleets.

In the Strait of Georgia, the *TI* analysis showed that while the overall trophic impact of marine mammals was near neutral (−0.83), there was a strong negative impact by fisheries on the food web (−7.79). This negative impact of fisheries affected almost all fish species in the ecosystem (Table S1), except for dogfish (*Squalus acanthias*), juvenile herring, and juvenile coho salmon, which seemed to benefit from fisheries. Marine mammals' negative impact on fish was always smaller than fisheries' impact on the same groups. Interestingly, the overall effect of marine mammal on fishing was positive for trawlers, gill nets, seiners and industrial fleets. Total marine mammals impact was also strongly negative for resident killer whales, seals and sea lions (Table S1).

Dynamic simulations

Different simulations were done for each model (eradication of all marine mammals, or of their component groups), and changes in biomass trends were recorded (see, e.g., Figures 3A and 3B). In both examples, the complete eradication of different groups of marine mammals created a change in the biomass trajectory of fish species, however this specific change was less than 15%, and could approach zero even after 50 years of simulations.

After simulating marine mammal extirpation in the eastern Bering Sea ecosystem, there was an increase in biomass (compared to the simulation without marine mammal extirpation) in the other demersal fish, deep pelagics, deep-water fish, jellyfish, and cephalopods, but all other groups showed a decrease in biomass (Figure 4A). Over a period of 51 years, there was an overall decrease of 6% of total biomass if marine mammals were eradicated ($B_{tot} = 316\ t{\cdot}km^{-2}$ with marine mammals, and $298\ t{\cdot}km^{-2}$ without them). This represents of course the complete extirpation of marine mammals biomass itself, but also seabirds, and other fish species such as small flatfish, as well as a critical biomass decrease for shallow pelagics (−99%), and epifauna (−97%). Our simulation also predicted that there were less fish to catch for the main species targeted by fisheries (adult pollock and shallow-water pelagics) without marine mammals (a decrease of 16% and 99%, respectively) (Figure 4A). Five fisheries out of eight suffered from a decrease in the biomass of their target species if there were no marine mammals in the ecosystem.

In the Gulf of St. Lawrence, when all seal and cetacean species were removed from the ecosystem, its structure changed. There was an explosion in the biomass of Greenland halibut, and an increase in large pelagic and demersal groups. Most groups that originally had lower biomasses appeared to increase their biomass after the extirpation of marine mammals. However, for commercially important groups, (e.g., adult cod, capelin, and small

Table 2. Mean trophic level of marine mammals' consumption (TL_Q), Mean trophic level of fisheries' catches (TL_C), primary production required (*PPR*), and overlap indices ($\alpha_{j,l}$) for marine mammals and fisheries, and connectance in our study areas.

Ecosystem model	TL_Q	TL_C	*PPR* Marine mammals' Q (% of total PP)	*PPR* Fisheries catch (% of total PP)	$\alpha_{j,l}$ per food type	$\alpha_{j,l}$ per trophic group	Connectance
Eastern Bering Sea	2.83	3.42	31.8	53.9	0.031	0.006	0.274
Gulf of St. Lawrence	3.24	3.71	9.8	18.4	0.161	0.034	0.298
Benguela	3.65	3.73	2.2	3.2	0.714	0.120	0.231
Eastern tropical Pacific	3.76	4.70	14.1	6.3	0.005	0.0003	0.218
North Sea	3.25	3.44	3.0	50.1	0.890	0.020	0.219
Gulf of Thailand	2.08	2.46	1.8	2.3	0.468	0.100	0.139
Strait of Georgia	3.36	3.25	6.0	6.7	0.163	0.024	0.250
Global average	**2.88**	**3.42**	**9.7**	**20.1**	**0.149**	**0.043**	-

The overlap index scales from 0 (no overlap) to 0.250 (identical resource). Connectance is an index of ecosystem complexity that represents the proportion of possible links between groups that are realized (links/species^2).

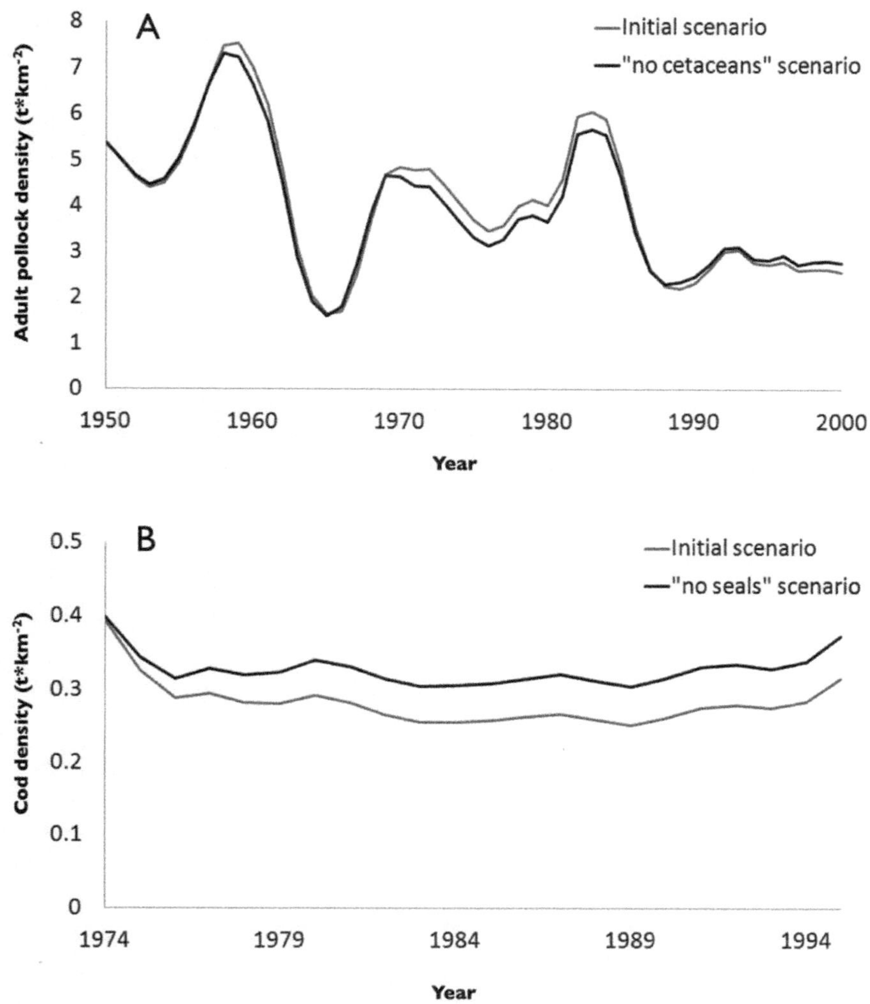

Figure 3. Change in commercially important fish biomass before (grey line) and after (black line) the simulated eradication of all marine mammals. Examples for pollock and cetaceans in the Eastern Bering Sea (A) and for cod and seals in the North Sea (B).

planktivorous pelagics), the increase was limited (Figure 4B). The most important fishery in terms of landings in the 1980s was for cod, redfish, small planktivorous pelagics (herring), and shrimp. Without marine mammals, most of these target species showed no significant change, or decreased slightly (Figure 4B).

On the one hand, in the Benguela system, the removal of the cetaceans and seals groups seemed to generate an explosion of sardine, as well as anchovy, snoek, other pelagics and seabirds. On the other hand, groups such as mesopelagics (−64%), other small pelagics (−27%), redeye (−20%), and pelagic demersals (−19%) showed decrease in biomass (Figure 4C). Increase in biomass was observed for snoek (47%), other large pelagics (28%), seabirds (27%), and large *M. capensis* (20%). The main targeted species in the Benguela ecosystem were anchovy, sardine, redeye (all caught by purse seine) and large deep-water Cape hake (caught by offshore trawl). When marine mammals were removed from the ecosystem, most of these commercially important fish ended up with less biomass than in the initial ecosystem, except for anchovy (Figure 4C). Out of 15 different fisheries in the Benguela ecosystem, eight underwent notable loss in the biomass of their target species after 25 years without marine mammals. Moreover, four of these were in the top-five fisheries with the most important catch at the beginning of the simulation.

In the eastern tropical Pacific, when marine mammals (here dolphins) were removed from the ecosystem, there was an increase in the biomass and hence predation of large fish such as wahoo (strongest increase in biomass; 248%), yellowfin tuna, bigeye, sharks, skipjack, albacore, bluefin tuna (*Thunnus orientalis*), and dorado (Figure 4D). The most important fish in terms of biomass (*Auxis* sp., flyingfish, and miscellaneous epipelagic fish) would decline if marine mammals were removed from the ecosystem (Figure 4D). However, commercially important species (especially bigeye, wahoo and skipjack), which tend to have lower biomasses, would benefit from the extirpation of marine mammals.

The Gulf of Thailand model showed that when all marine mammals were extirpated, this lead to an explosion in demersal benthivore species, but also to large variation in the biomass of small demersals, small pelagics and *Rastrelliger* spp. The remaining groups seemed to stabilize around equilibrium after the 24-year simulation. The most important groups in the Gulf of Thailand in terms of landings were 'trash fish', shellfish, shrimps, *Rastrelliger* spp., cephalopods, small pelagics, and crabs & lobsters. In a scenario without marine mammals in the ecosystem, most of these target species increased, except for shrimp, which was commercially very important (Figure 4E).

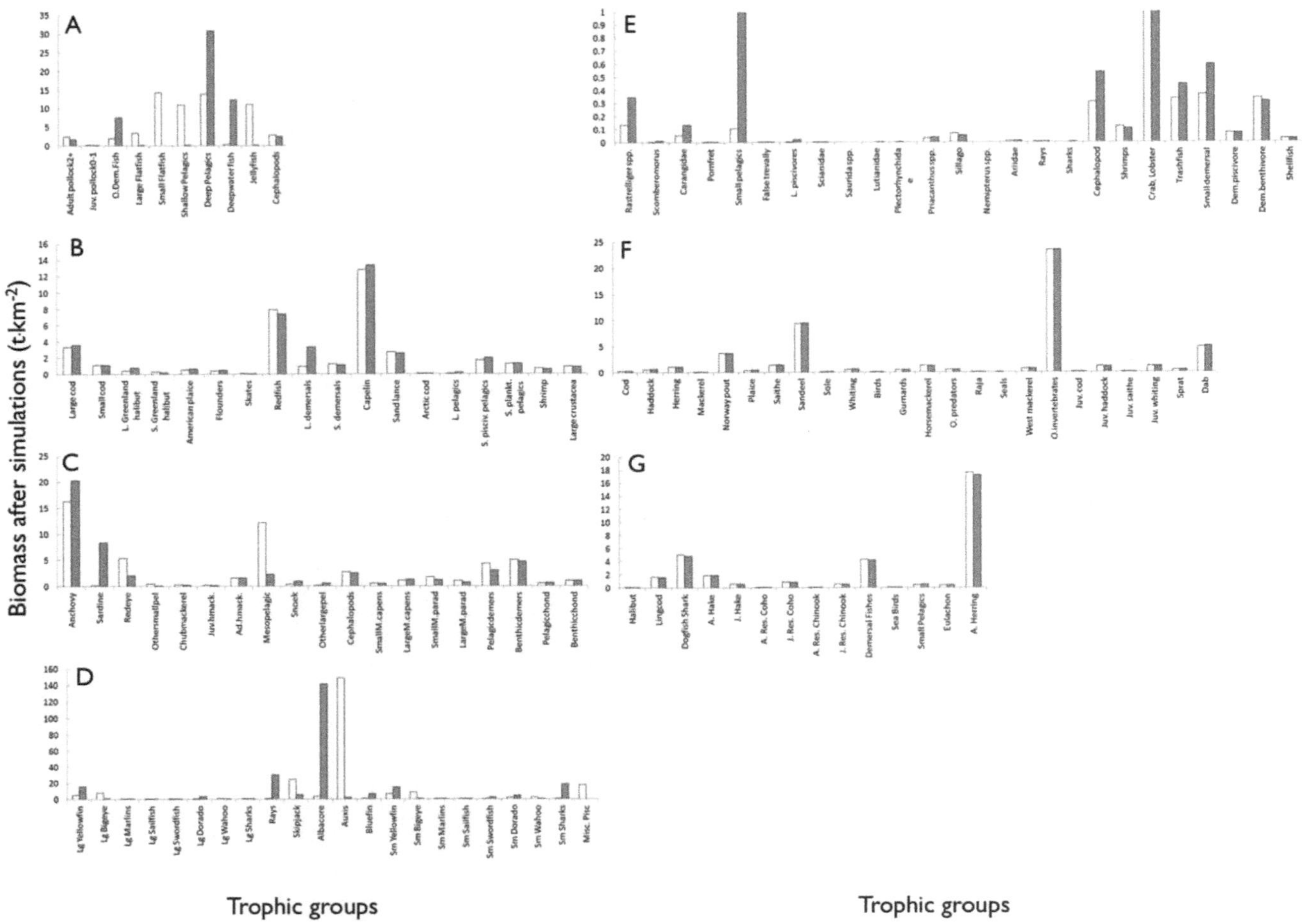

Figure 4. Biomass changes simulations with (white) and without (grey) marine mammals in the Eastern Bering Sea (A), Gulf of St. Lawrence (B), Benguela (C), Eastern tropical Pacific (D), Gulf of Thailand (E), North Sea (F), and Strait of Georgia (G) ecosystems.

In the North Sea model, after the simulated seal extirpation (Figure 4F), the dynamics remained approximately the same. No fish species showed a clear increase in biomass. Their biomass may have been replaced by cod and saithe, which showed an increase in biomass following the decline of seals (Figure 4F). The main species targeted by the North Sea fisheries were adult Norway pout, sandeel, sprat, saithe, herring, whiting, and haddock. When marine mammals were absent from the system, *Ecosim* predicted that these fisheries caught about the same amount of fish than if marine mammals were present in the system (the largest changes were increases of 11% and 14% for whiting and saithe, respectively). Out of four fisheries, one (seiners, catching herring and mackerel) was definitely decreasing in terms of catch if there were no marine mammals in the ecosystem. In general, all trophic groups stayed at approximately the same level of biomass, with or without marine mammals in the ecosystem (Figure 4F).

Finally, in the Strait of Georgia, at the end of the simulation extirpating marine mammals, there was strong variations in halibut biomass, and the biomass of small pelagics, jellyfish, eulachon (*Thaleichthys pacificus*), and adult hake (*Urophycis tenuis*) came close to zero. All other groups appeared to stabilize at lower levels at the end of the 50-years simulation. The main targeted species were herring, resident Chinook salmon, lingcod, and resident Coho salmon. When marine mammals were removed from the ecosystem, most of these commercially important fish ended up with more biomass than in the initial ecosystem, except for herring, which decreased by 13% (Figure 4G).

Differences between groups of marine mammals

In order to have noticeable effect on the biomass of commercially important fish, we would need to eradicate all marine mammals of the ecosystem (Table 3). However, even in that case, unexpected effects occurred, such as decreases of all commercially important fish in the Benguela and Strait of Georgia ecosystem. When removing only whale species, some ecosystems showed an increase in biomass for commercially important species, (e.g., Gulf of St. Lawrence), while some showed an (expected) increase in biomass, (e.g., predator fish and benthos in the Bering Sea). However, this effect was mainly done to the eradication of toothed whales, while baleen whales had almost no effect on these systems (Table 3). The eradication of marine mammals also had important impacts on fish that were not targeted by fisheries (Table 4). Indeed, most species that were not fished suffered from a decrease in biomass when marine mammals were removed from their respective ecosystem. This included depletion of forage fish in the eastern Bering Sea, and to a lesser extent a noticeable reduction of predatory fishes in the Gulf of St. Lawrence. Interestingly, the slight increases in biomass were more noted after eradicating baleen whales than any other marine mammal groups, while the most cases of a biomass decrease of non-

Table 3. Percentages of change in biomass for unfished trophic groups of prey, by different marine mammals groups, in the seven study ecosystems.

Ecosystem model	Trophic groups	No MM	No pinnipeds	No dolphins	No whales	No toothed whales	No baleen whales
Bering Sea	Predator fish	-	-	-	-	-	-
	Forage fish	−97.0	−38.1	-	−54.9	−36.0	−10.0
	Cephalopods	−15.9	−18.3	-	16.9	0.7	17.2
	Benthos	−14.9	11.0	-	−9.4	−7.4	3.6
	Plankton	1.1	0.2	-	0.4	0.2	0.1
Northern Gulf of	Predator fish	−18.3	−18.8	-	−2.2	-	-
St. Lawrence	Forage fish	−5.1	2.4	-	−2.7	-	-
	Cephalopods	-	5.7	-	-	-	-
	Benthos	0.02	−0.6	-	0.1	-	-
	Plankton	−0.03	−0.3	-	−0.1	-	-
Benguela	Predator fish	4.9	9.8	-	2.4	-	-
system	Forage fish	-	-	-	-	-	-
	Cephalopods	-	-	-	-	-	-
	Benthos	−2.9	−2.3	-	−0.6	-	-
	Plankton	−6.5	−5.4	-	−3.1	-	-
Eastern tropical	Predator fish	−1.9	-	−1.6	-	−0.4	0.03
Pacific	Forage fish	−0.7	-	0.1	-	−0.9	0.1
	Cephalopods	5.1	-	5.1	-	−0.2	0.2
	Benthos	0.03	-	−0.02	-	−0.1	0.04
	Plankton	0.2	-	−0.01	-	0.1	0.02
North Sea	Predator fish	-	-	-	-	-	-
	Forage fish	-	-	-	-	-	-
	Cephalopods	-	-	-	-	-	-
	Benthos	−0.1	−0.1	-	-	-	-
	Plankton	−0.1	−0.1	-	-	-	-
Gulf of	Predator fish	-	-	-	-	-	-
Thailand	Forage fish	-	-	-	-	-	-
	Cephalopods	−2.1	-	-	-	-	-
	Benthos	0.2	-	-	-	-	-
	Plankton	0.1	-	-	-	-	-
Strait of Georgia	Predator fish	14.1	14.4	-	-	−3.7	-
	Forage fish	5.4	2.4	-	-	2.3	-
	Cephalopods	7.4	5.7	-	-	1.3	-
	Benthos	−0.7	−0.6	-	-	−0.1	-
	Plankton	−0.1	−0.3	-	-	0.1	-

exploited species was when eradicating pinnipeds and dolphins (Table 4).

When all the simulations were pooled and analyzed by type of marine mammals (seals, dolphins, toothed whales, or baleen whales), the percentage of change in the biomass of commercially important fish (the groups with fishing mortality in the original models) increased differently according to the type of eradication simulated (Figure 5). When eradicating all marine mammals, the average response from all studied ecosystems showed the most important increase (Figure 5). However, the general response could vary from a slight decrease in biomass to an increase of about 20% of commercially important fishes. Similarly, when eradicating all whale species from the ecosystems, the overall response was a slight increase (less than 10%) in the biomass of commercially important fish. However, this was mainly driven by the removal of toothed whales, for which eradication leads to more chances of a positive effect (increase) on commercial fish species. When removing baleen whales from these systems, the response was almost null, but represented a negative effect on commercial fish biomass. Interestingly, the same effect was seen for seals, whose eradication would suggest an overall decrease of commercial fish biomass in most of the studied systems (Figure 5).

Mixed trophic impact *versus* catch or consumption

When mixed trophic impacts were plotted against consumption (for marine mammals) or catch (for fisheries), it appeared that marine mammals consumed generally less than fisheries catch, and that their *TI* was less negative than that of fisheries for the same

Table 4. Percentages of change in biomass for commercially important trophic groups of prey, by different marine mammals groups, in the seven study ecosystems.

Ecosystem model	Trophic groups	No MM	No pinnipeds	No dolphins	No whales	No toothed whales	No baleen whales
Bering Sea	Predator fish	160.2	39.0	-	18.5	13.3	4.1
	Forage fish	−21.0	−33.6	-	−48.4	−45.6	0.5
	Cephalopods	-	-	-	-	-	-
	Benthos	53.1	15.4	-	38.0	35.4	−3.9
	Plankton	-	-	-	-	-	-
Gulf of St.	Predator fish	19.7	8.3	-	6.7	-	-
Lawrence	Forage fish	4.2	0.7	-	−23.67	-	-
	Cephalopods	-	-	-	-	-	-
	Benthos	0.01	−0.1	-	0.1	-	-
	Plankton	−9.0	−6.5	-	−3.9	-	-
Benguela	Predator fish	−8.9	−7.1	-	−5.1	-	-
	Forage fish	−4.5	−1.5	-	11.4	-	-
	Cephalopods	−8.7	−11.1	-	−15.4	-	-
	Benthos	-	-	-	-	-	-
	Plankton	-	-	-	-	-	-
Eastern tropical	Predator fish	8.9	-	1.5	-	7.1	0.2
Pacific	Forage fish	-	-	-	-	-	-
	Cephalopods	-	-	-	-	-	-
	Benthos	-	-	-	-	-	-
	Plankton	-	-	-	-	-	-
North Sea	Predator fish	2.0	2.0	-	-	-	-
	Forage fish	1.5	1.5	-	-	-	-
	Cephalopods	-	-	-	-	-	-
	Benthos	-	-	-	-	-	-
	Plankton	-	-	-	-	-	-
Gulf of	Predator fish	90.1	-	-	-	-	-
Thailand	Forage fish	181.2	-	-	-	-	-
	Cephalopods	74.1	-	-	-	-	-
	Benthos	3.1	-	-	-	-	-
	Plankton	−18.6	-	-	-	-	-
Strait of Georgia	Predator fish	−7.5	−6.0	-	-	−2.2	-
	Forage fish	-	-	-	-	-	-
	Cephalopods	-	-	-	-	-	-
	Benthos	-	-	-	-	-	-
	Plankton	-	-	-	-	-	-

consumption or catch level (Figure 6). Moreover, the overall mixed trophic impacts of the marine mammals on the whole ecosystem became less negative with increasing consumption. This was a rather surprising result, to be discussed further below.

Discussion

The trophic impact of marine mammals on food webs is generally seen as a direct relationship between the predators and their prey. Many studies [59–62] have addressed the question of the trophic role of top predators without taking into account the indirect trophic effects that they can have on their prey. The results presented here suggest that marine mammals can have important indirect effects on trophic structure. Therefore, our analysis offers a new perspective on the function of these predators in marine food webs, and their interaction with fisheries. This simulation modeling exercise proved to be a very informative aspect of our study, because it allowed us to capture the dynamic behavior of the system and the way in which trophic interactions might reconfigure when marine mammals are removed from these ecosystems.

We clearly see from results presented above that a change in marine mammal biomass can lead to important alterations in the structure of the ecosystem. In a time where marine ecosystems are overexploited [63–64], polluted [65–67] and subject to climate change [68], improving our ability to understand ecological

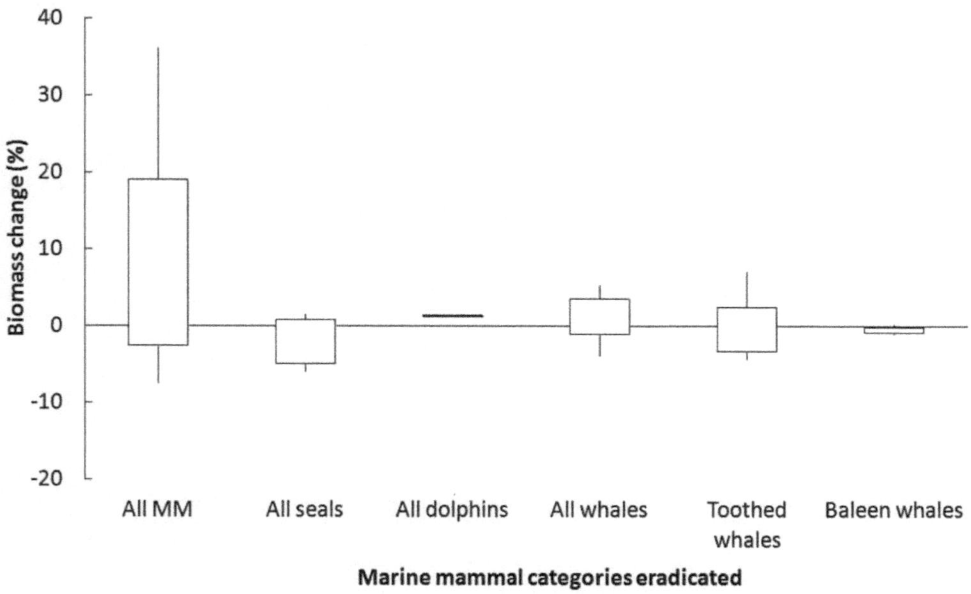

Figure 5. Total percentage of increase in the biomass of commercially important fishes when removing different groups of marine mammals. Boxes represent the lower 25% quartile up to the higher 75% quartile, while lines represent the smallest and the largest observations.

processes involving marine mammals and fisheries becomes crucial.

Strengths and weaknesses of the modeling approach

Studying marine mammals – fisheries interactions through a modelling approach can enhance our understanding of effect of relationships that would otherwise be very difficult to study. Our approach allowed to compare different ecosystems constructed using a common *EwE* framework. The level of detail included in the *Ecopath* model structure and equations is a real asset and represent a rigorous analytical framework [69].

There is a general, background controversy about appropriate levels of model complexity for the management of marine resources in the context of marine mammals – fisheries interactions [10,69–70]. While single-species approaches to modeling fish stocks are more or less bound to conclude that marine mammals are detrimental to the stocks, because such models cannot incorporate the effects of indirect interactions, more complex multi-species models (which potentially can capture the behavior of the whole system) are more difficult to parameterize; also, *EwE* depends greatly on assumptions and parameter estimateswhich often cannot be rigorously tested by comparing with data.

Our results and conclusions are the product of the quality of data sources, assumptions and results of the seven *EwE* models that were used for this study. The level of aggregation, for example, was different from one model to another, depending on the original authors' choices, and the context or ecological

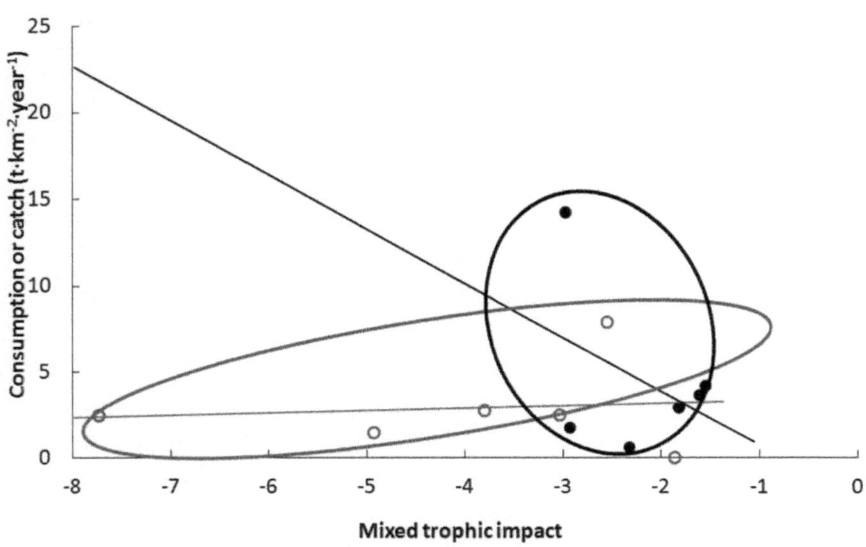

Figure 6. Total consumption by marine mammals (black dots) or total catch by fisheries (open dots) versus their respective overall mixed trophic impact for each studied ecosystem. Density ellipses represent the 90% confidence intervals.

question each model was built for. These differences in aggregation could lead to different prediction in dynamic models [71], and could be problematic for global comparisons. We did not re-aggregate the models but kept them with the same structure they had in their published version, mainly focusing on the fact that they covered marine mammal groups. This was therefore important to select the models that had the highest quality of data, and that showed a high accuracy in projecting biomass trends over time (i.e. they were fitted to time series in *Ecosim*).

Another potential issue with the way *EwE* models generally are constructed is that there is no possibility to discriminate between trophic levels that represent juveniles of larger fish or species that are truly occupying this trophic level at adult stage. Although this can be partly addressed with multi-stanza categorizations in *Ecosim* [34], most of the groups do not feature ontogenetic differences in the dynamics and diet. However, most of the marine mammals' diets used here are detailed to the species level, not often to the life stage or size of each prey. When diet information is entered in a non multi-stanza *Ecopath* model, a prey species, even juvenile, will be entered in the diet composition at the trophic level where the whole group (i.e., adult species) is. The groupings in *Ecopath* are done like this because information is based on adult diet. The juveniles are thus assumed to have the same diets than the adults, and their trophic levels represent the 'potential trophic level' they can reach at adult stage.

Some authors have warned modellers that caution must be taken in applying *EwE* models to marine mammal populations, because their life history is very different from most fish [69]. However here, the simulations we performed consisted in removing all marine mammals, and thus releasing their predation on other trophic groups in the ecosystem. The temporal variation of marine mammals' abundance itself was not addressed here (the simulated scenarios always presented how the system would react to the extirpation of marine mammal species, not with the possible causes and modalities of such extirpation). In *Ecosim* runs with original fishing effort (where marine mammals are present in ecosystems), we assume the vulnerabilities and the ecological reliability of each model is representative of the species dynamics and life cycles. The marine mammals groups present in the models are assumed to have realistic ecological parameters. However, the analysis done with these seven *Ecopath* models did not include any environmental effects that may affect ecosystem dynamics. We thus assumed that everything was explained by fisheries impact and species interactions, which is definitely not the case [72]. There is indeed strong evidence that climate change, for example, has an important impact on the availability of marine mammals' prey [73], as well as on their distribution and abundance [74–75].

One shortcoming of our approach is that we grouped all marine mammals together instead of analyzing pinnipeds, toothed whales, and baleen whales separately. Doing so, we masked the major differences in their feeding ecology. Also, analyzing marine mammals as a whole group could not show what effects their relative abundance have on the trophic impacts. The relative proportions of different marine mammal species would also affect the trophic level. For example, a large biomass of baleen whales, which would mainly consume krill, would significantly lower our estimate of trophic level of consumption. However, since we used the structure of the models directly as they were constructed, we could not separate these effects.

The calculated overlap index has the disadvantage of representing very large categories of species. Aggregation into large functional groups does not well represent the dynamics of the ecosystem. Thus, while the overlap index represents a good way to obtain a global and simple representation of the interaction between fisheries and marine mammals, the in-depth analysis of the structure of ecosystems remains crucial. The overlap index calculated at the species level showed that in the case of the North Sea, this can lead to opposite conclusions. Finally, the fact that the overlap index was calculated for marine mammals as a combined group presents, once again, a difficulty. Indeed, as most overlap with fisheries occurs among pinnipeds and dolphins, our estimates of the global overlap between fisheries and marine mammals may be underestimated.

The simulations are exploratory and represent extreme scenarios (the complete eradication of marine mammals). We did not explore any alternative or intermediate scenario. Consequently these last results should not be perceived as management strategies in any way.

Resource overlap and trophic levels

On a global scale, most food consumed by marine mammals consists of prey types that are not the main target of fisheries, and whales seem to consume most of their food in areas where commercial vessels do not fish [43]. In areas where competition between marine mammals and fisheries is evident (identified as hotspots of resource overlap by Kaschner and Pauly [43]), our results show that the resource overlap is indeed higher than the global average presented in Kaschner and Pauly [43]. However, most overlap appears to occur between fisheries and larger, deep-diving toothed whales [43], so when marine mammals are analyzed overall, their overlap is not as strong as may be expected.

Depending on the ecosystem, the overlap between marine mammals and fisheries index involves different food types. In the North Sea, the Benguela, and the Strait of Georgia systems, marine mammals and fisheries compete mainly for the 'miscellaneous fish'. This group includes demersal, benthic, benthopelagic and bathydemersal fish that are less than 150 cm, and pelagic fish that are between 60 cm and 150 cm [76]. For the purpose of this analysis, this prey group is clearly over-aggregated; and it is necessary to look at the composition of the 'miscellaneous fish' group in different ecosystems. In the North Sea, the Benguela and the Strait of Georgia ecosystems, there were important fisheries for larger fish [77–79], and the marine mammal species in these ecosystems are higher trophic-level predators, who mostly feed on these large fish [77–78,80]. These ecosystems are quite different in terms of structure from the Gulf of St. Lawrence, where the intense fishing activity in the 1980s has lead to the depletion of most groundfish stocks, leaving mainly smaller planktivorous fish and crustaceans for fisheries and marine mammals [32,81]. Consequently, in this ecosystem, small pelagics are the main overlapping resource. In the Gulf of Thailand, benthic invertebrates and miscellaneous fish are the food types that are most overlapping. Here again, the major development of trawl fisheries in the 1960s has lead to a shift to lower trophic levels such as 'trash fish' and shrimps [82], and most marine mammals are dolphins and whales that mainly eat smaller fish [83]. There may have been a time when fisheries and marine mammals did not overlap much in terms of food resources, but now that fisheries have moved down the food web [63], the target food types might have become very similar. Finally, in the eastern tropical Pacific Ocean, where the resource overlap is the lowest, there is not much competition between fisheries and marine mammal. Fisheries target mainly miscellaneous fish and large species such as tuna, and marine mammals feed mostly on small squids, mesopelagics and small pelagic fish. The number of trophic links in the ecosystem also has an effect on the resource overlap between marine mammal and fisheries. Indeed, food webs with lower connectance (less trophic links) tend to have higher overlap values.

Some ecosystems show major overlap between marine mammals and fisheries when analyzed by food types: the North Sea, Benguela and Gulf of Thailand most clearly. When the same analysis is done per trophic group (using the complete structure of the catch and the consumption matrix of marine mammals), the overlap index is reduced, but most ecosystems showing major overlap between marine mammals and fisheries remain the same. However, in the case of the North Sea, the overlap index calculated per trophic group is very low compared to the index calculated by food type, suggesting that there could be no overlap at the species level. However, diet studies on grey seals [84] show that sandeels, gadoids and other flatfish (targeted by fisheries) are important part of the diet. Grellier and Hammond [85] also showed that applying some digestion correction factors on grey seals diet in the North Sea reveal that some of its prey might in fact be similar to what is taken by the fisheries. This reduced overlap when we look at trophic group caught rather than food types might indicate that what seems to be similar in terms of type of fish caught by fisheries and eaten by marine mammals might in fact, when looked at it more specifically, be different species of fish. While *EwE* is suited to indicate general ecosystem properties, it is important to look at the species-specific results to have a clearer portrait of the situation.

Even at the trophic group level, the way the aggregations were done by the different authors who constructed the models seem to be important for our conclusions. Indeed, when fish species are aggregated onto trophic groups (i.e., large demersal feeders in the Gulf of St. Lawrence), chances are that even if this covers a group of species that are eaten by marine mammals and caught by fisheries, these could represent different fish species. Our study cannot address that level of detail, but this is definitely worth investigating. The real question in fisheries management is "are marine mammals stealing our fish?". To this, our study provides part of an answer, showing that marine mammals and fisheries do not have the same targets. And the closer we look at aggregation levels, the clearer it gets that they target different species of fish.

Overall, landings from global fisheries are shifting gradually from large piscivorous fish toward smaller invertebrates and planktivorous fish [63]. Our results further show that overall the trophic level of marine mammals prey is lower than the trophic level of the catch (2.88 *versus* 3.42). Thus, as fisheries continue to move further down in terms of the trophic level of species caught, the competition for food resources with marine mammals may become more important. In ecosystems such as the eastern tropical Pacific where the mean trophic level of the catch still is high, the overlap with marine mammals is negligible. Interestingly, in the Strait of Georgia the marine mammals' consume prey at a higher trophic level than is caught by the fisheries. This is because of transient killer whales, which mainly feed on pinnipeds, at a very high trophic level.

Lately, some studies have stressed that whales globally consume 3 to 5 times more marine fish and invertebrates annually than is fished for direct human consumption or for reduction into fish meal and oil [86]. This situation, it is alleged, is not "in balance" with the world's increasing need for a stable food supply. Such arguments are used extensively to justify whaling activity, as it is shown in this quote by Mr. Masayuki Komatsu, formerly the executive director of the Japanese Marine Fisheries Research and Development Department, to BBC on "the forces that drive Japanese whaling", 15 June 2006: *"Whale [are] abundant. The number of fish is falling while the number of whales is rising. Surely, the rapid increase in the whale population influences the level of fish stocks? We need to know more about it".*

Our results show that there is no clear and direct relationship between marine mammals' predation and the potential fish catch in the world's oceans. Many whales do eat fish, but the species that they eat are not necessarily targeted by fisheries. In fact the global overlap of food resources, representing the main 'hotspots' of competition between marine mammals and fisheries [43], is relatively low. Moreover, as the simulation results showed, it is not that clear whether the extirpation of marine mammals in ecosystems would even increase the biomass of the fish targeted by most fisheries.

Primary production required to sustain marine mammals and fisheries

In the ecosystems studied, the primary production required to sustain marine mammals represents an average of 9.7% of total primary production, less than half the PPR to sustain fisheries catch (20.1%). The latter value represents nearly three times the estimate by Pauly and Christensen [87] for global fisheries, as our analysis focuses on zones where fishing activity is intense.

In most ecosystems, PPR for marine mammals is lower than PPR for the catch, excepted in the eastern tropical Pacific Ocean. This model represents a very large area, and information about marine mammals (biomass, consumption rates, diet, production, etc.) is applied directly on the populations known to be within this area, while fisheries' effects might be more 'diluted' and less important in high sea.

The highest PPR for fisheries catch occurs in the North Sea (50%) and in the Eastern Bering Sea, (54%). PPR required by marine mammals is also the highest in the Bering Sea, followed by the eastern tropical Pacific. At the opposite end of the spectrum, the Benguela ecosystem, with the lowest PPR values for marine mammals as well as fisheries, is known to be a very productive ecosystem, due to the seasonal, wind-driven upwelling, and hence PPR is low [88].

Comparing the mixed trophic impact of different marine mammals types (seals, toothed whales, small cetaceans, baleen whales)

While there is a growing concern about the potential impact of marine mammal populations on fisheries catches [11,15,43,86,89], our results show that the highest overlap occurs between fisheries and pinnipeds or dolphins. Whales, especially baleen whales, have the lowest impact on the ecosystem, and reducing their population would not benefit fisheries in any way. Even if it is known that baleen whales do eat fish in some areas [90–92], their impact on the ecosystem is still minimal, and they don't seem to be a threat to fisheries, even in high overlap areas.

Comparing the mixed trophic impact of marine mammals and fisheries

The effect of marine mammals on their prey and consequently on available resources for fisheries is not only a direct predator-prey relationship. Rather their effect is also indirect, for example through feeding both on a prey and the competitors of the prey.

Even if negative for all studied ecosystems, the overall trophic impact of marine mammals on the different trophic groups of the ecosystem was always less strong than that of fisheries, except in the eastern tropical Pacific Ocean where fisheries target mainly large tunas (important predators of many trophic groups in the ecosystem), while marine mammals feed on a larger array of smaller prey [93].

For marine mammals, there is a paradoxical trend suggesting that the more they consume, the less they tend to reduce overall biomass. This is possible due to the fact that if marine mammals have to increase their consumption, they will feed on a wider array of prey and induce beneficial predation. In contrast, the mixed trophic impact of fisheries on the entire ecosystem is always strongly negative.

What if there were no marine mammals?

The mixed trophic impact routine of *EwE* provides a first overview of the negative or positive impacts that marine mammals have in the systems when they are at equilibrium, and considering indirect effects. This represents a first step in understanding their trophic role. The second step is to explore if these effects are creating any reconfiguration in terms of structure in the long term, if marine mammals are removed. When the extirpation of marine mammals was simulated, the biomass of other species in the food web also changed. In some ecosystems, commercially important species increased significantly after the eradication of marine mammals, (e.g., halibut and large pelagics in the Gulf of St. Lawrence, tuna species in the eastern tropical Pacific, tuna and pelagic species in the Gulf of Thailand, and cod and plaice in the North Sea). However, when all commercial species were considered, there was no obvious benefit for the fisheries. Indeed, total biomass, with no marine mammals in the ecosystem, remained generally and surprisingly similar or even decreased (as it is the case with the Eastern Bering Sea and the Gulf of St. Lawrence). Indeed, the extirpation of marine mammals may lead to reduced abundances of commercial important fish in some ecosystems. Cape hake, sardine, redeye in the Benguela upwelling, and herring in the North Sea and in the Strait of Georgia decreased when marine mammals were removed from these systems. In the case of the Gulf of Thailand, the Plectorhynchidae group became totally depleted when marine mammals were absent. On the other hand, when species or groups increased as a result of the extirpation of marine mammals in the ecosystem, these species or groups were not necessarily the most important commercially (deep-water fish, jellyfish and cephalopods in the Eastern Bering Sea; cephalopods, juvenile pelagics, or juvenile carangids (jacks) in the Gulf of Thailand). Finally, when commercially important species increased following the extirpation of marine mammals, it was not necessarily a stable equilibrium [94]. There might be more fish to catch, but once overfished, these ecosystems could become unstable and at risk of severe losses in biodiversity.

In certain areas, (e.g., in Eastern Canada), there has been a heated debate on culling marine mammals in an attempt to rebuild stocks of once commercially important fish species, notably of cod [95–96]. In that particular case, at least for the Gulf of St. Lawrence ecosystem, our results suggested that culling of marine mammals would not have led to recovery of the stocks of cod, nor otherwise benefited the commercial fishery. This corroborates the findings of Trzcinski et al. [97], who suggested that even the complete removal of grey seal predation in the eastern Scotian Shelf (Northwest Atlantic) would not assure the recovery of the cod population, given the high levels of other sources of natural mortality.

Are marine mammals a threat to fisheries?

Even in areas where there could be a competition between marine mammals and fisheries, the problem is mostly due to human use of marine resources. Over time, we have exploited and depleted the best marine resources, and now we are turning to what is left. In the process, we have moved from a zone of 'no conflict' to an area of higher overlap, as result of changing fisheries, and the collapse of overexploited large predatory fish.

At the International Whaling Commission, it has been proposed that baleen whales were a threat to fisheries, and that they should be culled. Regrettably, in this debate, it is difficult to assess whether it is based on any scientific evidence, considering the lack of evidence for existing large-scale competition between whales and fisheries [43], the well documented fact that the world's oceans increasingly are overexploited [63,98–102], and the unpredictable consequences of culling [103–105]. Moreover, the areas where this argument is used the most are the Caribbean and Northwest Africa (L. Morissette, unpublished data), two regions where baleen whales are breeding, and where they reduce their consumption rate to about 10% of what it is in their high latitude feeding areas [106].

Conclusions

Our analysis identified that marine mammals are important top predators in marine ecosystems, and that they play an important role in structuring the trophic relationships within food webs. Our results showed that even in hotspots of competition between marine mammals and fisheries, the overlap for food resources was lower than earlier presumed. Our results confirmed the findings of Kaschner and Pauly [43], who suggested that even the complete eradication of all marine mammals, from all oceans, would likely not increase fisheries catches. Hence, large-scale culling, as advocated in various Japanese studies (see, e.g., [86]) would probably not increase fisheries catches.

This study has focused on the top-down influences of marine mammals and fisheries on the fish species in different ecosystems. Although 'bottom-up' changes were not investigated here, their effects might just be additive and alter even more the structure of ecosystems. This is particularly true in the actual context of climate change, which can affect the productivity of the world's oceans [107–108]. There is still much debate about this idea and it will be important to find different ways of addressing this issue. The analysis presented here provided an insight into the problem, but further work on this needs to be pursued.

Supporting Information

Table S1 Trophic groups of the seven ecosystem models used in this study and how they fall into food types categories defined by Pauly et al. [75]: Non-marine mammal food (NM), miscellaneous fishes (MF), small pelagic fishes (SP), benthic invertebrates (BI), small squids (SS), large squids (LS), mesopelagic fishes (MP) large zooplankton (LZ), higher vertebrates (HV). The mixed trophic impact (MTI) of marine mammals and fisheries is given for each impacted trophic group.

Acknowledgments

This research was carried out as a contribution to the doctoral thesis of the first author at the Fisheries Centre of the University of British Columbia. Such a global approach would not have been possible without the precious collaboration of all the colleagues who made their model and data available for the analyses, and the Ecopath Consortium for development of the modeling techniques used.

Author Contributions

Conceived and designed the experiments: LM VC DP. Performed the experiments: LM. Analyzed the data: LM. Contributed reagents/materials/analysis tools: LM VC DP. Wrote the paper: LM.

References

1. Smith TD (1995) Interactions between marine mammals and fisheries: an unresolved problem for fisheries research. In: Bilx AS, Walløe L, Ultant Ø, editors. Whales, seals, fish and man. Waltham, MA: Elsevier Science. pp. 527–536.
2. Yodzis P (1998) Local trophodynamics and the interaction of marine mammals and fisheries in the Benguela ecosystem. J Anim Ecol 67: 635–658.
3. Gales N, Hindell M, Kirkwood R (2003) Marine mammals: fisheries, tourism and management issues. Melbourne, Australia: CSIRO publishing. 460 p.
4. Mackinson S, Blanchard JL, Pinnegar JK, Scott R (2003) Consequences of alternative functional response formulations in models exploring whale-fishery interactions. Mar Mamm Sci 19: 661–681.
5. DeMaster DP, Fowler CW, Perry SL, Richlen MF (2001) Predation and competition: the impact of fisheries on marine-mammals populations over the next one hundred years. J Mammal 82: 641–651.
6. Dans SL, Alonso MK, Pedraza NS, Crespo EA (2003) Incidental catch of dolphins in trawling fisheries off Patagonia, Argentina: Can population persist? Ecol Appl 13: 754–762.
7. Roman J, Palumbi SR (2003) Whales before whaling in the North Atlantic. Science 301: 508–510.
8. Bowen WD (1997) Role of marine mammals in aquatic ecosystems. Mar Ecol Prog Ser 158: 267–274.
9. Kaschner K, Watson R, Christensen V, Trites AW, Pauly D (2001) Modelling and maping trophic overlap between marine mammals and commercial fisheries in the North Atlantic. In: Zeller D, Watson R, Pauly D, editors. Fisheries impacts on North Atlantic ecosystems: catch, effort and national/regional data sets. Fisheries Centre Research Reports 9(3). pp. 35–45.
10. Plagányi ÉE, Butterworth DS (2002) Competition with fisheries. In: Perrin WF, Würsig B, Thewissen HGM, editors. Encyclopedia of Marine Mammals. San Diego: Academic Press. pp. 268–273.
11. Dillingham PW, Skalski JR, Ryding KE (2006) Fine-scale geographic interactions between Steller sea lion (*Eumetopias jubatus*) trends and local fisheries. Can J Fish Aquat Sci 63: 107–119.
12. Gerber LR, Morissette L, Kaschner K, Pauly D (2009) Should whales be culled to increase fishery yield? Science 323: 880–881.
13. Morissette L, Kaschner K, Gerber LR (2010) Ecosystem models clarify the trophic role of whales off Northwest Africa. Mar Ecol Prog Ser 404: 289–302.
14. Morissette L, Kaschner K, Gerber LR (2010) 'Whales eat fish'? Demystifying the myth in the Caribbean marine ecosystem. Fish Fish 11: 388–404.
15. Trites AW, Christensen V, Pauly D (1997) Competition between fisheries and marine mammals for prey and primary production in the Pacific Ocean. J Northw Atlant Fish Sci 22: 173–187.
16. Springer AM, Estes JA, van Vliet GB, Williams TM, Doak DF, et al. (2003) Sequential megafaunal collapse in the North Pacific Ocean; an ongoing legacy of industrial whaling? Proc Natl Acad Sci U S A 100: 12 223–12 228.
17. Estes JA, Tinker MT, Doroff AM, Burn DM (2005) Continuing Sea Otter population declines in the Aleutian Archipelago. Mar Mamm Sci 21: 169–172.
18. Kraus SD, Brown MW, Caswell H, Clark CW, Fujiwara M, et al. (2005) North Atlantic Right Whales in crisis. Science 309: 561–562.
19. Christensen L (2006) Marine mammal populations: reconstructing historical abundances at the global scale. Fisheries Centre Research Reports 14(9).
20. Matthiopoulos J, Smout S, Winship AJ, Thompson D, Boyd IL, et al.(2008) Getting beneath the surface of marine mammal – fisheries competition. Mamm Rev 38: 167–188.
21. Read AJ (2008) The looming crisis: interactions between marine mammals and fisheries. J Mamm 89: 541–548.
22. Northridge SP (1991) An updated world review of interactions between marine mammals and fisheries. FAO Fisheries Technical Paper 251, Supplement 1. Available: http://www.fao.org/docrep/003/T0452E/T0452E00.HTM. Accessed 2012 Aug 15.
23. MacKenzie BR, Alheit J, Conley DJ, Holm P, Kinze CC (2002) Ecological hypotheses for a historical reconstruction of upper trophic level biomass in the Baltic Sea and Skagerrak. Can J Fish Aquat Sci 59: 173–190.
24. Williams TM, Estes JA, Doak DF, Springer AM (2004) Killer appetites: assessing the role of predators in ecological communities. Ecology 85: 3373–3384.
25. Kaneko H, Lawler IR (2006) Can near infrared spectroscopy be used to improve assessment of marine mammal diets via fecal analysis? Mar Mamm Sci 22: 261–275.
26. Lawson JW, Stenson GB (1995) Historic variation in the diet of harp seals (*Phoca groenlandica*) in the northwest Atlantic. In: Blix AS, Ultang Ø. Walløe L, editors. Seals, whales, fish and man. Waltham, MA: Elsevier Science. pp. 261–269.
27. Bowen WD, Siniff DB (1999) Distribution, population biology, and feeding ecology of marine mammals. In: Reynolds JE, Rommels SA, editors. Biology of marine mammals. Washington, DC: Smithsonian Institution Press. pp. 423–484
28. Laake JL, Browne P, DeLong RL, Huber HR (2002) Pinniped diet composition: a comparison of estimation models. Fish Bull 100: 434–447.
29. Hume F, Hindell MA, Pemberton D, Gales R (2004) Spatial and temporal variation in the diet of a high trophic level predator, the Australian fur seal (*Arctocephalus pusillus doriferus*). Mar Biol 144: 407–415.
30. National Research Council (2003) The decline of Steller Sea Lion in Alaskan waters: Untangling food webs and fishing nets. Committee on the Alaska groundfish fishery and Steller Sea lions. Washington, DC: National Academies Press.
31. Zeller D, Pauly D (2005) Good news, bad news: global fisheries discards are declining, but so are total catches. Fish Fish 6: 156–159.
32. Morissette L, Hammill MO, Savenkoff C (2006) The trophic role of marine mammals in the Northern Gulf of St. Lawrence. Mar Mamm Sci 22: 74–103.
33. Morissette L (2009) An ecosystem approach to the 'whales eat fish' issue. In: Morissette L, Melgo JL, Kaschner K, Gerber L, editors. Fisheries Centre Research Reports 17(2). pp 3–5.
34. Christensen V, Walters CJ, Pauly D, Forrest R (2008) Ecopath with Ecosim version 6: user guide. Lenfest Ocean Futures Project, University of British Columbia, Vancouver. November 2008 edition.
35. Sparre P (1991) Introduction to multispecies virtual population analysis. ICES Marine Science Symposia 193: 12–21.
36. Morissette L (2001) Modélisation écosystémique du nord du Golfe du Saint-Laurent. M.Sc. thesis, Université du Québec, Rimouski, QC, Canada.
37. Plagányi ÉE (2007) Models for an ecosystem approach to fisheries. FAO Fisheries Technical Paper 477. 106 pp. Available: ftp://ftp.fao.org/docrep/fao/010/a1149e/a1149e.pdf. Accessed 2012 Aug 15.
38. Christensen V, Pauly D (1992) ECOPATH II-a software for balancing steady-state ecosystem models and calculating network characteristics. Ecol Modell 61: 169–185.
39. Allen KR (1971) Relation between production and biomass. J Fish Res Board of Can 28: 1573–1581.
40. Christensen V, Walters CJ (2004) Ecopath with Ecosim: methods, capabilities and limitations. Ecol Modell 172: 109–139.
41. Walters CJ, Christensen V, Pauly D (1997) Structuring dynamic models of exploited ecosystems from trophic mass-balance assessments. J. Fish Biol Fish 7: 139–172.
42. Pitcher TJ (2005) Simulating Antarctic ecosystems and fisheries: weapons of mass construction. In: Palomares MLD, Pruvost P, Pitcher TJ, Pauly D, editors. Modeling Antarctic marine ecosystems. Fisheries Centre Research Reports 13(7). pp. 3–4.
43. Kaschner K, Pauly D (2005) Competition between marine mammals and fisheries: Food for thought. In: Salem DJ, Rowan, AN, editors. The state of animals III: 2005. Washington, DC: Humane Society of the United States Press. pp. 95–117.
44. Fritz LW, Ferrero RC, Berg RJ (1995) The threatened status of the Steller sea lion, *Eumetopias jubatus*, under the Endangered Species Act: Effects on Alaska Groundfish Management. Mar Fish Rev 57: 14–27.
45. Loughlin TR, York A (2000) An accounting of the sources of Steller sea lion, *Eumetopias jubatus*, mortality. Mar Fish Rev 62: 40–45.
46. Punt AE, Butterworth DS (1995) The effects of future consumption by Cape fur seal on catches and catch rates of the Cape hakes. 4. Modelling the biological interaction between Cape fur seals *Arctocephalus pusillus pusillus* and the Cape hake *Merluccius capensis* and *Merluccius paradoxus*. S Afr J Mar Sci 16: 255–285.
47. Yodzis P (2001) Must top predators be culled for the sake of fisheries? Trends Ecol Evol 16: 78–84.
48. Morissette L (2005) Addressing uncertainty in marine ecosystems modeling. In Levner E, Linkov I, Proth J-M, editors. Strategic management of marine ecosystems. NATO Science Series: IV: Earth and environmental sciences, Volume 50. Heidelberg, Germany: Springer-Verlag. pp. 127–142.
49. Pauly D, Watson R (2005) Background and interpretation of the 'Marine Trophic Index' as a measure of biodiversity. Philos Trans R Soc Lond B Biol Sci 360: 415–423.
50. Lindeman RL (1942) The trophic-dynamic aspect of ecology. Ecology 23: 399–418.
51. Odum WE, Heald EJ (1975) The detritus-based food web of an estuarine mangrove community. In Wiley M., editor. Estuarine research, Vol. 1. New York: Academic Press. pp 265–286.
52. Pauly D, Trites AW, Capuli E, Christensen V (1998). Diet composition and trophic levels of marine mammals. ICES J Mar Sci 55: 467–481.
53. Froese R, Pauly D (2012) FishBase, version (08/2012). Available: http://www.fishbase.org. Accessed 2012 Aug 15.
54. Ulanowicz RE, Puccia CJ (1990) Mixed trophic impacts in ecosystems. Coenoses 5: 7–16.
55. Walters C. Kitchell JF (2001) Cultivation/depensation effects on juvenile survival and recruitment: Implications for the theory of fishing. Can J Fish Aquat Sci 58: 39–50.
56. Walters CJ, Christensen V, Martell SJ, Kitchell JF (2005) Possible ecosystem impacts of applying MSY policies from single-species assessments. ICES J Mar Sci 62: 558–568.
57. Vasconcellos V, Mackinson S, Sloman K, Pauly D (1997) The stability of trophic mass-balance models of marine ecosystems: a comparative analysis. Ecol Modell 100: 125–143.
58. Patterson K (1992) Fisheries for small pelagic species: an empirical approach to management targets. Rev Fish Biol Fish 2: 321–338.
59. Parsons TR (1992) The removal of marine predators by fisheries and the impact of trophic structure. Mar Pol Bull 25: 51–53.

60. Kenney RD, Scott GP, Thompson TJ, Winn HE (1997) Estimates of prey consumption and trophic impacts of cetaceans in the USA northeast continental shelf ecosystem. J Northw Atl Fish Sci 22: 155–171.
61. Lesage V, Hammill MO, Kovacs KM (2001) Marine mammals and the community structure of the Estuary and Gulf of St. Lawrence, Canada: evidence from stable isotope analysis. Mar Ecol Prog Ser 210: 203–221.
62. Tjelmeland S, Lindstrøm U (2005) An ecosystem element added to the assessment of Norwegian spring-spawning herring: implementing predation by minke whales. ICES J Mar Sci 62: 285–294.
63. Pauly D, Christensen V, Dalsgaard J, Froese R, Toreres F Jr (1998) Fishing down marine food webs. Science 279: 860–863.
64. Lotze HK, Lenihan HS, Bourque BJ, Bradbury RH, Cooke RG, et al. (2006) Depletion, degradation, and recovery potential of estuaries and coastal seas. Science 312: 1806–1809.
65. Angel MV (1995) Deep sea pollution – who cares? IEE Colloquium on "Pollution of Land, Sea and Air: An Overview for Engineers" 165: 9/1–9/5.
66. Sindermann CJ (1996) Ocean pollution: effects on living resources and humans. Boca Raton, FL: CRC Press. 304 p.
67. Clark RB (2001) Marine pollution. New York: Oxford University Press. 248 p.
68. Harley CDG, Randall HA, Hultgren KM, Miner BG, Sorte CJB, et al. (2006) The impacts of climate change in coastal marine systems. Ecol Lett 9: 228–241.
69. Plagányi ÉE, Butterworth DS (2004) A critical look at the potential of Ecopath with Ecosim to assist in practical fisheries management. Afr J Mar Sci 26: 261–287.
70. Whipple SJ, Link JS, Garrison LP, Fogarty MJ (2001) Models of predation and fishing mortality. Fish Fish 1: 22–40.
71. Essington TE (2007) Evaluating the sensitivity of a trophic mass-balance model (Ecopath) to imprecise data inputs. Can J Fish Aquat Sci 64: 628–637.
72. Christensen V, Walters CJ (2011) Progress in the use of ecosystem models for fisheries management. In: Christensen V, Maclean J, editors. Ecosystem Approaches to Fisheries: a Global Perspective. Cambridge: Cambridge University Press. pp. 189–205.
73. Hardwood J (2001) Marine mammals and their environment in the twenty-first century. J Mammal 82: 630–640.
74. Simmonds MP, Issac SJ (2007) The impacts of climate change on marine mammals: Early signs of significant problems. Oryx 41: 19–26.
75. Cheung WWL, Lam VWY, Sarmiento JL, Kearney K, Watson R, et al. (2009) Projecting global marine biodiversity impacts under climate change scenarios. Fish Fish 10: 235–251.
76. Pauly D, Trites AW, Capuli E, Christensen V (1995) Diet composition and trophic levels of marine mammals. ICES Counc Meet Pap 1995/N:13.
77. Christensen V, Beyer JE, Gislason H, Vinter M (2002) A comparative analysis of the North Sea based on Ecopath with Ecosim and multispecies virtual population analysis. In: Christensen V, Reck G, Maclean JL, editors. Proceedings of the INCO-DC Conference Placing Fisheries in their Ecosystem Context, Galàpagos Islands, Ecuador, 4–8 December 2000. ACP-EU Fisheries Research Report. pp. 39.
78. Shannon LJ, Christensen V, Walters CJ (2004) Modelling stock dynamics in the southern Benguela ecosystem for the period 1978–2002. Afr J Mar Sci 26: 179–196.
79. Wallace SS (1998) Changes in human exploitation of marine resources in British Columbia (pre-contact to present day). In: Pauly D, Pitcher T, Preikshot D, Hearne J, editors. Back to the future: reconstructing the strait of Geortia ecosystem. Fisheries Centre Research Reports 6(5). pp. 58–64.
80. Dalsgaard J, Wallace SS, Salas S, Preikshot D (1998) Mass-balance model reconstructions of the Strait of Georgia: the present, one hundred and five hundred years ago. In: Pauly D, Pitcher T, Preikshot D, Hearne J, editors. Back to the future: reconstructing the Strait of Georgia ecosystem. Fisheries Centre Research Reports 6(5). pp. 72–86.
81. Savenkoff C, Swain DP, Hanson JM, Castonguay M, Hammill MO, et al. (2007) Effects of fishing and predation in a heavily exploited ecosystem: Comparing periods before and after the collapse of groundfish in the southern Gulf of St. Lawrence (Canada). Ecol Modell 204: 115–128.
82. Gulland JA (1983) World resources of fisheries and their management. In: Kinne O, editor. Marine Ecology, vol. V, part 2. Hoboken, NJ: John Wiley and Sons Ltd. pp. 839–1061
83. Perrin WF, Reeves RR, Dolar MLL, Jefferson TA, Marsh H, et al. (2005) Report of the second workshop on the biology and conservation of small cetaceans and dugongs of Southeast Asia. CMS Technical series publication No. 9. UNEP/CMS secretariat, Bonn, Germany.
84. Hammond PS, Hall AJ, Prime JH (1994) The diet of grey seals around Orkney and other islands and mainland sites in North-Eastern Scotland. J Appl Ecol 31: 340–350.
85. Grellier K, Hammond PS (2006) Robust digestion and passage rate estimates for hard parts of grey seal (*Halichoerus grypus*) prey. Can J Fish Aquat Sci 63: 1982–1998.
86. Anonymous (2001) Increasing competition between fisheries and whales. Japan's whale research in the Western North Pacific (JARPN II). Fisheries Agency.
87. Pauly D, Christensen V (1995) Primary production required to sustain global fisheries. Nature 374: 255–257.
88. Shannon LJ, Moloney CL, Jarre A, Field JG (2003) Trophic flows in the southern Benguela during the 1980s and 1990s. J Mar Sys 39: 83–116.
89. Harwood J, Croxall JP (1988) The assessment of competition between seals and commercial fisheries in the North sea and Antarctic. Mar Mamm Sci 4:13–33.
90. Lindstrøm U, Haug T, Rottingen I (2002) Predation on herring, *Clupea harengus*, by minke whales, *Balaenoptera acutorostrata*, in the Barents Sea. ICES J Mar Sci 59: 58–70.
91. Lydersen C, Weslawski JM, Oritsland NA (2006) Stomach content analysis of inke whales *Balaenoptera acutorostrata* from the Lofoten and Vesteralen areas, Norway. Ecography 14: 219–222.
92. Lawson JW, Gosselin JF (2008) Don't Ignore The Whales: Cetacean Biomass Consumption estimates Based On The Recent TNASS Aerial Survey of Atlantic Canada. Presented during NAFO symposium "The role of marine mammals in ecosystems in the 21st century", 29 September–1 October, 2008, Dartmouth, NS, Canada.
93. Olson RJ, Watters GM (2003) A model of the pelagic ecosystem in the eastern tropical Pacific ocean. Inter-Amer Trop Tuna Comm 22: 133–218.
94. Berkeley SA, Hixon MA, Larson RJ, Love MS (2004) Fisheries sustainability via protection of age structure and spatial distribution of fish populations. Fisheries 29: 23–32.
95. Dwyer E (1999) Empty nets: Shad fisherman say catches have fallen off dramatically and they're blaming the harbour seals. Saint John Times Globe, June 14, 1999.
96. Fisheries Resource Conservation Council (2002) 2002/2003 conservation requirements for groundfish stocks in the Gulf of St. Lawrence. Report to the Minister of Fisheries and Oceans, April 2002. FRCC.2002.R.2.
97. Trzcinski MK, Mohn R, Bowen WD (2006) Continued decline of an Atlantic cod population: How important is gray seal predation? Ecol App 16: 2276–2292.
98. Jackson JB, Kirby MX, Berger WH, Bjorndal KA, Botsford LW, et al. (2001) Historical overfishing and the recent collapse of coastal ecosystems. Science 293: 629–638.
99. Pauly D, Christensen V, Guénette S, Pitcher TJ, Sumaila UR, et al. (2002) Towards sustainability in world fisheries. Nature 418: 689–695.
100. Baum JK, Myers RA, Kehler DG, Worm B, Harley J, et al. (2003) Collapse and conservation on shark populations in the Northwest Atlantic. Science 299: 389–392.
101. Christensen V, Guenette S, Heymans JJ, Walters CJ, Watson R, et al. (2003) Hundred-year decline of North Atlantic predatory fishes. Fish Fish 4: 1–24.
102. Myers RA, Worm B (2003) Rapid worldwide depletion of predatory fish communities. Nature 423:280–283.
103. Paine RT, Tegner MJ, Johnson EA (1998) Compounded perturbations yield ecological surprises. Ecosystems 1: 535–545.
104. Scheffer M, Carpenter S, Foley JA, Folke C, Walker B (2001) Catastrophic shifts in ecosystems. Nature 413: 591–596.
105. Pace ML, Cole JJ, Carpenter SR, Kitchell JF (1999) Trophic cascades revealed in diverse ecosystems. Trends Ecol Evol 14: 483–488.
106. Lockyer C (1981) Growth and energy budgets of large baleen whales from the southern hemisphere. In: FAO Fisheries Series 5, Mammals in the Sea. pp. 379–488.
107. Loreng H (2004) Marine Systems: The impact of climate change (ACIA Chapter 9). ACIA International Scientific Symposium on Climate Change in the Arctic, Reykjavik (Iceland), 9–12 Nov 2004.
108. Sanae C, Tadokoro K, Sugisaki H, Saino T (2006) Effects of decadal climate change on zooplankton over the last 50 years in the western subarctic North Pacific. Glob Chang Biol 12: 907–920.
109. Morissette L, Despatie S-P, Savenkoff C, Hammill MO, Bourdages H (2003) Data gathering and input parameters to construct ecosystem models for the northern Gulf of St. Lawrence (mid-1980s). Can Tech Rep Fish Aquat Sci 2497.
110. FAO/FISHCODE (2001) Report of a bio-economic modeling workshop and a policy dialogue meeting on the Thai demersal fisheries in the Gulf of Thailand held at Hua Hin, Thailand, 31 May–9 June 2000. FI: GCP/INT/648NOR: Field Report F-16 (En). Rome, FAO.
111. Martell SJ, Beattie AI, Walters CJ, Nayar T, Briese R (2002) Simulating fisheries management strategies in the Gulf of Georgia ecosystem using Ecopath with Ecosim. In Pitcher T, Cochrane K, editors. The use of ecosystem models to investigate multispecies management strategies for capture fisheries. Fisheries Centre Research Reports 10(2). pp. 16–23.

12

Significant Increase in Ecosystem C Can Be Achieved with Sustainable Forest Management in Subtropical Plantation Forests

Xiaohua Wei[1], Juan A. Blanco[2]*

1 Department of Earth and Environmental Sciences, University of British Columbia, Kelowna, British Columbia, Canada, 2 Departmento de Ciencias del Medio Natural, Universidad Pública de Navarra, Pamplona, Navarra, Spain

Abstract

Subtropical planted forests are rapidly expanding. They are traditionally managed for intensive, short-term goals that often lead to long-term yield decline and reduced carbon sequestration capacity. Here we show how it is possible to increase and sustain carbon stored in subtropical forest plantations if management is switched towards more sustainable forestry. We first conducted a literature review to explore possible management factors that contribute to the potentials in ecosystem C in tropical and subtropical plantations. We found that broadleaves plantations have significantly higher ecosystem C than conifer plantations. In addition, ecosystem C increases with plantation age, and reaches a peak with intermediate stand densities of 1500–2500 trees ha^{-1}. We then used the FORECAST model to simulate the regional implications of switching from traditional to sustainable management regimes, using Chinese fir (*Cunninghamia lanceolata*) plantations in subtropical China as a study case. We randomly simulated 200 traditional short-rotation pure stands and 200 sustainably-managed mixed Chinese fir – *Phoebe bournei* plantations, for 120 years. Our results showed that mixed, sustainably-managed plantations have on average 67.5% more ecosystem C than traditional pure conifer plantations. If all pure plantations were gradually transformed into mixed plantations during the next 10 years, carbon stocks could rise in 2050 by 260.22 TgC in east-central China. Assuming similar differences for temperate and boreal plantations, if sustainable forestry practices were applied to all new forest plantation types in China, stored carbon could increase by 1,482.80 TgC in 2050. Such an increase would be equivalent to a yearly sequestration rate of 40.08 TgC yr^{-1}, offsetting 1.9% of China's annual emissions in 2010. More importantly, this C increase can be sustained in the long term through the maintenance of higher amounts of soil organic carbon and the production of timber products with longer life spans.

Editor: Dafeng Hui, Tennessee State University, United States of America

Funding: Funding for this research was provided by: 1) Forest Carbon Counting Method Project (XDA05060101) of the Chinese Academy of Science Strategic and Flagship Program on the Chinese Ecological and Engineering Projects and their Contributions to Carbon; and 2) BC-China Innovation and Commercialization Strategic Development Program (ICSD-2007-Wei-Kimmins). The funders had no role in study design, data collection and analysis, decision to publish, or preparation of the manuscript.

Competing Interests: The authors have declared that no competing interests exist.

* E-mail: juan.blanco@unavarra.es

Introduction

Reforestation efforts are promoted globally to meet the growing demand of forest products, especially in developing countries. Subtropical forests are therefore promoting the maintenance of forest cover in the region and the production of goods that improve the livelihoods of local communities. Subtropical forests cover 7–10% of the global land area, and store 40–50% of carbon present in terrestrial vegetation [1]. However, subtropical forests also account for most of current deforestation, with accumulated losses in the 1996–2010 period of ~100 million ha [2]. Because forests are a major carbon pool, their management is crucial to develop successful policies for climate change mitigation. Carbon storage versus production of timber and non-timber forest products are seen sometimes as incompatible objectives to be implemented at the same time in a forest, but this view is being challenged [3]. Clarifying if subtropical forest plantations can support multi-objective forestry and estimating how much carbon could be stored in working forest plantations would help to maximize the outcome of forest management plans while helping to reach a more sustainable development of the subtropical regions.

New planted forests are typically established on sites that have rapid growth and access to processing facilities and growing markets. Hence, very large areas of new forest plantations have been established in subtropical countries, particularly in Latin America and parts of Asia. In spite of such rapid growth, only 3% of the world forest land is covered with productive forest plantations. However, this area expanded by 2 million ha annually in the 1990s and by 2.8 million ha in the 2000s [4].

While much attention and discussion have been devoted on controlling and reverting deforestation, how to enhance forest carbon stocks by improving forest ecosystem condition has received less emphasis until recently [5]. Reforestation has been suggested not only as an effective way to restore degraded ecosystems, but also as a way to mitigate elevated atmospheric CO_2, hence contributing towards the reduction of climate change [3,6]. Whether and how these efforts can maximize and sustain forest C storage largely remains unexamined [7,8,9]. Recent decisions by the U.N. Framework Convention on Climate Change have encouraged consideration of other ecosystem services while implementing REDD+ (Reducing greenhouse gas Emissions from Deforestation and forest Degradation) projects for forest carbon sequestration and storage. Therefore, selecting the most appropriate approaches for implementing multipurpose forest management in planted forests is just as important as growing more forests [10]. However, few incentives and policy guidelines are in place to encourage growing better forests globally, perhaps due to the lack of a clear understanding of long-term benefits from growing better forests as well as operational difficulties.

Traditional forest management practices in forest plantations, particularly in developing countries, usually involve application of short rotations, usage of monoculture and exotic species, and high-level removal of biomass, with the main purpose of maximizing short-term economic gains [11]. The ecological problems associated with those traditional forestry practices have been well documented: declines in long-term productivity [12,13], biodiversity loss [14], increased susceptibility to insects and diseases [15], physicochemical changes in forest soils [16], and losses of other ecological services [17]. In the context of climate change mitigation, traditional forestry practices may lead to short-term carbon sequestration in wood products, but those gains are unlikely sustainable due to losses in ecosystem C linked to the above-mentioned ecological problems. Sustainable forestry practices, however, could have an important role in mitigating climate change impacts, if they are well designed and implemented [3]. For example, transformation of conifer monocultures into mixed conifer-broadleaved plantations has been considered as an efficient strategy to sustain forest productivity and restore degraded forests [18].

China has both subtropical and temperate forests, and aggressive reforestation policies have turned it into one of the five most forest-rich countries in the world, accounting for the largest gain in forested areas globally [12,19]. However, most of ecosystems reforested during the past decades in China have been by monocultures dominated by coniferous species. It has been widely reported that timber yield declines and degradation in soil fertility in coniferous plantations over several continuous rotations (see [20], and references therein). On the other hand, in subtropical China, evergreen broadleaved forests are the main native ecosystem type in the region, with a clear potential capacity for carbon sequestration [11]. Among the native species, the broadleaf *Phoebe bournei* (Hemsley) Yang is one of the most valuable tree species in this subtropical region because of its high-quality wood properties, with significant economic and ecological benefits [21]. Sharing the same distribution area, the conifer Chinese fir (*Cunninghamia lanceolata* (Lamb.) Hook. is the most important commercial forest species in subtropical China. It has been widely planted in the southeast provinces of China [20,22], becoming one of the most abundant planted species in the world [23].

The traditional management of Chinese fir plantations is representative of the type of management usually carried out in

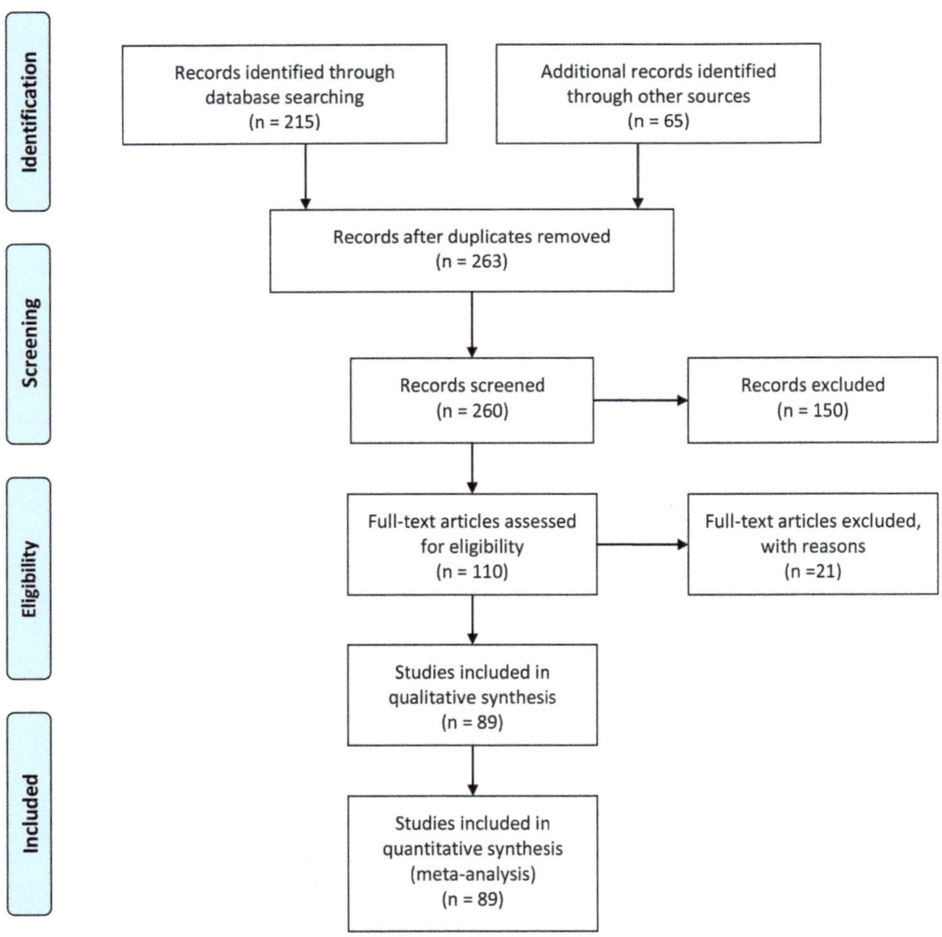

Figure 1. PRISMA flow diagram for the creation of the database of carbon pools in tropical and subtropical forest plantations.

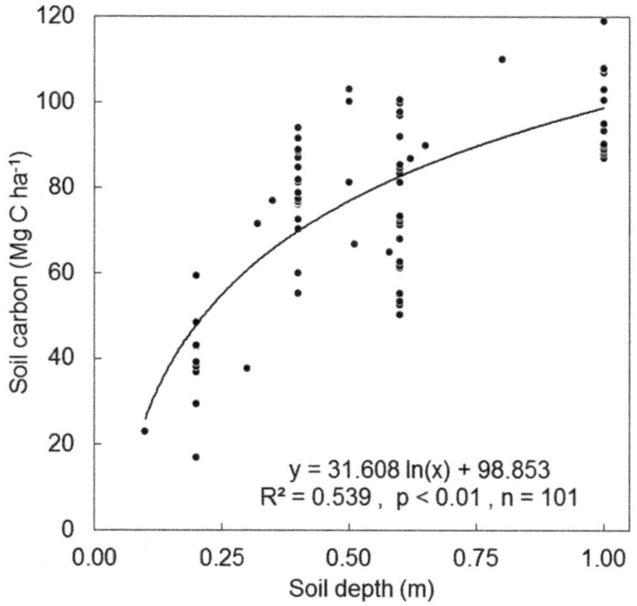

Figure 2. Values of soil carbon *vs.* soil depth in the reviewed literature.

many other tropical and subtropical forest regions around the world. After harvesting and slash-burning the native mixed evergreen broad-leaved forests, pure plantations of fast-growing conifers were established. The plantation sites used to be abandoned after one or two rotations and allowed to regenerate naturally to mixed species stands by stump sprouting and natural seeding, which then acted as a fallow period to restore the site [24]. However, since the 1950s, the plantation area of Chinese fir has been enlarged, and this species has been repeatedly planted on the same sites without intercropping or periods of fallow. Local foresters have generally used a 25-year rotation, with variation from 20 to 30 years depending on soil nutrient abundance [25]. A trend towards rotation length shortening is being observed in the latest years, driven by increasing demand for timber products fueled by economic development and population increase in China. Continuous cultivation of Chinese fir at the same site has resulted in a well-known ecological problem: yield decline over consecutive rotations [20,26]. With these conditionings, an important question has recently been raised in Chinese forestry with implications for the management of forest plantations in other subtropical regions of the world: can forest plantations be managed to produce forest products but in a way that also maintains or even increases ecosystem C? If so, can the differences in ecosystem C between traditional and sustainable management be quantified?

To answer these questions, in this work we tested the hypothesis that subtropical tree plantations accumulate less carbon under traditional exploitative management than under sustainable forest management, which is more similar to natural stand development. We also aim to quantify the potential increase in ecosystem C at a regional level. To test this hypothesis we first reviewed available information on C pools in subtropical forest plantations and related them to factors defining their management (species type, rotation length, stand density). Also, the literature review allowed us to identify the management prescriptions with the highest carbon storage, and to define realistic sustainable management scenarios. Finally, we estimated potential C sequestration gains through time at a regional level by comparing ecosystem C values under traditional and sustainable management regimes in subtropical plantations in China, simulating plantation growth with the forest ecosystem model FORECAST [27].

Materials and Methods

We first carried out a literature review to test if differences in ecosystem C among subtropical plantation types exist, and also to identify the management factors that have maximized ecosystem C in those plantations in the past. Then, we used ecological modelling to estimate the average differences between traditional forest management and sustainable forest management, with consideration of the factors identified in the literature review. As ecological models need to be calibrated for specific ecosystem types [28], we used Chinese fir plantations (*Cunninghamia lanceolata* (Lamb.) Hook.) located in the subtropical region of China as a case study. Continuous repetitions of traditional practices over several rotations (monoculture, short rotations, whole-tree harvesting, and slash burning) have resulted in severe yield decline [26]. Among various possible reasons, N limitation by nutrient exports and resource competition between tree seedlings and understory are the most important factors [20]. Therefore, there is a need to improve the management of this species to establish a more sustainable and multi-purpose forestry in China.

Literature Review of Carbon in Subtropical Plantations

We conducted a structured search of published scientific documents to define the observed range in carbon storage in the ecosystem, as well as in the belowground and aboveground fractions, and their relationships with plantation type (broadleaf or coniferous), plantation age, and stand density. We searched in the on-line databases Google Scholar (by Google Inc. http://scholar.google.com), ISI Web of Knowledge (by Thomson Reuters, http://www.isiwebofknowledge.com), China Academic Journal Database (by China National Knowledge Infrastructure, http://cnki.en.eastview.com/kns50), CAB Abstracts (by the Centre for Agricultural Bioscience International, http://www.cabdirect.org), and the databases hosted by the Canadian Forest Service Bookstore (http://bookstore.cfs.nrcan.gc.ca), the University of British Columbia's library (http://www.ubc.ca/library), and the Public University of Navarre's library (http://www1.unavarra.es/biblioteca).

We divided the population of collected documents into two groups: 1) documents reporting C pools in broadleaf or other non-coniferous forests, and 2) documents describing coniferous or mixed conifer-broadleaf plantations. The final number of documents and sites used for further analysis after a standardized selection process [29] were, for broadleaves 38 documents and 138 sites; whereas for conifers 51 documents and 160 sites (Figure 1; see the full reference list and values of carbon and management variables for each site in Material S1). For each site, we gathered data on plantation age, density, aboveground C density (MgC ha^{-1}), belowground C density, and ecosystem C density. Aboveground C was calculated as tree C plus understory C (if reported). Belowground C carbon was calculated as C in tree roots (and in plant roots if reported), litter C and mineral soil C. Values from graphs were extracted using DataThief III [30]. Original values for mineral soil C were often not directly comparable, as they were measured for different soil depths. To standardize soil carbon values, we modified the method by [31] and used bulk density and soil C concentration to calculate and plot soil C content versus soil depth, obtaining a significant correlation between both variables (Figure 2). We then estimated the

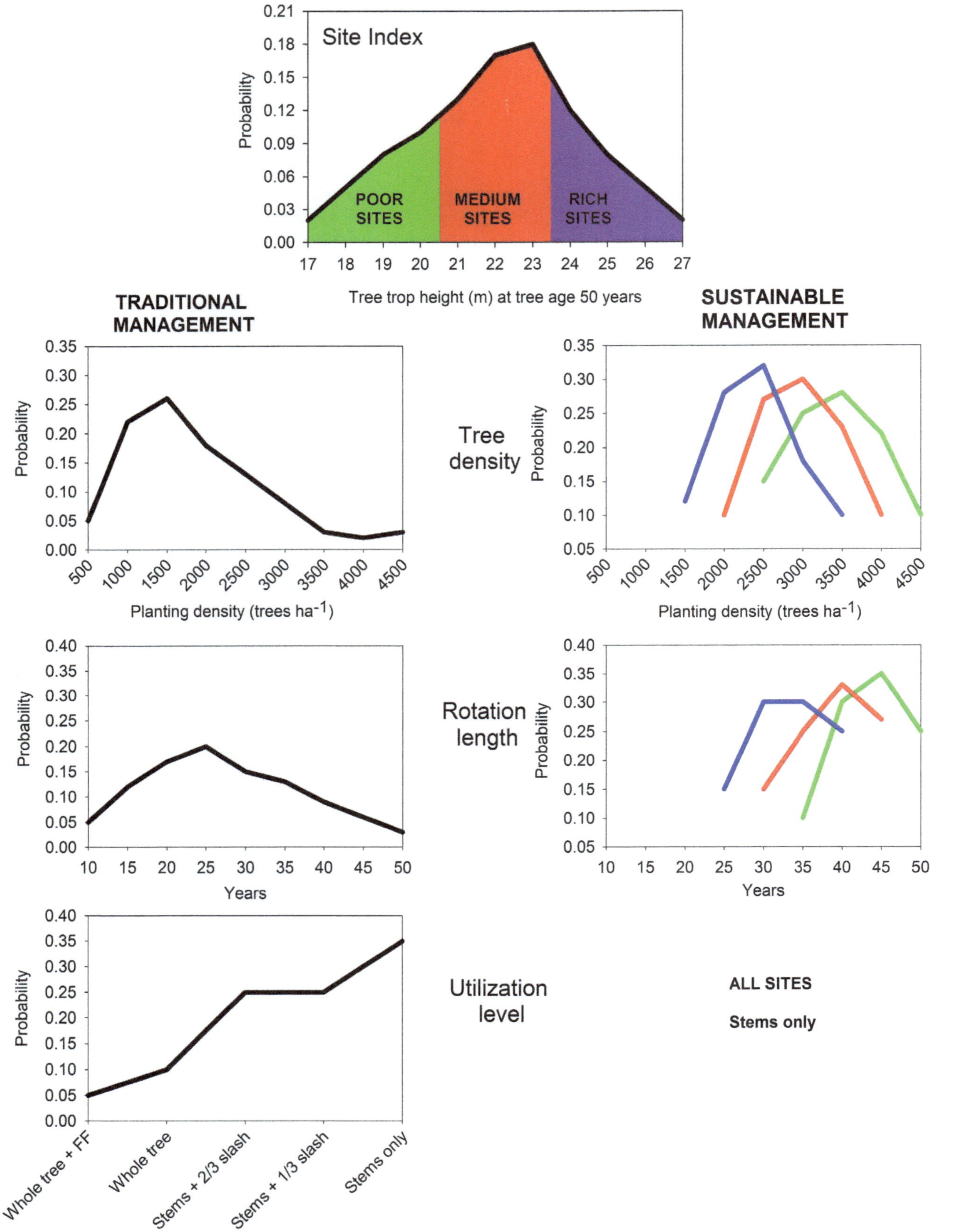

Figure 3. Probability distributions of the four variables defining the management scenarios. Different color lines correspond to the different site qualities as defined in the top graph. FF: Forest floor removal.

Table 1. Values used to calibrate FORECAST parameters related to Chinese-fir.

Parameter	Unit	Rich site	Poor site
Chinese fir parameters			
Nitrogen concentration in needles young/old/dead	%	1.53/1.36/1.13	1.21/1.11/0.93
Nitrogen concentration in stem sapwood/heartwood	%	0.14/0.03	0.12/0.03
Nitrogen concentration in bark live/dead	%	0.44/0.27	0.37/0.24
Nitrogen concentration in branches live/dead	%	0.67/0.52	0.55/0.47
Nitrogen concentration in root sapwood/heartwood	%	0.37/0.06	0.35/0.06
Nitrogen concentration in fine roots live/dead	%	1.17/0.97	0.96/0.79
Shading by maximum foliage biomass	% of full light	8	30
Soil volume occupied at maximum fine root biomass	%	100	95
Efficiency of N root capture	%	98	100
Retention time for young/old foliage/dead branches	years	1/2/40	1/2/40
Fine roots turnover	$year^{-1}$	0.95	1.35
Decomposition rates			
Sapwood (by litter age)	% $year^{-1}$	1–5 years (2.0); 6–10 years (10.0); 11–15 years (30.0); 16–20 years (20.0); > 20 years (4.0)	
Heartwood	% $year^{-1}$	1–10 years (0.4); 11–15 years (10.0); 16–25 years (15.0); 25–40 years (10.0); >40 years (2.0)	
Bark	% $year^{-1}$	1–5 years (2.0); 6–20 years (12.0); 20–40 years (20.0); >40 years (4.0)	
Branches and large roots	% $year^{-1}$	1–5 years (10.0); 6–10 years (45.0); 11–15 years (35.0); >15 years (4.0)	
Needles (poor site)	% $year^{-1}$	1–2 years (20.0); 3–5 years (30.0); 6–10 years (40.0); >10 years (2.0)	
Needles (good site)	% $year^{-1}$	1–2 years (27.0); 3–5 years (30.0); 6–10 years (40.0) >10 years (3.0)	
Fine roots	% $year^{-1}$	1–2 years (30.0); 3–4 years (50.0); >4 years (9.0)	

Decomposition rates indicate the mass loss in one year as a fraction of the initial mass at that year. See text for list of bibliographical sources for model calibration.

standardized mineral soil carbon content at 60 cm depth as the value that would correspond if soil carbon distribution in each soil followed a relationship with soil depth parallel to the regression line.

We tested the equality of means for aboveground, belowground and ecosystem C between the two types of plantations using Welch's modified *t*-test, (a robust test that does not assume equality of variances; [32]), setting significance level at 0.05. We also carried out regressions (weighted by the error associated to each measure) of carbon density vs. plantation age and stand density, lumping together data from conifer and broadleaf plantations as initial analysis did not show improvement of the regression results by splitting the data into species groups or location (data not shown). All the statistical analyses where carried out with Sigmaplot 10.0 (Systat Software Inc.), JMP v.5.0.1 (SAS Institute, NC, USA), and S-PLUS 6.1 (Insightful Corp.).

Simulating Ecosystem C in Planted Forests with FORECAST

In the second part of our research, we used the results from the literature review to define probability distributions for the two main factors defining forest plantation management (stand density and rotation length) that maximized ecosystem C, and we assumed that they are indicators of more sustainable forest management. Probability distribution of site quality was defined using the distribution map of site qualities for Chinese fir in south-eastern China [33]. Probability distributions of planting density, rotation length and utilization level for traditional management were also created for both sets of simulations after consultations with panels of local experts composed by academics, forest officers, and forest managers (Figure 3). The panel of local experts indicated the need for creating different management scenarios based on site quality for sustainable management. Hence, probability distributions of tree density and rotation length were different in each site quality for sustainable management, but they were the same in all sites under traditional management.

Using the probability functions, we constructed 200 Monte Carlo projections for each set of simulations (traditional management and sustainable forest management). Each run simulated a management scenario (a random combination of site quality, rotation length, and stand density). In traditional management, all trees planted were Chinese fir. In sustainable management, probability distributions show total tree density, which was composed by Chinese fir and *Phoebe bournei* in a 1:1 ratio, the most suitable species mix for carbon sequestration [21].

We then simulated each scenario for 120 years with the ecosystem model FORECAST, which is a stand-level model specially designed for assessing the effects of various forest management practices on forest carbon storage and sequestration. FORECAST has been successfully tested and applied in many world forests before, including tropical and subtropical forests [28].

FORECAST is a management-oriented, deterministic, non-spatial stand-level forest growth and ecosystem dynamics simulator that operates at annual time steps. The model has been described in detail before [27,28], and therefore only a summary of the main driving function used to calculate tree growth is provided in Material S1. The model uses a mass balance approach to estimate ecosystem nutrient circulation, and how nutrient availability limits tree growth together with available light in the canopy.

FORECAST has three application stages: 1) assembling calibration data and generating historical rates of key ecosystem processes; 2) model initialization by establishing the ecosystem condition for the beginning of a simulation run; and 3) simulation of tree and plant growth.

Input data came from sites covering the observed range of Chinese-fir plantation growth, qualitatively described as poor, medium and good (Table 1). The sites were scaled quantitatively as 17, 21 and 27 (from poor to good, based on canopy top height in m at age 50 years) to represent a relative index of tree growth required for extrapolation within the model. Tree data on biomass, mortality, stand density, tree height, canopy height, nutrient concentrations of live tissues and other data were based on reported values for similar Chinese-fir plantations [25,34,35,36,37,38,39,40,41,42,43].

Model calibration data for *Phoebe bournei* forests (Table 2) came from literary sources and field data [21]. They included data of biomass accumulation and stand density [44,45,46,47,48,49,50], decomposition rates [51,52,53,54], soil nutrients under different site conditions [55,56,57], and photosynthetic efficiency [58,59,60,61]. Understory was simulated as grass and shrub complexes [20]. Published data were used to characterize shrub biomass, height, tissue nutrient concentrations and other relevant data [62,63,64,65] (Table 3).

Data describing soil processes and nutrient inputs in precipitation and slope seepage, mineral soil cation and anion exchange capacities, humus mass, nutrient concentrations in litterfall, litter decomposition rates and others. were derived from literature [38,39,40,66,67,68,69,70,71] (Table 3). To establish initial site conditions the model was run in set-up mode, forcing the model to match the observed site conditions [72,73]. Initial conditions were created for each site quality, by running several cycles of tree growth ending with stand-replacing windthrow, as typhoons are the most common natural disturbance in this region [74]. A library of initial conditions was created with initial soil organic matter values (humus+litter) ranging from 90.9 to 296.3 Mg ha^{-1} for the poorest and richest sites, respectively.

Published field data on several chronosequences at different site qualities were used to evaluate FORECAST performance for Chinese fir and *Phoebe bournei* plantations (see Material S1). After obtaining a satisfactory model performance, we run the model for the 400 randomly-generated scenarios described above. A sub-set of documents gathered during the literature review dealing with Chinese fir plantations was also used to evaluate the model performance for ecosystem C.

Results

From our review, 23.9% of the studies on bradleaves used local native species, but *Populus* spp. (19.6%), *Tectionia sp.* (13.1%) and *Eucalyptus* spp. (8.10%) were the most common commercial species used. For conifers, 81.4% of the studies reported data from *Cunninghamia* sp. plantations, and 14.3% came from *Pinus* spp. plantations. Our results showed that subtropical broadleaves plantations can, on average, sustain higher carbon densities than subtropical conifer plantations (Figure 4). The differences, however, are mostly caused by differences in belowground C, whereas differences among plantations types in aboveground C are non-significant. Not surprisingly, the trend in subtropical forest plantations is to increase ecosystem carbon storage as plantation age increases. This is basically the result of increase in tree size,

Table 2. Values used to calibrate FORECAST parameters related to *Phoebe bournei.*

Parameter	Unit	Rich site	Poor site
Chinese fir parameters			
Nitrogen concentration in leaves young/old/dead	%	1.35/0.92/0.69	1.15/0.80/0.67
Nitrogen concentration in stem sapwood/heartwood	%	0.18/0.04	0.16/0.04
Nitrogen concentration in bark live/dead	%	0.55/0.47	0.49/0.33
Nitrogen concentration in branches live/dead	%	0.33/0.29	0.28/0.26
Nitrogen concentration in root sapwood/heartwood	%	0.37/0.29	0.34/0.27
Nitrogen concentration in fine roots live/dead	%	0.81/0.57	0.65/0.56
Shading by maximum foliage biomass	% of full light	8	25
Soil volume occupied at maximum fine root biomass	%	100	100
Efficiency of N root capture	%	100	100
Retention time for young/old foliage/dead branches	years	1/0.5/30	1/0.5/30
Fine roots turnover	year^{-1}	1.00	1.15
Decomposition rates			
Sapwood (by litter age)	% year^{-1}	1–3 years (0.1); 4–15 years (2.0); 15–20 years (12.0); >20 years (9.0)	
Heartwood	% year^{-1}	1–3 years (0.1); 4–15 years (2.0); 15–20 years (12.0); 20–40 years (9.0); > 40 years (2.0)	
Bark	% year^{-1}	1–10 years (20.0); 11–20 years (40.0); 21–30 years (35.0); >30 years (2.0)	
Branches and large roots	% year^{-1}	1–5 years (15.0); 6–15 years (35.0); 16–20 years (45.0); >20 years (15.0)	
Leaves (poor site)	% year^{-1}	1–2 years (45.0); 3–5 years (35.0); >6 years (20.0)	
Leaves (good site)	% year^{-1}	1–2 years (50.0); 3–5 years (40.0); >6 years (20.0)	
Fine roots	% year^{-1}	1–5 years (45.0); >6 years (35.0)	

Decomposition rates indicate the mass loss in one year as a fraction of the initial mass at that year. See text for list of bibliographical sources for model calibration.

Table 3. Values used to calibrate FORECAST parameters related to the shrub complex and soil processes.

Shrub complex parameters	Unit	Rich site	Poor site
Nitrogen concentration in leaves live/dead	%	1.68/1.38	1.12/0.76
Nitrogen concentration in stems live/dead	%	0.48/0.14	0.32/0.10
Nitrogen concentration in rhizomes live/dead	%	1.14/1.05	0.76/0.70
Nitrogen concentration in roots live/dead	%	1.10/0.84	0.73/0.56
Shading by maximum foliage biomass	% of full light	0.20	0.45
Soil volume occupied at maximum fine root biomass	%	60	50
Efficiency of N root capture	%	99	99
Transfer from live to dead stem/rhizomes/roots	% year^{-1}	20/20/30	20/20/40
Retention time for foliage	years	1	1
Decomposition rates		Litter age in years (decomposition rate in %)	
Foliage (poor site)	% year^{-1}	1 year (80.0); 2–3 years (60.0); 4–5 years (50.0); >5 years (2.0)	
Foliage (rich site)	% year^{-1}	1 year (95.0); 2–3 years (60.0); 4–5 years (50.0); >5 years (2.0)	
Stems & roots	% year^{-1}	1 year (20.0); 2–3 years (30.0); 4–5 years (40.0); >5 years (2.0)	
Soil parameters		**Rich site**	**Poor site**
Nitrogen concentration in slow/fast humus	%	3.00/1.40	3.0/1.40
Decomposition rate slow/fast humus	% year^{-1}	0.15/2.00	0.15/2.00
CEC soil (CEC humus)/AEC	kg N ha^{-1}	80.0 (0.2)/20.0	40.0 (0.2)/5.0
Atmospheric deposition/non-symbiotic fixation	kg N ha^{-1} year^{-1}	4.9/1.0	4.9/1.0
Non-symbiotic N fixation rate	Kg N ha^{-1}	2.5	3.0

Decomposition rates indicate the mass loss in one year as a fraction of the initial mass at that year.

and accumulation of litter and coarse woody debris biomass. This trend, however, seems to slow down over time, with little expected gains beyond plantation age ~90, as described by the significant exponential model (Figure 5). Ecosystem C initially increases as plantation density increases, reaching a peak at about 1500–2500 stems ha^{-1}. For higher densities the trend is to reduce ecosystem C (Figure 5).

During the simulation runs, the most frequent scenario for sustainable forest management was 3000 trees ha^{-1} (1500 Chinese fir, 1500 *P. bournei*), rotation length of 40 years and stem-only harvesting. For traditional management, the most frequent scenario was 1500 trees ha^{-1} (Chinese fir only), rotation length of 25 years, stem plus slash removal, and site preparation by understory burning. Comparing these two scenarios when run for 240 years, we can see how plantations managed for sustainable, long-term goals, maintain ecosystem productivity higher and during longer periods that traditional intense management.

Yearly net primary productivity under traditional management quickly drops in each consecutive rotation, and as a consequence tree biomass (and the associated merchantable volume) also decreases rapidly over time (Figure 6). In addition, there is an increasingly stronger competition with understory. Although with traditional management higher peaks of available soil N are achieved than in the sustainable management, the total amount of ecosystem N decreases over time. On the other hand, under sustainable management higher levels of soil organic matter are sustained (Figure 6).

When all the inputs and outputs are accumulated through the 240 years of simulation, it is clear how current management is depleting ecosystem N reserves. On the other hand, sustainable management can maintain and even increase N pools (Figure 7). Traditional management removes N from the ecosystems during the site preparation (directly through slash-burning and indirectly by erosion caused by heavy machinery and losses of plant cover), with the export of harvested products, and by N leaching. Under sustainable forest management, slash-burning is not used and erosion is greatly reduced as practices affecting soil stability and plant cover are minimized. N removal in harvested products is also lower as only stems (with low N concentrations) leave the plantations. On the other hand, as N reserves under sustainable management are higher, leaching losses are also higher, just because the soil N pool is bigger than in traditional management. All flows considered, N outputs with current practices are not only higher, but N inputs are also smaller, as there is less N fixation due to the reduced biological activity in sites of lower productivity. Also, because any given stand gets less N inputs by seepage as it is surrounded by stands that also have lower N capitals.

When simulating how the C pools in Chinese fir stands would evolve over the 21st century, our results showed that if sustainable forestry practices were introduced at a rate of 10% of the total planted area, the region would keep on average a higher and sustained level of carbon storage over the long term, compared to traditional management (Figure 8). At the beginning of our scenarios a large overlapping range for ecosystem C between traditional and sustainable plantations can be seen (grey area). Such fact indicates that some time is needed for the sustainable management to significatively reduce C loses compared to traditional management. This is not surprising as productivity losses are accumulative and occur mostly at the change of rotation. However, by the middle and until the end of the simulated 120 years the overlap has been greatly reduced. By that time, only the sites with the highest C contents and managed traditionally would be comparable to the poor sites under sustainable management. Such increase in carbon density could be 12% by 2020. This difference would grow over time, reaching 17%, 52%, 79%, and 67% greater carbon density under sustainable forest management

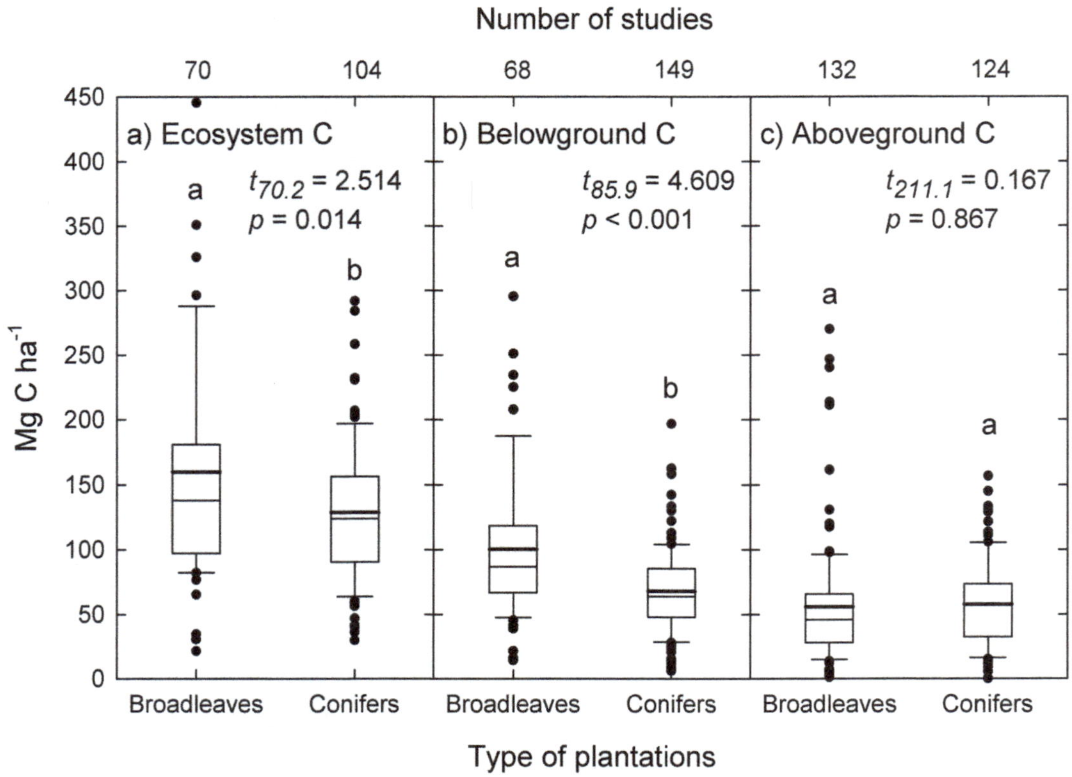

Figure 4. Distribution of C storage in published bibliography for different subtropical plantation types. The thick horizontal line indicates the average of all observations. Letters indicate significant differences with Welch's *t* at $P<0.05$. See Material S1 for the complete list of references used.

after 25, 50, 75 and 100 years, respectively. The average weighted carbon density increase over the 100-year simulation period was 67.5%.

As a way to validate our estimations at the stand level, when analyzing the average and 95% confidence intervals for the whole set of projected ecosystem C values they were always inside the range of reported values for Chinese-fir plantations (Figure 8), providing confidence on our estimations.

Stand-level differences between management regimes would have large regional implications. Extrapolated to the whole area dedicated to Chinese fir plantations (7,046,520 ha [11]), together with the plan of substituting traditional by sustainable management implemented at a ratio of 10% of the land every year, the increase in ecosystem C stored in the mixed Chinese fir plantations could reach 261.2 TgC by 2050 (Figure 8). After that year, the differences in ecosystem carbon storage between management options show a temporal decline. This is as a consequence of the

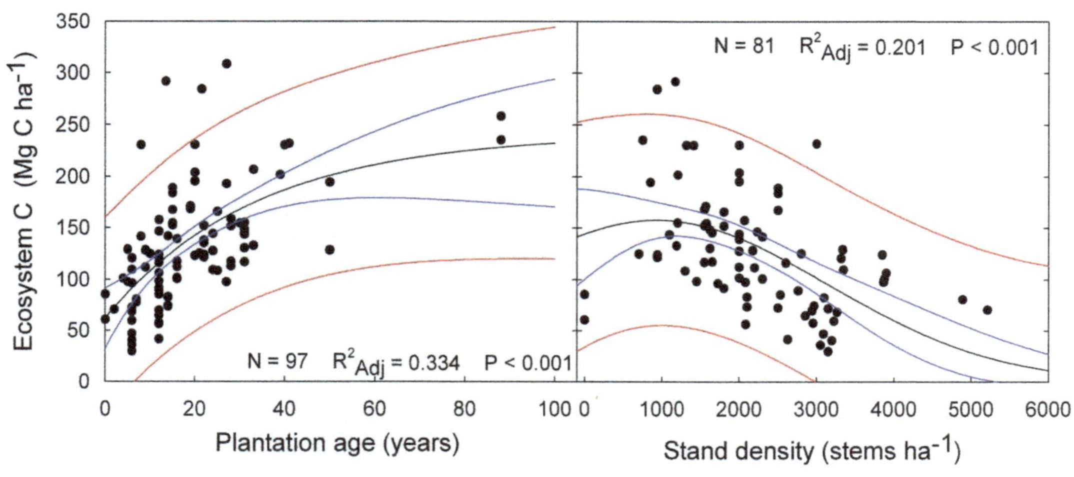

Figure 5. Carbon stocks in subtropical plantations for different A) rotation ages, and B) stand densities. Black lines indicate the best-fit model, blue lines 95% confidence interval and red lines 95% prediction intervals.

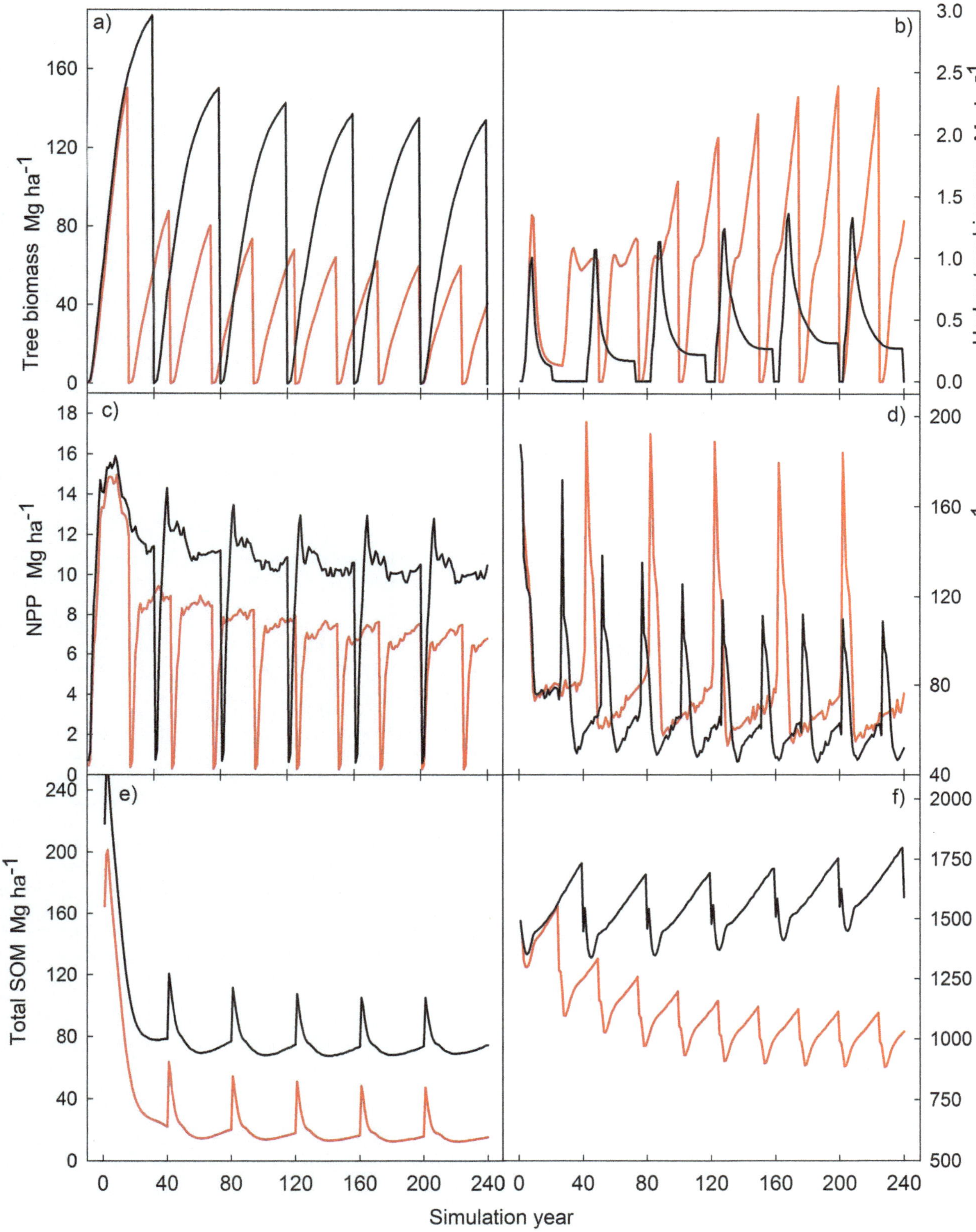

Figure 6. Change in ecological variables in Chinese fir plantations in 240 years of simulation under the average plan two types of management. Current (red line) 25-year rotation, whole-tree harvesting, 1500 trees ha^{-1} Chinese fir planted; and close-to-nature management (black line) 40-year rotations, stem-only harvesting, 1500 trees ha^{-1} Chinese fir +1500 trees ha^{-1} *Phoebe bournei.*

harvesting of sustainable forest plantations primarily clustered around the years 40 to 60. The second rotation growth would again produce an increase in stored C, which would remain above 200 TgC after 120 years of simulation.

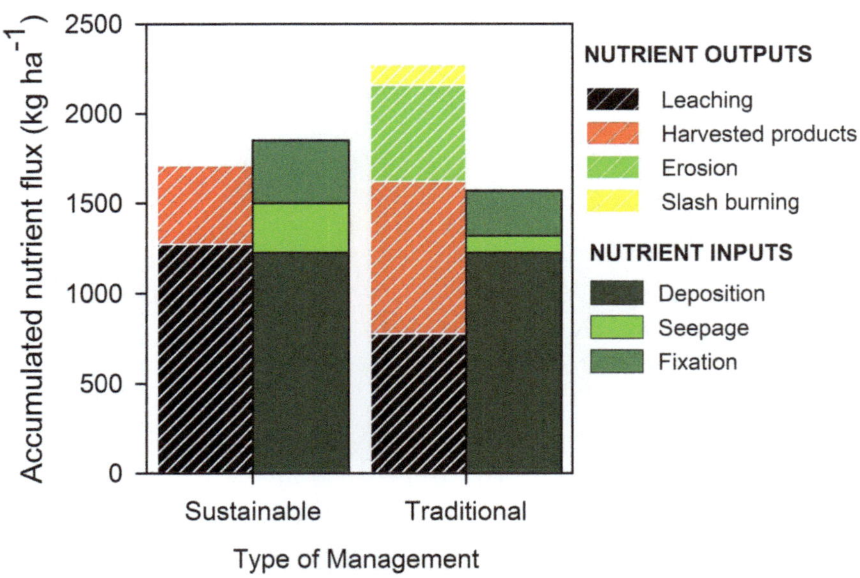

Figure 7. Accumulated N inputs (solid bars) and outputs (dashed bars) in Chinese subtropical plantations after 240 years of management under two most typical management regimes for traditional and sustainable plantations.

Discussion

Subtropical plantations have been described as the most effective for C sequestration [75,76,77]. In subtropical regions, local broadleaved species are usually planted, but the available data on C stocks in subtropical plantations came mostly from the five genuses more widely planted around the globe: *Eucalyptus, Populus, Tectonia, Pinus* and *Cunninghamia*. These are the most valuable among the fast-growing species currently used, and our results show that selecting among the species type (broadleaves or conifers) can produce significant differences in C pools depending on plantation type and management regime. Ecosystem C increases with tree age in subtropical plantations, as these systems develop ecological structures that make them closer to natural forests over time [78]. Older plantation age has also been associated to higher soil carbon levels [79]. Therefore, increasing rotation length should be encouraged when implementing sustainable forestry plans for increasing carbon stocks, as well as generating more biophysical structure for higher niche diversity and its associate biodiversity.

[80] have pointed out that density plays an important role in the productivity of mixed stands, with higher productivities in mixed stands than in monocultures at medium to high densities (>1500 trees ha^{-1}). Our results also indicate that density should be maintained at medium values, avoiding excessive inter-tree competition for resources such as light, water and nutrients that could reduce stand productivity, but also avoiding low densities that may leave some resources unused by trees. Our estimates also fit with several field experiments in mixed plantations of Chinese fir with other trees species, which have consistently shown higher biomass and C density in mixed stands with a tree density ratio of 1:1 for Chinese fir/companion species, for total densities in the 2000–3000 trees ha^{-1} range [81,82,83,84].

Our results indicate that managing plantations designed for multiple values could sustain higher carbon storage potentials if they include broadleaves, are managed for rotations lengths of 25–50 years, and have planting densities of 1500–2500 stems ha^{-1}. Our estimations corroborate previous field research that have reported maximum biomass values in mixed Chinese fir plantations for similar rotation and density ranges, reporting little gain or even biomass decrease with plantation densities above 3000 trees ha^{-1} [82,85]. In this regard, forest plantations seem to have important functional similarities with native forests, where intermediate tree densities have been reported as the ones reaching the carrying capacity of the ecosystem, without surpassing it [86].

Differences in C density among management types can be better understood by comparing the typical (most frequent in our set of simulations) management regimes for traditional and sustainable forest management practices. Depletion in ecosystem N reserves reduces ecosystem productivity, which in turn produces a long-term decrease in tree biomass [87]. This productivity reduction is much more pronounced under traditional practices, as shorter rotation lengths accumulate higher and faster nutrient exports. In addition, short rotations and lower stand densities produce an under-canopy environment with more light available and therefore more suitable for understory growth (Figure 6). In these plantations, understory grass and shrubs usually sprout from rhizomes and other underground structures, rather than from seeds [20]. This phenomenon provides a competitive advantage against planted tree seedlings, which is exacerbated by the advantage of understory's lower nutrient needs in increasingly nutrient-poor sites as a consequence of intensive organic matter extraction after harvesting [88], slash burning [89], and N leaching [90] (Figure 7).

Reductions in soil organic matter and its associated sources of nutrients, water retention and physical properties would also be more severe under traditional management [16], leading to the already reported yield decline in Chinese fir plantations [26]. On the other hand, our results show that using longer rotations, moderately higher stand densities, and mixing conifers and broadleaves would prevent site productivity decline. With more sustainable management, the ecosystem would maintain and even can increase N reserves by keeping a positive N balance, an indicator of ecological sustainability of forest management in the long-term [26,91].

Theoretical forest ecology indicates that in tree species mixtures, if each species has different growing patterns (differential shade

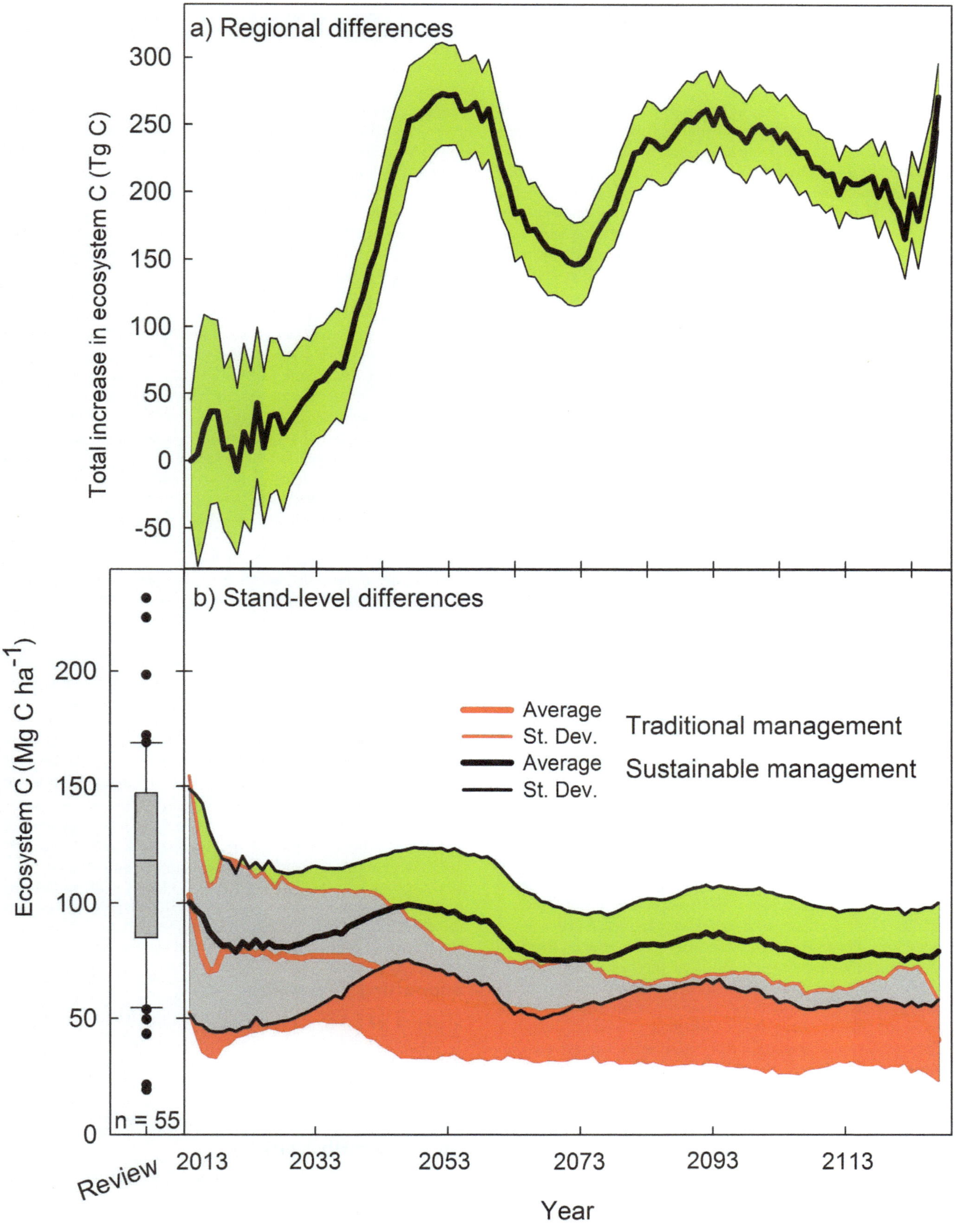

Figure 8. Total carbon gains when switching from current traditional intensive forestry to sustainable forest management at an annual rate of 10% of total area dedicated to Chinese-fir plantations (panel A). Ecosystem carbon density in Chinese fir plantations under current intensive management (red lines), and sustainable management (green lines) (panel B). Thick lines represent average and color areas

represent one standard deviation. The grey area indicates values of ecosystem C that can be achieved by either sustainable or traditional management plans. The box-plot in the bottom-left summarizes the values obtained from a review of 55 published peer-reviewed papers on carbon stocks in Chinese fir plantations (see Material S1 for the complete reference list).

tolerance, physical separation of canopies, phenological differences, successional separation, and differences in soil resource utilization), inter-specific competition may be reduced. Growing evidence indicates a generalized pattern in mixed forests, in which niche separation at both canopy and root levels leads to a more complete use of site resources [80,92,93,94]. This phenomenon has already been reported in mixed tropical plantations [95,96,97]. Therefore, resources in the mixed stand are used more efficiently, hence total stand productivity increases. This phenomenon is due to two facts: complementary resource use between species and facilitative improvement in nutrition in mixtures with a nitrogen-fixing species [93,95,98].

Our results fit with this theoretical background, showing that in Chinese fir plantations, implementing mixtures with *Phoebe bournei* will not only increase productivity, but it will also prevent SOM and N losses and their associated yield decline over time. The additional increase in sequestered ecosystem C is another positive effect of species mixtures in these sites. Similar results have also been reported when simulating boreal [99], temperate [73] and tropical [100] forests.

Forest management can greatly affect net C exchange with the atmosphere, not only because it changes ecosystem C pools, but also because the C removed from the ecosystem during harvesting is stored in wood products, such as paper, cardboard, furniture, structural timber, etc. [101]. We have shown that subtropical plantations managed under sustainable forest management can store more ecosystem carbon. However, their net effect on C sequestration depends on forest types, their management regimes, the type of wood products produced, and the efficiency of biomass conversion, as well as assumptions about how the wood and wood residues will substitute for other products with greater GHG emissions and for fossil energy [102].

In Chinese fir plantations under traditional management, most of the production goes to the market as small diameter trunks. Over time, harvested sizes become increasingly smaller as plantation sites lose productivity due to the nutrient losses and competition with understory. However, using a more sustainable forest management, harvesting rates are reduced as it takes more time to complete the harvesting turn, but on the other hand harvested tree sizes are bigger and stay bigger through several rotations. These wood products are expected to reach better prices and last longer because they will be used more as structural timber and high quality furniture [103]. The increased ecosystem C storage, combined with the longer lifespan of higher quality wood products, could compensate the reduced harvesting rates compared to traditional management. Although a life-cycle analysis is needed to test this hypothesis, previous work has shown that secondary forests allowed growing rather than being harvested and their wood products stored could result in lower net CO_2 emissions [104].

Effects of forest management actions on carbon should be examined for large areas and over long time periods [105]. In the case of China, government's commitment to increase forest cover by 40 million ha with new plantations may not have an impact on carbon storage as important as intended unless better forest management is also implemented [106]. In this context, applying a more ecologically-based forestry could make a significant difference. Assuming those new 40 million ha were evenly distributed among all Chinese forest types, there could be 17.55 million ha of new subtropical plantations by 2020 [8]. If the differences between sustainable and traditional forestry were similar to the ones we have estimated for Chinese fir plantations, a more ecologically-based forestry implemented in Chinese subtropical plantations could add 650.8 TgC by 2050.

Previous research on long-term C storage under sustainable forest management regimes in Chinese temperate forests have estimated ecosystem C densities of 600–750 MgC ha^{-1} in poplar (Dai *et al.* under review), 450–800 MgC ha^{-1} in spruce [107], and 200–325 MgC ha^{-1} larch plantations [108]. In all cases, estimates C densities for temperate plantations were higher than our estimates for subtropical plantations. Therefore, it seems reasonable to assume that differences between traditional and sustainable forest management regimes in temperate plantations are at least similar to the ones calculated for subtropical plantations. Under this assumption, C storage in Chinese plantations could potentially be increased by 1,482.8 TgC by 2050. These estimations could even be conservative, as temperate plantations have on average higher levels of ecosystem carbon stocks [109].

In terms of emissions, applying sustainable forestry to all new Chinese plantations could potentially lead to an average net increase in stored C of 0.04 PgC y^{-1} in 2050, equivalent to offsetting 1.91% of China's annual carbon emissions in 2010 [110]. More importantly, this carbon increase would be sustainable with opportunities to maintain other ecological services and functions linked to increased biodiversity and less degraded forest soil [3].

In conclusion, our results show that applying a more ecologically-based forestry would have positive impacts not only on long-term plantations productivity, but it can also be a better strategy to implement climate change mitigation programs. Carbon stored in plantations could be further increased if sustainable forestry practices were combined with other practices not examined here, such as genetic improvement of trees, landscape design, etc., although the ecological sustainability of such practices should also be assessed. Finally, socio-economic incentives for farmers to extend rotations and to market non-conifer timber and other products would also be important.

Supporting Information

Material S1 Additional information describing the FORECAST model, its validation for subtropical forests in China, and the list of references used for the review on carbon pools in subtropical planted forests.

Author Contributions

Conceived and designed the experiments: XW JAB. Performed the experiments: JAB. Analyzed the data: XW JAB. Contributed reagents/materials/analysis tools: XW JAB. Wrote the paper: XW JAB.

References

1. Lewis SL, López-González G, Sonké B, Affum-Baffoe K, Baker TR, et al. (2009) Increasing carbon storage in intact African tropical forests. Nature 457: 1003–1007.
2. Food and Agriculture Organization (2012) State of World's Forests 2012. Rome: Food and Agriculture Organization.
3. Paquette A, Messier C (2010) The role of plantations in managing world's forests in the Anthropocene. Fron Ecol Environ 8: 27–34.
4. Kirilenko AP, Sedjo RA (2007) Climate change impacts on forestry. PNAS 104, 19697–19702.
5. Rhemtulla JM, Mladenoff DJ, Clayton MK (2009) Historical forest baselines reveal potential for continue carbon sequestration. PNAS 105: 6082–6087.
6. Kimmins JP, Blanco JA, Seely B, Welham C, Scoullar K (2008) Complexity in Modeling Forest Ecosystems; How Much is Enough? For Ecol Manage 256: 1646–1658.
7. Niu XZ, Duiker SW (2006) Carbon sequestration potential by afforestation of marginal agricultural land in the Midwestern U.S. For Ecol Manage 223: 415–427.
8. State Forestry Administration People's Republic of China (2009). People's Republic of China forestry outlook study. FAO Working Paper No. APFSOS II/WP/2009/11. Bangkok.
9. Chen X, Zhang X, Zhang Y, Wan C (2009) Carbon sequestration potential of the stands under the Grain for Green program in Yunnan Province, China. For Ecol Manage 258: 199–206.
10. Nabuurs GJ et al. (2007) Forestry. In: Metz, B., Davidson, O.R., Bosch, P.R., Dave, R., Meyer, L.A. (eds) Climate Change 2007: Mitigation. Contribution of Working Group III to the Fourth Assessment Report of the Intergovernmental Panel on Climate Change. Cambridge: Cambridge University Press.
11. Wei X, Blanco JA, Jiang H, Kimmins JP (2012) Effects of nitrogen deposition on carbon sequestration in Chinese fir forests. Sci Tot Environ 416: 351–361.
12. Food and Agriculture Organization (2001) Biological Sustainability of productivity in successive rotations. Report based on the work of J. Evans. Forest Plantation Thematic Papers, Working Paper 2. Forest Resources Development Service, Forest Resources Division. Rome: Food and Agriculture Organization.
13. Carson KM, Curran LM, Asner GP, Pittman AM, Trigg SM, et al. (2013) Carbon emissions from forest conversion by Kalimantan oil palm plantations. Nature Clim Change 3: 283–287.
14. Gibson L, Lee TM, Koh LP, Brook BW, Gardner TA, et al. (2011) Primary forests are irreplaceable for sustaining tropical biodiversity. Nature 478: 378–381.
15. Jactel H, Brockerhoff EG (2007) Tree diversity reduces herbivory by forest insects. Ecol Let 10: 835–848.
16. Liao C, Luo Y, Fang C, Chen J, Li B (2012) The effects of plantation practice on soil properties based on the comparison between natural and planted forests: a meta-analysis. Global Ecol Biogeogr 21: 318–327.
17. Constanza R, D'arge R, De Groot R, Farber S, Grasso M, et al. (1997) The value of the world's ecosystem services and natural capital. Nature 387: 253–260.
18. Lo YH, Lin YC, Blanco JA, Yu CH, Guan BT (2012) Moving from ecological conservation to restoration: an example from central Taiwan, Asia. In: Blanco JA, Lo YH (Eds.) Forest Ecosystems: more than just trees. Rijeka: InTech. 339–354.
19. Lewis SL (2006) Tropical forests and the changing Earth system. Phil Trans R Soc Lond B 261: 195–210.
20. Bi J, Blanco JA, Kimmins JP, Ding Y, Seely B, et al. (2007) Yield decline in Chinese Fir plantations: A simulation investigation with implications for model complexity. Can J For Res 37: 1615–1630.
21. Wang W, Wei X, Liao W, Blanco JA, Liu Y, et al. (2013) Evaluation of the effects of forest management strategies on carbon sequestration in evergreen broad-leaved (*Phoebe bournei*) plantation forests using FORECAST ecosystem model. For Ecol Manage 300: 21–32.
22. Lu ZH, Wu G, Ma X, Bai GX (2002) Current situation of Chinese forestry tactics and strategy of sustainable development. J For Res 13: 319–22.
23. Food and Agriculture Organization (2006) Global planted forests thematic study: results and analysis, by D. Del Lungo, J. Ball, J. Carle. Planted Forests and Trees Working Paper 38. Rome: Food and Agriculture Organization.
24. Yu XT (1997) (Ed.) Silviculture of Chinese Fir. Fuzhou: Science and Technology Press of Fujian. 250 pp. (in Chinese).
25. Wu ZL (1984) Chinese-fir China Forestry. Beijing: Publishing House. (in Chinese).
26. Tian DL, Xiang WH, Chen XY, Yan WD, Fang X, et al. (2011) A long-term evaluation of biomass production in first and second rotations of Chinese fir plantations at the same site. Forestry 84: 411–418.
27. Kimmins JP, Mailly D, Seely B (1999) Modelling forest ecosystem net primary production: the hybrid simulation approach use in FORECAST. Ecol Model 122: 195–224.
28. Kimmins JP, Blanco JA, Seely B, Welham C, Scoullar K (2010) Forecasting Forest Futures: A Hybrid Modelling Approach to the Assessment of Sustainability of Forest Ecosystems and their Values. London: Earthscan. 218 p.
29. Moher D, Liberati A, Tetzlaff J, Altman DG, The PRISMA Group (2009) Preferred Reporting Items for Systematic Reviews and Meta-Analyses: The PRISMA Statement. PLoS Med 6(6): e1000097. doi:10.1371/journal.pmed1000097.
30. Tummer B (2006) DataThief III v1.6. Available: http://datathief.org. Accessed: 22 October 2013.
31. Guo LB, Gifford M (2002) Soil carbon stocks and land use change: a meta analysis. Global Chan Biol 8: 345–360.
32. Quinn GP, Keough MJ (2002) Experimental design and data analysis for biologist. Cambridge: Cambridge University Press.
33. Yu XT (1999) Chinese fir: In Chinese Forests Forests Editorial Committee (Eds) (China Forestry Publishing House, Beijing) 1–29 (in Chinese).
34. Pan WC, Tian DL, Lei ZX, Kang WX (1983) Studies on the nutrient cycling in the Chinese-fir plantations: (II) Contents, accumulation rate and biological cycling of nutrient elements in the fast-growing Chinese-fir forest in the hill regions. J Central South For Inst 3: 1–12 (in Chinese).
35. Shao L (1992) Chinese fir. In: Sheng W, Chen B, Xu H, Xu X (Eds) Research on Site Degradation of Timber Plantation. Beijing: Chinese Science & Technology Press. 87–92 (in Chinese).
36. Yang YS, He ZM, Qiu RH, Yu XT, Huang BL (1999) Growth pattern of 29-year-old Chinese-fir grown from seed in different rotations. Sci Silvae Sin 35: 32–36 (in Chinese).
37. Xiao W, Nie D, Zhang J (1999) Study on biomass and energy use efficiency of the stands of *Cunninghamia lanceolata* in China. For Res 3: 253–258.
38. Liu F, Luo R, Jiang J (1991) Quantities and effects of various phosphorus compounds in soil under Chinese-fir plantation. J Nanjing For Univ 4: 7–12 (in Chinese).
39. Tian DL, Zhao K (1989) Studies on the litter in a Chinese-fir plantation ecosystem: (I) Amounts, composition and dynamics of litter. J Central South For Inst 9(Sup): 38–43 (in Chinese).
40. Tian DL (1994) Study on bio-geochemistry cycling in Chinese-fir ecosystem. In: Forestry Ministry of China (Ed.) Studies on Forest Ecosystems in China. Harbin: Publishing House of Northeast Forestry Univ. 136–145 (in Chinese).
41. Zhong AL, Hsiung WY (1993) Evaluation and diagnosis of tree nutritional status in Chinese-fir (*Cunninghamia lanceolata* (Lamb) Hook) plantations, Jiangxi, China. For Ecol Manage 62: 245–270.
42. Zhou X (1994) Long-Term Research On China's Forest Ecosystems. Harbin: Northeast Forestry Univ. Press. 320 pp. (in Chinese).
43. Zhou H (1999) Study on productivity of mixed forest planted on site of continuously planted Chinese-fir. J Fujian College For 2: 165–169. (in Chinese).
44. Du J, Lu Y (2009) Study on growth regularity of *Phoebe zhennan* plantation. J Zhejiang For Sci Technol 29: 9–12 (in Chinese).
45. Ma M, Jiang H, Liu Y (2008). Biomass, carbon content, carbon storage and their vertical distribution of *Phoebe bournei* artificial stand. Scien Silv Sin 44: 34–39 (in Chinese).
46. Peng L (2003) Biomass and productivity in 35 year old *Phoebe bournei* plantation. J Fujian College For 23: 128–131 (in Chinese).
47. Peng L (2008) Preliminary study on the biomass of *Phoebe bournei* plantation with different density and site condition. J Fujian For Sci Technol 35: 15–23 (in Chinese).
48. Wu W (2009) Study on stand structure and growth law of near mature *Phoebe bournei* plantation. Develop. For Sci Technol 23: 75–78 (in Chinese).
49. Lu M (2006) Study On Biomass And Nutrient Of Phoebe Bournei Plantation In Heterogeneity Spatiality Of Density And Site. PhD Thesis. Fujian: Fujian Agriculture and Forestry Univ (in Chinese).
50. Liao H, Di D, Zhang C, et al. (1989) Planting density management to *Cinnamoum camphora* and *Phoebe bournei*. J. Fujian College For. 9: 80–84 (in Chinese).
51. Cheng Y (2003) Study On Litter Fall Decomposing Dynamic And Nutrients Releasing Rule Of Phoebe Bournei. PhD Thesis. Fujian: Fujian Agriculture and Forestry Univ. (in Chinese).
52. Guo Y (2007) Decomposition and the nutrient dynamics of *Phoebe bournei* leaf litter. J Fujian College For 27: 199–202 (in Chinese).
53. Lin KM, Zhang ZQ, Cao GQ, He ZM, et al. (2006) Decomposition characteristics and its nutrient dynamics of leaf litter mixtures of both Chinese fir and *Phoebe bournei*. Acta Ecol Sin 26: 2732–2738 (in Chinese).
54. Lin KM, Zhang ZQ, Ye FM, Lin Y, Li QS. (2010) Dynamic analysis of decomposition characteristics and content change of nutrient elements of leaf litter of *Cunninghamia lanceolata*, *Phoebe bournei* and *Schima superba* under *C. lanceolata* artificial forest. J Plant Resour Environ 19: 34–39 (in Chinese).
55. Peng L (2008) Study on nutrient elements of Phoebe bournei plantation in the different site conditions. J Fujian For Sci Tech 35: 10–15 (in Chinese).
56. Chen S (2010) Effect of different site conditions on the growth of *Phoebe bournei*. J. Fujian College For. 30: 157–160 (in Chinese).
57. Liu Q (2009) Study of soil fertility of *Phoebe bournei* planted forest. J Jiangxi For Sci Technol 1: 31–33 (in Chinese).
58. Du J, Fan ZX, Ye DY, Lu CT (2010) The predicting models of crown volume and LAI for *Phoebe zhennan* plantation. J Zhejiang For Sci Technol 30: 37–41 (in Chinese).

59. Hu J, Liu G (2010) Preliminary investigation on photosynthetic characteristics of saplings of *Phoebe zhennan* and *Machilus leptophylla*. J Anhui Agri Univ 37: 541–546 (in Chinese).
60. Ren CH, Chen H, Zhou LH (1990) Study on L.A.I. (Leaf Area Index) of *Phoebe bournei* stand in Youxi county. J Fujian College For 10: 67–71 (in Chinese).
61. Wu Z, Chen S (2004) The effect of light on growth of *Phoebe bournei* plantation. J Fujian College For 24: 371–373 (in Chinese).
62. Fan X, Ma X, Fu R, Liu A (2001) Comparative study on underground vegetation develop of different generation plantations of Chinese-fir. For Res 14: 11–19 (in Chinese).
63. Lin KM, Yu XT, Hong W, Huang BL (2001) Effect of minor vegetation on soil fertility in Chinese-fir plantation. Scient Silv Sin 37: 94–98 (in Chinese).
64. Xiang WH, Tian DL, Yan WD, Kang WX, Fang X (2003) Biomass and nutrient content of vegetation on fallow site after clear-cutting of Chinese-fir plantation. Acta Ecol Sin 23: 695–702 (in Chinese).
65. Yan W, Tian D, Jiao X (2003) A study on biomass dynamics and distribution of undervegetation in the secondary generation of Chinese-fir plantation in Hui Tong. For Res 16: 82–86 (in Chinese).
66. Zhou X, Luo R, Ye J (1991) Effect of continuous cropping with Chinese-fir upon soil nutrients and its feedback. J Nanjing For Univ 3: 47–52 (in Chinese).
67. Tian DL, Zhu XP, Cai BY, Gao YM (1989) Studies on the litter in a Chinese-fir plantation ecosystem: (II) Nutrient contents and decomposition rate of litter. J Central South For Inst 9(Sup): 45–55 (in Chinese).
68. Liao LP (2000) Decomposition of leaf litter of Chinese-fir in mixture with major associated broad-leaved plantation species. Acta Phytoecol Sin 24: 27–33 (in Chinese).
69. Ding YX, Tian Y, Qi L (1999) A testing simulation with FORECAST on long-term productivity of Chinese-fir plantations. For Studies China 1: 34–38 (in Chinese).
70. Ding YX, Chen JL (1995) Effect of continuous plantation of Chinese-fir on soil fertility. Pedosphere 5: 57–66.
71. Yang YS, Chen GS, Huang BL (2000) Variation in the soil water and nutrients between different rotation stands of Chinese-fir. J Nanjing For Univ 24: 25–28 (in Chinese).
72. Seely B, Welham C, Kimmins JP (2002) Carbon sequestration in a boreal forest ecosystem: results from the ecosystem simulation model, FORECAST. For Ecol Manage 169: 123–135.
73. Blanco JA, Seely B, Welham C, Kimmins JP, Seebacher TM (2007) Testing the performance of FORECAST, a forest ecosystem model, against 29 years of field data in a *Pseudotsuga menziesii* plantation. Can J For Res 37: 1808–1820.
74. Guan BT, Wrigth WE, Chung CH, Chang ST (2012a) ENSO and PDO strongly influence Taiwan spruce height growth. For Ecol Manage 267: 50–57.
75. Masera OR, Garza-Caligaris JF, Kanninen M, Karjalainen T, Liski J, et al. (2003) Modeling carbon sequestration in afforestation, agroforestry and forest management projects: the CO2FIX V.2 approach. Ecol Model 164: 77–199.
76. Cannell MGR (2003) Carbon sequestration and biomass energy offset: Theoretical, potential and achievable capacities globally, in Europe and the UK. Biomass Bioener 24: 97–116.
77. Van Minnen JG, Strngers BJ, Eickhout B, Swart RJ, Leemans R (2008) Quantifying the effectiveness of climate change mitigation through forest plantations and carbon sequestration with an integrated land-use model. Carbon Balance Manage 3: 3.
78. Blanco JA (2012) Forests may need centuries to recover their original productivity after continuous intensive management: an example from Douglas-fir. Sci Total Environ 437: 91–103.
79. Marín-Spiotta E, Sharma S (2013) Carbon storage in successional and plantation forest soils: a tropical analysis. Global Ecol Biogeogr 22: 105–117.
80. Amoroso MM, Turnblom MM (2006) Comparing productivity of pure and mixed Douglas-fir and western hemlock plantations in the Pacific Northwest. Can J For Res 36: 1484–1496.
81. Wu Z (2005) Study on growth effect of the mixed forest of *Cunninghamia lanceolata* and *Phoebe bournei*. J Fujian College For 25(2): 142–146.
82. Duan W, Zhu B, Lu JL, Lu R, Yang CJ, et al. (2008) Carbon Storage Effects in Artificial Mixed Forest of *Cryptomeria fortune* - *Cunninghamia lanceolata* in MontaneAreas of Western Sichuan Province of China. Sichuan Environment 27(5): 57–63 (in Chinese).
83. Zhang G (2009) Growth and biomass in the Chinese fir mixtures. Anhui Agri. Sci. Bull. 15(21): 131–132.
84. Lin T (2011) Mixed Forests of *Michelia macclurei* and *Cunninghamia lanceolata* in Coastal Mountains Land of Southern Fujian. Jiangxi Forestry Sci Tech 3: 16–19.
85. Chen XY (2009) Stand biomass and its distribution of mixed forests of *Alnus crematogyne* and *Cunninghamia lanceolata*. Jiangxi Forest Science and Technology 1, 14–17.
86. Keith H, MaKey B, Berry S, Lindenmayer D, Gibbons P (2010) Estimating carbon carrying capacity in natural forest ecosystems across heterogeneous landscapes: addressing sources of error. Glob Change Biol 16: 2971–2989.
87. Berthrong ST, Jobbagy EG, Jackson RB (2009) A global-meta-analysis of soil exchangeable cation, pH, carbon and nitrogen with afforestation. Ecol Appl 19: 2228–2241.
88. Minghe L, Ritchie GA (1999) Eight hundred years of clonal forestry in China: I. Traditional afforestation with Chinese fir (*Cunninghamia lanceolata* (Lamb.) Hook.). New For 18: 131–142.
89. Zhang XQ, Kirschbaum MKU, Hou Z, Guo Z (2004) Carbon stock changes in successive rotations of Chinese-fir (*Cunninghamia lanceolata* (Lamb.) Hook) plantations. For Ecol Manage 202: 131–147.
90. Blanco JA, Wei X, Jiang H, Jie CY, Xin ZH (2012) Impacts of enhanced nitrogen deposition and soil acidification on biomass production and nitrogen leaching in Chinese fir plantations. Can J For Res 42: 437–450.
91. Blanco JA, Zavala MA, Imbert JB, Castillo FJ (2005) Sustainability of forest management practices: Evaluation through a simulation model of nutrient cycling. For Ecol Manage 213: 209–228.
92. Kelty MJ (1992) Comparative productivity of monocultures and mixed-species stands. In: Kelty, M.J., Larson, B.C., Oliver, C.D. (Eds.), The Ecology and Silviculture of Mixed-Species Forests. Kluwer Academic Publishers, Dordrecht, Boston, 125–141.
93. Pretzsch H, Bielak K, Block J, Bruchwald A, Dieler J, et al. (2013) Productivity of mixed versus pure stands of oak (*Quercus petraea* (Matt.) Liebl. and *Quercus robur* L.) and European beech (*Fagus sylvatica* L.) along an ecological gradient. Eur J For Res 132: 263–280.
94. Primicia I, Camarero JJ, Imbert JB, Castillo FJ (2013) Effects of thinning and canopy type on growth dynamics of *Pinus sylvestris*: inter-annual variations and intra-annual interactions with microclimate. Eur J Forest Res 132: 121–135.
95. Bauhus J, Khanna PK, Menden N (2000) Aboveground and belowground interactions in mixed plantations of *Eucalyptus globulus* and *Acacia mearnsii*. Can J For Res 30: 1886–1894.
96. Erskine PD, Lamb D, Bristow M (2006) Tree species diversity and ecosystem function: Can tropical multi-species plantations generate greater productivity? For Ecol Manage 233: 206–210.
97. Forrester DI, Bauhus J, Cowie AL, Vanclay JK (2006) Mixed-species plantation of *Eucalyptus* with nitrogen fixing trees: a review. For Ecol Manage 233: 211–230.
98. Kelty MJ (2006) The role of species mixtures in plantation forestry. For Ecol Manage 233: 195–204.
99. Seely B, Welham C, Blanco JA (2010) Towards the application of soil organic matter as an indicator of ecosystem productivity: Deriving thresholds, developing monitoring systems, and evaluating practices. Ecol Ind 10: 999–1008.
100. Blanco JA, González E (2010) Exploring the sustainability of current management prescriptions for Pinus caribaea plantations in Cuba: a modelling approach. J. Trop. For. Sci. 22: 139–154.
101. Fahey TJ, Woodbury PB, Battles JJ, Goodale CL, Hamburg SP, et al. (2010) Forest carbon storage: ecology, management, and policy. Front Ecol Environ 8: 245–252.
102. Eriksson E, Gillespie AR, Gustavsson L, Langvll O, Olsson M, et al. (2007) Integrated carbon analysis of forest management practices and wood substitution. Can J For Res 37: 671–681.
103. Lippke B, Oneil E, Harrison R, Skog K, Gustavsson L, et al. (2011) Life cycle impacts of forest management and wood utilization on carbon mitigation: knowns and unknowns. Carbon Manage 2: 303–333.
104. Perez-Garcia J, Lippke B, Comnick J, Manriquez C (2005) An assessment of carbon pools, storage, and wood products market substitution using life-cycle analysis results. Wood Fiber Sci 37: 140–148.
105. McKinley DC, Ryan MG, Birdsey RA, Giardina CP, Harmon ME, et al. (2011) A synthesis of current knowledge on forests and carbon storage in the United States. Ecol Appl 21: 1902–1924.
106. Liu P, Yin R (2012) Sequestering carbon in China's forest ecosystems: potential and challenges. Forests 3: 417–430.
107. Jie C, Jiang H, Zhou G, Wei X, Blanco JA, et al. (2011) Simulating the carbon storage of spruce forests based on the FORECAST model and remotely sensed data. Proceedings - 2011 19th International Conference on Geoinformatics, Geoinformatics 2011. Article number 5981581. ISBN: 978-161284848-8. DOI: 10.1109/GeoInformatics.2011.5981581.
108. Qi G (2011) Carbon pool and sequestration potential of larch plantations in Northeast China. PhD thesis. Shenyang: Chinese Academy of Sciences.
109. Keith H, Mackey BG, Lindenmayer DB (2009) Re-evaluation of forest biomass carbon stocks and lessons from the world's most carbon-dense forests. PNAS 106: 11635–11640.
110. Guan D, Liu Z, Geng Y, Lindner S, Hubacek K (2012b) The gigatonne gap in China's carbon dioxide inventories. Nature Clim Change 2: 672–675.

13

Generalized Additive Models Used to Predict Species Abundance in the Gulf of Mexico: An Ecosystem Modeling Tool

Michael Drexler*, Cameron H. Ainsworth

College of Marine Science, University of South Florida, Saint Petersburg, Florida, United States of America

Abstract

Spatially explicit ecosystem models of all types require an initial allocation of biomass, often in areas where fisheries independent abundance estimates do not exist. A generalized additive modelling (GAM) approach is used to describe the abundance of 40 species groups (i.e. functional groups) across the Gulf of Mexico (GoM) using a large fisheries independent data set (SEAMAP) and climate scale oceanographic conditions. Predictor variables included in the model are chlorophyll *a*, sediment type, dissolved oxygen, temperature, and depth. Despite the presence of a large number of zeros in the data, a single GAM using a negative binomial distribution was suitable to make predictions of abundance for multiple functional groups. We present an example case study using pink shrimp (*Farfantepenaeus duroarum*) and compare the results to known distributions. The model successfully predicts the known areas of high abundance in the GoM, including those areas where no data was inputted into the model fitting. Overall, the model reliably captures areas of high and low abundance for the large majority of functional groups observed in SEAMAP. The result of this method allows for the objective setting of spatial distributions for numerous functional groups across a modeling domain, even where abundance data may not exist.

Editor: Andrew Davies, Bangor University, United Kingdom

Funding: The authors would like thank the funding sources: Florida Sea Grant, Gulf of Mexico Research Initiative, C-IMAGE (Center for Integrated Modeling and Analysis of the Gulf Ecosystem) consortium, and the National Marine Fisheries Service/USF College of Marine Science Marine Resource Assessment fellowship. The funders had no role in study design, data collection and analysis, decision to publish, or preparation of the manuscript.

Competing Interests: The authors have declared that no competing interests exist.

* E-mail: mdrexler@mail.usf.edu

Introduction

The need for ecosystem-based approaches to fisheries management has been widely recognized throughout the world [1]. Marine ecosystem models (e.g., Ecospace, Atlantis, InVitro, OSMOSE, Gadget, IBEM, etc.) are becoming an important tool in achieving those goals as they incorporate predator-prey dynamics and environmental interactions in a spatially explicit context. Spatially explicit models allow managers to better understand certain ecosystem processes, but they require large amounts of data in comparison to models that assume homogeneous space. One example of these additional requirements is that these models require an initial spatial allocation of functional group biomass or abundance. It is not straightforward to develop biomass distribution grids due to the lack of comprehensive stock assessments outside a handful of commercially valued species and there is a particular lack of spatial distribution data from international waters. In most cases, this limits the development of ecosystem models to those areas that are rich in fisheries independent data. Efforts have been made to extrapolate data from limited spatial areas to larger scales using a variety of methods including interpolation over arbitrarily assigned regions [2] and similarity matrices [3]. Generalized additive modeling offers an objective way to predict abundance or biomass according to the known ecology of the animals over broad geographic areas. Generalized additive models (GAMs) [4] are a semi-parametric approach to predicting non-linear responses to a suite of predictor variables. In general, GAMs can be used to identify optimal conditions for a given species using environmental variables (e.g., depth and temperature) in order to predict the likelihood that a given species would inhabit a particular environment, or their abundance [5,6,7,8]. The outputs of these models are often used to interpolate species distributions at high resolution within coarsely sampled areas [9,5,7]. Model testing and sensitivity analysis can also help identify influential environmental variables and their corresponding range of influence [9,5]. In comparative studies, GAMs have often been shown to perform as well or better than other types of predictive models based on environmental conditions [8,10,11]. Despite their acknowledgment as a proven tool for ecological analyses, albeit with some caveats [11], few studies, have applied the method to make predictions outside of a sampled area.

Fisheries independent sampling efforts typically result in many zero observations for any given species, particularly in surveys that cover a broad range of habitats or depths. To deal with this problem a number of approaches have been developed to fit these types of data including lognormal delta distributions [12,13], delta method approximation of variance [14], and zero inflated distributions [15,16]. All of these methods can be applied to either generalized linear models, or generalized additive models. The latter of the two allows for greater flexibility in the model fitting. Despite these advanced methods dealing with zero inflation, Warton (2005) [17] found that in most cases a negative binomial was sufficient to model data with many zeros. In this paper we utilized a negative binomial GAM to predict the relative abundance of functional groups across shelf areas of the entire Gulf of Mexico (GoM) including Mexican and Cuban waters and areas where fisheries independent surveys do not exist, based on environmental and habitat predictors.

The purpose of this study was to find a single parsimonious framework to predict the distribution of multiple functional groups within an ecosystem. This framework can then be utilized in the objective distribution of relative abundance for a spatially explicit Atlantis ecosystem model of the GoM (Atlantis-GoM) in preparation for NOAA's Integrated Ecosystem Assessment and other applications (Atlantis; [18,19]). In order to provide better spatial management of fisheries, a better understanding of the ecosystem-wide influences on that stock requires modeling of the entire ecosystem and not just those parts that are adequately sampled. We validate the model by predicting the distribution of pink shrimp (*Farfantepenaeus duroarum*) throughout the Gulf of Mexico and compare the model performance of both the aggregated results used in ecosystem models and the high-resolution gridded values taken directly from the GAM. While numerous species were considered in the model fitting, only the summer abundance of pink shrimp is illustrated here. Pink shrimp were chosen as an example species because they were well represented in the available observational data set used in model training and their distribution is strongly correlated with environmental and habitat predictor variables.

Methods

Study Area

The Gulf of Mexico (GoM) is one of the world's 64 Large Marine Ecosystems [20]. This ecosystem spans tropical and subtropical climates and is enveloped by the economic exclusive zones (EEZ) of the United States, Mexico, and Cuba. The EEZ of the United States alone supports 25 million recreational fishing trips [21] and a commercial fisheries harvest in excess of one million tonnes per year [22]. Gulf shrimp remain one of the most important fisheries in the region with combined landings value of 368 million dollars [23]. In this article, we estimate the abundance and spatial distribution of 40 functional groups (groups of species aggregated according to niche similarities). The single-species functional group of pink shrimp is illustrated here in detail as a case study. The range of pink shrimp extends through the entire GoM coastal waters [24]. Nearly 94% of the pink shrimp harvested in the GoM are landed on the west coast of Florida [23] where they are particularly abundant. Known "hot spots" for pink shrimp include the area surrounding the Dry Tortugas as well as the eastern coast of the Golfo de Campeche (Figure 1). The highest adult abundance is found between 9 and 44 m of water [25,26].

SEAMAP Groundfish survey

The Southeast Area Monitoring and Assessment Program (SEAMAP) is a multiagency fisheries independent data collection program coordinated by the Gulf States Marine Fisheries Commission [27]. Groundfish surveys are conducted on an annual, and sometimes seasonal, basis using a 40-ft otter trawl throughout the northern GoM. The general area surveyed by SEAMAP includes most the continental shelf up to 200 m depth, but only within the territorial waters of the United States in the northern Gulf of Mexico (Figure 1). Survey data was extracted from the public SEAMAP database [28] and aggregated by functional groups. Sampling effort was estimated as the total area swept of each SEAMAP tow using the Euclidean distance between start and end points and an assumed 40-ft trawling width.

Environmental Conditions

Predictor variables included in the model were surface chlorophyll *a* (chl *a*), sediment type, bottom dissolved oxygen (DO), bottom temperature, and depth (Table 1). These variables were chosen due to the wide spatial coverage throughout the Gulf of Mexico. Sediment type was divided into the following categories: mud, sand, gravel, and rock. A 0.1^0 gridded map of seasonal environmental parameters was made for each season (Winter: Jan-Mar, Spring: Apr-Jun, Summer: Jul-Aug, Fall: Sep-Dec) with data collected from the following sources. Measurements of bottom temperature and DO at the maximum depth recorded for each grid point were extracted for each season from the National Oceanographic Data Center (NODC) regional climatology database [29]. Surface chl *a* measurements were obtained by averaging the seasonal composites of MODIS – Terra satellite measurements and the NASA Ocean Biogeochemical Model from 2005–2010 accessed through the GIOVANNI portal (http://disc.sci.gsfc.nasa.gov/Giovanni/overview/index.html). A continuous raster of bathymetry was derived from the SRTM30_PLUS global bathymetry grid [30], which was accessed from the Gulf of Mexico Coastal Ocean Observing System (gcoos.tamu.edu). The best available data on bottom sediment type, dbSEABED2006 [31,32] does not provide complete coverage for the entire Gulf of Mexico, although the area sampled by SEAMAP has substantial coverage. A nearest neighbor function was executed on a 0.1^0 grid using the natural neighbor function in GIS v10.0 [33] for those grid points lacking any sediment data. Incomplete and low resolution environmental data was subjected to a spline interpolation using the same grid and GIS software in order to provide a contiguous surface from which to make model predictions. The seasonal environmental conditions grid was then overlaid with the SEAMAP starting locations for a given season. These environmental grids were used both in the fitting of the model and in and in predicting abundance distributions for unsampled areas.

Model description

A GAM approach was used to predict relative abundance of Atlantis-GoM functional groups in shelf areas across the entire Gulf of Mexico based on estimates of abundance and regional oceanographic conditions occurring at SEAMAP trawl locations. Due to the large number of zero observations and the need for a single parsimonious model to make predictions for a large number of functional groups, a GAM was developed using a negative binomial distribution with an offset for effort [15]. Prior to the fitting of the final model, the data was randomly split into training and test sets: 2/3 of the data was for training and 1/3 for model validation. Once validated, the full set of pink shrimp summer abundance data was then used to fit the model and make the final predictions for the 'summer' season. All models were fit using the 'mgcv' package in the R version 2.14.0 environment [34] following the equation:

$$g(\eta)\tilde{s}(\text{depth}) + s(\text{chla}) + s(\text{temp}) + s(\text{oxygen}) + \text{factor}(\text{sedimenttype}) + offset(g(effort))$$

where η represents the expected abundance resulting from the generalization of the predictor terms according to the link function *g*. The abundance data was modeled using a negative binomial distribution with a log link function, including an offset, with

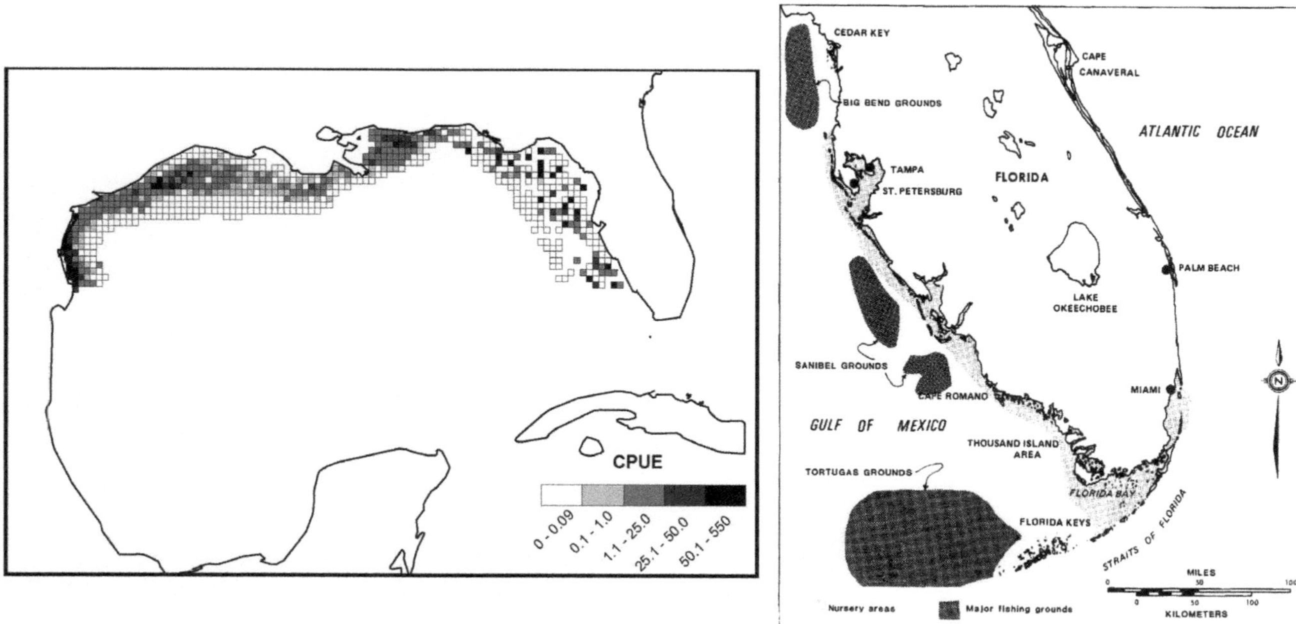

Figure 1. Pink shrimp abundance data. Historical fishing grounds of pink shrimp (*Farfantepenaeus duorarum*) off of the west coast of Florida in 1983 (left) and the observed abundance sampled from SEAMAP sampling locations from 1987 to 2009 in terms of individuals caught per one-hour tow of 40-ft shrimp trawl (right). Reproduced with permission from Bielsa et al. (1983).

equivalent link function, to allow for variations in effort. Function *s* is a thin plate regression spline fit to a given environmental parameter. The smoothness selection was fit using a spline-based penalized likelihood estimation. Theta parameters and weighted penalties were determined by Un-Biased Risk Estimator (UBRE) which is similar to an AIC rescaled [34]. Estimation of the theta parameter was limited to a range of 1–10. An extra penalty was applied to each parameter as the smoothing parameter approached zero, allowing the complete removal of a term from the model when the smoothing parameter is equal to zero. This extra penalty allows for partially automated model selection and is especially useful given the models broad application to numerous functional groups.

The initial training model fits were evaluated by analyzing the total deviance explained and UBRE score. Model performance was evaluated by predicting the abundance at each of the data points in the test data set, given the environmental conditions at that point. The predictions were then compared to the observed values at each respective point. Plots of the predicted vs. observed abundance were made for each group, and a least squares regression was used to evaluate any trends. Once the predictions of the training/test model were compared and deemed suitable for analysis, the entire data set was then used to predict abundance from the environmental conditions within each 0.1^0 grid cell for depths up to 200 m. To illustrate the usefulness of this method as a way of initializing spatially explicit ecological models, the results of the pink shrimp distributions were averaged according to the Atlantis-GoM spatial polygons and displayed with their associated 95% confidence intervals.

Results

For pink shrimp, the model described 45.5% (UBRE = 1.6) of the deviance. The models for the remaining functional groups described between 10% and 83% of the deviance with a median value of 33.6% (Table S1 in File S1). All of the functional groups observed in SEAMAP trawls also had a positive slope of the

Table 1. Data sources for model.

Environmental Parameter	Data Source	Resolution	Manipulations
Abundance	Southeast Area Monitoring and Assessment Program (SEAMAP)	varies	Standardized to area swept centered around each starting point
Surface chlorophyll *a* (chl *a*)	MODIS-Terra 4km Satellite -NASA Giovanni Portal	4km	NA
Sediment	dbSEABED – GoM Data Altas	low	Nearest neighbor interpolation
Bottom Temperature	NODC/GoM regional Climatology	0.1^0	Spline interpolation
Bottom Dissolved Oxygen (DO)	NODC/GoM regional Climatology	1.0^0	Spline interpolation
Depth	DOC/NOAA/NESDIS/NGDC – GoM Data Atlas	1.85 km	NA

List of data sources used in the model, the resolution of that data, and any manipulations that were required to attain a contiguous surface with which to make model predictions.

observed/predicted line indicating that we can reliably predict low and high density areas at least qualitatively. All of the continuous predictors were found to have a smoothing term significantly different from zero ($p<0.001$) and thus contributed to the model fit for pink shrimp. This was also true for most demersal invertebrate groups. However, the significance of the predictors from other groups not commonly selected by a benthic trawl varied widely.

Individual parameter values were in general agreement with the habitat suitability index model derived by Mulholland (1984) [24]. Mulholland reports catches of shrimp over a wide range of temperatures (5°–38°C) with the highest density of catch from 20–38°C. The curve fit to the modeled distribution found the highest abundance to be in the range of 18–32°C (Figure 2). Values higher than 38°C likely reflect the error associated with the interpolation of environmental data. Temperatures lower than 15°C had a negative effect on the expected abundance.

Mulholland [24] also predicts sandy-silt and silty-sand to be the sediment types with the highest suitability for pink shrimp, followed by hard bottom. The lowest suitability was in soft bottom. The modeled data predicted significantly greater densities of shrimp on sand and rock habitats ($p<0.05$; Figure 3). Depth only had a slight positive effect on the estimated abundance up to 30–40 m; any depth greater than this had a negative effect on the estimated abundance. Chlorophyll *a* concentrations were inversely correlated to pink shrimp abundance for values up to 15 mg/m^3 at which point the excepted abundance dropped sharply. The influence of dissolved oxygen did not fluctuate greatly across the range of values between 2.0–4.5 mL/L. Dissolved oxygen values greater than 6 had a positive effect on the model, but were not common in the seasonal averages of the environmental conditions. Therefore, the fitted model reflects previous research pertaining to the habitat preference for pink shrimp with regard to temperature

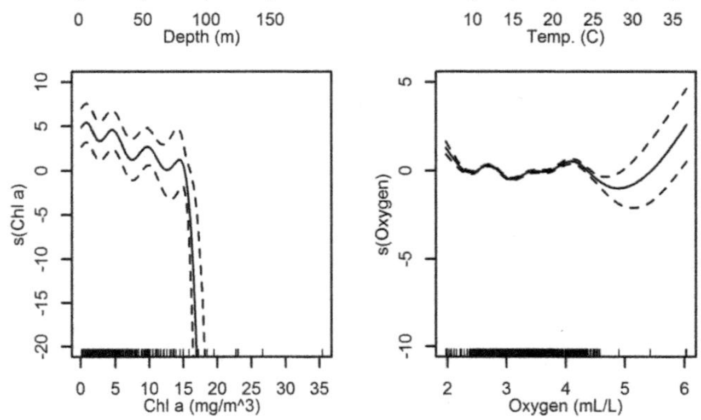

Figure 2. Model fits. Smoothed curve of the additive effect to the estimated abundance of pink shrimp for the individual environmental parameters in the GAM. Dotted lines represent 95% confidence intervals, marks along the lower axis represent a single observation. A straight line represents an additive effect of zero.

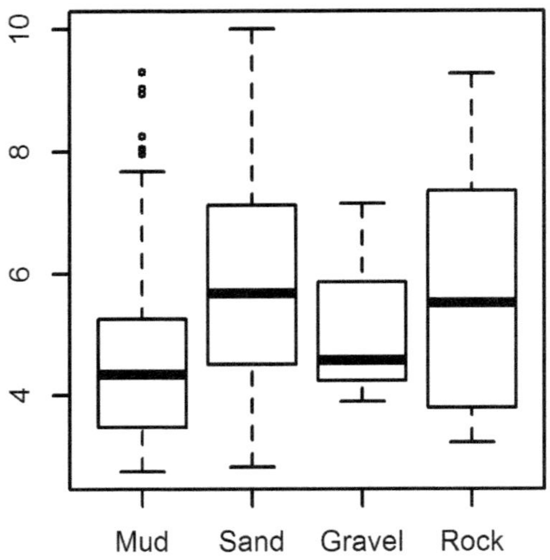

Figure 3. Sediment type vs. pink shrimp abundance. Natural log distribution of pink shrimp abundance for each sediment type. Since we are primarily interested in which category of sediment type is suitable for pink shrimp the zeros have been removed for display purposes.

and sediment type and introduces some additional suitability parameters.

Using the test set of SEAMAP data, the ability of the model to predict the response was evaluated. In general, the model predicted a higher mean abundance for those stations where high abundances were observed (Figure 4). Although less than ideal, ideal being equivalent 1:1 slope, the difference captures the general trend of the data. In every functional group assessed using this model, the slope of the least square line was greater than zero.

The model predictions were then aggregated by polygon and compared to the mean abundance occurring within those polygons with fisheries independent data. The aggregated predictions were in better agreement with the observed data with a normal distribution of residuals around the least square line. However, like the point estimates, the intercept was greater than zero, and may indicate an overestimation of abundances close to zero.

The spatial distribution of the gridded values, predict high abundance along the entire mid-depth portion of the West Florida Shelf, and some additional hotspots near the Dry Tortugas, Louisiana-Texas border, Texas-Mexico border, and on the north-western Campeche Bank (Figure 5). The highest abundance was found in the areas north of the Florida Keys/Dry Tortugas with abundances approaching 1.2 million shrimp per grid cell. Abundances near the Florida panhandle were two orders of magnitude less than those found near the Florida Keys. The hotspots around the Texas-Mexico border and Campeche Bank were a similar order of magnitude less than those near the Dry Tortugas although those distributions were patchy and smaller in area.

The Atlantis-GoM polygon spatial distributions reflect the general distribution of pink shrimp in the 0.1^0 gridded results (Figure 5,6). Highest abundance occurs at mid depth over the West Florida Shelf and Dry Tortugas, and the nearshore polygons along the western gulf. The comparison of the observed to predicted values of the aggregated polygons were in better agreement than the higher resolution gridded values (Figure 4). The observed mean of those polygons with SEAMAP data (3.52) was found to be significantly different from the predicted mean (6.62) through a paired t-test ($t=3.90$, $p=0.0002$). The 95% confidence intervals were inversely related to the mean abundance, and ranged between 0 and 1.65 on a log scale for the

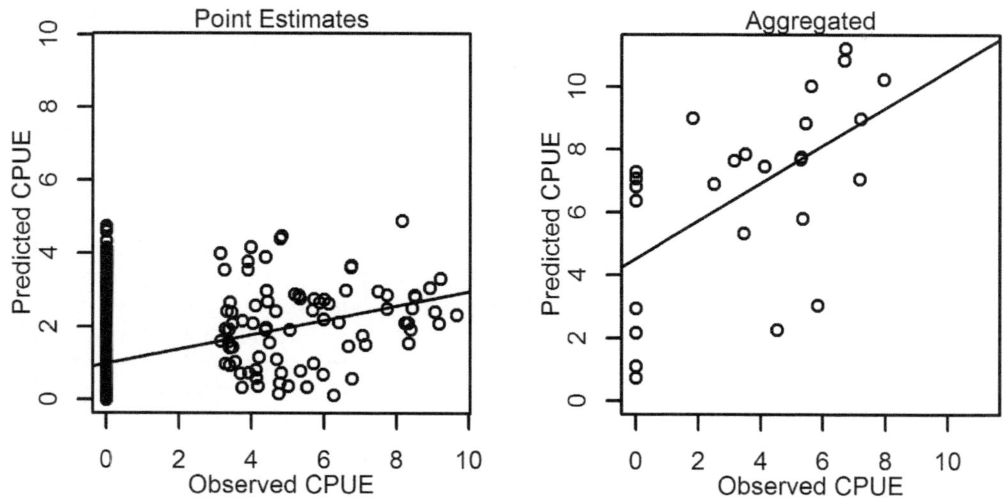

Figure 4. Model performance. Comparison of model predictions to observed ln(abundance) estimated from the environmental conditions of the test data set and compared to the observed values. The resulting scatterplots and least-squared line of fit is shown for both the grid estimates from the model and those estimates aggregated to the Atlantis-GoM polygon level.

majority of the cells. The few polygons with very few data points (estuarine and deep water polygons) had much higher confidence intervals approaching 12 orders of magnitude, and the estimates are not reliable.

Discussion

The method described here provides a standardized way to generate abundance distributions for models that 1) require comprehensive spatial distributions for 2) a large number of species 3) but are limited in terms of fisheries independent data. Therefore it is an ideal supporting application for spatial ecosystem models. In this paper we extend the use of the GAM approach to make predictions of abundance based on the environmental conditions beyond the sampling domain of the data used to fit the model. We also show how the data derived in this model will be incorporated into the Gulf of Mexico Atlantis ecosystem model in Figure 6. The absolute predictions for individual functional groups may not be precise using the limited abundance data set from SEAMAP employed here. Also, the

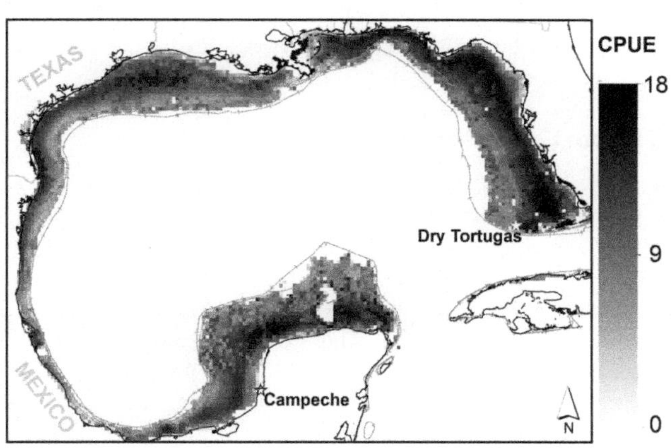

Figure 5. Modeled pink shrimp abundance for the entire GoM. Results of the pink shrimp GAM predicting estimate of abundance. Bathymetric contours of 50 m, 100 m, and 200 m also shown. CPUE expressed as ln(ind./km^2).

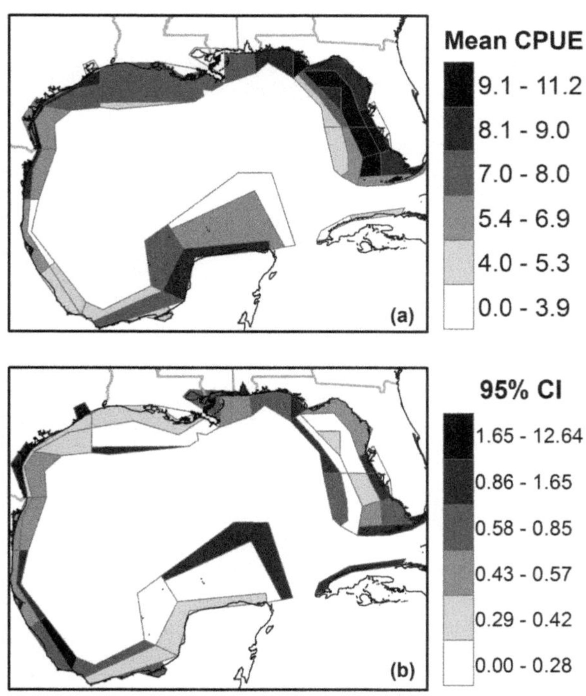

Figure 6. Aggregated model results. Example of the spatial aggregation that can be performed from the GAM predictions of pink shrimp ln(abundance). Mean CPUE ln (ind./km^2) derived from Figure 6 is calculated according to (a) the box geometry of the Gulf of Mexico Atlantis ecosystem model and (b) the associated 95% confidence intervals (+/− ln(CPUE)) for each box. The few boxes with a confidence interval greater than 1.65 did not have high spatial coverage and the results should be thrown out.

environmental data used in making predictions is averaged over time and space, missing environmental extremes that may have significant influence on species distributions. Future studies might test environmental data with a higher temporal resolution, given the data is available. Also, the presence of spatial autocorrelation on a regional basis may be addressed by splitting the training/test data into regional blocks and examining the residuals in the remaining sub-regions. In our case we considered this approach impractical due to data scarcity and the application of the modeled results to a course resolution ecosystem model.

The high degree of correlation between model results and the validation data set indicates that we can reliably predict qualitative differences between low- and high-biomass regions, especially in terms of relative abundance. Regardless of the degree of precision provided by this model, the results offer a vast improvement over the assumption of a homogenous distribution of a population commonly used in stock assessments. Further, when aggregated to the level used for spatially explicit ecosystem models, the model provides a better fit to the observed data points. Therefore, the information supplied by this modelling framework can be used to initialize the spatial distribution of species for dynamic ecosystem models whose spatial distributions will settle to a new, but related, equilibrium at run-time. This proof of concept application can be improved as additional CPUE and environmental data become available. Additional data could be incorporated from fisheries-dependent data such as spatially referenced commercial catch statistics and observer data. Coupling these models with spatially explicit estimations of pressures can ultimately determine the absolute contributions of these predictors on species abundance in lieu of pressure.

We demonstrated the utility of the model by predicting areas of high abundance of pink shrimp near the historical fishing areas where no observational abundance data was available. The model results were also in general agreement with previous research regarding the suitability of each parameter used. Thus the extension of this model to the entirety of the southern Gulf of Mexico should provide reasonable estimates of abundance. It should be noted that although pink shrimp occur throughout the Gulf of Mexico, one would expect the species composition of other groups of animals to vary with latitude. While the use of multispecies functional groups occupying similar niches does provide some added flexibility to the final predictions, the inclusion of a latitude parameter could explicitly differentiate these variations in species abundance. Unfortunately, no data is available for latitudes south of Texas-Mexico border to the west, and south of Florida to the east. Thus, including this terms will limit the predictive capabilities to only those latitudes where data exist, undermining the purpose of the model.

While the magnitude of abundance from a benthic trawl may be expected to reflect the absolute abundance of pink shrimp, estimating the abundance of other functional groups, particularly non-demersal and non-benthic species using the same gear will be subject to a catchability bias. There was adequate data to predict the distributions of 40 of the 90 Atlantis-GoM functional groups. Those groups which are less vulnerable to benthic trawling gear, such as large sharks, greater amberjack, king mackerel, spanish mackerel, and squid, did not show a close relationship between observed and predicted abundance values (Figure S1 in File S1). Abundance estimates for these functional groups may be unreliable until additional abundance data from other sampling gear types can be incorporated into the training data set. Abundance was better predicted for slower moving, smaller, mainly demersal species that would be selected by a benthic trawler, such as benthic grazers, bivalves, blue crab, flatfish, pink shrimp, and sessile filter feeders. It is worth re-iterating that predictions of these functional groups were limited by the depth range sampled by SEAMAP trawling. For this reason, the current application should only be used to set distributions on the continental shelf. However, the substitution of a general set of predictors suitable to the pelagic or deepwater environments should yield plausible results.

Despite the positive relationship between the predicted and observed values, none of the least squares lines approached a slope of one. Therefore, the model tends to overestimate abundance where no catch was observed. However, the model did manage to consistently detect a lowered combined abundance at sites where no catch was observed. This bias is likely related to the fact that these simple environmental indices are not satisfactory to explain a portion of the variability in the distribution of these functional groups. The model tends to fail at low population densities that may be heavily influenced by minute differences in oceanographic conditions, patchiness, or other unexplained variability. These differences will not be captured by the regional/seasonal environmental data used in this study. The environmental variables chosen for this study were done so for their wide spatial distribution spanning the entire modelling domain. Given this caveat, these high resolution errors become less important when used to initialize a course resolution model such as the Atlantis-GoM; where polygons are on the order of thousands of square kilometers (Figure 6). The final gridded spatial distributions for all 40 functional groups can be found in Figure S2 in File S1.

Aggregating the results to a courser resolution, i.e. the Atlantis-GoM polygons, allows us to average out the variance, as the large polygons will tend to be closer to the global average. The general distribution of pink shrimp is adequately represented by the Atlantis-GoM polygons (Figures 5, 6). This aggregation provided a better fit to the observational data than the individual points (Figure 4). The significant differences detected between the overall polygon means suggests, as in the gridded model, that we may still be slightly overestimating the total abundance across the entire system. As stated before, this bias may be related to the fact that we are fitting our model to seasonal data. The same reasoning for the bias in the gridded values also applies here. GAMs require many degrees of freedom and introduce a tradeoff between including addition degrees of freedom by aggregating data across seasons or spatial scales, or more highly resolved data with low predictive power to which we chose the prior.

In conclusion, this paper notes the utility of GAMs beyond the common applications of identifying influential environmental variables and interpolating abundance and biomass within sampling regions. For applications like initializing biomass distributions in spatial ecosystem models, where wide spatial and taxonomic coverage is desirable but the benefit of high precision estimates is lost at run-time, these statistical approaches hold unrealized potential.

Supporting Information

File S1 Table S1, Summary of individual model performance in terms of deviation explained for every Atlantis-GOM functional group observed during SEAMAP sampling from 2005–2010. Figure S1, Combined model fits of the observed (x-axis) versus predicted (y-axis) values of data for all 40 functional groups estimated from this model. The log-log line of least squares is plotted for visualization. Those functional groups with a slope less than or equal to zero ('deepwater fish' and 'large sharks') are not reliable and should be estimated with a separate set of parameters. Figure S2, Spatial distribution maps predicted by the combined GAM model for all 40 functional groups observed in the SEAMAP database. Grey scale represents the log transformed abundance per square kilometer of each functional group.

Acknowledgments

We would like acknowledge all of the efforts that were invested in generating the large data sets which made this study possible. First we would like to thank the Southeast Area Monitoring and Assessment Program (SEAMAP) for the routine sampling efforts to collect fisheries independent throughout the Gulf of Mexico, David Chagaris for his help navigating the database and Bob McMichael for the opportunity to participate in the SEAMAP cruises. We also acknowledge all the efforts to compile regional information pertaining to the Gulf of Mexico including: the usSEABED/dbSEABED team for compiling offshore surficial sediments in the region, the National Oceanographic Data Center (NODC), and the Gulf of Mexico Data Atlas for organizing this information.

Author Contributions

Conceived and designed the experiments: MD CA. Performed the experiments: MD. Analyzed the data: MD. Contributed reagents/materials/analysis tools: MD CA. Wrote the paper: MD CA.

References

1. Pikitch EK, Santora C, Babcock E, Bakun A, Bonfil R, et al. (2004) Ecosystem-based fishery management. Science 305: 346–347.
2. Brand EJ, Kaplan IC, Harvey CJ, Levin PS, Fulton EA, et al. (2007) A spatially explicit ecosystem model of the California Current's food web and Oceanography. U.S. Department of Commerce., NOAA Tech Memo, NMFS-NWFSC-84, 145 p.
3. Ainsworth CH, Kaplan IC, Levin PS, Cudney-Bueno R, Fulton EA, et al. (2011) Atlantis model development for the northern Gulf of California. U.S. Department of Commerce., NOAA Tech Memo, NMFS-NWFSC-110, 293 p.
4. Hastie T, Tibshirani R (1986) Generalized Additive Models. Statistical Science, 1, 297–310.
5. Stoner AW, Manderson JP, Pessutti JP (2001) Spatially explicit analysis of estuarine habitat for juvenile winter flounder: combining generalized additive models and geographic information systems. Mar Ecol Prog Ser 213: 253–271.
6. Maravelias CD (1999) Habitat selection and clustering of a pelagic fish: effects of topography and bathymetry on species dynamics. Can J Fish Aquat Sci 56: 437–450.
7. Yee TW, Mitchell ND (1991) Generalized additive models in plant ecology. Journal of Vegetation Science 2: 587–602.
8. Walsh W, Kleiber P (2001) Generalized additive model and regression tree analysis of blue shark (*Prionace glauca*) by the Hawaii-based longline fishery. Fish Res 53: 115–131.
9. Katsanevakis S, Maravelis CD (2009) Bathymetric distribution of demersal fish in the Aegean and Ionian Seas based on generalized additive modeling. Fish Sci 75(1): 13–23.
10. Moisen GC, Frescino TS (2002) Comparing five modelling techniques for predicting forest characteristics. Ecol Modell 157: 209–225.
11. Guisan A, Edwards TC, Hastie T (2002) Generalized linear and generalized additive models in studies of species distributions: setting the scene. Ecol Modell 157: 89–100.
12. Aitchison J, Brown JAC (1957) The Lognormal Distribution. Cambridge University Press, Cambridge, UK.
13. Pennington M (1983) Efficient estimators of abundance, for fish and plankton surveys. Biometrics 39: 281–286.
14. Stefánsson G 1996) Analysis of Groundfish survey adundance data: combining the GLM and delta approaches. ICES J. Mar Sci 53: 577–588.
15. Barry SC, Welsh AH (2002) Generalized additive modelling and zero inflated count data. Ecol Modell 157: 179–188.
16. Minami M, Lennert-Cody CE, Gao W, Román-Verdesoto M (2007) Modeling shark bycatch: The zero-inflated negative binomial regression model with smoothing. Fish Res 84: 210–221.
17. Warton DI (2005) Many zeros does not mean zero inflation: comparing the goodness-of-fit of parametric models to multivariate abundance data. Environmetrics 16: 275–289.
18. Fulton EA, Parslow JS, Smith ADM, Johnson CR (2004) Biogeochemical marine ecosystem models. 2. The effect of physiological data on model performance. Ecol Modell 173: 371–406.
19. Fulton EA, Smith ADM, Smith DC (2007) Alternative management strategies for Southeast Australian Commonwealth Fisheries: Stage 2: Quantitative Management Strategy Evaluation. Australian Fisheries Management Authority. Fisheries Research and Development Corporation.
20. Sherman K, Hempel G (Editors) (2008) The UNEP Large Marine Ecosystem Report: A perspective on changing conditions in LMEs of the world's Regional Seas. UNEP Regional Seas Report and Studies No. 182. United Nations Environment Programme. Nairobi, Kenya.
21. NOS (2008) Gulf of Mexico at a Glance: U.S. National Ocean Service, Department of Commerce, National Oceanic and Atmospheric Administration. Washington, D.C.
22. Vidal L, Pauly D (2004) Integration of subsystems models as a tool toward describing feeding interactions and fisheries impacts in a large marine ecosystem, the Gulf of Mexico. Ocean Coast Manag 47: 709–725.
23. NOAA (2010) Annual Commercial Landing Statistics. National Marine Fisheries Service. National Oceanic and Atmospheric Administration. Electronic database. Available at http://www.st.nmfs.noaa.gov/st1/commercial/landings/annual_landings.html. Date accessed: June 15, 2012.
24. Mulholland R (1984) Habitat suitability index models: pink shrimp. U.S. Fish and Wildlife Service FWS/OBS-82/10.76, 17 p.
25. GMFMC (2006) Options paper Amendment 15 to the shrimp fishery management plan (including environmental impact statement and regulatory impact review). Gulf of Mexico Fishery Management Council. Tampa, Florida.
26. Bielsa LM, Murdich WH, Labisky RF (1983) Species profiles: life histories and environmental requirements of coastal fishes and invertebrates (south Florida) – pink shrimp. U.S. Fish and Wildlife Service. FWS/OBS-82/11.17. U.S. Army Corps of Engineers, TR EL-82-4. 21 p.
27. GSMFC (2011) SEAMAP environmental and biological atlas of the Gulf of Mexico, 2009. Gulf States Marine Fisheries Commission.
28. Rester JK (2011) SEAMAP Environmental and Biological Atlas of the Gulf of Mexico, 2009. Gulf States Marine Fisheries Commission. Ocean Springs, MS.
29. Boyer TP, Antonov JI, Baranova OK, Garcia HE, Johnson DR, et al. (2009) World Ocean Database 2009. Levitus, S., Ed., NOAA Atlas NESDIS 66, U.S. Government Printing Office, Wash., D.C., 216 p.
30. Becker JJ, Sandwell DT, Smith WHF, Braud J, Binder B, et al. (2009) Global Bathymetry and Elevation Data at 30 Arc Seconds Resolution: SRTM30_PLUS, Marine Geodesy 32(4), 355–371.
31. Buczkowski BJ, Reid JA, Jenkins CJ, Reid JM, Williams SJ, et al. (2006) usSEABED: Gulf of Mexico and Caribbean (Puerto Rico and U.S. Virgin Islands) offshore surficial sediment data release: U.S. Geological Survey Data Series 146, version 1.0.
32. Jenkins C (2011) Dominant Bottom Types and Habitats In Gulf of Mexico Data Atlas [Internet]. Stennis Space Center (MS): National Coastal Data Development Center.
33. ESRI (2011) ArcGIS Desktop: Release 10. Redlands, CA: Environmental Systems Research Institute.
34. Wood S (2006) Generalized Additive Models: An Introduction with R. Chapman and Hall/CRC.

14

Responses of Terrestrial Ecosystems' Net Primary Productivity to Future Regional Climate Change in China

Dongsheng Zhao, Shaohong Wu*, Yunhe Yin

Institute of Geographical Sciences and Natural Resources Research, Chinese Academy of Sciences, Anwai, Beijing, China

Abstract

The impact of regional climate change on net primary productivity (NPP) is an important aspect in the study of ecosystems' response to global climate change. China's ecosystems are very sensitive to climate change owing to the influence of the East Asian monsoon. The Lund–Potsdam–Jena Dynamic Global Vegetation Model for China (LPJ-CN), a global dynamical vegetation model developed for China's terrestrial ecosystems, was applied in this study to simulate the NPP changes affected by future climate change. As the LPJ-CN model is based on natural vegetation, the simulation in this study did not consider the influence of anthropogenic activities. Results suggest that future climate change would have adverse effects on natural ecosystems, with NPP tending to decrease in eastern China, particularly in the temperate and warm temperate regions. NPP would increase in western China, with a concentration in the Tibetan Plateau and the northwest arid regions. The increasing trend in NPP in western China and the decreasing trend in eastern China would be further enhanced by the warming climate. The spatial distribution of NPP, which declines from the southeast coast to the northwest inland, would have minimal variation under scenarios of climate change.

Editor: Ben Bond-Lamberty, DOE Pacific Northwest National Laboratory, United States of America

Funding: This study was supported by the National Scientific Technical Supporting Programs during the 12th Five-year Plan of China (2012BAC19B04, 2012BAC19B10). The funders had no role in study design, data collection and analysis, decision to publish, or preparation of the manuscript.

Competing Interests: The authors have declared that no competing interests exist.

* E-mail: wush@igsnrr.ac.cn

Introduction

The impact of climate change on ecosystems is an important topic that has elicited substantial interest across the world [1]. Analysis of recorded temperature in the past century shows a global surface average temperature rise of approximately 0.74°C. Based on the mean temperature recorded from 1980 to 1999, the global surface average temperature is projected to increase by about 1.1°C to 6.4°C by 2100 [2]. China's climate, dominated by the East Asian monsoon, is extremely sensitive to global change [3]. Following projections from numerous general circulation models (GCMs), China would experience obvious climate changes in the future, including increase in average temperature, frequent occurrences of extreme climatic events, spatial and temporal heterogeneity in enhancing precipitation, and enlargement of its arid [4]. These changes in the climate can induce substantial variations in the composition, structure, and function of terrestrial ecosystems, thus inducing changes in ecosystem services, which are closely associated with the living environment of human beings and socioeconomic sustainability.

The impact of climate change on the ecosystem is poorly quantified because long-term in situ measurements are very sparse, and remote sensing techniques are only partially effective [2,5]. Recent studies on the interaction between terrestrial ecosystems and climate change have primarily focused on enhancing the simulation of ecosystem models. Considering that ecosystem models can simulate not only the interaction among ecological processes but also the feedback between climate and ecosystems, these models are of paramount importance in understanding energy balance as well as the water and carbon cycles in an ecosystem [6–9]. Net primary productivity (NPP) is the rate at which carbohydrates accumulate in a plant's tissues [10,11]. NPP is not only an important index to describe an ecosystem's structure and function but also a key element in describing carbon sequestration in an ecosystem during climate change [12].

Many models, including equilibrium biogeography and biogeochemical models, have been used in previous studies to simulate the impact of climate change on NPP in China. For example, Ni et al. employed an equilibrium biogeography model called BIOME3 to simulate changes in NPP under climate change scenarios [13]. The researchers found that NPP would increase in China from 2070 to 2099. Results from a process-based biogeochemical model (InTEC) indicated that the average forest NPP might be reduced from 2091 to 2100 under climate change, thereby inhibiting the CO_2 fertilizing effect in plants [14]. Ji et al. found that the NPP of terrestrial ecosystems in China would decrease from 1991 to 2100 based on simulations by an atmospheric–vegetation interaction model (AVIM2) [15]. These discrepancies in NPP trends can be partially attributed to ecosystem models because most of them disregard the role of vegetation dynamics in the carbon cycle during climate change. Dynamic vegetation models, including vegetation dynamics and biogeochemical processes, have supplied us with an effective approach to project transient responses of the ecosystem to rapid climate change [16,17]. Previous studies were almost entirely based on climate scenarios generated by GCMs with a resolution of 200 km to 300 km, which is too coarse for studies at regional or national scales. Regional climate models (RCMs) may be a better alternative in overcoming these shortcomings considering that these models provide greater spatial detail through dynamic downscaling of the GCM output [18].

Climate changes in China have significant regional differences; thus, the responses of terrestrial ecosystems to climate change vary in different regions. This study simulated the NPP of ecosystems in China under regional climate change scenarios (A2, B2, and A1B) based on a modified Lund–Potsdam–Jena Dynamic Global Vegetation Model (LPJ-DGVM). The temporal and spatial changes in the NPP of ecosystems in different regions were examined according to eco-regions. The purpose of the study was to elucidate the impacts of climate change on ecosystems at a regional scale and provide scientific basis for local adaptation and mitigating strategies.

Methods and Data

1. Methods

LPJ-DGVM [19] is an integrated dynamic biogeography–biogeochemistry model developed based on an earlier equilibrium model, BIOME3. LPJ-DGVM is constructed in a modular framework, which combines process-based representations of terrestrial vegetation dynamics and land–atmosphere carbon and water exchanges. LPJ-DGVM explicitly considers key ecosystem processes such as vegetation growth, mortality, carbon allocation, and resource competition. Vegetation structure and composition are described by ten plant functional types (PFTs) (Table 1), which are distinguished according to their phenology, physiology, physiognomy, disturbance response attributes, and bioclimatic constraints. Gross primary production is computed based on a coupled photosynthesis–water balance scheme established through canopy conductance. Net primary production is calculated by subtracting autotrophic respiration. The sequestrated carbon is stored in seven PFT-associated pools representing leaves, sapwood, heartwood, fine roots, fast and slow decomposition in the aboveground litter pool, and a below-ground litter pool. The decomposition rates of soil and litter organic carbon depend on soil temperature and moisture. Model input includes monthly mean air temperature, total precipitation, number of wet days and percentage of full sunshine, annual CO_2 concentration, and soil texture class. The full description of LPJ-DGVM was provided by Sitch et al. [19]; thus, only a short overview is provided in this paper. The LPJ-DGVM model is a typical dynamic vegetation model that has been widely utilized to study terrestrial ecosystem dynamics and interactions between climate change and ecosystems at global, regional, and site scales [8,9,16,17,20]. The version employed in the present study includes improved hydrology by Gerten et al. [20].

In our previous study [21], LPJ-DGVM was carefully modified by adding shrub and cold grass PFTs, which were parameterized based on various inventory and observational data, with respect to the characteristics of ecosystems in China. The results of the simulations by the modified LPJ-DGVM for China (LPJ-CN) were validated with data sets obtained from the sites [21]. The simulated NPP results from LPJ-CN matched the observed data ($R^2 = 0.64$, $P<0.01$), which was better compared with the original LPJ-DGVM data ($R^2 = 0.10$) employed by Ni [22]. Therefore, LPJ-CN is assumed to be appropriate for simulating NPP in China.

Considering that the East Asian monsoon caused by differences in the heat-absorbing capacity of the continent and the ocean dominates the climate in China, climatic types vary from tropical in the south to cold temperate in the north and from humid in the east to dry in the west. Diverse climates and complex topography result in high biodiversity in China. The macro-spatial distribution of ecosystems forms different eco-regions with unique characteristics. Ecosystem responses to climate change in different eco-regions vary. According to the ecosystem regionalization scheme of China by Zheng et al. [23], China can be divided into eight eco-regions, namely, cold temperate humid region (I), temperate humid/sub-humid region (II), north semi-arid region (III), warm temperate humid/sub-humid region (IV), subtropical humid region (V), tropical humid region (VI), northwest arid region (VII), and Tibetan Plateau region (VIII). The impact of climate change on NPP was discussed in this paper according to the eco-region scheme.

According to the impact degree of future climate change on NPP, the change would be slight if the absolute change in NPP compared with the baseline term is <20%; the change would be moderate at 20% to 40%, the change would be severe at 40% to 60%, and the change would be extremely severe at over 60%. The slopes of linear regressions, obtained using the least squares method, were utilized to assess variation trends in the time series. Positive slopes indicate increasing trends, whereas negative slopes indicate decreasing trends. The significance levels of the trends were assessed by the non-parametric Mann-Kendall test.

2. Data

2.1 Climatic data. The data on climate scenarios utilized in this study were provided by the climate change research group of the Institute of Environment and Sustainable Development in Agriculture, Chinese Academy of Agricultural Sciences. Based on the greenhouse gas emission scenarios of the IPCC Special Report on Emission Scenarios (SRES) [24], the group produced high-resolution ($0.5° \times 0.5°$) climate data scenarios for late 21st century China through the Providing Regional Climate for Impacts Studies (PRECIS) system [25]. PRECIS was validated by applying reanalysis data derived from the European Centre for Medium-Range Weather Forecasts (ECMWF) as lateral boundary conditions; PRECIS was found to simulate terrestrial climate change in China effectively [26]. Projected climatic data, including three SRES emission scenarios (A2, B2, and A1B), for the period 1961 to 2100 were employed in the present study.

Compared with the baseline climate (1961–1990), the annual mean temperature of China (Fig. 1a) is projected to increase from 3.26°C (B2) and 4.64°C (A1B), with the highest increase in northwest and northeast China and the lowest increase in southeast China. The total annual precipitation across China (Fig. 1b) is projected to increase from 8% (B2) and 21% (A2), with the highest increase in northwestern China. A slight decrease in precipitation would occur in northeast China.

The periods were divided into baseline period (1961–1990), near term period (1991–2020), middle term period (2021–2050), and long term period (2051–2080). Each term was discussed according to the average of 30 years.

2.2 Soil data. This research adopted soil texture data from the map of soil texture types (1:14,000,000) [27], which contains information on the proportions of mineral grains in the top soil and geographical distribution of the different soil texture types. The soil textures were reclassified as clay, silt, sand, silty sand, sandy clay, silty clay, and clay with silt and sand according to the FAO classification standard for soil texture to address the data input requirements of LPJ-CN [28]. The soil data were then transformed into the ArcInfo grid format and resampled to the spatial resolution of $0.5° \times 0.5°$.

3. Modeling Protocol

LPJ was operated for a "spin up" period of 900 years to achieve equilibrium in stable vegetation structures and C pools. The model

Table 1. PFTs Bioclimatic limiting factors used by LPJ-CN.

PFTs	T_c min (°C)	T_c max (°C)	GDD min (°C)	T_w max (°C)
Tropical broad-leaved evergreen forest	*12.0* (15.5)			
Tropical broad-leaved raingreen forest	*12.0* (15.5)			
Temperate needle-leaved evergreen forest	−2.0	22.0	900	
Temperate broad-leaved evergreen forest	*0* (3.0)	*14.0* (18.8)	*1500* (1200)	
Temperate broad-leaved summergreen forest	−17.0	*.0* (15.5)	*1500* (1200)	
Boreal needle-leaved evergreen forest	−32.5	*−25.0* (2.0)	600	23.0
Boreal needle-leaved summergreen forest		*−2.0* (2.0)	350	23.0
Boreal broad-leaved summergreen forest		*−15.0* (2.0)	350	23.0
Shrub		−5.0	350	
Cold grass		−12.0		12
Temperate grass		*10.0* (15.5)		
Tropical grass	*15.0* (15.5)			

Note: The parameter values in round brackets given by Sitch et al. [19] are replaced by italic values in this study. Shrub and cold grass are newly added PFTs in LPJ. T_c min the minimum coldest monthly mean temperature for survival; T_c max the maximum coldest monthly mean temperature; GDD min the minimum growing degree days of over 5°C; T_w max the upper limit of temperature of the warmest month.

was thereafter subjected to transient climate. Observed atmospheric CO_2 concentration was applied to the ecosystem model simulation for the period of 1961 to 1990. In the projected simulation, the atmospheric CO_2 concentration would remain at the value of 1990. The scheme was designed to eliminate the fertilizing effects of atmospheric CO_2 concentration on the ecosystem.

Results

1. Variation Trend of NPP in Different Eco-regions

Figure 2 illustrates the inter-annual variations of average NPP in eight eco-regions under A2, B2, and A1B scenarios from 1991 to 2100. In the future, NPP may exhibit a significant decreasing trend in all eco-regions except in the Tibetan plateau and northwest arid regions. This projection is associated with the spatial heterogeneity of climate change and sensitivity of the ecosystem to climate change. In cold temperate humid region, the NPP trend can be distinguished in two periods: before and after 2030. From 1991 to 2030, NPP would remain unchanged but would significantly decrease after 2030 (Fig. 2). NPP in the temperate humid/sub-humid region is projected to decline at a rate greater than 1.5 g C m^{-2} yr^{-1} and would demonstrate a large decadal fluctuation particularly in the B2 and A1B scenarios. The largest decrease is projected to occur in the warm, temperate humid/sub-humid region then in the tropical humid and sub-tropical humid regions (Table 2). The various climate trends in the three climate scenarios can produce slight differences in the NPP trend for each eco-region. Considering that the average NPP in the three eco-regions is higher than that in the other eco-regions, the rapid decrease could have a significant contribution to the decline of NPP in entire China. Increase in NPP may occur in the Tibetan plateau and northwest arid regions due to warming climate. The largest increase in NPP may be found in the Tibetan plateau region, with an average NPP of 1.7 g C m^{-2} yr^{-1} for the

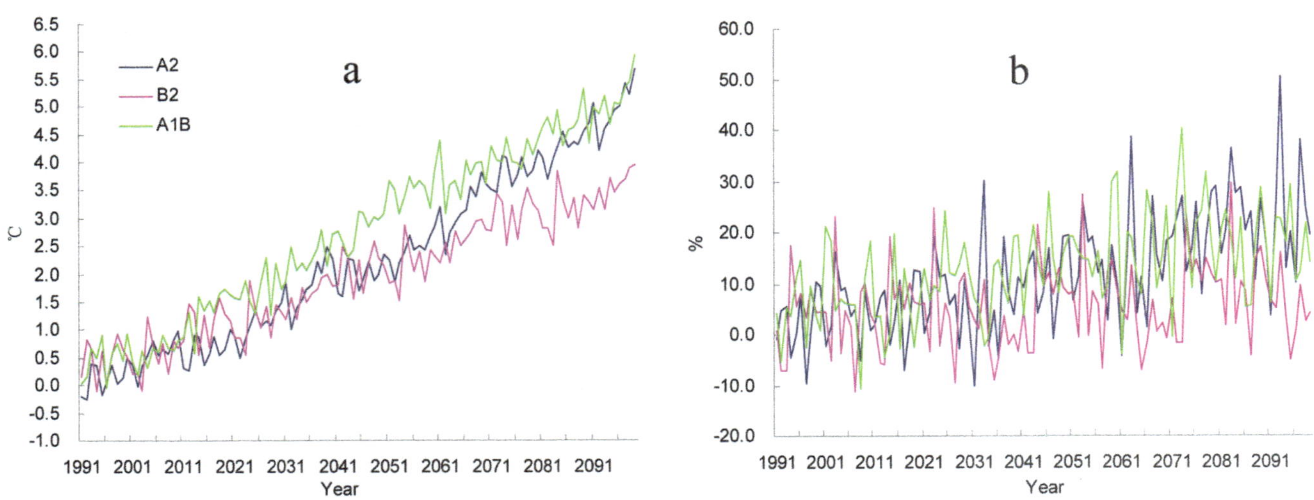

Figure 1. Inter-annual variations of temperature (a) and precipitation (b) anomalies averaged across China (relative to the reference averages from 1961 to 1990, projected by the PRECIS under SRES A2, B2, and A1B scenarios).

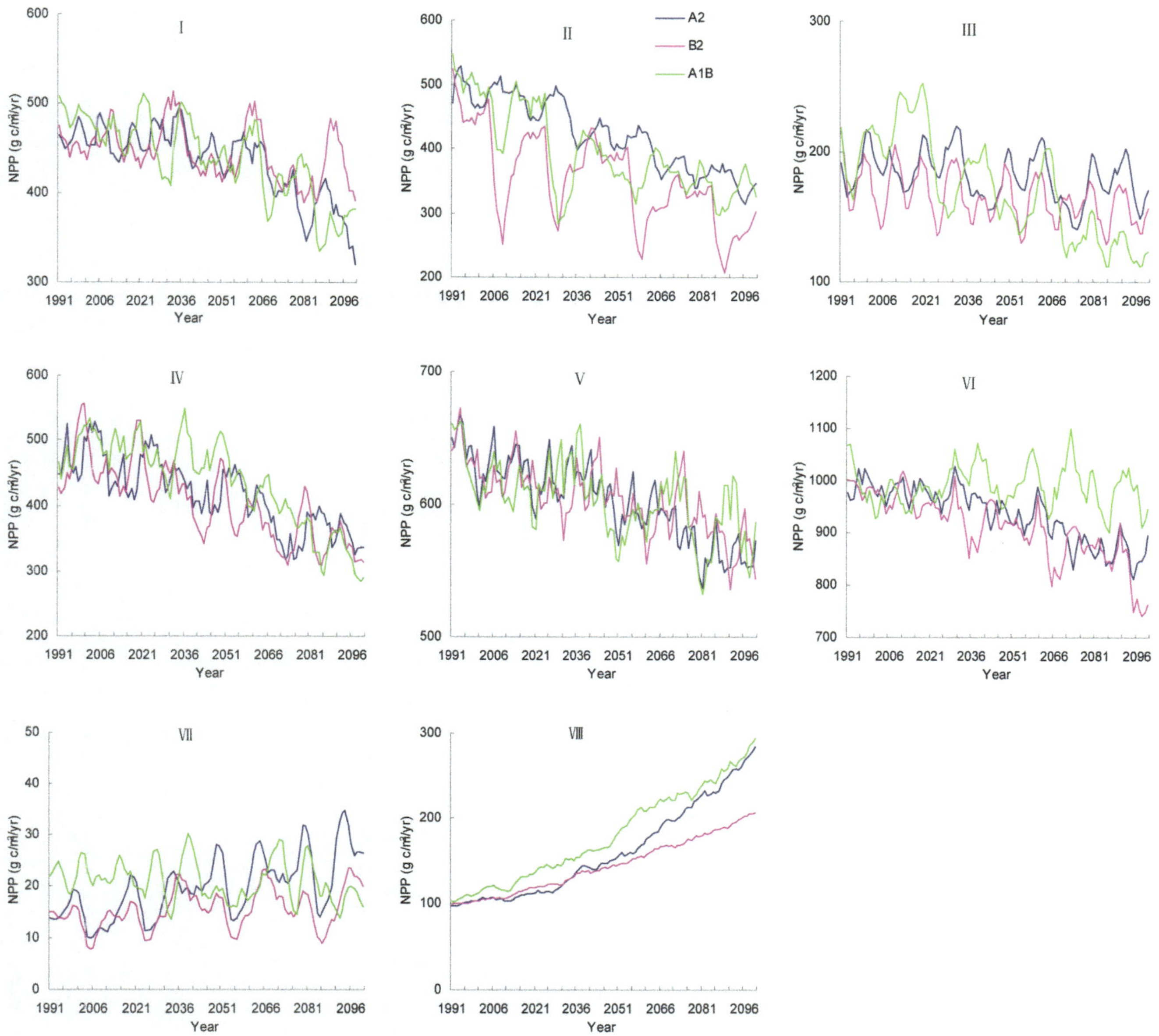

Figure 2. Inter-annual variation of NPP in different eco-regions (I refers to cold temperate humid region; II, temperate humid/sub-humid region; III, north semi-arid region; IV, warm temperate humid/sub-humid region; V, subtropical humid region; VI, tropical humid region; VII, northwest arid region; VIII, Tibetan Plateau region.) from 1991 to 2100 under SRES A2, B2 and A1B scenarios.

A2 and A1B scenarios. The NPP increment in the northwest arid region may be slight; approximately 0.13 g C m^{-2} yr^{-1} ($R^2=0.46$, $P<0.001$) for A2 and 0.14 g C m^{-2} yr^{-1} ($R^2=0.14$, $P<0.001$) for B2. The annual variations in NPP are projected to exhibit a mild decreasing trend in the A1B scenario, although this projection did not pass the statistical tests.

2. Spatial Variation of NPP

Figure 3 shows the spatial pattern of NPP in near term and its variation compared with the baseline term. The NPP in the majority of the ecosystems is projected to increase in near term in over 60% of the total land area of China under the A2, B2, and A1B scenarios. A slight increase in NPP is likely to be predominant. Over-moderate increment may exist only in a few ecosystems concentrated in the Tibetan Plateau and northwest arid region. A few ecosystems may experience a decrease in NPP particularly under the A1B scenario, which shows the largest area of NPP decrement amounting to approximately 37% of the overall territorial ecosystems. A slight decrement dominates the largest areas of ecosystems with decreasing NPP, accounting for 62%, 64%, and 84% in the A2, B2, and A1B scenarios, respectively. Only a few ecosystems are likely to have a rate of decrement over moderate levels; these ecosystems are mainly distributed in the north semi-arid and northwest arid regions, particularly in the A2 scenario.

The trends of NPP, decreasing in east China and increasing in west China, are more obvious in the middle term compared with the near term (Fig. 4). A slight NPP decrement may dominate China's ecosystems in the A2, B2, and A1B scenarios, occupying 25% and 26% of the overall territorial ecosystem in the A2 and B2 scenarios, respectively, which mainly occurs in the sub-tropical humid and warm temperate humid/sub-humid regions. In

Table 2. NPP Slopes (g C m^{-2} yr^{-1}) in different eco-regions and its coefficient of determination under different climate scenarios.

Eco-regions	A2		B2		A1B	
	Slope	R^2	Slope	R^2	Slope	R^2
I	−0.95	0.61	−0.37	0.17	−1.11	0.64
II	−1.63	0.87	−1.53	0.50	−1.59	0.60
III	−0.19	0.11	−0.24	0.18	−0.94	0.62
IV	−1.36	0.66	−1.49	0.64	−1.76	0.71
V	−0.83	0.74	−0.60	0.55	−0.53	0.36
VI	−1.43	0.75	−1.65	0.70	−0.05	0.01
VII	0.13	0.46	0.04	0.14	−0.03*	0.01
VIII	1.66	0.97	1.00	0.99	1.69	0.98

Note: *means the significance level is P<0.5, or else P<0.001.

addition, ecosystems with over-moderate NPP decrement are limited to the temperate humid/sub-humid, north semi-arid, and northwest arid regions. Severe and extremely severe NPP decrements primarily exist in the central temperate humid/sub-humid region and the adjacent areas between the north semi-arid and northwest arid regions. The ecosystems with NPP increment are likely to decline, with approximately 53% of the total ecosystem in the B2 and A1B scenarios and 59% in the A2 scenario. In the B2 scenario, the area with a slight increment in NPP is reduced to 21% of the overall ecosystem, which is smaller than the value in the near term. Ecosystems with over-moderate NPP increment are still concentrated in the Tibetan Plateau and northwest arid region where the absolute NPP value is only 60 g C m^{-2}, a value that is considerably lower than those in the other regions.

Figure 5 illustrates the NPP distribution in the long term climate and change in NPP relative to the baseline term climate. In the A1B, A2, and B2 scenarios, the ecosystems with NPP decrement are expected to occupy 47%, 46%, and 52% of total area in China, respectively, which means that nearly half of the ecosystems in China may be exposed to the adverse effects of climate change. An NPP decrement may occur primarily in eastern China, which has an inherently high NPP. An NPP increment may occur in western China, which has low NPP. Ecosystems with slight NPP decrement would still predominate in China and would retain spatial patterns similar to that in the middle term. Areas with over-moderate NPP decrement are likely to expand further and are projected to become distributed in the temperate humid/sub-humid and warm temperate humid/sub-humid regions. Ecosystems may be severely impacted by climate change in the A1B scenario, and the area with the extremely severe NPP decrement would be greater than 7% of the overall territorial ecosystems. This projection is expected to occur in most parts of northern China. On the contrary, the NPP increment

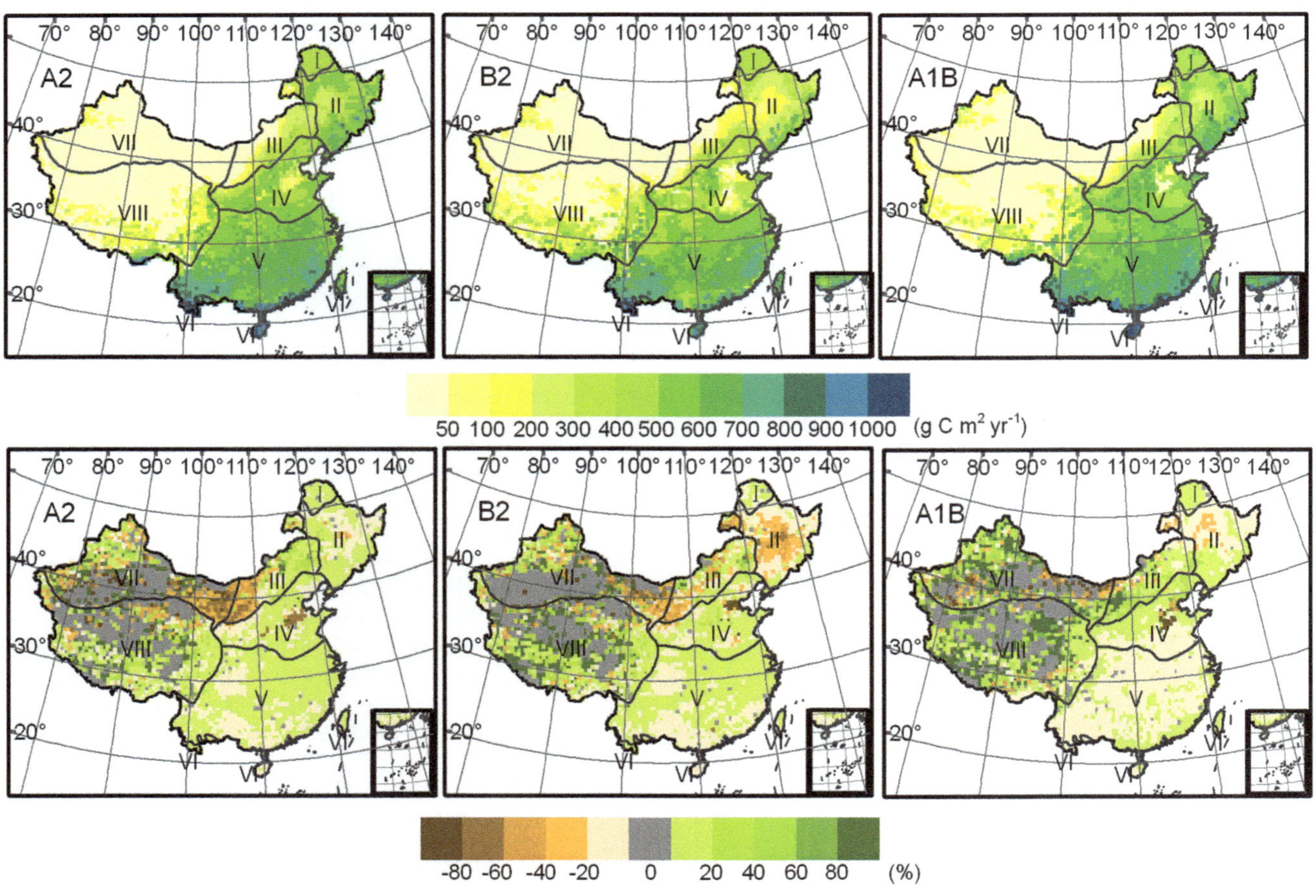

Figure 3. NPP spatial pattern (g C m^2 yr^{-1}) (upper panel) in the near term and its anomalies (%) (lower panel) with baseline over China modeled through climate change projections for SRES A2, B2, and A1B scenarios.

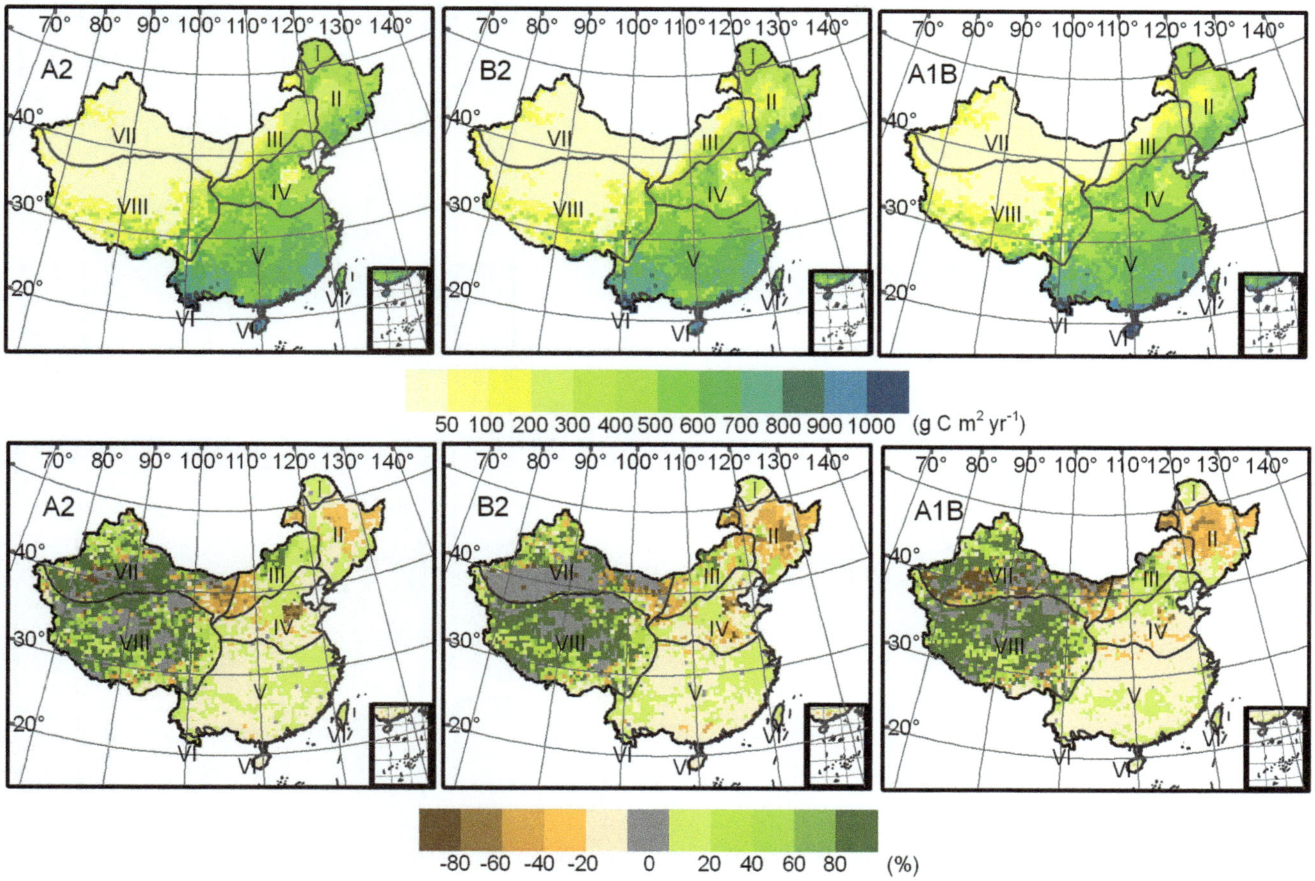

Figure 4. NPP pattern (g C m^2 yr^{-1}) (upper panel) in the midterm and its anomalies (%) (lower panel) with baseline over China modeled through climate change projections for SRES A2, B2, and A1B scenarios.

would be mainly concentrated in western China, particularly in the Tibetan Plateau region where the average NPP increment may reach approximately 60%. The NPP decrement in East China and increment in West China may be enhanced further in terms of spatial variation.

Discussion

1. Effects of Climate Change on NPP

The projected climate change characterized by warm and wet climate may cause variations in NPP in most areas of China according to our simulated results. Regional differences are mainly associated with the combinations of variations in temperature and precipitation in all three scenarios. The NPP decrease in cold temperate humid regions may be associated with the rising temperature because a slight increase in temperature can benefit the boreal forest growth, whereas a significant increase in temperature may reduce the water effectiveness through thawing frozen soil and enhancing evapotranspiration, which in turn can result in negative effects for the boreal forest growth [15,16,29]. The large annual variation in NPP in the temperate humid/sub-humid region can be attributed to frequent drought derived from the climate scenarios, particularly in B2. The decrement in precipitation can impair the photosynthesis rate of vegetation in the model because of reduced stomatal conductance [30,31].

The NPP decrement in warm temperate, sub-tropical, and tropical humid regions can be explained by increased water stress as a consequence of increased evapotranspiration that is not counterbalanced by an increase in rainfall. Piao et al. [29] also found that increase in temperature alone does not benefit vegetation NPP in temperate and tropical ecosystems because of water control. Studies on the historical tree ring records of tropical forests and temperature [32,33] suggest that water availability is a key limiting factor that controls vegetation NPP. Given that water limits photosynthesis in plants in the northwest arid region where plants are more sensitive to precipitation variation than temperature variation, increasing precipitation can result in an increase in carboxylation efficiency, which is the primary cause of enhanced productivity [30]. The NPP in the A1B scenario exhibits a different trend from that in A2 and B2. This different trend could have been caused by different precipitation trends. Before 2030, the A1B scenario has a greater amount of rainfall than the other two scenarios; however, the amount of rainfall would remain unchanged or even decrease after 2031. The relatively low temperature in the Tibetan Plateau limits plant growth, but warming may solve this limitation, thus leading to a general increase in NPP [13,34].

The results of this research are generally consistent with the results of previous studies [15,14,28,35]. However, some discrepancies exist. Our study suggests that NPP in eastern China, which is covered with forests, is likely to decrease even with the influence of climate change. This is similar to the result obtained by Ju et al. [14]. The spatial change in NPP, namely, a decrease in eastern China and an increase in western China under the climate change scenarios, is consistent with Wu et al. [35] and Ji et al.'s [15] results. The NPP changes in several sensitive areas located at the

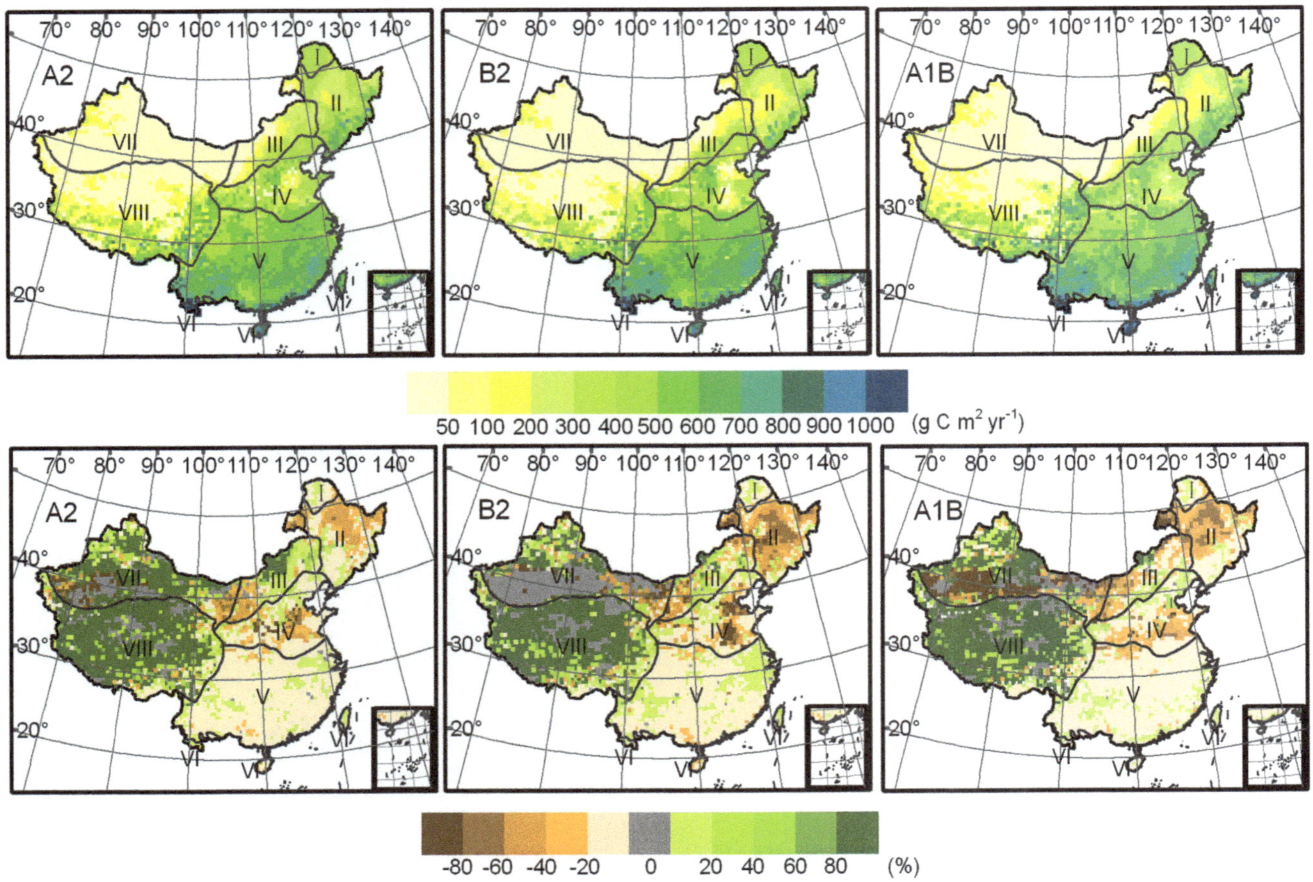

Figure 5. NPP pattern (g C m^2 yr^{-1}) (upper panel) in the long term and its anomalies (%) (lower panel) with baseline over China modeled through climate change projections for SRES A2, B2, and A1B scenarios.

temperate humid/sub-humid, warm temperate humid/sub-humid, and north semi-arid regions are different from the results of previous studies [15,35]. These areas have an increasing incidence of extreme climatic events; therefore, biogeography and biogeochemical models cannot simulate the effects of extreme climatic events on ecosystems. Dynamic vegetation models can capture these effects, leading to discrepancies in results.

2. Uncertainties in the Study

LPJ-CN, a process-based ecosystem model, was applied in this study to simulate NPP changes in China under scenarios A2, B2, and A1B. The study examined the NPP variation in three projected periods, which improves the understanding of the impact of the different warming levels. However, several uncertainties underlying future climate change and ecosystem responses exist. The trends of climate change in China are different in various climatic scenarios. However, we only probed the possibilities for the IPCC SRES A2, B2, and A1B scenarios. Given that the climatic system is complicated and nonlinear and China's landform is very complex with altitudes varying from −50 m to 8843 m, GCMs are often unable to depict the accurate spatial patterns of the climatic factors employed in ecosystem modeling with low resolution. Although the simulation capability of PRECIS with boundary condition from the general circulation model (GCMs)-HadAM3 [25] has been improved in this study by enhancing spatial resolution, the aforementioned problems still cannot be addressed entirely to eliminate uncertainties. Moreover, some uncertainties remain with regard to the parameterization and process representations in the LPJ-DGVM model [36]. For example, the model does not provide an explicit representation of the nitrogen cycle; some scientific assumptions are unreasonable, and some lagging responses and self-adaptive mechanisms of the ecosystem to climate change are omitted in the LPJ-DGVM model. These issues have attracted the attention of researchers around the world, but there are insufficient studies to support the quantification of these processes in the model.

The LPJ-CN model predicts the temporal and spatial change of ecosystem NPP according to potential natural vegetation, which is free from the influence of anthropogenic activities. However, the great expansion in human activity can definitely have a major impact on ecosystem dynamics [37]. Liu et al. reported that approximately 5.2 Mha of grassland and woodland in China were converted to cropland in the 1990s because of reclamation activities [38]. Piao et al. found that the large-scale reforestation and afforestation programs since the 1980s in China have resulted in the increase of forest biomass carbon stocks [39]. Considering the lack of information on the amount and spatial patterns of land use in the future, the accurate estimation of the magnitude of NPP change associated with land use remains a challenge [40]. These uncertainties can be addressed by integrating human disturbances in future studies.

Conclusions

We utilized a dynamic vegetation model, LPJ-CN, to simulate the temporal and spatial responses of ecosystem NPP to climate change in 21st century China under A2, B2, and A1B scenarios. Our results indicate that rising temperature and slightly changing precipitation would lead to a serious impact on the natural ecosystem's NPP in general. The near-term, mid-term, and long-term impact of climate change may become worse gradually. In the near term, the impact of climate change on the ecosystem may be minimal; more than half of the ecosystems may benefit from the change. However, an adverse impact may occur in the middle term. In the long term, more than half of ecosystems may be exposed to adverse effects. NPP enhancement may appear mainly in the Tibetan Plateau and northwest arid region. The other eco-regions may be dominated by NPP increase. Although NPP may increase by a relatively high percentage in the Tibetan Plateau and northwest arid region, it cannot generate a significant influence on the overall NPP distribution in China because of its low initial productivity level. Therefore, the spatial distribution of NPP, which decreases from the southeast coast to the northwest inland, would not be altered under the climate change scenarios. The response of China's terrestrial ecosystems to climate change will contribute to our understanding of the vulnerability and adaptability of ecosystems to climate change in a regional scale. The result obtained in this study can provide a basis for environmental policy making.

Acknowledgments

We thank Prof. Yinlong Xu from Institute of Environment and Sustainable Development in Agriculture, Chinese Academy of Agriculture Sciences, for providing climate scenarios data.

Author Contributions

Conceived and designed the experiments: SW. Performed the experiments: DZ. Analyzed the data: DZ. Wrote the paper: DZ YY.

References

1. Heimann M, Reichstein M (2008) Terrestrial ecosystem carbon dynamics and climate feedbacks. Nature 451, 289–292.
2. IPCC (2007) Summary for Policymakers. In: Climate Change 2007: Impacts, Adaptation and Vulnerability. Contribution of Working Group II to the Fourth Assessment Report of the Intergovernmental Panel on Climate Change, M.L. Parry, O.F. Canziani, J.P. Palutikof, P.J. Van der Linden and C.E. Hanson, editors, Cambridge, Cambridge University Press.
3. Ding YH, Johnny CLC (2005) The East Asian summer monsoon: an overview. Meteorol Atmos Phys 89, 117–142.
4. Taskforce on China's National Assessment Report on Climate Change (2007) China's National Assessment Report on Climate Change. Science Press: Beijing.
5. Valentini R, Matteucci G, Dolman AJ, Schulze ED, Rebmann C, et al. (2000) Respiration as the main determinant of carbon balance in European forests. Nature 404: 861–865.
6. Tian H, Melillo JM, Kicklighter DW, Pan S, Liu J, et al. (2003) Regional carbon dynamics in monsoon Asia and its implications for the global carbon cycle. Global Planet Change 37: 201–207.
7. Koca D, Smith B, Sykes MT (2006) Modeling regional climate change effects on potential natural ecosystems in Sweden. Climatic Change 78: 381–401.
8. Morales P, Hick T, Rowell DP, Smith B, Sykes MT (2007) Changes in European ecosystem productivity and carbon balance driven by regional climate model output. Global Change Biol 13: 108–122.
9. Doherty RM, Sitch S, Smith B, Lewis SL, Thornton PK (2010) Implication of future climate and atmospheric CO_2 content for regional biogeochemistry, biogeography and ecosystem services across East Africa, Global Change Biol 16: 617–640.
10. Odum EP (1959) Fundamentals of ecology. Philadelphia: W.B. Saunders Co.
11. Lieth H, Whittaker RH (1975) Primary Productivity of the biosphere. New York: Springer-Verlag Press.
12. Nemani RR, Keeling CD, Hashimoto H (2003) Climate-driven increases in global terrestrial net primary production from 1982–1999. Science 300: 1560–1563.
13. Ni J (2000) A simulation of biomes on the Tibetan Plateau and their responses to global climate change. Mt Res Dev 20: 80–89.
14. Ju WM, Chen JM, Harvey D, Wang S (2007) Future carbon balance of China's forests under climate change and increasing CO_2. J Environ Manage 85: 538–562.
15. Ji JJ, Huang M, Li KR (2009) Prediction of carbon exchanges between China terrestrial ecosystem and atmosphere in 21st century. Sci China Ser D 51(6): 885–898.
16. Cramer W, Bondeau A, Woodward FI, Prentice IC, Betts RA, et al. (2001) Global response of terrestrial ecosystem structure and function to CO_2 and climate change: Results from six dynamic global vegetation models. Global Change Biol 7: 357–373.
17. Bachelet D, Neilson RP, Hickler T, Drapek RJ, Lenihan JM, et al. (2003) Simulating past and future dynamics of natural ecosystems in the United States. Global Biogeochem Cy 17(2), 1045, doi:10.1029/2001GB001508.
18. McGregor JL (1997) Regional climate modeling. Meteorol Atmos Phys 63: 105–117.
19. Sitch S, Smith B, Prentice IC. Arneth A, Bondeau A, et al. (2003) Evaluation of ecosystem dynamics, plant geography and terrestrial carbon cycling in the LPJ dynamic global vegetation model. Global Vegetation Model. Global Change Biol 9: 161–185.
20. Gerten D, Schaphoff S, Haberlandt U, Lucht W, Sitch S (2004) Terrestrial vegetation and water balance-Hydrological evaluation of a dynamic global vegetation model. J Hydrol 286: 249–270.
21. Zhao DS, Wu SH, Yin YH (2011) Variation trends of natural vegetation net primary productivity in China under climate change scenario. Chinese J Appl Ecol 22(4): 897–904.
22. Ni J (2003) Net primary productivity in forests of China: scaling-up of national inventory data and comparison with model predictions. Forest Eco Manag 176: 485–495.
23. Zheng D (1996) The system of physico-geographical regions of the Qinghai-Xizang (Tibet) Plateau. Sci China Ser D 39: 410–417.
24. Nakicenovic etal (2000) Special Report on Emissions Scenarios (SRES). Cambridge: Cambridge University Press.
25. Jones R, Noguer M, Hassell D, Hudson D, Wilson S, et al. (2004) Generating high resolution climate change scenarios using PRECIS. Met Office Hadley Centre, Exeter UK.
26. Xu YL, Jones R (2004) Validating PRECIS with ECMWF reanalysis data over China. Chinese J Agrometeorol 25(1): 5–9.
27. Zhang SH, Peng GB, Huang M (2004) The feature extraction and data fusion of regional soil textures based on GIS techniques. Clim Environ Res 6: 65–79.
28. Ni J, Sykes MT, Prentice IC, Cramer W (2001) Modeling the vegetation of China using the process-based equilibrium terrestrial biosphere model BIOME3. Global Ecol Biogeogr 9: 463–479.
29. Piao SL, Ciais P, Friedlingstein P, de Noblet-Duconudre N, Cadule P, et al. (2009) Spatiotemporal patterns of terrestrial carbon cycle during the 20th century. Global Biogeochem Cy 23, GB4026, doi:10.1029/2008GB003339.
30. Hou YY, Liu QH, Yan H, Tian GL (2007) Variation trend of China terrestrial vegetation net primary productivity and its responses to climate factors in 1982–2000. Chinese J Appl Ecol 18(7): 1546–1553.
31. Yang YH, Fang JY, Ma WH, Guo DL, Mohammat A (2010) Large-scale pattern of biomass partitioning across China's grassland. Global Ecol Biogeogr 19: 268–277.
32. Clark DA, Piper SC, Keeling DC, Clark DB (2003) Tropical rain forest tree growth and atmospheric carbon dynamics linked to interannual temperature variation during 1984–2000. Pro Natl Acad Sci U S A 100: 5852–5857.
33. Liang EY, Shao XM, Kong ZC, Lin JX (2003) The extreme drought in the 1920s and its effect on tree growth deduced from tree ring analysis: A case study in north China. Ann Forest Sci 60: 145–152.
34. Zhao DS, Wu SH, Yin YH, Yin ZY (2011) Vegetation distribution on Tibetan Plateau under climate change scenario. Reg Environ Change 11: 905–915.
35. Wu SH, Yin YH, Zhao DS, Huang M, Shao XM, et al. (2010) Impact of future climate change on terrestrial ecosystems in China. Int J Climatol 30: 866–873.
36. Zaehle S, Sitch S, Smith B, Hatterman F (2005) Effects of parameter uncertainties on the modeling of terrestrial biosphere dynamics. Global Biogeochem Cy 19, GB3020, doi:10.1029/2004GB002395.
37. Bondeau A, Smith PC, Zaehle S, Schaphoff S, Lucht W, et al. (2007) Modelling the role of agriculture for the 20th century global terrestrial carbon balance. Global Change Biol 13: 679–706.
38. Liu JY, Liu ML, Tian HQ, Zhuang DF, Zhang ZX, et al. (2005) Spatial and temporal patterns of China's cropland during 1990–2000: An analysis based on Landsat TM data. Remote Sens Environ 98: 442–456.
39. Piao SL, Fang JY, Ciais P, Peylin P, Huang Y, et al. (2009) The Carbon balance of terrestrial ecosystems in China. Nature 458: 1009–1013.
40. Houghton RA (2007) Balancing the global carbon budget. Ann Rev Earth Pl Sc 35: 313–347.

Detecting the Differences in Responses of Stomatal Conductance to Moisture Stresses between Deciduous Shrubs and *Artemisia* Subshrubs

Qiong Gao[1], Mei Yu[2]*, Chan Zhou[3]

1 Landscape Analysis and Modeling Group, National Laboratory of Earth Surface Processes and Resources Ecology, Beijing Normal University, Beijing, China, 2 Institute for Tropical Ecosystem Studies, University of Puerto Rico - Rio Piedras, San Juan, Puerto Rico, United States of America, 3 College of Life Sciences, Liaoning University, Shenyang, China

Abstract

Shrubs and subshrubs can tolerate wider ranges of moisture stresses in both soil and air than other plant life forms, and thus represent greater nonlinearity and uncertainty in ecosystem physiology. The objectives of this paper are to model shrub/subshrub stomatal conductance by synthesizing the field leaf gas exchanges data of 24 species in China, in order to detect the differences between deciduous shrubs and *Artemisia* subshrubs in their responses of stomatal conductance to changes in the moisture stresses. We revised a model of stomatal conductance by incorporating the tradeoff between xylem hydraulic efficiency and cavitation loss risk. We then fit the model at the three hierarchical levels: global (pooling all data as a single group), three functional groups (deciduous non-legume shrubs, deciduous legume shrubs, and subshrubs in *Artemisia* genus), and individual observations (species × sites). Bayesian inference with Markov Chain Monte Carlo method was applied to obtain the model parameters at the three levels. We found that the model at the level of functional groups is a significant improvement over that at the global level, indicating the significant differences in the stomatal behavior among the three functional groups. The differences in tolerance and sensitivities to changes in moisture stresses are the most evident between the shrubs and the subshrubs: The two shrub groups can tolerate much higher soil water stress than the subshrubs. The analysis at the observation level is also a significant improvement over that at the functional group level, indicating great variations within each group. Our analysis offered a clue for the equivocal issue of shrub encroachment into grasslands: While the invasion by the shrubs may be irreversible, the dominance of subshrubs, due to their lower resistance and tolerance to moisture stresses, may be put down by appropriate grassland management.

Editor: Gil Bohrer, The Ohio State University, United States of America

Funding: The National Science Foundation of China grants # 41171445 and 41321001, National Lab of Earth Surface Processes and Resources Ecology grant # 2012-TDZY-31, National Basic Research Project # 2014CB954303. The funders had no role in study design, data collection and analysis, decision to publish, or preparation of the manuscript.

Competing Interests: The authors have declared that no competing interests exist.

* E-mail: meiyu@ites.upr.edu

Introduction

Dynamic global vegetation models (DGVMs) and regional ecosystem models have been making projections of structures and functions in response to changes in climate and anthropogenic activities [1,2]. However, there existed great disagreements among the models, largely due to the uncertainties in physiological parameters that control the carbon assimilation into and emission out of ecosystems [3]. Ecosystem model parameters are often derived from plant functional traits, which have been recognized as important links to ecosystem functions, structures, and adaptations [4,5]. Relationships among maximum leaf photosynthesis, maximum stomatal conductance, specific leaf area, leaf life span, leaf size, and leaf nitrogen have been found from the global leaf traits dataset by means of meta-analyses [6,7,8,9]. Shrubland ecosystem physiology in the sub-humid, semiarid, and arid regions involves great uncertainty because of the superior tolerance to moisture stresses, morphological plasticity, and adaptability [10].

Shrubs have been reported to invade grasslands in many places over the world [10,11,12,13,14,15]. Many causal factors have been hypothesized to trigger the processes of shrub encroachment. Among these hypotheses, increased drought frequency and shifted rainfall seasonality/intensity have been considered as major drivers [16,17]. These hypotheses are based on a common assumption that shrubs are more tolerant to drought than grasses. The parameterized stomatal model used in the Patch Arid Land Simulator (PALS) [18,19] showed that the shrub stomata can tolerate much more severe soil water stress than the grass stomata. When soil moisture is ample, the grasses showed greater stomata conductance than the shrubs.

Shrubs tend to have deep roots, high ratio of leaf to sapwood areas, low vertical shading, and strong morphological plasticity to adapt to variations in climate and soils [20]. The same shrub species can be phreatophyte in sandy soils, but xerophyte in heavy clayey soils to maintain active when soil water potential lower than −5 MPa [21,22].

Models of stomatal conductance are important tools to quantify leaf production processes of shrubs, since leaf photosynthesis is co-limited by stomatal conductance and biochemical carboxylation [23]. Many stomatal models exist in the literature, with different mechanistic/empirical assumptions and treatments. Ball et al. [24] proposed a simple empirical model of stomatal conductance as a function of net photosynthesis rates and relative humidity on leaf surface. Another model was developed to count the composite effects of net photosynthesis, vapor pressure deficit, and CO_2 pressure on stomatal conductance [25]. More mechanistic model of stomatal conductance that considered the transient response of stomatal conductance to changes in driving variables was developed by Buckley et al. [26,27] Incorporating stomatal conductance with hydrological structure and function of tree canopy showed complex interactions between crown structure and hydrological function [28].

A semi-mechanistic model was developed by Gao et al. [29] to calculate stomatal conductance as a function of soil water potential, vapor pressure deficit in air, intercellular CO_2 concentration, and light irradiance. The model was revised [30]to consider the cavitation loss of xylem conductance and hydrological capacitance, the model considered the cavitation loss of xylem hydraulic conductivity linearly dependent on xylem water potential, so that

$$K_{soil-to-leaf} = g_p(1+\zeta\psi_x)$$

where $K_{soil-to-leaf}$ is the apparent soil-to-leaf conductance, g_p is the maximum conductance, and ζ is the parameter signifying the linear dependence of $K_{soil-to-leaf}$ on ψ_x, the xylem water potential. However, recent prevailing literatures on functional anatomy of plant xylem of various life forms revealed that the loss of xylem hydraulic conductivity due to cavitation and embolism is significantly related to the vessel diameters [31,32,33,34,35,36,37,38], despite of some equivocal results with insignificant relationship from a few studies [39,40]. According to the prevailing literatures, vessels with larger diameter provide greater hydraulic conductivity, however, larger vessels are more prone to cavitation damage by negative pressure inside. Therefore, there exists a tradeoff between the hydraulic conductivity and its sensitivity to cavitation loss. The results of these experiments also indicated that the whole-plant (soil-to-leaf) conductivity decays exponentially with the xylem pressure, rather than the linear assumption in the previous stomatal models.

In this paper, we revised the model of stomatal conductance by Gao et al. [29,30] according to the prevailing literatures to reflect the tradeoff between the hydraulic efficiency and cavitation resistance. We then fit the revised model using the field diurnal measurements of leaf gas exchange of 24 shrub species in China to test the hypothesis that shrub functional groups differ significantly in their stomatal responses to changes in source (soil) and air moisture stresses. The results indicated that the three shrub functional groups (deciduous non-legume, deciduous legume, and *Artemisia* subshrubs) were significantly different in stomatal behavior. The sensitivities of stomatal conductance of the three shrub functional groups to relevant driving variables were calculated and discussed in the context of experimental evidences.

Methods

The Model

We revised the model of stomatal conductance by Gao et al. [30] so that the dependence of soil-to-leaf hydraulic conductance ($K_{soil\text{-}to\text{-}leaf}$) on xylem water potential (ψ_x) is hyperbolic rather than linear, and that the loss of xylem hydraulic conductance depends on the maximum soil-to-leaf conductivity (g_p). Specifically,

$$K_{soil-to-leaf} = \frac{g_p}{1-\lambda\psi_x} \tag{1}$$

where λ is a parameter. Moreover, we assumed λ is proportional to g_p, i.e.$\lambda = C_\lambda g_p$, so that

$$K_{soil-to-leaf} = \frac{g_p}{1-C_\lambda g_p\psi_x} \tag{2}$$

where C_λ is a constant across all species and all levels of the analysis. This formulation allowed us to balance the hydraulic conductance and the xylem safety, hence the greater the gp, the more vulnerable the xylem system.

In addition to the above changes, the osmotic adjustment is now dependent on net photosynthesis A_n rather than light intensity I_p [41], so that the osmotic pressure of plant leaf and guard cells, π, is

$$\pi = \pi_0\left(1+\xi\frac{A_n}{K_i+A_n}\right)\left(\frac{C_0}{C_i}\right) \tag{3}$$

where π_0 is the baseline osmotic pressure (MPa), C_i and C_0 are the intercellular and reference CO_2 concentrations, respectively, and ξ and K_i are parameters. The model now takes similar treatment as Ball et al. [24,25].

The assumptions in the previous model [30] include: stomatal conductance, g_s, is proportional to the leaf turgor pressure P_{turgor},

$$g_s = K_\psi P_{turgor} = K_\psi(\pi+\psi_x) \tag{4}$$

where K_ψ is the apparent compliance of guard cell structure; and transpiration is the product of stomatal conductance and the scaled vapor pressure deficit (D), i.e. the absolute vapor pressure deficit (VPD) divided by air pressure (P).

The mass balance between leaf transpiration and soil-to-leaf flow gives the equation

$$\frac{g_p}{1-C_\lambda g_p\psi_x}(\psi-\psi_x) = K_\psi(\pi+\psi_x)D \tag{5}$$

where ψ is the soil water potential. Solving the Equation (5) for ψ_x and substituting the result in (4) give the following model

$$g_s = \frac{b-\sqrt{b^2-4ac}}{2a} \tag{6}$$

where

$$a = C_\lambda D \tag{7}$$

$$b = 1+\left(1/g_p+C_\lambda\pi\right)K_\psi D \tag{8}$$

$$c = K_\psi(\pi+\psi) \tag{9}$$

Equations (3, 6–9) constitute the complete model with 3 basic parameters of K_ψ,g_p, and C_λ, and three osmotic adjustment parameters of π_0, K_i, and ξ.

Data Collection and Preparation

This analysis used the field diurnal gas exchange data of shrub leaves in northern China (Table 1). Part of the data were collected by the authors and colleagues during 2002–2007, and others were reported in the literature from 1990 to 2003. All the field measurements were done in public lands with free access to researchers. No specific permissions were required since the experiments had no obvious impacts on the sites and did not involve any endangered or protected species. The data from literature were read from tables and charts. Measurement on one species at a site is called an *observation*, and an observation may be done in multiple days. Each observation produced a data table with a number of variables (columns) including stomatal conductance/resistance, net photosynthesis, air pressure, and vapor pressure deficit, at leaf surface. Each record (row) in the data table represented one replicate of measurements on one plant leaf surface at specific hour of the day. There were a total of 43 observations within 80 diurnal measurements involving 24 species in this analysis. Table 1 also listed the approximate geographic location, elevation, instrument used, and data source. Data measured by the authors and colleagues are provided in Table S1.

To facilitate the analysis, all datasets of the diurnal measurements were converted to the formats and units used by the portable photosynthesis system Licor-6400 (Licor, Nebraska, USA). To avoid the problem of repeated measurements on the same plant leaf, the records of all data tables were averaged for each leaf at a particular hour.

Procedures of Analysis

The observations were grouped into three functional groups: non-legume deciduous shrubs (DCDS), deciduous legume shrubs (LEGM, mostly in *Caragana* genus), and subshrubs (SUBS, *Artemisia* genus). The LEGM group, with their potential nitrogen fixation capability, may be advantageous over the DCDS when nitrogen is limiting. The subshrubs have characteristics of both woody and herbaceous plants (woody based but herbaceous primary growth).

The above model of stomatal conductance was fitted to the data at three hierarchical levels: observations (OBS, species cross sites), functional groups (PFT), and global (GLB, pooling all data into a single group). Parameter C_λ is treated as a super parameter for all the species and observations, so that the tradeoff between conductivity and xylem safety is realized. Model parameters were estimated by means of Bayesian Inference using Gibbs sampling with Markov Chain Monte Carlo (MCMC), implemented in the WinBUGS program [42,43] run within the free software of R [44].

Fitting the stomatal conductance model required the data for soil water potential which is not available. To get around the problem, we made an additional assumption that soil water potential is approximately constant for each day and the daily soil water potentials were estimated at the observation level. The assumption behind this treatment is that minimizing the difference between the model and the data should also allow us to estimate both model parameters and the unknown daily soil water potential [29,30]. The estimated daily soil water potentials from the observation level were used at the levels of GLB and PFT.

MCMC requires specification of prior probability distributions for the parameters. On the other hand, it is generally difficult for the iterations to converge to the physiologically meaningful results because of the nature of nonlinear models. We handled this difficulty by providing priors of uniform distributions within the specified physiologically meaningful ranges. This allowed us to place the following bounds to the parameters: $K_\psi \in [0.01,1.5]$ mol H_2O m^{-2} s^{-1} MPa^{-1}, $g_p \in [0.1,60]$mol H_2O m^{-2} s^{-1} MPa^{-1}, $K_i \in [0.02,500]$ μmol CO_2 m^{-2} s^{-1}, $C_\lambda \in [1,10]$ $mmol^{-1}$ m^2 s, $\pi_0 \in [0.01,2.0]$ MPa, and $\xi \in [0.1,500]$. The parameter C_λ was only fitted at the observation level, so that the parameter at the other two levels (PFT and GLB) was fixed at the mean value from the observational fit. Since the previous studies have recognized the higher drought tolerance of shrubs than other functional types, we placed the bounds of daily soil water potential as $\psi \in [-0.03, -3.5]$ MPa. These bounds were realized in the specification of uniform prior distribution. For example, a statement in the WINBUGS program, ~dunif (0.01, 1.5) specifies that the prior for the parameter K_ψ is a uniform distribution between 0.01 and 1.5 mol m^{-2} s^{-1} MPa^{-1}. For each level of the hierarchical analysis, the MCMC was iterated for 5,000 times so that almost all parameters can converge.

Results

The Statistics of the Fitted Models and Parameters

Summary statistics (Table 2) for the analysis at the three levels indicate that from global to shrub functional groups to individual observations, the deviances of the models decrease, while the number of parameters increases. The deviance for stomatal model is negative because most measured and predicted stomatal conductance are smaller than 1. Specifically, the deviance decreased from −1,423 to −1,804 to −3,981, and the standard error of the residual decreased from 0.170 to 0.158 to 0.095 mol m^{-2} s^{-1}, for GLB, PFT, and OBS levels, respectively. Smaller deviances indicate better fits of the models to the data, in the price of increased model complexity with more parameters. The difference in deviances between any two hierarchical levels has been shown to follow the χ^2 distribution with the degrees of freedom equal to the difference in number of parameters between the two levels. Significant differences in model quality between the two hierarchical levels should be indicated in a significantly large χ^2 value. The deviance tests (Table 2) among the three levels, viz. PFT vs. GLB, and OBS vs. PFT, indicate that the model for shrub functional groups (PFT) is a significant improvement over the global model (GLB), and that there is significant difference in behavior of stomatal conductance among the three shrub functional groups. Similarly, the model at the observation level is significantly better than the models at the level of functional groups. The trend of improved model fit from global to functional groups to observations is further demonstrated in the increased correlations between the model predictions and measurements (Table 2). Specifically the Pearson's correlation coefficient between the measured and predicted stomatal conductance increased from 0.61 to 0.70 to 0.90 (Fig. 1) as the analyses went from global to functional groups to observation levels. The plot of the model predicted against the measured stomatal conductance (Fig. 1) showed the progressive improvement of the fitting from coarse to fine granularity.

The obtained parameters of the stomatal model at the observation level (Table 3) showed wide ranges of variations across species and sites. Differences in plant and leaf ages, water and nutrient status, instruments used, completeness and accuracy of the data, timing of measurements, and instability of nonlinear algorithms, might all have contributed to these large variations. The super parameter C_λ is 3.13 ± 1.66 (m^2 s mol^{-1}).

Despite that we used the independent uniform prior distributions for the model parameters, we detected significant negative correlations between the fitted parameters K_ψ and g_p ($r = -0.44$, $p = 0.002$), and between π_0 and μ ($r = -0.44$, $p = 0.0015$), and positive correlations between parameters K_ψ and π_0 ($r = 0.33$, $p = 0.021$), between g_p and K_i ($r = 0.44$, $p = 0.073$). The negative

Table 1. List of data sources. PFT – plant functional type, N – number of records.

Observation	Species	PFT	N	Latitude	Longitude	Elevation (m)	Instrument	Reference
1	*Chimonanthus praecox* (L.) Link	DCDS	24					[59]
2	*Mussaenda esquirolli* Levl.	DCDS	47	109.61	26.85	500	LI-6400	Measured
3	*Syringa pekinensis* Rupr.	DCDS	53	115.48	40.02	1100	LI-6400	
4	*Lespedeza bicolor* Turcz.	LEGM	44	115.48	40.02	1100	LI-6400	
5	*Vitex negundo* Linn.	DCDS	53	115.48	40.02	1100	LI-6400	
6	*Caragana pygmaea* (Linn.) DC.	LEGM	176	112.7	42.71	1100	LI-6400	
7	*Caragana korshinskii* Kom.	LEGM	72	112.7	42.71	1100	LI-6400	
8	*Caragana intermedia* Kuang et H. C. Fu	LEGM	72	112.7	42.71	1100	LI-6400	
9	*Artemisia frigida* Willd.	SUBS	122	112.7	42.71	1100	LI-6400	
10	*Artemisia frigida* Willd.	SUBS	82	112.7	42.71	1100	LI-6400	
11	*Artemisia frigida* Willd.	SUBS	124	112.7	42.71	1100	LI-6400	
12	*Caragana microphylla* Lam.	LEGM	144	115.47	42.12	1350	LI-6400	
13	*Artemisia frigida* Willd.	SUBS	169	115.47	42.12	1350	LI-6400	
14	*Salix psammophila* C. Wang et Ch. Y. Yang	DCDS	36	109.19	39.49	1300	LI-6400	
15	*Artemisia ordosica* Krasch.	SUBS	36	109.19	39.49	1300	LI-6400	
16	*Hedysarum fruticosum* Pall.	LEGM	72	109.19	39.49	1300	LI-6400	
17	*Caragana intermedia* Kuang et H. C. Fu	LEGM	72	116.73	43.55	1200	LI-6400	
18	*Salix psammophila* C. Wang et Ch. Y. Yang	DCDS	72	116.73	43.55	1200	LI-6400	
19	*Lespedeza potaninii* Vass.	LEGM	69	116.73	43.55	1200	LI-6400	
20	*Hippophae rhamnoides* Linn.	DCDS	72	116.73	43.55	1200	LI-6400	
21	*Artemisia oxycephala* Kitag.	SUBS	32	116.7	43.63	1200	CI-301 PS	[60]
22	*Paeonia suffruticosa* Andr.	DCDS	36	115.42	39.97	100	CI-301PS	[61]
23	*Caragana intermedia* Kuang et H. C. Fu	LEGM	23	109.85	39.03	1300	LI-6000	[62]
24	*Artemisia ordosica* Krasch.	SUBS	24	109.85	39.03	1300	LI-6000	
25	*Lonicera maackii* (Rupr.) Maxim.	DCDS	13	116.37	39.93	50	CI-301 PS	[63]
26	*Lagerstroemia indica* Linn.	DCDS	13	116.37	39.93	50	CI-301 PS	
27	*Viburnum rhytidophyllum* Hemsl.	DCDS	13	116.37	39.93	50	CI-301 PS	
28	*Sambucus williamsii* Hance	DCDS	13	116.37	39.93	50	CI-301 PS	
29	*Artemisia frigida* Willd.	SUBS	7	120.75	42.88	500	CI-301 PS	[64]
30	*Caragana microphylla* Lam.	LEGM	7	120.75	42.88	500	CI-301 PS	
31	*Hippophae rhamnoides* Linn.	DCDS	48	109.25	36.71	1350	LI-6400	Measured
32	*Caragana korshinskii* Kom.	LEGM	36	109.25	36.71	1350	LI-6400	
33	*Lespedeza daurica* (Laxm.) Schindl.	LEGM	22	109.25	36.71	1350	LI-6400	
34	*Caragana korshinskii* Kom.	LEGM	18	104.85	37.45	1300	LI-6200	[65]
35	*Artemisia ordosica* Krasch.	SUBS	18	104.85	37.45	1300	LI-6200	
36	*Hedysarum scoparium* Fisch. et Mey.	LEGM	13	104.85	37.45	1300	Li-6200	[66]
37	*Caragana korshinskii* Kom.	LEGM	13	104.85	37.45	1300	Li-6200	[67]
38	*Artemisia ordosica* Krasch.	SUBS	13	104.85	37.45	1300	Li-6200	
39	*Caragana korshinskii* Kom.	LEGM	21	104.95	37.33	1300	CI-301PS	[68]
40	*Caragana intermedia* Kuang et H. C. Fu	LEGM	21	109.19	39.49	1300	CI-301PS	
41	*Caragana microphylla* Lam.	LEGM	21	120.92	42.38	1300	CI-301PS	
42	*Alhagi sparsifolia* Shap.	DCDS	42	86.2	38.37	1400	LI-6400	[69]
43	*Calligonum caput-medusae* Schrenk	DCDS	42	86.2	38.37	1400	LI-6400	

correlation between K_ψ and g_p indicated that the stiffness of guard cell structure and efficiency of hydraulic conductance exhibit some kind of coordination. The more efficient vertical xylem transport, the stiffer the guard cell structure to keep the stomata open. It also reflects the balance between stomatal sensitivity to turgor pressure and xylem sensitivity to cavitation.

The obtained soil water potentials at the observation level (Fig. 2) were plotted as histograms of the three functional types. We found the mean and standard deviations of the soil water

Table 2. Deviances calculated by the WINBUGS, and the Chi-square tests of deviances among the three levels (Global, Functional group, and Observation).

Analysis Level	Global	Functional group	Observation
Deviance	−1,423	−1,804	−3,981
Deviance Information Criterion	−1,485	−1,789	−24,249
Number of Parameters	5	15	296
Standard deviation of error	0.170	0.158	0.095
Difference in deviance	378		2,897
Difference in number of parameters	10		281
p-value of χ^2 test	<0.0001		<0.0001
Correlation between measured and predicted stomatal conductance	0.61	0.70	0.90

potentials were -0.94 ± 0.49, -1.08 ± 0.66, and -0.73 ± 0.41 MPa for DCDS, LEGM, and SUBS, respectively. The t test to compare the means between LEGM and SUBS resulted in a p-value of 0.024, and the comparison between DCDS and SUBS resulted in a p-value of 0.13. The result indicates that the fitted soil water potentials for the subshrubs are significantly higher than the LEGM shrubs. The means of the soil water potential are not significantly different from each other between the LEGM and DCDS groups. The correlations among the model parameters reflect the complex interactions among various components involved in water transportation from soil to plant leaves [28].

At the functional group level (Table 4), the fitted mean compliance K_ψ varied from 0.16 to 0.70 to 1.45, reflecting the decreased stiffness of guard cell structure from DCDS to LEGM to SUBS, respectively. However, the maximum potential soil-to-leaf conductance is the greatest for SUBS (5.66 mmol m^{-2} s^{-1} MPa^{-1}), but the smallest for LEGM (2.05 mmol m^{-2} s^{-1} MPa^{-1}), with DCDS in between (2.57 mmol m^{-2} s^{-1} MPa^{-1}).

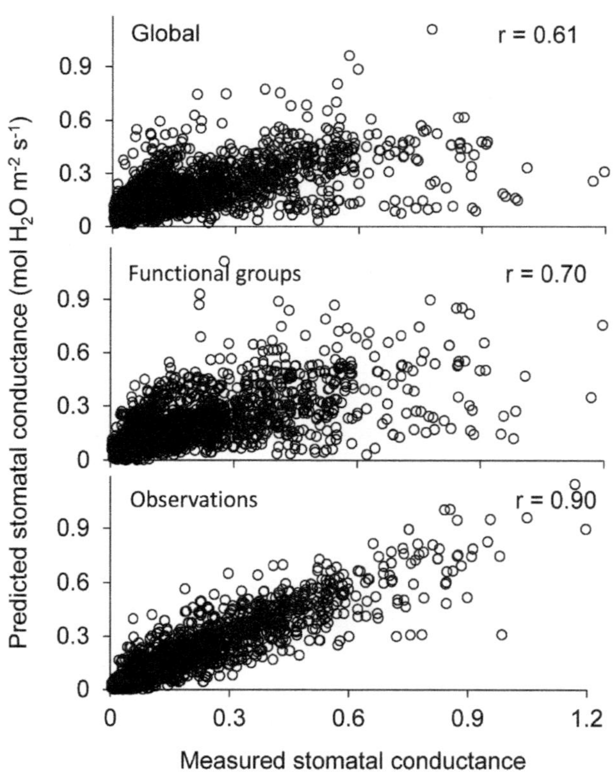

Figure 1. Model predicted vs. measured stomatal conductance at the three hierarchical levels.

While π_0 is the baseline osmotic pressure, the combination of K_i and ξ determines the extent and sensitivity of osmotic adjustment. The three functional groups have similar π_0 in the range of 1.87 to 1.94 MPa. The DCDS group is shown to have the smallest mean K_i value (198.8 μmol CO_2 m^{-2} s^{-1}), whereas the SUBS has K_i of 319.3. The SUBS has the lowest ξ (9.9) followed by 23.8 for DCDS and 34.1 for LEGM. Referring to Equation (3), the efficiency of osmotic adjustment is largely determined by the ratio of ξ to K_i since K_i is much greater than the maximum net assimilation (<20 μmol m^{-2} s^{-1}) in this case. The obtained model parameters at the global level (Table 4) mostly fall in the range of those at the level of functional groups, except for π_0 of 1.76, which is smaller than any of the three functional groups.

Behavior of the Stomatal Model

Based on the model parameters at the levels of PFT and GLB, calculated stomatal conductance was plotted as functions of soil water potential and dimensionless vapor pressure deficit (Fig. 3), with leaf net assimilation fixed at 6.5 μmol m^{-2} s^{-1} (approximately the mean value of net photosynthesis for these groups). Sharp comparison among the three shrub functional groups exists. The lower right corners of the panels depict the maximum stomatal conductance at favorable moisture conditions with $\psi = -0.033$ MPa (approximately field capacity) and $D = 0.0035$. At temperature of 30°C at the sea level, D of 0.0035 means 91% of relative humidity. At these conditions, the subshrub has the highest stomatal conductance of 1.7 mol m^{-2} s^{-1} followed by the 1.1 of the LEGM group. However, the DCDS group has the much smaller stomatal conductance of 0.44 mol m^{-2} s^{-1}. The GLB has the maximum stomatal conductance of 0.83 mol m^{-2} s^{-1}, slightly smaller than the average of the three functional groups. The ranking of these values is largely determined by the K_ψ parameter (Table 4). When moisture conditions are favorable, greater guard cell compliance implies greater aperture of stomata, hence greater stomatal conductance.

The subshrub group has lower tolerance to soil water stress than the two deciduous shrub groups, as the stomatal conductance of SUBS closes at the soil water potential of −2.2 MPa. In contrast, the stomatal conductance of DCDS and LEGM groups decrease to zero at −3.2 and −3.5 MPa, respectively. However, the subshrub group has higher tolerance to vapor pressure deficit than the two deciduous shrub groups (Fig. 3). When the soil water

Table 3. Obtained parameters of the stomatal model at individual observation level.

Observation	PFT	K_ψ	g_p	K_i	π_0	ξ
1	DCDS	0.088	33.11	295.4	0.95	149.6
2	DCDS	0.552	28.10	326.8	1.30	9.8
3	DCDS	0.380	20.08	324.1	1.02	99.0
4	LEGM	0.454	18.14	330.1	1.02	79.8
5	DCDS	0.297	26.79	329.7	1.02	69.1
6	LEGM	0.591	10.95	330.1	1.54	2.9
7	LEGM	0.639	17.16	333.9	1.04	12.0
8	LEGM	0.734	10.57	329.9	1.20	11.9
9	SUBS	0.228	25.90	345.8	1.41	26.0
10	SUBS	0.508	22.16	338.7	1.09	27.4
11	SUBS	1.084	11.64	345.1	1.51	2.9
12	LEGM	0.927	1.94	362.5	1.39	88.7
13	SUBS	1.461	5.06	369.7	1.95	15.7
14	DCDS	1.034	2.77	302.1	1.39	45.8
15	SUBS	0.986	1.47	227.2	1.16	135.1
16	LEGM	0.576	9.83	41.4	0.94	5.6
17	LEGM	1.034	1.66	67.7	1.26	38.3
18	DCDS	0.460	19.17	308.3	1.12	36.5
19	LEGM	1.307	3.90	219.8	1.74	34.2
20	DCDS	0.526	6.42	45.3	1.04	49.3
21	SUBS	0.282	30.17	316.3	1.54	8.8
22	DCDS	0.263	20.60	304.4	1.06	191.7
23	LEGM	1.407	2.82	26.5	1.35	3.6
24	SUBS	1.482	8.28	331.1	1.98	1.7
25	DCDS	0.322	20.99	305.4	0.71	149.0
26	DCDS	0.345	19.93	311.0	0.76	151.9
27	DCDS	0.311	22.57	308.1	0.77	148.7
28	DCDS	0.180	27.62	313.2	0.85	140.8
29	SUBS	0.311	22.94	314.0	0.82	152.0
30	LEGM	0.262	25.18	305.6	0.78	162.8
31	DCDS	1.090	0.87	27.4	1.43	10.6
32	LEGM	0.845	1.71	319.5	1.55	113.5
33	LEGM	0.246	23.21	271.1	1.21	130.4
34	LEGM	0.565	26.70	325.6	1.31	33.9
35	SUBS	0.599	8.05	303.0	1.17	158.1
36	LEGM	0.551	38.58	320.1	1.38	30.9
37	LEGM	0.452	50.90	341.6	1.69	51.5
38	SUBS	0.807	55.83	314.9	1.86	7.4
39	LEGM	0.236	18.59	298.9	1.15	205.1
40	LEGM	0.636	5.95	323.0	1.43	87.9
41	LEGM	0.157	23.07	317.9	1.27	150.3
42	DCDS	0.745	2.56	325.9	1.50	119.3
43	DCDS	0.407	12.50	309.5	1.32	150.7
Mean		0.613	19.06	283.9	17.36	76.7
Standard deviation		0.373	13.96	93.0	12.95	63.5
Maximum		1.482	62.13	369.7	55.83	205.1
Minimum		0.088	1.04	26.5	0.87	1.7

C_λ is 3.13±1.66 m^2 s mol^{-1}. Units of the parameters: K_ψ, mol m^{-2} s^{-1} MPa^{-1}, g_p, mmol m^{-2} s^{-1} MPa^{-1}, K_i, μmol CO_2 m^{-2} s^{-1}, π_0, MPa, and ξ is dimensionless.

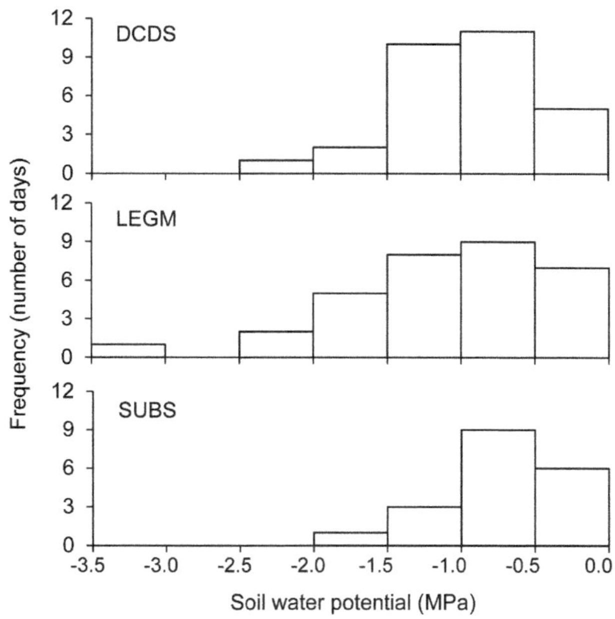

Figure 2. Histograms of the fitted soil water potentials for the three functional groups.

potential is high (−0.033 MPa) but the scaled vapor pressure deficit is great (0.05), the subshrub group maintains greater stomatal conductance (0.23 mol m^{-2} s^{-1}) than the two deciduous shrub groups (approximately 0.14 mol m^{-2} s^{-1}).

The tolerance to soil water stress is primarily determined by the osmotic pressure and the compliance of guard cell structure. The SUBS has low tolerance to soil water stress largely because of the greatest K_ψ which makes the stomata sensitive to changes in soil water stress.

The spacing and slope of the contour lines (Fig. 3) represent the relative sensitivities of stomatal conductance with respect to soil water stress and vapor pressure deficit. The response of stomatal conductance to vapor pressure deficit is determined by the ratio of

Table 4. Parameters of the stomatal model estimated by the WinBUGS at the global (GLB) and the functional group levels.

Parameter		GLB	DCDS	LEGM	SUBS
K_ψ	mean	0.38	0.16	0.70	1.42
	STD	0.10	0.03	0.27	0.15
g_p	mean	3.60	2.57	2.05	5.66
	STD	3.70	1.19	0.90	2.37
K_i	mean	202.6	198.8	284.0	319.8
	STD	129.5	142.5	130.0	122.8
π_0	mean	1.76	1.91	1.94	1.87
	STD	0.39	0.08	0.17	0.23
ξ	mean	25.5	23.8	34.1	9.9
	STD	48.6	15.9	17.3	11.5
N_{day}	80		24	36	20
N	2119		576	916	627
N_{obs}	43		15	18	10

N is the number of data points involved in the WinBUGS calculation. N_{obs} is the number of observations, and N_{day} the number of diurnal days of measurements.

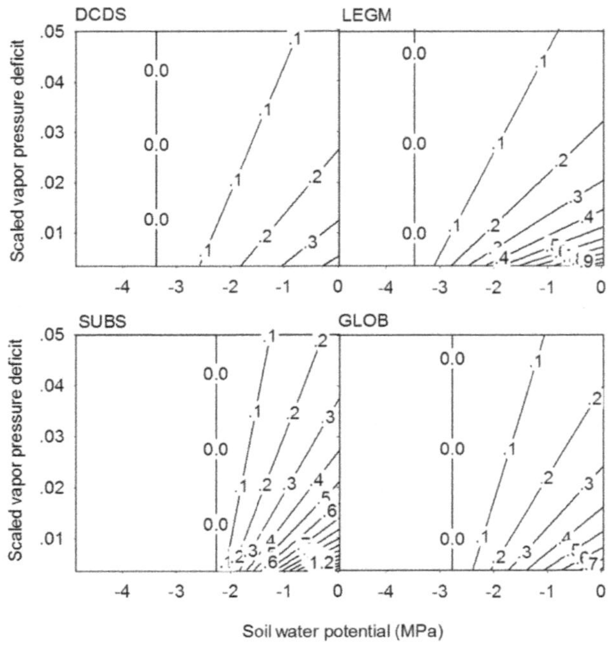

Figure 3. Stomatal conductance (contour lines, mol H_2O m^{-2} s^{-1}) as functions of soil water potential and scaled vapor pressure deficit for the three shrub functional groups, calculated based on the parameters obtained at the level of functional groups: the deciduous non-legume shrubs (DCDS), the deciduous legume shrubs (LEGM), and the *Artemisia* subshrubs (SUBS). The stomatal conductance calculated from the global level model (GLOB) is also plotted as the comparison. Net assimilation A_n is fixed at 6.5 μmol CO_2 m^{-2} s^{-1}.

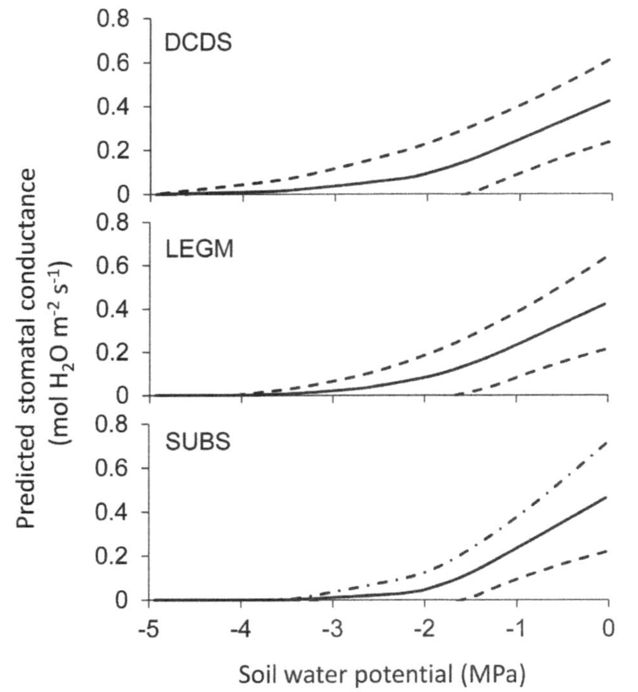

Figure 4. Predicted stomatal conductance as functions of soil water potential based on the parameters at the observation level. The scaled vapor pressure deficit is fixed at 0.03 and the net assimilation at 6.5 μmol CO_2 m^{-2} s^{-1}, for all individual observations. The solid line stands for the mean, and the broken lines indicate one standard deviation below and above the mean.

K_ψ to $K_{soil-to-leaf}$, so that greater K_ψ but smaller $K_{soil-to-leaf}$ makes the stomatal conductance more sensitive to VPD. A smaller $K_{soil-to-leaf}$ tends to cause insufficient water supply from roots to leaves under greater VPD, and the insufficient water supply decreases the leaf xylem water potential, turgor pressure, and stomatal conductance. A greater K_ψ will make stomata more sensitive to the decrease in the turgor pressure, so that the stomatal conductance decreases faster with VPD. The LEGM and SUBS groups have high sensitivity to vapor pressure deficit, largely because the former has the smallest g_p (maximum $K_{soil-to-leaf}$), and the latter has the greatest K_ψ values. The DCDS group has a small g_p (2.57 mmol m^{-2} s^{-1} MPa^{-1}), however, it also has a small K_ψ (0.16 mol m^{-2} s^{-1} MPa^{-1}), so that the stomatal conductance is not as sensitive as LEGM or SUBS.

We also used the fitted parameters at the observation level to calculate stomatal conductance of all the observations, and plotted the arithmetic means and one standard deviation below and above against soil water potential (Fig. 4) and scaled vapor pressure deficit (Fig. 5). The results confirmed the findings at the level of functional groups. The average stomatal conductance of DCDS, LEGM, and SUBS close approximately at -4, -3.6, and -3 MPa, respectively, so that the tolerance and sensitivity to soil water stress and vapor pressure deficit of the three functional groups follow the same ranks in Fig. 3.

Finally the comparison of mean $K_{soil-to-leaf}$ as a function of xylem water potential based on the parameters obtained at the observation level (Fig. 6) shows similar characteristics for the three functional groups. However, the calculated $\psi_{x,50}$ values,at which $K_{soil-to-leaf}$ decreased by a half of its maximum value, are -12.3,

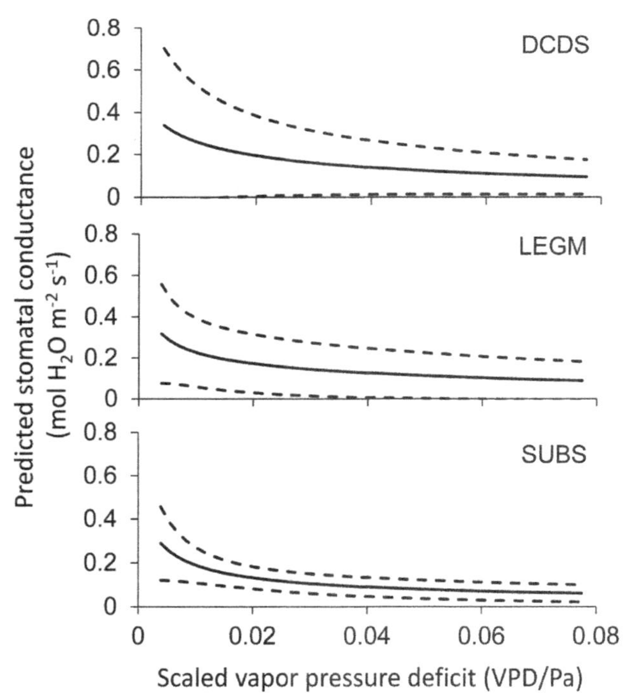

Figure 5. Predicted stomatal conductance as functions of scaled vapor pressure deficit based on the parameters at the observation level. The soil water potential is fixed at -1.0 MPa for all individual observations, and net assimilation is treated as the same as in Fig. 3. The solid line stands for the mean, and the broken lines indicate one standard deviation below and above the mean.

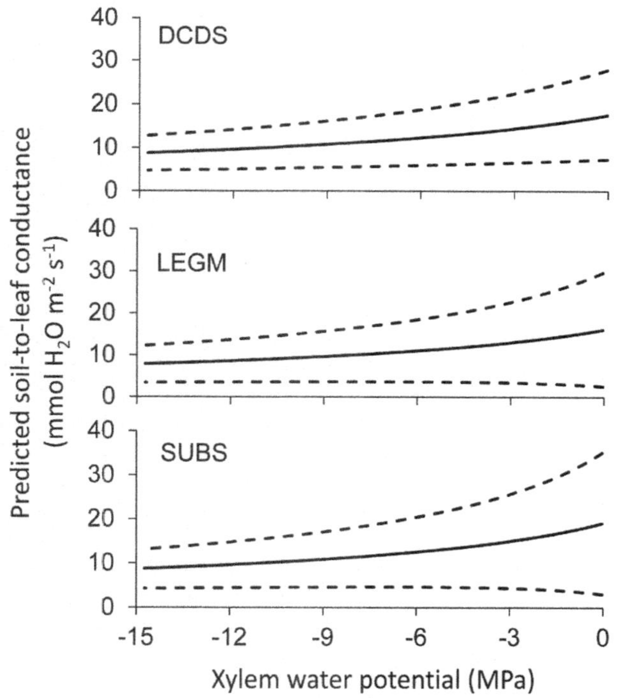

Figure 6. Predicted soil-to-leaf conductance as functions of xylem water potential based on the parameters at the observation level. The solid line stands for the mean, and the broken lines indicate one standard deviation below and above the mean.

−14.0, and −16.5 MPa for SUBS, LEGM, and DCDS groups, respectively, indicating that the xylem vessels of the two deciduous shrubs (DCDS and LEGM), on average, are slightly stronger than those of the SUBS.

Discussion

Experimental Evidences Connected to the Results

The behavior of stomatal conductance, driven by soil moisture and vapor pressure deficit, depends largely on the stiffness of guard cell structure, and the hydraulic conductance of plant xylem which transports water from soil to leaves. A number of literatures reported the findings of the xylem hydraulic conductivity of various plant life forms.

Kocacinar and Sage [45] experimentally measured xylem hydraulic conductivities of 16 shrub species in west America and central Asia, and found the mean hydraulic conductivities are 3.24×10^{-4} and 0.46×10^{-4} kg $m^{-1}s^{-1}$ MPa^{-1} for C_3 and C_4 species, respectively. If flow paths are about 2 m long, then the hydraulic conductance would be 9.0 and 1.3 mmol m^{-2} s^{-1} MPa^{-1} for C_3 and C_4 species, respectively. The synthesis of measurements on leaf-specific hydraulic conductivity of various plant growth forms [46] reveals that the leaf-specific whole plant conductance is in the range between 0.2 and 20 mmol m^{-2} s^{-1} MPa^{-1}, and that the desert subshrub tends to have higher hydraulic conductance than other growth forms. The measurement of vessel density, diameter, and leaf-specific hydraulic conductance of three *Caragana* species in northern China [47] showed that their soil-to-leaf hydraulic conductance varied between 0.5 to 20 mmol m^{-2} s^{-1} MPa^{-1}. The data of Iovi et al. [48] for the Mediterranean species showed that soil-to-leaf conductance varies within a wide range up to 40 mmol m^{-2} s^{-1} MPa^{-1}, and the herbaceous group has much higher hydraulic conductance than other groups. They also found a negative exponential decay of the hydraulic conductance with respect to decreased xylem water potential.

Our analysis at the observation level yielded g_p (maximum potential soil-to-leaf conductivity) in the range from 1.04 to 62.13 mmol m^{-2} s^{-1} MPa^{-1}, and the parameters of the three shrub functional groups are in the range from 2.05 to 5.66 mmol m^{-2} s^{-1} MPa^{-1}, comparable to the above experimental findings. Greater g_p is an indicator of greater vessel lumen diameter, but not an indicator of greater vessel density, as the negative correlation between the two quantities found by Poorter et al. [49]. However, the greater diameter of the vessel members in general means higher potential loss of hydraulic conductivity due to cavitation and embolism of vessel members under xylem tension. This tradeoff is reflected in our obtained C_λ parameter which makes the loss of conductivity proportional to the maximum conductivity. For example, the SUBS group has much higher hydraulic conductance, so that the decrease in conductance with xylem water stress is faster than the other two groups. The result is similar to those found by Iovi et al. [48].

Using finite element analysis, Cooke et al. [50] indicated that a typical stomata aperture width increases from 7 to 15 μm when turgor pressure inside guard cells increases from 0 to 700 kPa. If we assume that stomatal conductance is directly proportional to stomatal aperture and that stomatal apertures of 7 and 15 μm approximately correspond to typical stomatal conductance of 0.3 and 1.0 mol m^{-2} s^{-1}, respectively, an approximation of the compliance would be $(1.0–0.3)/0.7 = 1.0$ mol m^{-2} s^{-1} MPa^{-1}. The model parameter K_ψ in our analysis is a product of this compliance and stomata density, so that the apparent compliance K_ψ should have more variation range if the variation in stomatal density is considered. We found this parameter varies between 0.088 and 1.482 with a mean of 0.613 at the observation level, and between 0.16 to 1.42 mol m^{-2} s^{-1} MPa^{-1} at the functional group level. The result is comparable to Cook's finding.

From the findings by Leishman et al. [51,52], we know that the non-legume deciduous shrubs have lower stomatal conductance than the other two groups. The most deciduous shrubs in this study (both legume and non-legume) were found as xerophyte growing in relatively shallow soils with high water use efficiency. Their instantaneous photosynthesis can be high when soil water is abundant. However, due to the long-term drying conditions, they have to invest large amount of the assimilated products on constructing xylem for small-diameter vessels with lower g_p, which allows them to sustain the excessive drought conditions. Most legume species are thin-leaved *Caragana* species, which might have something to do with their relatively greater K_ψ than the non-legume deciduous group.

Finally, the SUBS group (*Artemisia*), with their less lignified stems and leaves, is shown to have the greatest compliance, and the largest potential hydraulic conductance than the two deciduous shrub groups. Thus the subshrubs are shown here to be typical drought-avoiding in most occasions. In this sense, the subshrubs share some characteristics of herbaceous plants, with stomatal conductance more responsive to variation of moistures in soil and air [18,48]. In other words, the subshrubs behave as an opportunist, so that they have a large stomatal conductance to accommodate the great photosynthesis when moistures are abundant in soil, but choose to close the stomata under severe drought conditions.

Implication for Macro-scale Ecosystem Dynamics

The dynamics of leaf stomatal conductance is coupled with plant morphological structure and other variables such as leaf nutrient status and boundary layer thickness [28,30]. The present model should be applied and further tested within more complex ecosystem models to improve its connection with external variables.

Our hierarchical analysis showed that the functional group model is a significant improvement over the global level model, and the model at the observation level is a significant improvement over that at the functional group level. This result suggests the parameters at three levels can be used in ecosystems modeling at large, intermediate, and small scales. At regional and global scales in which we need to place all shrubs in one group, the global level parameters can be used. For small patch-scale ecosystems, parameters at the observation level are more appropriate. Modeling at intermediate mesoscales (watershed or landscape) may necessitate distinguishing the properties among shrubs types, thus the results at the level of functional groups may be applicable.

Our findings at the functional group level offered a clue to the equivocal issue in shrub-grass interaction. The process of shrub encroachment into grasslands has been understood as involving nonlinear processes in soil and plant communities, with strong positive feedbacks that consequently lead to the domination of shrubs in the previous grasslands [53]. The current understanding is that once shrubs get established and dominated in the grasslands, it is impossible for the process to reverse to restore the grasslands because of the altered competition at the community scale. However, recent studies gave a number of exceptional contradictory cases. In a 16-year exclosure (excluding from livestock grazing) established at the Erdos Sandland Ecosystem Station in northern China, it was found that a previously dominating shrub (*Artemisia ordosica* Krasch) decreased by 90% [54], but grasses and forbs increased substantially inside the exclosure. Another example is the *Artemisia frigida* Willd., a commonly observed shrub species in northern China grasslands. The proliferation of this species in the so-called typical steppes has been considered a result of overgrazing [55]. Enclosure studies in northern China showed that this shrub can be put down with appropriate management. Several studies showed that the enclosures with managed livestock grazing kept the dominance of the species less than 8% in the enclosures of 5, 14, and 25 years, with comparison to the 30% dominance in the heavily grazed sites [56,57,58]. These results contradict the current understanding of the irreversibility of shrub invasion. However, we noticed that the species in the above case studies are *Artemisia* subshrubs. Our results showed that there is a significant difference in stomatal behavior between the subshrubs and the deciduous shrubs. The deciduous shrubs have great advantages in resistance and tolerance to soil water stress over the grasses, which may contribute to their irreversible encroachment into grasslands. The subshrubs are less resistant and tolerant to soil moisture stress than the shrubs, and thus are less advantageous in the competition with the grasses. Our hypothesis is that when the grazing pressure is heavy, the soil cannot hold enough water because of the decreased grass roots biomass. The subshrubs start to gain dominance. However, when grazing pressure is reduced, root biomass start to increase, allowing fast infiltration of water. Consequently the soil layers tend to hold more water to allow better grass growth, and the advantage of the subshrubs over the grasses is weakened. Therefore, with appropriate management, the invasion of the subshrubs into grasslands might be reversed and the health of the grassland ecosystems might be recovered. This hypothesis has to be tested in more strictly designed experiments.

Supporting Information

Table S1 Data measured by the authors group with Licor 6400 portable gas analyzer.

Acknowledgments

We appreciate Drs. Hongmei Xu, Yinghui Liu, Zhicai Zhang, Junshan Liu, and many others for the help in the field work.

Author Contributions

Conceived and designed the experiments: QG MY. Performed the experiments: QG MY CZ. Analyzed the data: QG MY. Contributed reagents/materials/analysis tools: QG MY. Wrote the paper: QG MY.

References

1. Cramer W, Bondeau A, Woodward FI, Prentice IC, Betts RA, et al. (2001) Global response of terrestrial ecosystem structure and function to CO2 and climate change: results from six dynamic global vegetation models. Global Change Biology 7: 357–373.
2. Gao Q, Yu M, Wang J, Jia H, Wang K (2004) Relationships between regional primary production and vegetation patterns. Ecological Modelling 172: 1–12.
3. Purves D, Pacala SW (2008) Predictive models of forest dynamics. Science 320: 1452–1453.
4. Hooper DU, Chapin FSI, Ewel JJ, Hector A, Inchaustl P, et al. (2005) Effects of biodiversity on ecosystem functioning: a consensus of current knowledge. Ecological Monographs 75: 3–35.
5. Yu M, Gao Q (2011) Leaf-traits and growth allometry explain competition and differences in responses to climate changes in a temperate forest landscape: a simulation study. Annuals of Botany.
6. Schulze ED, Kelliher FM, Korner C, Lloyd J, Leuning R (1994) Relationships among maximum stomatal conductance, ecosystem surface conductance, carbon assimilation rate, and plant nitrogen nutrition: A global ecology scaling exercise. Annual Review of Ecology and Systematics 25: 629–660.
7. Reich PB, Walters MB, Ellsworth DS (1997) From tropics to trundra: Global convergence in plant functioning. Proceedings of the National Academy of Sciences of the United States of America 94: 13730–13734.
8. Wright IJ, Ackerly DD, Bongers F, Harms KE, Ibarra-Manriquez G, et al. (2007) Relationships among ecologically important dimensions of plant trait variation in seven Neotropical forests. Annals of Botany 99: 1003–1015.
9. Medlyn BE, Badeck FW, De Pury DGG, Barton CVM, Broadmeadow M, et al. (1999) Effects of elevated CO_2 on photosynthesis in European forest species: a meta-analysis of model parameters. Plant, Cell and Environment 22: 1457–1495.
10. Reynolds JF, Smith DMS, Lambin EF, Turner BLI, Mortimore M, et al. (2007) Global desertification: building a science for dryland development. Science 316: 847–851.
11. Tong C, Yang JR, Yong WY, Yong SP (2002) Spatial pattern of steppe degraddation in Xilin River basin of Inner Mongolia. Journal of Natural Resources 17: 571–578.
12. Wigley BJ, Bond WJ, Hoffman MT (2009) Bush encroachment under three contrasting land-use practices in a mesic South African savanna. Afr J Ecol 47: 62–70.
13. Xiong XG, Han XG, Bai YF, Pan QM (2003) Increased distribution of Caragana microphylla in rangelands and its causes and consequences in Xilin River Basin. Acta Prataculturae Sinica 12: 57–62.
14. Archer S (1994) Woody plant encroachment into southwestern grasslands and savannas: rate, patterns and proximate causes. In: Vavra M, Laycock W, Pieper R, editors. Ecological Implications of Livestock Herbivory in the West. Denver, Colorado, USA.: Society for Range Management. 13–68.
15. Archer S, Schimel DS, Holland EA (1995) Mechanisms of Shrubland Expansion - Land-Use, Climate or Co-2. Climatic Change 29: 91–99.
16. Grover HD, Musick HB (1990) Shrubland encroachment in southern New Mexico, USA: An analysis of desertification processes in the American southwest. Climatic Change 17: 305–330.
17. Neilson RP (1986) High-resolution climatic analysis and southwest biogeography. Science 232: 27–34.

18. Kemp PR, Reynolds JF, Pachepsky Y, Chen J-L (1997) A comparative modelling study of soil water dynamics in a desert ecosystem. Water Resources Research 33: 73–90.
19. Reynolds JF, Kemp PR, Ogle K, Fernandez RJ, Gao Q, et al. (2006) Modeling the unique attributes of arid ecosystems: Lessons from the Jornada Basin. In: Havstad K, Huenneke LF, Schlesinger WH, editors. Structure and Function of Chihuahuan Desert Ecosystem, The Jornada Basin Long-Term Ecological Research Site. London: Oxford University Press.
20. Sperry JS, Hacke UG (2002) Desert shrub water relations with respect to soil characteristics and plant functional types. Functional Ecology 16: 367–378.
21. Nilsen ET, Sharifi MR, Rundel PW (1983) Diurnal and seasonal water relations of the desert phreatophyte Prosopis glandulosa (honey mesquite) in the Sonoran Desert of California. Ecology 64: 1381–1393.
22. Pelaez DV, Boo RM (1987) Plant water potential for shrubs in Argentina. Journal of Range Management 40: 6–9.
23. Farquhar GD, Shark TD (1982) Stomatal conductance and photosynthesis. Annual review of Plant Physiology 33: 317–345.
24. Ball JT, Woodrow IE, Berry JA (1987) A model predicting stomatal conductance and its contribution to the control of photosynthesis under different environmental conditions. In: Biggins J, editor. Progress in Photosynthesis Research. Dordrecht: Martinus Nijhoff Publishers. 221–224.
25. Leuning R (1995) A critical appraisal of a combined stomatal-photosynthesis model for C_3 plants. Plant, Cell and Environment 18: 339–355.
26. Buckley TN, Mott KA, Farquhar GD (2003) A hydromechanical and biochemical model of stomatal conductance. Plant, Cell and Environment 26: 1767–1785.
27. Buckley TN, Turnbull TL, Adams MA (2012) Simple models for stomatal conductane derived from a process model: cross-validation against sap flux data. Plant and Cell Physiology 35: 1647–1662.
28. Bohrer G, Mourad H, Laursen TA, Drewry D, Avissar R, et al. (2005) Finite element tree crown hydrodynamics model (FETCH) using porous media flow within branching elements: A new representation of tree hydrodynamics. Water Resources Research 41: doi:10.1029/2005WR004181.
29. Gao Q, Zhao P, Zeng X, Cai X, Shen W (2002) A model of stomatal conductance to quantify the relationship between leaf transpiration, microclimate and soil water stress. Plant, Cell and Environment 25: 1373–1381.
30. Gao Q, Zhang X, Xu H, Huang Y (2005) Modeling seasonal and diurnal dynamics of stomatal conductance of plants in a semiarid environment. Functional Plant Biology 32: 583–598.
31. Hoffmann WA, Marchin RM, Abit P, Lau OL (2011) Hydraulic failure and tree dieback are associated with high wood density in a temperate forest under extreme drought. Global Change Biology 17: 2731–2742.
32. Fichot R, Chamaillard S, Depardieu C, Le Thiec D, Cochard H, et al. (2011) Hydraulic efficiency and coordination with xylem resistance to cavitation, leaf function, and growth performance among eight unrelated Populus deltoidesx-Populus nigra hybrids. Journal of Experimental Botany 62: 2093–2106.
33. Damunupola JW, Ratnayake K, Joyce DC, Irving DE (2011) Characterisation of xylem conduits and their possible role in limiting the vase life of cut Acacia holosericea (Mimosaceae) foliage stems. Functional Plant Biology 38: 614–623.
34. Choat B, Medek DE, Stuart SA, Pasquet-Kok J, Egerton JJG, et al. (2011) Xylem traits mediate a trade-off between resistance to freeze-thaw-induced embolism and photosynthetic capacity in overwintering evergreens. New Phytologist 191: 996–1005.
35. Tombesi S, Johnson RS, Day KR, DeJong TM (2010) Relationships between xylem vessel characteristics, calculated axial hydraulic conductance and size-controlling capacity of peach rootstocks. Annals of Botany 105: 327–331.
36. Jacobsen AL, Esler KJ, Pratt RB, Ewers FW (2009) Water stress tolerance of shrubs in Mediterranean-type climate regions: Convergence of fynbos and succulent karoo communities with california shrub communities. American Journal of Botany 96: 1445–1453.
37. Reyes-Santamaria I, Terrazas T, Barrientos-Priego AF, Trejo C (2002) Xylem conductivity and vulnerability in cultivars and races of avocado. Scientia Horticulturae 92: 97–105.
38. Maherali H, Delucia EH (2000) Xylem conductivity and vulnerability to cavitation of ponderosa pine growling in contrasting climates. Tree Physiology 20: 859–867.
39. Lopez OR, Kursar TA, Cochard H, Tyree MT (2005) Interspecific variation in xylem vulnerability to cavitation among tropical tree and shrub species. Tree Physiology 25: 1553–1562.
40. Jacobsen AL, Pratt RB, Ewers FW, Davis SD (2007) Cavitation resistance among 26 chaparral species of southern California. Ecological Monographs 77.
41. Emanuel RE, Epstein HE, McGlynn BL, Welsch DL, Muth DJ, et al. (2010) Spatial and temporal controls on watershed ecohydrology in the northern Rocky Mountains. Water Resources Research 46.
42. Lunn DJ, Thomas A, Best N, Spiegelhalter D (2000) WinBUGS - a Bayesian modelling framework: concepts, structure, and extensibility. Statistics and Computing 10.
43. Ntzoufras I (2009) Bayesian Modeling Using WinBUGS; Giudici P, Givens GH, Mallick BK, editors. Hoboken, New Jersey: John Wiley & Sons.
44. R-Development-Cor-Team (2008) R: A language and environment for statistical computing. Vienna, Austria: R Foundation for Statistical Computing.
45. Kocacinar F, Sage RF (2004) Photosynthetic pathway alters hydraulic structure and function in woody plants. Oecologia 139: 214–223.
46. Mencuccini M (2003) The ecological significance of long-distance water transport: short-term regulation, long-term acclimation and the hydraulic costs of stature across plant life forms. Plant Cell and Environment 26: 163–182.
47. Li J, Gao Y-B, Zheng Z-R, Gao Z-l (2008) Water relations, hydraulic conductance, and vessel features of three Caragana species of the Inner Mongolia Plateau of China. Botanical Studies 49: 127–137.
48. Iovi K, Kolovou C, Kyparissis A (2009) An ecophysiological approach of hydraulic performance for nine Mediterranean species. Tree Physiology 29: 889–900.
49. Poorter L, McDonald I, Alarcon A, Licona J-C, Pena-Claros M, et al. (2010) The imprtance of wood traits and hydraulic conductance for the performance and life history strategies of 42 rainforest tree species. New Phytologist 185: 481–492.
50. Cooke JR, DeBaerdemaeker JG, Rand RH, Mang HA (1977) A finite element shell analysis of guard cell deformations. Transactions of ASAE 19: 1107–1121.
51. Leishman MR, Haslehurst T, Ares A, Baruch Z (2007) Leaf trait relationships of native and invasive plants: community- and global-scale comparisons. New Phytologist 176: 635–643.
52. Wright IJ, Reich PB, Westoby M, Ackerly DD, Baruch Z, et al. (2004) The worldwide leaf economics spectrum. Nature 428: 821–827.
53. Schlesinger WH, Reynolds JF, Gunningham GL, Hueneke LF, Jarrell WM, et al. (1990) Biological feedbacks in global desertification. Science 247: 1043–1048.
54. Xiong HQ, Duan JY, Zhang XS (2011) Effects of grazing exclusion on plant community characteristics in a degraded Mu Us Sandland, Inner Mongolia, northern China. Ecology and Environmental Sciences 20: 233–240.
55. Wang S, Li Y, Wang Y, Han YH (1998) Succession of Artemisia frigida rangeland and multivariation analysis under different stocking rates in Inner Mongolia. Acta Agrestia Sinica 6: 299–305.
56. Shan G, Xu Z, Ning F, Ma Y, Li L (2008) Influence of exclosure year on community structure and species diversity on a typical steppe. Acta Prataculturae Sinica 17: 1–8.
57. Liu ZG, Li ZQ (2006) Plant biodiversity of Aretemisia frigida communities on degraded grasslands under diferent grazing intensities after thirteen–year enclosure. Acta Ecologica Sinica 26: 475–482.
58. Wang J, Yang C, Wang TJ (2006) Changes of model of *Artemisia frigida* populations under the disturbance of grazing. Acta Ecologica Sinica 26: 960–965.
59. Li J, Liu YD, Cheng GX, Chen J, Zhu JY (2000) A preliminary study on the diurnal variation of net photosynthetic rate and transpiration rate for Chimonanthus praecox. Guihaia 20: 52–58.
60. Cui XY, Chen ZZ, Du ZC (2000) Photosynthetic characteristics of a semi-arid sandy grassland community in Inner Mongolica. Acta Phytoecologica Sinica 24: 541–546.
61. Zhang SR, Ma KP, Chen LZ (2003) Response of photosynthetic plasticity of Paeonia suffruticosa to changed light environments. Environmental and Experimental Botany 49: 121–133.
62. Zhou HY, Huang ZC (1996) Seasonal variation of diurnal leaf photosynthesis and transpiration of major plant species in Maowusu sandy region. Acta Phytoecologica Sinica 20: 120–131.
63. Leng PS, Yang XH, Hu Y, Zhu HY (2000) Characterstics of Photosynthesis and transpiration of five gardening trees. Journal of Beijing Agricultural College 15: 13–18.
64. Zhou HY (2000) Physioecological characteristics of four dominant plant species in Kerqin sandy land. Chinese Journal of Applied Ecology 11: 587–590.
65. Feng JC, Zhang CL, Huang ZC, Liang HG (1997) Plant leaf gas exchange study under different environmental conditions. Northwest Botanica Sinica (Chinese) 17: 135–141.
66. Zhang LP, Teng YW, Wang XP, Huang ZC, Liu XM (1996) Periodic oscillations in stomatal conductance of a zerophyte Hedysarum scoparium. Journal of Lanzhou University (Natural Science) 32: 128–131.
67. Zhang LP, Wang XP, Liu LC, Huang ZC, Liu XM (1998) Study on gas exchange characteristics of main constructive plants Artemisia ordosica and Caragana korshinskii in Shapotou region. Acta Ecologica Sinica 18: 133–137.
68. Zhou HY, Zhang JG, Long LQ, Zhao L (2001) Photosynthesis of several dominant shrubs of genus Caragana L. in the ecotone of northern China. Journal Of Desert Research 21: 227–231.
69. Deng X, Li XM, Zhang XM, Ye WH (2002) A study of the gas exchange characteristics of four desert plants. Acta Phytoecologica Sinica 26: 605–612.

16

California Coastal Upwelling Onset Variability: Cross-Shore and Bottom-Up Propagation in the Planktonic Ecosystem

Fanny Chenillat[1,2]*, Pascal Rivière[1], Xavier Capet[3], Peter J. S. Franks[4], Bruno Blanke[2]

1 Laboratoire des Sciences de l'Environnement Marin (LEMAR), CNRS/UBO/IRD/IFREMER, Institut Universitaire Européen de la Mer (IUEM), Plouzané, France, **2** Laboratoire de Physique des Océans (LPO), CNRS/IFREMER/IRD/UBO, Université de Bretagne Occidentale, Brest, France, **3** Laboratoire d'Océanographie et du Climat (LOCEAN), CNRS/UPMC/IRD/MNHN, Institut Pierre Simon Laplace (IPSL), Paris, France, **4** Integrative Oceanography Division, Scripps Institution of Oceanography, University of California San Diego, La Jolla, California, United States of America

Abstract

The variability of the California Current System (CCS) is primarily driven by variability in regional wind forcing. In particular, the timing of the spring transition, *i.e.*, the onset of upwelling-favorable winds, varies considerably in the CCS with changes in the North Pacific Gyre Oscillation. Using a coupled physical-biogeochemical model, this study examines the sensitivity of the ecosystem functioning in the CCS to a lead or lag in the spring transition. An early spring transition results in an increased vertical nutrient flux at the coast, with the largest ecosystem consequences, both in relative amplitude and persistence, hundreds of kilometers offshore and at the highest trophic level of the modeled food web. A budget analysis reveals that the propagation of the perturbation offshore and up the food web is driven by remineralization and grazing/predation involving both large and small plankton species.

Editor: Steven J. Bograd, National Oceanic and Atmospheric Administration/National Marine Fisheries Service/Southwest Fisheries Science Center, United States of America

Funding: The authors have no support or funding to report.

Competing Interests: The authors have declared that no competing interests exist.

* E-mail: fannychenillat@gmail.com

Introduction

The high biological productivity and fisheries activity of coastal upwelling systems is characterized by long-term (decadal) variability. Correlations of this biological variability in the California Current System (CCS) with North Pacific climate modes such as PDO (Pacific Decadal Oscillation) and ENSO (El Niño Southern Oscillation) [1,2,3,4] have failed to capture the decadal variability of salinity, nutrients and chlorophyll (Chl-a); the mechanisms driving the biological changes remain unclear. The recently introduced NPGO (North Pacific Gyre Oscillation) [5] is the second dominant mode of variability of sea surface height anomalies (SSHa) in the Northern Pacific and is associated with changes in strength of the central and eastern parts of the North Pacific Gyre. Interannual changes in the NPGO have been shown to explain a significant fraction of the long-term variability of salinity, nutrient and Chl-a in the CCS [5,6,7].

Variations in the NPGO correlate with changes in the strength and timing of wind-driven upwelling in the CCS. Upwelling in the central and northern CCS exhibits a strong seasonal cycle in response to seasonally varying equatorward winds. At the spring transition – the onset of upwelling – winds become predominantly upwelling-favorable, usually sometime between January and April, with some latitudinal and interannual variability. *Chenillat et al.* [8] showed that a positive phase of the NPGO is characterized by strong alongshore winter winds leading to an early spring transition, with the opposite patterns during a negative NPGO. These changes in the onset of upwelling change the nutrient input and consequent phytoplankton bloom, which in turn influence the rest of the coastal trophic web [9,10,11,12].

Only a few studies have focused on the mechanisms that communicate changes in the timing of the upwelling onset across trophic levels (*e.g.*, *Dorman et al.* [13] for krill; *Ji et al.* [14], and references therein). A bottom-up process can certainly be anticipated, but its expression in time and space and the way it affects trophic links are unknown. Building on new findings about the seasonal expression of the NPGO and its effect on the onset of upwelling [8], we investigate the influence of changes in the timing of the spring transition on the structure and functioning of the California Current planktonic ecosystem. In particular, we examine the cross-shore dependence of the ecosystem response to a time lag in the onset of wind-driven coastal upwelling.

We use a regional hydrodynamic model forced by two synthetic wind climatologies that only differ in winter, corresponding to early or late upwelling onset, such as those typically arising in NPGO+ or NPGO- conditions [8]. This model is coupled to an ecosystem model composed of several phytoplankton and zooplankton size classes. In principle, this model has enough complexity to reproduce the two types of food chains observed in the CCS: a short coastal chain characterized by large organisms, and a longer offshore chain composed of a more diverse set of organisms [15]. This allows us to explore how the perturbation in the onset of upwelling propagates both in space (across shore) [16] and through the food chain.

Materials and Methods

Physical model

The hydrodynamic model is the Regional Ocean Modeling System (ROMS), a three dimensional, free-surface, hydrostatic, eddy-resolving primitive equation ocean model based on the Boussinesq approximation and hydrostatic vertical momentum balance [17]. ROMS has been used successfully to model the North Pacific and particularly the CCS dynamics [5,18,19,20,21]. The configuration of the model is the same as in *Capet et al.* [22], except that the horizontal resolution is 15 km. In the vertical 32 σ-coordinate levels are irregularly spaced with a higher resolution near the surface to adequately resolve the upper ocean physics and ecosystem dynamics. The grid is rotated to follow the general orientation of the California coastline and covers the entire CCS, from the coast to 1000 km offshore and from Baja California (24°N) to Vancouver Island (50°N). The bathymetry is derived from etopo2 (http://www.ngdc.noaa.gov/mgg/global/etopo2.html) following the procedure described by *Penven et al.* [23].

Monthly mean climatologies are used to force the model at the surface and at its lateral boundaries. This is sufficient for our study and consistent with the horizontal resolution of our configuration: the ocean response to synoptic winds can be large within a few tens of kilometers from the shore (succession of upwelling and relaxation phases, sea breeze effects [24]) but this wind variability is not key to the system functioning on regional scales of 100 km or more [18]. It is worth noting here that high temporal resolution atmospheric forcing fields tend to have detrimental coastal biases [20,22] with possible implications for the system dynamics [25]. Moreover, using climatological forcing simplifies the construction of the two wind seasonal climatologies typical of early and late upwelling onset. To do so, we manipulate a QuikSCAT wind climatology for the period 2000–2008 by rearranging monthly mean fields as described in ***Table 1***. For the two particular rearrangements we have retained, the spring transition indices (the time of year when winds become upwelling favorable, [26]) differ by approximately 3 weeks, and the annual mean upwelling indices differ by 11% [27]. The rearrangements were chosen such that these differences are typical of those found when comparing NPGO+ and NPGO- winter wind conditions off central California [8]. NPGO fluctuations at regional scale are mainly characterized by a modification of winter winds, with no significant heat flux differences, which could be checked using NCEP reanalysis for the period 1958–2008. Thus, we expect our numerical experiments to be relevant to study NPGO impacts on the CCS ecosystem, despite some degree of idealization.

Table 1. Rearrangement of monthly climatological QuikSCAT wind data to build synthetic monthly climatologies representative of early and late upwelling onset.

New climatology months	Early upwelling onset	Late upwelling onset
December	December QuikSCAT	November QuikSCAT
January	January QuikSCAT	December QuikSCAT
February	March QuikSCAT	December QuikSCAT
March	April QuikSCAT	January QuikSCAT

An equivalent rearrangement was performed for SST. The first column indicates the month of the new climatology, and the second and third columns indicate the corresponding month of the QuikSCAT climatology used to build the early and late upwelling onset climatology, respectively. All the other months in the new climatologies are identical to those of the QuikSCAT climatology.

Ecological model

The physical model is coupled to the NEMURO (North Pacific Ecosystem Model for Understanding Regional Oceanography) lower trophic level ecosystem model [28], adapted to the North Pacific Ocean. This ecosystem model has been widely used in the CCS [29,30,31,32]. NEMURO consists of 11 state variables: nitrate (NO_3), ammonium (NH_4), silicic acid ($Si(OH)_4$), a small and a large phytoplankton (PS, representing non-siliceous phytoplankton; PL, representing diatoms), a small, a large, and a predatory zooplankton (ZS, representing microzooplankton; ZL, representing copepods; ZP, representing euphausiids), particulate and dissolved organic nitrogen (PON and DON), and particulate silica (opal). This model is based on two biogeochemical cycles, for nitrogen and silicon. Nitrogen is found in all living compartments (PS, PL, ZS, ZL, ZP), whereas silica is used by diatoms (PL) and transits to its consumers, ZL and ZP. The ratio Si:N is fixed in this model.

Coupled with a realistic hydrodynamical model, this ecosystem model is complex enough to reproduce the cross-shore gradient observed in biological fields, and in particular the existence of two distinct ecosystems: nearshore, a short food chain characterized by large organisms; offshore, a long food chain composed of organisms of every size [15]. As in many studies [33,34], biogeochemical parameters were adjusted for a constant temperature, chosen as 10°C [30]. The complexity added by temperature dependent processes, was deemed unnecessary in the context of this study, in particular because NPGO has no major impact on temperature off California [5]. We used the phytoplankton parameters tuned for the CCS by *Li et al.* [31]. The zooplankton parameters were designed for the CCS by comparison with other planktonic model studies [29,33,34,35]. We also used a Holling type III formulation for grazing, which is numerically more stable than the default Ivlev formulation used commonly in NEMURO ([36,37], and references therein). Note that *Li et al.* [32] compared Ivlev and Holling type III grazing formulations in the CCS and showed that they gave statistically indistinguishable results. A compilation of the main biological parameters is given in ***Table 2***. For more details about NEMURO, the reader may refer to *Kishi et al.* [28].

Initial conditions and monthly-averaged boundary conditions for the biological fields were taken from the high resolution OFES model (Ocean general circulation model For the Earth Simulator [38,39,40]) run with a 0.1° horizontal resolution. The OFES model includes a simple ecosystem model (Nutrient-Phytoplankton-Zooplankton-Detritus, or NPZD type) over the whole Pacific domain, integrated numerically from 2000 to 2007. However, the OFES-NPZD model is composed of only 4 biological components, whereas NEMURO has several nutrients, phytoplankters, zooplankters, and detritus. With the guidance of open ocean observations [41,42], 2/3 (1/3) of OFES phytoplankton was assigned to NEMURO PS (PL), and each zooplankton class of NEMURO received one third of the total OFES zooplankton. Boundary and initial conditions for silicic acid were provided by nitrate profiles taken from the OFES-NPZD model and adjusted by a mean ratio Si:N of about 2.0, in accordance with the Levitus World Ocean Atlas [43].

Set of simulations

A 30-year spin-up simulation was performed with the physical model only (SP-PHY), and we have verified that the end of this run reached a statistical equilibrium. The final state provides an initial

Table 2. Parameters and terms used in NEMURO model.

Symbol	Definition	Value			Unit	Source
	Parameters for phytoplankton	**PS**	**PL**			
V_{maxj}	Maximum photosynthetic rate	0.4	1.0		d^{-1}	[31]
K_{NO3}	Half saturation constant for nitrate	1.0	3.0		mmolN m^{-3}	[28]
K_{NH4}	Half saturation constant for ammonium	0.1	0.3		mmolN m^{-3}	[28]
$K_{Si(OH)4}$	Half saturation constant for silicate	-	4.0		mmolN m^{-3}	[31]
ψ	Ammonium inhibition coefficient	1.5	4.6		(mmolN $m^{-3})^{-1}$	[28]
α	Initial slope of photosynthesis-irradiance curve	0.014	0.028		$(Wm^{-2})^{-1}$ d^{-1}	[31]
β	Photoinhibition coefficient	0.001	0.008		$(Wm^{-2})^{-1}$ d^{-1}	[31]
Mor_{P0}	Mortality rate at 0°C	0.0585	0.0290		(mmolN $m^{-3})^{-1}$ d^{-1}	[28]
	Pameters for zooplankton	**ZS**	**ZL**	**ZP**		
$Gr_{max/PS}$	Maximum grazing rate on PS at 0°C	1.0	0.2	-	d^{-1}	[29,33,34,35]
$G_{rmax/PL}$	Maximum grazing rate on PL at 0°C	-	1.0	1.0	d^{-1}	[29,33,34,35]
$G_{rmax/ZS}$	Maximum grazing rate on ZS at 0°C	-	0.5	0.4	d^{-1}	[29,33,34,35]
$G_{rmax/ZL}$	Maximum grazing rate on PL at 0°C	-	-	0.8	d^{-1}	[29,33,34,35]
K_{PS2Z}	Half saturation constant for PS	1.0	1.0	-	(mmolN $m^{-3})^2$	[29,33,34,35]
K_{PL2Z}	Half saturation constant for PL	-	1.0	1.0	(mmolN $m^{-3})^2$	[29,33,34,35]
K_{ZS2Z}	Half saturation constant for ZS	-	1.0	1.0	(mmolN $m^{-3})^2$	[29,33,34,35]
K_{ZL2Z}	Half saturation constant for ZL	-	-	1.0	(mmolN $m^{-3})^2$	[29,33,34,35]
ψ_{PL}	ZP on PL grazing inhibition coefficient	-	-	4.605	$(mmolN^{-1})^{-1}$	[28]
ψ_{ZS}	ZP on ZS grazing inhibition coefficient	-	-	3.010	$(mmolN^{-1})^{-1}$	[28]
Mor_{Z0}	Mortality rate at 0°C	0.0585	0.0585	0.0585	(mmolN d^{-3} $m^{-3})^{-1}$ d^{-1}	[28]
$Alpha_Z$	Assimilation efficiency	0.70	0.70	0.70	-	[28]
$Beta_Z$	Growth efficiency	0.30	0.30	0.30	-	[28]
	Other parameters					
g_T	Temperature coefficient set to 10°C	2			-	[30]
att_w	Light Extinction Coefficient of Sea Water	0.0554			m^{-1}	[31]
att_P	Self Shading Coefficient	0.0249			mmolN m^{-3} m^{-1}	[31]
Nit_O	Nitrification rate at 0°C	0.03			d^{-1}	[28]
$VP2N_O$	Decomposition rate at 0°C, PON to NH_4	0.10			d^{-1}	[28]
$VP2D_O$	Decomposition rate at 0°C, PON to DON	0.10			d^{-1}	[28]
$VD2N_O$	Decomposition rate at 0°C, DON to NH_4	0.02			d^{-1}	[28]
$VP2Si_O$	Decomposition rate at 0°C, Opal to $Si(OH)_4$	0.01			d^{-1}	[28]
RSiN	Si :N ratio	1.0			-	[31]
setVP	Sinking velocity of PON	40.0			m d^{-1}	[28]
setVO	Sinking velocity of Opal	40.0			m d^{-1}	[28]

Bibliographic sources are indicated in the table. References [29,33,34,35] are based on ecosystem models published data that focus on upwelling system.

condition for the second spin-up that couples biology and physics (SP-BIOPHY). SP-BIOPHY was run for 12 years. These two spin-up simulations were forced by synthetic SST and QuikSCAT wind data given by the average of the early and late upwelling onset climatologies. Next, we ran two 24-year long experiments for the early upwelling onset (NPGO+ like) and late upwelling onset (NPGO− like) scenarios (respectively EXP-EU and EXP-LU). The only differences between these twin experiments were the climatological wind conditions that were representative of the effect of NPGO on winter wind upwelling regime; these two wind fields differ only during wintertime from December to March. The two wind climatologies are based on the study of *Chenillat et al.* [8] and were constructed by combining the original monthly mean wind fields of the QuikSCAT climatology as described in ***Table 1***. The alongshore components of these two wind climatologies differ by 11% off the Central California Current System. The whole spin-up procedure is summarized in ***Figure 1***.

Results and Discussion

Model evaluation

We only aim at qualitatively assessing the model skill because our model, forced by climatological atmospheric fluxes, lacks several important features (synoptic forcing variability, resolution

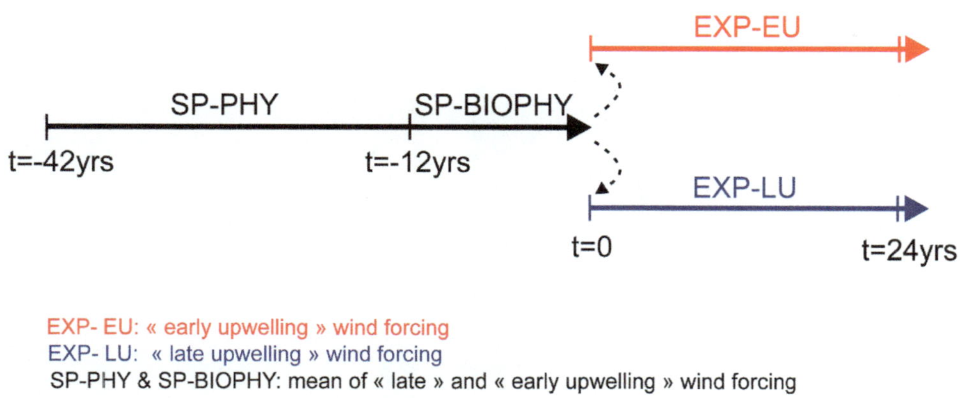

Figure 1. Experimental layout. Times are in years. EXP-EU: early upwelling forcing winds. EXP-LU: late upwelling forcing winds. SP-PHY and SP-BIOPHY: mean forcing winds.

of fine scale frontal processes [44,45]) and because the observations used for comparisons do not span over the same periods.

The evaluation of the model is based on averages calculated over the final 6 years of SP-BIOPHY, except for eddy kinetic energy (EKE) which requires a longer time series for its computation, with high-frequency outputs for velocity. Therefore, EKE is evaluated in a 30-year-long simulation run of the physical model (using the same forcing as SP-PHY).

The annual mean climatology EKE is a good proxy for the intensity of mesoscale turbulence. The modeled EKE (***Fig. 2a***) is compared with MADT altimetry data (http://www.aviso.oceanobs.com/duacs/) between 1992 and 2009 (***Fig. 2b***). The model EKE has maximum values around 120 cm^2/s^2 in the central CCS, about 300 km from the coast, and minimum values along the coast, similar to the observations. The maximum values of altimetry-derived EKE are higher than the model (200 cm^2/s^2) and extend over a broader area, which indicates that the model underestimates EKE. This is a direct consequence of the model resolution that is too coarse to fully resolve mesoscale and submesoscale dynamics in this system [22].

To assess the mean surface circulation in the CCS we compare the annual climatological mean model SSH (***Fig. 3a***) with equivalent AVISO data from 1992–2009 (***Fig. 3b***). The cross-shore gradient (from −0.2 m at the coast to +0.2 m offshore) and the placement of the isolines are in fair agreement with the observations, which shows that the intensity of our California Current is realistic. The values nearshore are more negative in the observations than in the model, but altimetric sea level measurement should be used with caution close to shore [46].

Figure 4 shows the annual climatological mean model Chlorophyll-a (Chl-a) concentration at 10 m depth and climatological Sea-WIFS (Sea-Viewing Wide Field-of-view Sensor) data averaged over 1997–2005. The model phytoplankton biomass (sum of PS and PL, in mmol N m^{-3}) is converted to Chl-*a* using a C:N ratio of 106:16, and a conversion from C to Chl-*a* based on *Cloern et al.* [47]. The comparison reveals a general agreement between the model and the observations in the central and southern CCS in a 500 km strip along the coast despite a small negative bias of the model very nearshore and far offshore. The main discrepancy is found in the northern CCS with a more systematic negative bias in modeled surface Chl-*a* concentration. Note that our study focuses on central California, where the model best reproduces the observed chlorophyll distribution. Seasonal profiles in the central CCS, averaged from Point Conception

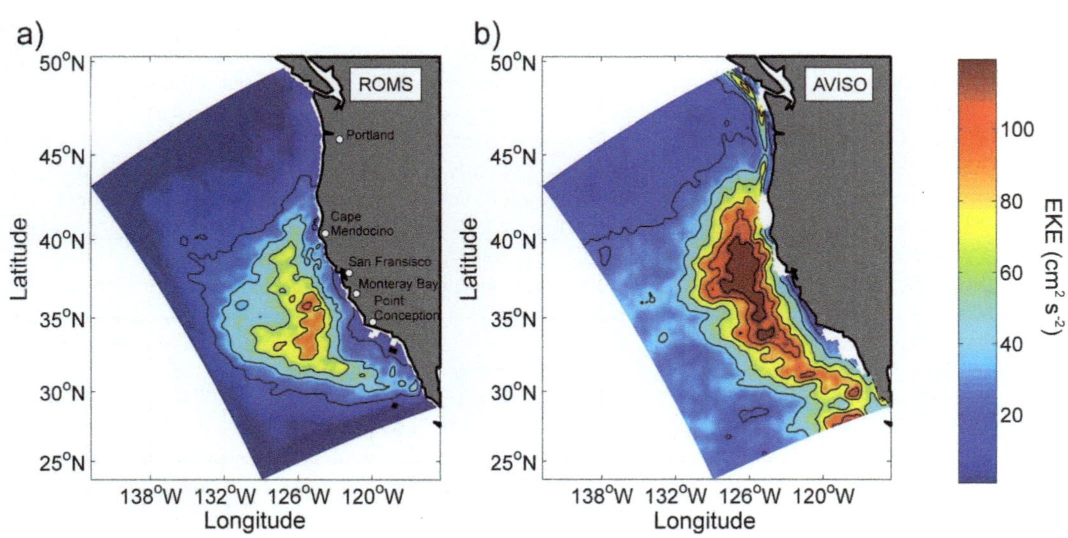

Figure 2. Annual mean eddy kinetic energy (EKE) from (a) the model and (b) satellite-derived SSH observations (AVISO, 1992–2009). Units: cm^2/s^2. The contour interval is 20 cm^2/s^2.

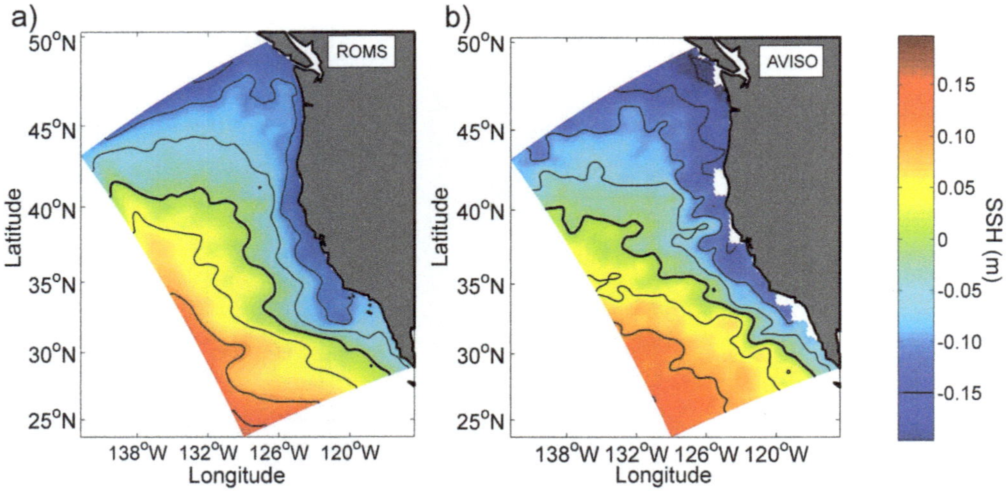

Figure 3. Annual mean sea surface height from (a) the model and (b) satellite-derived SSH observations (AVISO, 1992–2009). Units: m. The contour interval is 0.05 m and the 0 m contour is in bold.

(34.5°N) to Cape Mendocino (40°N), confirm the model bias in Chl-a nearshore, mainly in spring (***Figs. 4c–f***). The far offshore negative bias is most prominent in winter and fall, whereas the surface Chl-*a* concentration is in better agreement with observations during spring and summer.

Finally, we examine the vertical structure of the model (***Fig. 5***) using CalCOFI data averaged over 1949–2000. Along line 70 (south of Monterey), from the coast to 350 km offshore a general agreement is found between model and CalCOFI annual climatological mean temperature, nitrate and Chl-*a* climatologies, including the cross-shore structure of the 13°C isotherm depth (***Figs. 5a–b***), the 20 mmol N m^{-3} isoline (***Figs. 5c–d***), and the presence of a subsurface maximum in Chl-*a* (***Figs. 5e–f***). Some differences are also evident: offshore, the model overestimates Chl-*a* by a factor of 2 at the surface and by a factor of 3 around the depth of the subsurface maximum (70 m). These differences coincide with a model underestimation of subsurface nitrate concentrations. Close to the coast, the model 12°C isotherm is deeper than in the observations, which suggests that the model fails to accurately represent the pathway and/or transformation of upwelling waters.

Overall and despite some weaknesses, the model shows evident skill and realism for both its physical and biogeochemical components. It reproduces a realistic CCS general circulation (for the mean and mesoscale dynamics) and it is able to simulate a seasonal cycle and a spatial cross-shore structure of "bulk phytoplankton" that resemble the observations, at least off central California. In the next section, we analyze the functioning of the modeled biological system and its perturbation by a change in the timing of the onset of upwelling.

Ecosystem response to differential onset of upwelling

We focus on the central CCS, from Point Conception (34.5°N) to Cape Mendocino (40°N). Diagnoses of tracer concentration and tracer budgets are carried out in two subregional boxes. The nearshore box is bounded by the latitudes 34.5°N and 40°N, the coastline, and a line that follows the mean orientation of the central coast but is located 150 km offshore. The offshore box is defined similarly, but within two lateral limits situated at 300 and 450 km from the coastline. The vertical extension of both boxes is from the surface down to 100 m (or the ocean bottom in areas shallower than 100 m). The ecosystem response will be described in terms of nitrate and biomass concentrations expressed as nitrogen concentrations. We estimate the statistical significance of the differences between the twin resulting climatologies with a non-parametric test [48]. The role of the other components, ammonium and dissolved or particulate nitrogen (involved in the recycling loop), will be discussed in the following sections. For space constraints and clarity, the silicon cycle will be ignored. However, we have verified that, at least in the model we use, it does not play a key role in the ecosystem response to a differential upwelling onset.

For the nearshore domain, the annual mean cycle of all model variables (***Fig. 6***) agrees with the observed central California upwelling dynamics and biogeochemistry, regardless of the details of the winter wind. PL, ZL and ZP (***Figs. 6d, f and g***) dominate the nearshore ecosystem biomass with a marked bloom in late winter/early spring, a peak of biomass in early summer, and a return to lower values afterwards. Their evolution roughly mirrors that of SSH (***Fig. 6a***), but with smoother changes. The peak for ZP (***Fig. 6g***) lags the ZL peak by about a month, and the PL peak by about 2 weeks, in qualitative agreement with a bottom-up forcing of the ecosystem. PS and ZS (***Figs. 6c and e***) are present in much smaller concentrations. The PS bloom appears slightly later (2 weeks) than the PL bloom, and is followed by a smooth peak of ZS in early fall. The evolution of ZS is in quadrature with the evolution of large zooplankton ZL and ZP: its minimum is in spring when the concentrations of the larger plankton size classes increase the fastest, and vice versa in fall.

The mean annual cycles of biogeochemical tracer concentration in the offshore box are 2 to 5 times smaller than in the nearshore box, with a less pronounced (***Figs. 6c*** to ***f***) or even reversed (NO_3) seasonal cycle (***Fig. 6b***), except for ZP (***Fig. 6g***). The nearshore box minus the offshore box differences are consistent with tracer maxima originating nearshore and moving offshore with some attenuation. Other more subtle differences deal with the biomass of small (***Figs. 6c and e***) and large (***Figs. 6d, f and g***) plankton size classes. Nearshore, the ecosystem biomass is dominated by large plankton species, whereas the offshore trophic chain is longer and consists of a roughly equal proportion of small and large organisms.

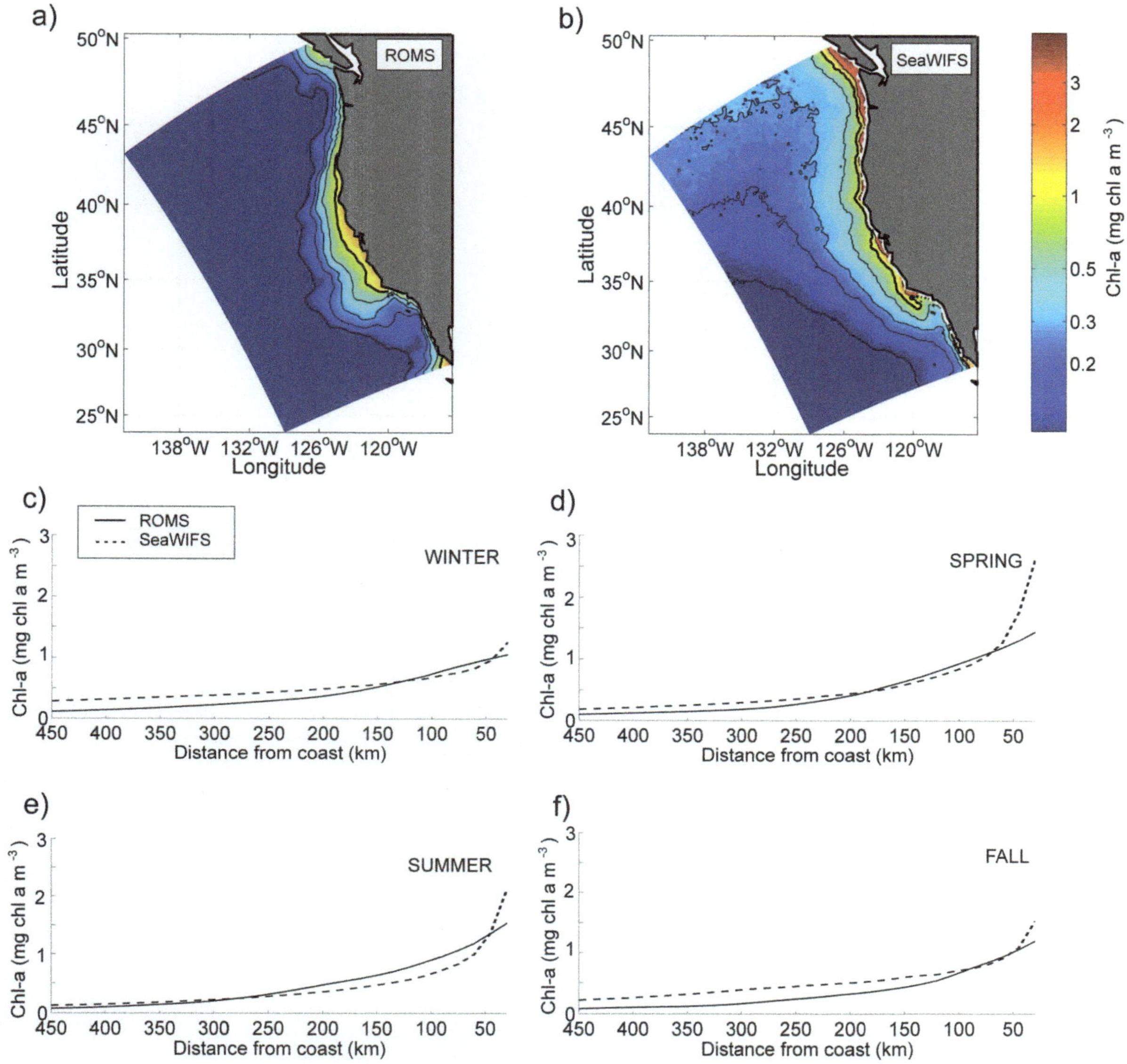

Figure 4. Annual mean surface Chlorophyll-*a* from (a) the model and (b) observations (SeaWIFS, 1997–2005). Seasonal cross-shore profiles of surface Chlorophyll-a averaged for the central CCS (from 34.5°N to 40°N) for the model (solid line) and the observations (SeaWIFS, 1997–2005; dashed line) for (c) winter (January–February–March), (d) spring (April–May–June), (e) summer (July–August–September), and (f) fall (October–November–December). Units: mg Chl-a m^{-3}.

We now study the system response to a perturbation in the onset of upwelling from monthly climatologies of the last 12 years of the EXP-LU and EXP-EU simulations.

The modeled sea level seasonal cycle is smooth enough to define the upwelling onset as the time at which the nearshore sea level starts decreasing. This definition gives a statistically significant 10-day difference between the two simulations (***Fig. 6a***). Wind differences also directly translate into a larger annual stock of nearshore nitrate in the "early upwelling" (NPGO+) simulation. A statistically significant difference of 15% is observed during December-April (***Fig. 6b***); the change is only 5% when averaged over the entire year. An impact on plankton nearshore biomasses is observed from January to July. Differences in nearshore biomass of PL (7% in January and 20% in March), ZL (8% and 28%) and ZP (3% and 21%) (***Figs. 6d, f and g***) are consistent with the propagation of the perturbation up the food chain. Interestingly, the small nearshore plankton PS and particularly ZS (***Figs. 6c and e***) are noticeably less affected by the perturbation.

Overall, the winter wind perturbation has a very limited impact on the nearshore ecosystem beyond July. However, a noteworthy consequence of the winter wind perturbation is the presence of a reverse trend in nearshore biological/nutrient concentrations in late summer compared with the beginning of the year. Although the trend is of limited amplitude and tends to be smoothed by spatial averaging, it is apparent for NO_3 from July to October (***Fig. 6b***). In other words, stronger winter winds lead to reduced nitrogen concentrations about 6 months later in conjunction with a sea level rise (***Fig. 6a***) mainly in July and August. Indeed, the response of a linear eastern-boundary system to periodic winds includes free Rossby waves that would propagate offshore for some months and lead to a rise in sea level [49]. Our non-linear CCS undergoes, on average, a similar SSH evolution and a depression

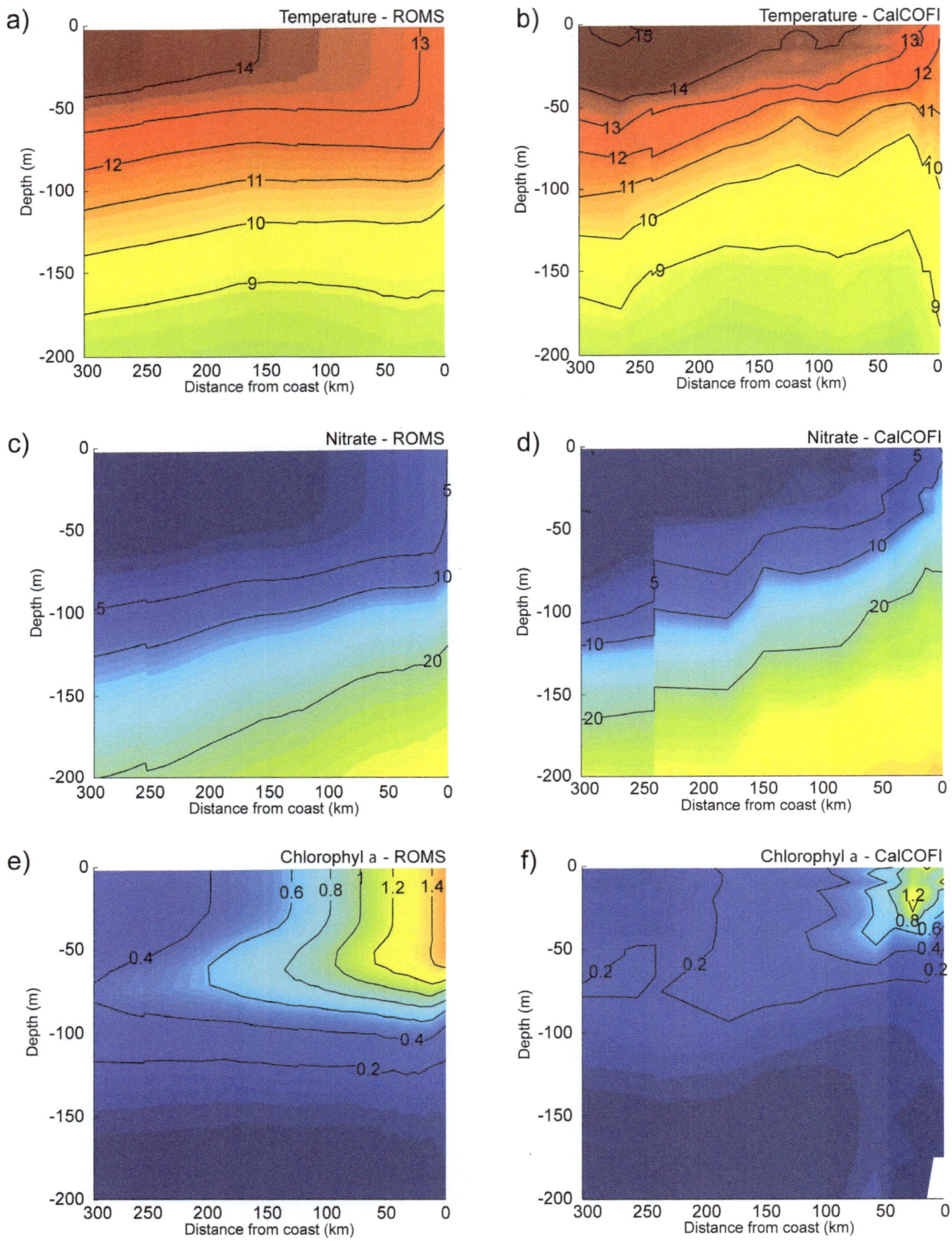

Figure 5. Comparison along CalCOFI line 70 of mean vertical sections for the model (left) and observations (right): (a) and (b) temperature (°C); (c) and (d) nitrate concentration (mmol N m^{-3}); (e) and (f) Chlorophyll-*a* (mg Chl-a m^{-3}).

of the upper thermocline (10 to 20 m) that reduces the nutrient input to nearshore areas (not shown).

The effect of the wind perturbation on the offshore ecosystem is distinct and arguably more important. In terms of duration, the increase in biogeochemical concentrations is notable from March

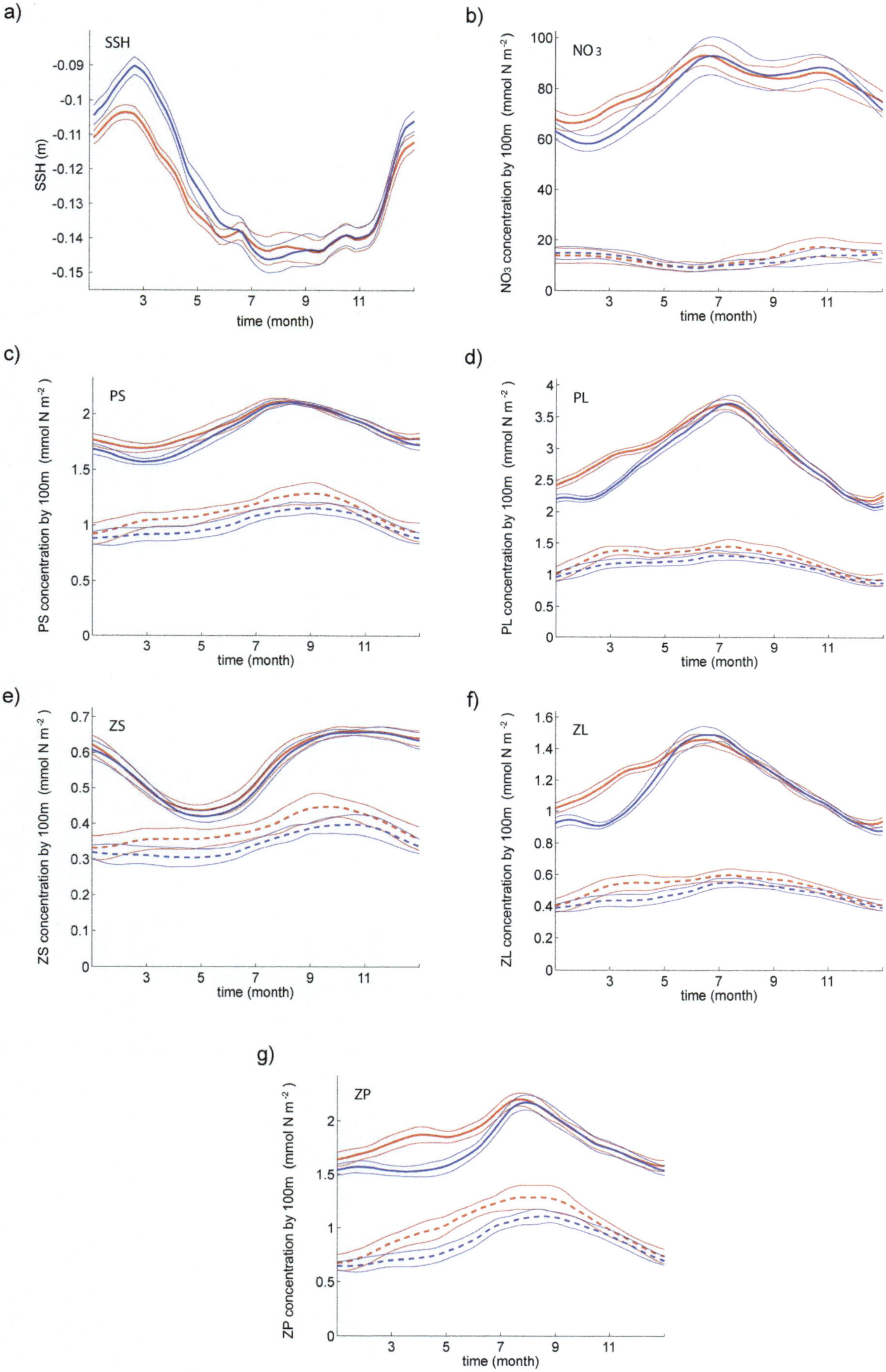
a)
SSH
SSH (m)
-0.09
-0.1
-0.11
-0.12
-0.13
-0.14
-0.15
3 5 7 9 11
time (month)
b)
NO3
NO3 concentration by 100m (mmol N m-2)
100
80
60
40
20
0
3 5 7 9 11
time (month)
c)
PS
PS concentration by 100m (mmol N m-2)
2
1.5
1
0.5
0
3 5 7 9 11
time (month)
d)
PL
PL concentration by 100m (mmol N m-2)
4
3.5
3
2.5
2
1.5
1
0.5
0
3 5 7 9 11
time (month)
e)
ZS
ZS concentration by 100m (mmol N m-2)
0.7
0.6
0.5
0.4
0.3
0.2
0.1
0
3 5 7 9 11
time (month)
f)
ZL
ZL concentration by 100m (mmol N m-2)
1.6
1.4
1.2
1
0.8
0.6
0.4
0.2
0
3 5 7 9 11
time (month)
g)
ZP
ZP concentration by 100m (mmol N m-2)
2
1.5
1
0.5
0
3 5 7 9 11
time (month)

Figure 6. Time evolution of key modeled variables in the early upwelling (red) and the late upwelling (blue) scenario, in the central CCS. (a) Nearshore SSH (in m) averaged over a 50 km wide strip along the coast. (b–g) Biological variables (in mmol N m^{-2}) integrated from the surface to 100 m depth and averaged over the nearshore region (0–150 km offshore, bold solid line) and over the offshore region (300–450 km offshore, bold dashed line): (b) nitrate; (c) PS; (d) PL; (e) ZS; (f) ZL; (g) ZP. Gray-shadowed areas show the months when winds differ between both scenarios. The thin lines represent the year-to-year variability diagnosed during the 12-year-long scenarios (standard deviation).

to late October, *i.e.*, for most of the year. In particular, the ZP biomass (***Fig. 6g***) is very sensitive to winter winds, with a maximum difference between the two simulations of about 35% in May. Nutrients are a notable exception: they remain strongly depleted throughout the year (***Fig. 6b***).

These results indicate an offshore propagation of the nearshore nutrient input perturbation, in qualitative agreement with the conceptual view of the mixed-layer conveyor model proposed by Botsford et al. [16] (see their Fig. 1). As it propagates offshore at a speed of a few cm/s (not shown), the perturbation also propagates up the trophic chain. We now characterize this cross-shore and bottom-up propagation by studying the functioning of the system by means of tracer budgets.

Cross-shore difference of biological tracer budget

We analyze the propagation of the effects of the winter wind perturbation in space (from nearshore to offshore) and across trophic levels (bottom-up). We do so by studying the physical and biological terms (flux divergences and source/sink terms) that compose the budget of every biological component of the ecosystem within the box framework defined in section 4. Both seasonal (***Figs. 7*** and ***8***) and 12-year averaged (***Fig. 9***) fluxes are shown. For sake of simplicity, only the most important fluxes are presented.

Seasonal biological budgets. The origin of the perturbation can be traced back in the nearshore nitrate budget (***Fig. 7a***): the advection term is larger from January to March under stronger winter upwelling winds. The difference arises primarily from vertical advection (not shown). Over the rest of the year, the nitrate flux divergence remains positive, but some compensation occurs due to the cross-shore (mainly advective) nitrate flux out of the nearshore box in the early upwelling onset case. The advective flux perturbation results mainly in an uptake perturbation: the total phytoplankton uptake increases continuously from January with early upwelling winds, whereas it is constant until the beginning of March with late upwelling winds (***Fig. 7a***). Noticeable nearshore uptake differences are limited to the first half of the year.

We now turn to nearshore bulk phytoplankton production (***Fig. 7b***). We recall that phytoplankton growth is the ombination of new production (nitrate uptake) and recycled production (ammonium uptake). These two processes are not separately shown in ***Figure 7***, but can be easily estimated as follows: new production is reflected by nitrate uptake (***Fig. 7a***), and recycled production is reflected by the difference between phytoplankton growth (***Fig. 7b***) and new production. Approximately half of the nearshore phytoplankton growth is associated with ammonium uptake, *i.e.*, with the recycling loop; the other half is associated with nitrate uptake, *i.e.*, with new production. Moreover, the wind-induced perturbation is of the same order for ammonium and nitrate uptake.

Offshore, total nitrate uptake differences (about 20%) are spread over most of the year (***Fig. 7a***). Unlike the nearshore region, the offshore winter wind-induced perturbation is more important for ammonium uptake than for nitrate uptake. Indeed, the difference between the two upwelling scenarios is about 60% for the total uptake, estimated over the entire year (***Fig. 7b***), which is much more than the 20% difference obtained for nitrate uptake (***Fig. 7a***). Thus, in the offshore region, an early upwelling stimulates recycled production more efficiently than new production.

For the phytoplankton and zooplankton budgets, the consequences of the upwelling onset perturbation are mainly expressed through the physiological source/sink terms (***Figs. 7b*** and ***8***) while advection plays a modest role (not shown, but its role as a nearshore and offshore source will be discussed below in the context of the annual budgets). For the total phytoplankton, the

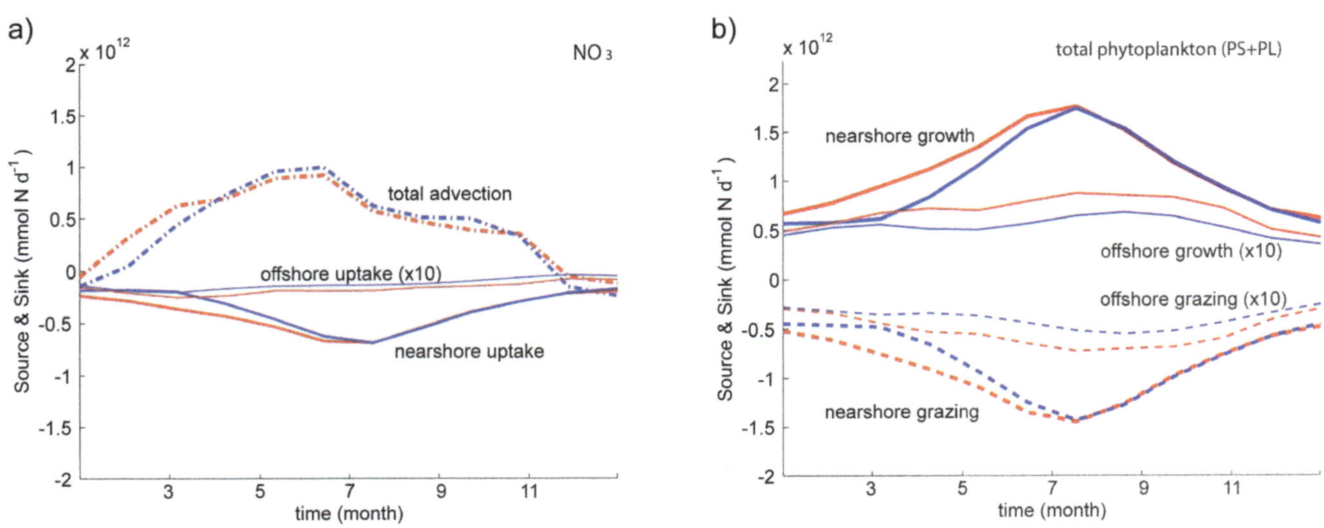

Figure 7. Time evolution of nitrogen sources and sinks (in mmol N d^{-1}) in the early upwelling (red) and late upwelling (blue) scenarios: (a) nearshore total advection of nitrate, and nearshore and offshore uptake by total phytoplankton; (b) nearshore and offshore total phytoplankton growth and grazing. The offshore fluxes are rescaled by the factor 10 for clarity. Gray-shadowed areas show the months when winds differ between both scenarios.

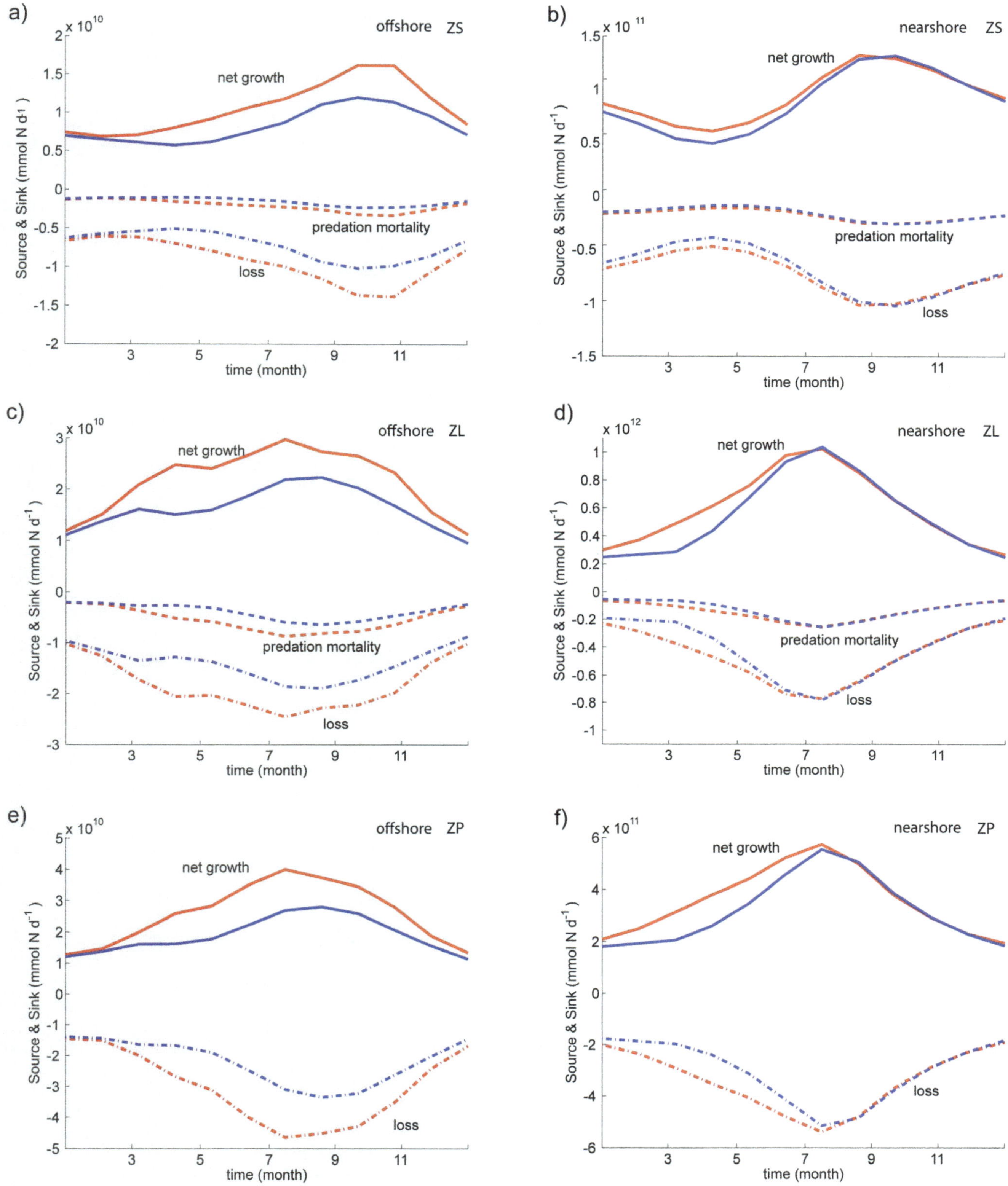

Figure 8. Time evolution of biological sources and sinks (in mmol N d^{-1}) in the early upwelling (red) and late upwelling (blue) scenarios, both nearshore (right) and offshore (left): (a) and (b) small zooplankton (ZS) fluxes; (c) and (d) large zooplankton (ZL) fluxes; (e) and (f) small zooplankton (ZS) fluxes. Gray-shadowed areas show the months when winds differ between both scenarios.

growth perturbation is largely compensated by changes in zooplankton grazing (***Fig. 7b***), a signature of propagation of the perturbation up the trophic chain. For all nearshore/offshore zooplankton classes, wind-induced perturbations in net growth are compensated by predation mortality (~20–25%) and loss (75–80%), the latter being a mixture of excretion, egestion and natural

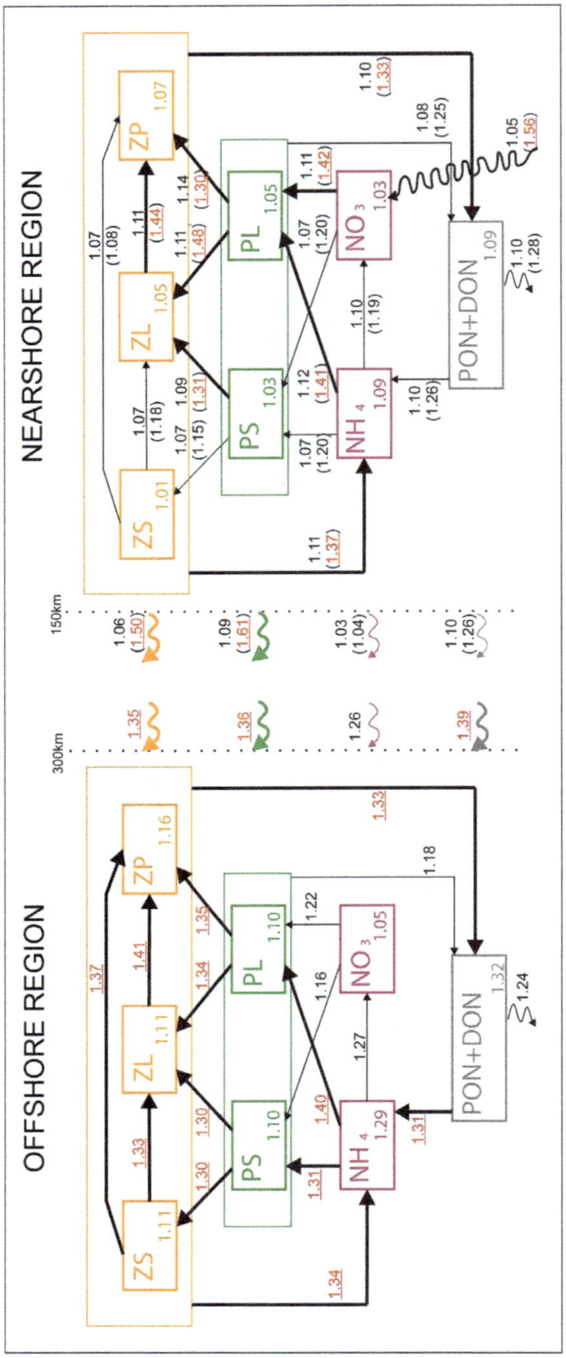

Figure 9. Biological (biomasses and fluxes) and advective (fluxes) budget for biological tracers in the nearshore and offshore boxes. Numbers in the main boxes indicate the ratios of the early upwelling scenario annual mean biomasses to the late upwelling scenario annual mean biomasses. Numbers above arrows indicate the ratios of the early upwelling scenario annual means fluxes to the late upwelling scenario annual means fluxes. Note that for the nearshore region, numbers in parentheses indicate the ratios of the winter mean fluxes. Ratios underlined in red above bold arrows emphasize 30% or larger differences between early and late upwelling scenarios.

mortality (about 45–50%, 35–40% and 10–20% of the loss, for both small and large zooplankton; 35–40%, 25–30% and 25–40% for predator zooplankton).

A careful examination of the small and large zooplankton budgets for the nearshore domain (***Fig. 8b–d***) indicates that small zooplankton fluxes are one order of magnitude smaller than the large zooplankton fluxes. We also note that small zooplankton predation mortality is only weakly affected by the winter wind perturbation. This confirms the nearshore dominance of the short food chain involving large species, both in terms of biomass (***Fig. 6***) and inter-size class fluxes (***Fig. 8***). In contrast, in the offshore region, the small and large zooplankton fluxes are of the same order and are equally affected by the winter wind perturbation. Offshore, the entire trophic web is modified by the perturbation. These points are explored in more detail in the next subsection.

Annual mean biological budgets. The effects of a winter wind perturbation are discussed using annual mean biological tracers budgets including advection and biological sources and sinks.

Nearshore (***Fig. 9***), a winter wind perturbation has a weak effect on the annual-mean budget, with a less than 10% flux increase (and a less than 7% biomass increase) in the early upwelling wind scenario compared to the late upwelling wind scenario. As described above, the nearshore ecosystem perturbation has a limited duration, roughly from January to March when winds differ (see the difference ratios over that period also given in ***Fig. 9***), and the ecosystem (mainly the large plankton species) is stimulated by early upwelling winds. During the perturbation, the fluxes involving PL, ZL and ZP increase by more than 30% compared to the late upwelling case. Interestingly, the most sensitive food chain is PL to ZL to ZP, with a short-term increase of more than 40%. This chain is stimulated by a statistically significant increase of both new and regenerated production, in similar proportion. The remineralization loop is also enhanced, mainly via zooplankton losses to NH_4 and PON/DON. This stimulation of the food chain has important consequences for the cross-shore export of biomass, which increases strikingly (~50%) over the same period. On the other hand, the export of nutrients (nitrate, ammonium) is not affected.

Nearshore the modeled food chain involving small plankton species (PS and ZS) is weakly sensitive to the wind perturbation, as one would expect given the local dominance of large cells. The only notable perturbation of the nearshore fluxes concerns the grazing of PS by ZL in winter; this perturbation is unable to propagate up the small plankton chain because the ZL grazing pressure diverts it.

The effect of the wind perturbation on the offshore budget is noticeably larger than on the nearshore budget (***Fig. 9***). Most annual mean fluxes are >30% larger in the early upwelling wind case. Both small and large species are involved in the response to the wind perturbation: primary production, grazing, and predation fluxes increase by more than 30% in the early upwelling wind case (***Fig. 9***). The ecosystem stimulation by winter winds benefits ZP the most via predation. It is important to note that the stimulation of the entire food chain is primarily induced by a boost in regenerated production (30% increase for PS and 40% for PL), whereas the new production is only increased by 16% and 22% respectively. This offshore system perturbation stems largely from enhanced cross-shore advection of nearshore material: the physical transport is increased significantly (more than 30%) for biomass, PON and DON, and to a lesser extent for NO_3 (26%).

Overall, this analysis demonstrates the importance of cross-shore transport of biomass and the recycling loop. Both processes strongly shape the ecosystem response to the winter wind perturbation. As it propagates offshore, the perturbation at low trophic levels is "recycled" and increasingly migrates up the food chain.

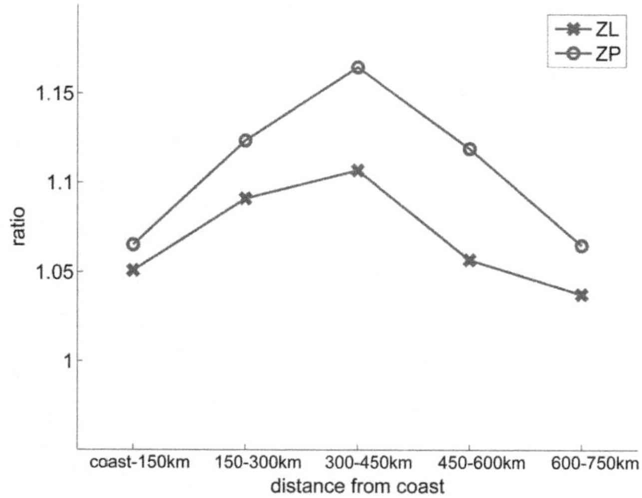

Figure 10. Ratio of annual mean biomasses (see caption of Figure 9 for details) as a function of the distance from the coast, and computed in successive 150 km-wide boxes. The ratios are given both for the large zooplankton and the zooplankton predator.

There is a limit to how far the perturbation moves offshore. Annual climatological mean ZL and ZP biomass ratios (between early- and late-upwelling scenarios) are the largest for the box situated 300 to 450 km offshore (***Fig. 10***). This maximum can be explained as follows: nearshore nutrients are abundant, and a reduction or increase of their concentration has less effect on production than farther offshore where nutrients are more strongly limiting (this was checked by computing the phytoplankton nutrient limitation terms). On the other hand, at a distance from the coast of the order of 1000 kilometers (see ***Fig. 10***), the perturbation is no longer felt because the enrichment of coastal origin plays a negligible role in the local ecosystem dynamics.

Several processes contribute to the cross-shore profile of the ecosystem response (***Fig. 10***): mean/eddy cross-shore advection, which propagates the anomaly offshore; alongshore advection, which advects the anomaly southward; and vertical export through sinking of particulate organic nitrogen (and possibly through subduction), which progressively attenuates the signal as it moves offshore. The representation of these processes and their relative importance depend to some degree on the model parameter settings. The horizontal grid resolution (15 km) is too coarse to simulate frontal processes realistically, so subduction is probably underestimated. The sinking velocities for particulate carbon are high (40 m/day see ***Table 2***) even though, despite large uncertainties [50], values around 10 m/day are often used in upwelling ecosystem studies [33,34,35,51]. Our choice is conservative: it limits the offshore propagation of coastal perturbations, and avoids underestimation of both vertical sinking and subduction, which would have raised doubts about the cross-shore structure of the response to wind anomalies. We ensured that the cross-shore response of large and predator zooplankton was robust with respect to the parameters that control the recycling loop (nitrification, remineralization, and decomposition rates).

It is clear that eddies are important for the biogeochemical functioning of the California Current (see for instance [52]), and we have shown that offshore propagation of perturbations to the nearshore ecosystem is an important determinant of offshore ecosystem variability. We note, however, that the mesoscale activity in our numerical simulations does not change appreciably from one run to another (see Appendix S1 for a more detailed analysis).

Conclusion

In this numerical study, we investigated the CCS ecosystem response to a time lag or lead in the onset of upwelling, as observed, for instance, in conjunction with NPGO variability. By analyzing the response in terms of structural changes (in biogeochemical tracer concentrations) and also in terms of dynamics (biogeochemical fluxes), we are able to clarify the mechanisms involved in driving the ecosystem response. The numerical approach relies on a biological and physical coupled model, forced by synthetic winds (derived from QuikSCAT climatology) that differ during winter (December to March). Despite some deficiencies, the model qualitatively represents the essential features of the CCS planktonic ecosystem (cross-shore gradient, seasonal cycle and depth distribution).

The intensity of early onset of upwelling winds exerts a major impact on the planktonic ecosystem, both in the nearshore region, but more importantly offshore (typically 300–500 km from the coast) where the perturbation is stronger, more persistent, and more complex. Nearshore, the ecosystem perturbation is undetectable in all biogeochemical fields after July. Before July, the effect of early upwelling winds simply arises from enhanced vertical nitrate fluxes that stimulate phytoplankton primary production and, in turn, zooplankton population growth. Overall, the changes in the nearshore ecosystem with respect to wind and upwelling changes (11% difference in Ekman transport) are modest, with less than a 7% difference in annual-average biomass between the two analyzed solutions (***Fig. 9***).

Offshore, where the biomass and biogeochemical tracer concentrations are much lower, the winter wind perturbation is felt most of the year and its relative amplitude is a factor 2 or more, greater than the nearshore. Most importantly, the highest trophic level (predator zooplankton) is by far the most impacted. The processes responsible for the offshore response are subtle and involve a combination of i) cross-shore transport which advects the coastal perturbation offshore at speeds consistent with an Ekman drift, ii) remineralization cycles which provide an increasingly large fraction of the nutrients available for primary production, and iii) relatively efficient transmission up the food chain through successive grazing/predation stages in which both small and large zooplankton participate.

There is thus a simultaneous propagation of the coastal upwelling/nitrogen perturbation in space (mainly cross-shore, with a modest contribution of eddies in our model) and from the bottom to the top of the trophic chain. During this propagation, which occurs over several months, the remineralization loop is very active on time scales of a few days. A total nitrogen (organic plus inorganic) perturbation climbs up the food chain while being transported offshore by the Ekman conveyor belt. A schematic view summarizes this propagation in ***Figure 11***. Leaks do exist (via sinking and alongshore advection by the California Current), but the typical length scale over which nearshore upwelled nitrogen exits the system is large enough for the perturbation to be felt hundreds of kilometers from the coast.

Our conclusions must be considered with caution as our study lacks realism in several respects. However, we believe they provide valuable original insight concerning ecosystem dynamics and response to large-scale wind anomalies. A horizontal model resolution of about 5 km would allow better realization of mesoscale eddies. We anticipate that at such resolution their effect would be increased: Gruber *et al.* [52] found that eddies reduce

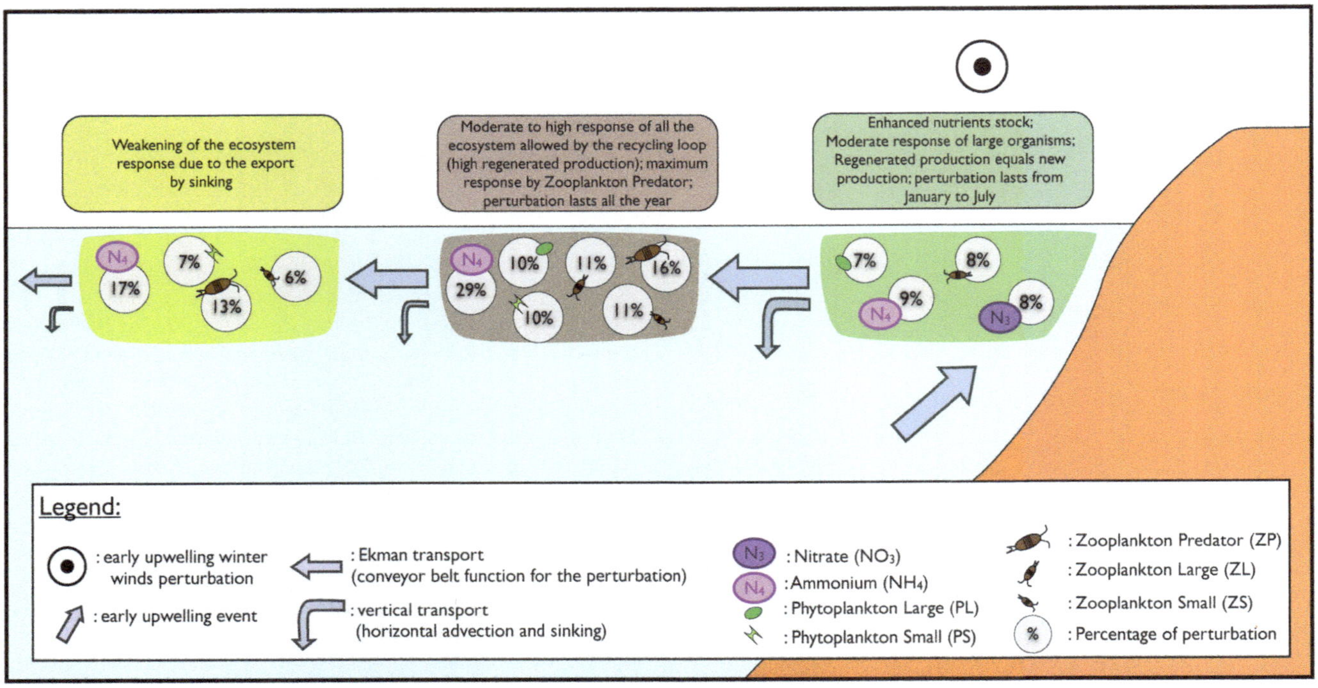

Figure 11. Schematic view of the upwelling ecosystem response to a perturbation that consists in early coastal upwelling winter-winds. This sketch illustrates the simultaneous propagation of the coastal upwelling/nitrogen perturbation cross-shore in space and bottom-up across the trophic chain. Relative changes in physical and biological processes affected by the perturbation are presented. 1) In the nearshore region (green box) the response is moderate and characterized by an increase of upwelled nitrates input and a modest increase of large organisms' biomass (mainly large phytoplankton PL and large zooplankton ZL). Moreover, in this region, the perturbation is no more detectable after July. 2) In the Offshore region (grey box), the response to the perturbation is larger and visible at all trophic levels and most importantly at the highest level (predator zooplankton ZP). This is induced by a wind-driven cross-shore transport of the coastal perturbation that recycles en route. We note a high perturbation of ammonium concentrations and the activation of the recycling loop. In this region, the perturbation lasts much longer and persists all the year. 3) In the far offshore region (yellow box), the perturbation is still non negligible with a main signature at the highest trophic level, despite a significant weakening of the ecosystem response because of vertical export by sinking.

new primary production by as much as 70% in a region located 100 to 500 km off central California. Hence, better-resolved eddies would presumably further enhance the offshore propagation of the perturbation. In contrast, a more accurate representation of submesoscale frontal processes would lead to an increased nutrient export away from the euphotic zone that would probably attenuate the ecosystem perturbation more rapidly [52,53]. A 1 km horizontal resolution simulation would be required to include both of these competing effects and determine their relative importance in the context of winter-wind perturbation propagation. Forcing variability at synoptic scales would also be important in that case; unfortunately, such forcing is not yet available. Lastly the ecosystem model used in this study included only minimal ecosystem structure. Hence, it provides a simplified view of the highly complex local ecosystem: it ignores size-structured relationships among organisms, size distribution of particles and aggregation dynamics known to play an important role in export production. It also includes a very simplified parameterization of higher trophic level predation and ignores iron limitation.

Higher trophic levels than those explicitly represented here tend to have marked preferences for the cross-shore location of their habitat. These preferences are seen as a tradeoff between food availability, optimal temperature for growth, predation level, and possibly other factors. More precisely, in the present system, food availability is higher nearshore, optimal temperatures for growth – while species-dependent – tend to be found offshore, and predation mortality (in particular at the larval stage) is higher in the macrozooplankton-rich nearshore waters. Several key CCS species such as the Pacific Sardine are found some distance away from the more eutrophic coastal zone. Our study suggests enhanced sensitivity of the offshore zone to winter wind anomalies as a possible new element to the habitat selection trade-off, with a late onset of upwelling having long-lasting detrimental effects. This would need to be tested with observations in which, unfortunately, other processes complicate the ecosystem response to a lag or lead in the upwelling onset timing. The late upwelling observed in 2005 (corresponding to a strong negative NPGO index) had profound consequences for the ecosystem [7]. In particular, off Oregon, *Mackas et al.* [54] reported nearshore zooplankton biomass anomalies persisting many months after the winds returned to normal. They hypothesized that the seasonal life history strategies of zooplankton may have affected their response to wind anomalies. A Lagrangian individual-based submodel would be needed to take zooplankton behavioral traits into account. It would also permit more elaborate analyses of nearshore versus offshore ecosystem sensitivity, through biogeochemical budgets along water parcel or animal trajectories.

Supporting Information

Appendix S1 In this appendix we give a detailed description of the cross-shore transport. We quantify the eddy and mean flow contributions to the advective flux divergence in the tracer budgets: for nitrate concentration and for biomass.

Acknowledgments

The altimeter products were produced by Ssalto/Duacs and distributed by Aviso, with support from CNES (http://www.aviso.oceanobs.com/duacs/). Numerical simulations were performed on the computers of the "Pôle de Calcul Intensif pour la Mer", Ifremer. We also thank Camille Mazoyer for her help in implementing the NEMURO module into ROMS-AGRIF. Authors acknowledge the INSU-CNRS CYBER-LEFE program through the TWISTED and TANGGO actions, the French Ministry of Research, the Université Européenne de Bretagne (UEB), and NSF CCE-LTER.

Author Contributions

Conceived and designed the experiments: FC PR XC PJSF BB. Performed the experiments: FC. Analyzed the data: FC PR XC. Contributed reagents/materials/analysis tools: FC PR XC. Wrote the paper: FC PR XC PJSF BB.

References

1. Roemmich D, McGowan J (1995) Climatic warming and the decline of zooplankton in the California current. Science 267: 1324–1326.
2. Hare SR, Mantua NJ, Francis RC (1999) Inverse production regimes: Alaska and West Coast Pacific salmon. Fisheries 24 (1): 6–14.
3. Chavez FP, Ryan J, Lluch-Cota SE, Ñiquen M (2003) From anchovies to sardines and back: multidecadal change in the Pacific Ocean. Science 299: 217–221.
4. Legaard KR, Thomas AC (2006) Spatial patterns in seasonal and interannual variability of chlorophyll and sea surface temperature in the California Current. J Geophys Res 111: C06032.
5. Di Lorenzo E, Schneider N, Cobb KM, Franks PJS, Chhak K, et al. (2008) North Pacific Gyre Oscillation links ocean climate and ecosystem change. Geophys Res Lett 35 (8): 1–6.
6. Di Lorenzo E, Fiechter J, Schneider N, Bracco A, Miller AJ, et al. (2009) Nutrient and salinity decadal variations in the central and eastern North Pacific. Geophys Res Lett 36 (14): 2003–2008.
7. Sydeman WJ, Thompson SA (2010) The California Current Integrated Ecosystem Assessment (IEA), Module II: Trends and Variability in Climate-Ecosystem State. NOAA Tech. Rep., NMFS SWFSC Environmental Research Division, (January), 1–59.
8. Chenillat F, Rivière P, Capet X, Di Lorenzo E, Blanke B (2012) North Pacific Gyre Oscillation modulates seasonal timing and ecosystem functioning in the California Current upwelling system, Geophys Res Lett 39: L01606.
9. MacIsaac JJ, Dugdale RC, Barber RT, Blasco D (1985) Primary production cycle in an upwelling center. Deep-Sea Res II 32(5A): 503–529.
10. Menge BA, Daley BA, Wheeler PA, Strub PT (1997) Rocky intertidal oceanography : An association between community structure and nearshore phytoplankton concentration. Limonol Oceanogr 42 (1): 57–66.
11. Snyder MA, Sloan LC, Diffenbaugh NS, Bell JL (2003) Future climate change and upwelling in the California Current. Geophys Res Lett 30 (15): 1–4.
12. Barth JA, Menge BA, Lubchenco J, Chan F, Bane JM, et al. (2007) Delayed upwelling alters nearshore coastal ocean ecosystems in the northern California current. Proc Natl Acad Sci USA 104 (10): 3719–24.
13. Dorman JG, Powell TM, Sydeman WJ, Bograd SJ (2011) Advection and starvation cause krill (*Euphausiapacifica*) decreases in 2005 Northern California coastal populations : Implications from a model study. Geophys Res Lett 38 (4): 1–5.
14. Ji R, Edwards M, Mackas DL, Runge JA, Thomas AC (2010) Marine plankton phenology and life history in a changing climate : current research and future directions. J Plankton Res 32 (10): 1355–1368.
15. Landry MR (1977) A review of important concepts in the trophic organization of pelagic ecosystems.Helgoländer Wissenschaftliche Meeresuntersuchungen 30 (1–4): 8–17.
16. Botsford LW, Lawrence CA, Dever EP, Hastings A, Largier J (2006) Effects of variable winds on biological productivity on continental shelves in coastal upwelling systems. Deep-Sea Res II 53: 3116–3140.
17. Shchepetkin AF, McWilliams JC (2005) The regional oceanic modeling system (ROMS) : a split-explicit, free-surface, topography following- coordinate oceanic model. Ocean Modell 9 (4): 347–404.
18. Marchesiello P, McWilliams JC, Shchepetkin AF (2003) Equilibrium Structure and Dynamics of the California Current System. J Phys Oceanogr 33 (4): 753–783.
19. Di Lorenzo E (2003) Seasonal dynamics of the surface circulation in the Southern California Current System. Deep-Sea Res II 50: 2371–2388.
20. Capet X, Marchesiello P, McWilliams JC (2004) Upwelling response to coastal wind profiles.Geophys Res Lett 31 (13): 1–4.
21. Di Lorenzo E, Miller AJ, Schneider N, McWilliams JC (2005) The Warming of the California Current System: Dynamics and Ecosystem Implications. J Phys Oceanogr 35 (3): 336–362.
22. Capet X, Colas F, McWilliams JC, Penven P, Marchesiello P (2008) Eddies in eastern boundary subtropical upwelling systems, in Ocean Modeling in an Eddying Regime. Geophys Monogr Ser, vol. 177, edited by M. W. Hecht and H. Hasumi, pp. 131–147, AGU, Washington, D. C.
23. Penven P, Marchesiello P, Debreu L, Lefèvre J (2007) Software tools for pre- and post-processing of oceanic regional simulations. Environmental Modelling & Software 23 (5): 660–662.
24. Woodson CB, Washburn L, Barth JA, Hoover DJ, Kirincich AR, et al. (2009) Northern Monterey Bay upwelling shadow front: Observations of a coastally and surface-trapped buoyant plume. J Geophys Res: Oceans 114 (12): 2156–2202.
25. Castelao RM, Barth JM (2007) The Role of Wind Stress Curl in Jet Separation at a Cape. J Phys Oceanogr 37 (11): 2652–2671.
26. Schwing FB, O'Farrell M, Steger JM, Baltz K (1996) Coastal Upwelling indices west coast of North America. NOAA Tech. Rep., NMFS SWFSC NMFS SWFSC 231: 144p. NOAA, Seatlle, Wash.
27. Bakun A (1973) Coastal upwelling indices, west coast of North America, 1946–71. U.S. Dep. Commer., NOAA Tech. Rep., NMFS SSRF-671, 103 p, NOAA, Seatlle, Wash.
28. Kishi M, Kashiwai M, Ware D, Megrey B, Eslinger D, et al. (2007) NEMURO - a lower trophic level model for the North Pacific marine ecosystem. Ecol Modell 202 (1–2): 12–25.
29. Rose K, Megrey B, Werner F, Ware D (2007) Calibration of the NEMURO nutrient–phytoplankton–zooplankton food web model to a coastal ecosystem: Evaluation of an automated calibration approach. Ecol Modell 202 (1–2): 38–51.
30. Wainwright T, Feinberg L, Hooff R, Peterson WT (2007) A comparison of two lower trophic models for the California Current System. Ecol Modell 202 (1–2): 120–131.
31. Li QP, Franks PJS, Landry MR, Goericke R, Taylor AG (2010) Modeling phytoplankton growth rates and chlorophyll to carbon ratios in California coastal and pelagic ecosystems. J Geophys Res 115 (G4): 1–12.
32. Li QP, Franks PJS, Landry MR (2011) Microzooplankton grazing dynamics : parameterizing grazing models with diluation experiment data from the California Current Ecosystem. Mar Ecol Prog Ser 438: 59–69.
33. Powell TM, Lewis CVW, Curchitser E, Haidvogel DB, Hermann AJ, et al. (2006) Results from a three-dimensional, nested biologicalphysical model of the California Current System and comparisons with statistics from satellite imagery. J Geophys Res 111 (C7): 1–14.
34. Chai F, Dugdale R, Peng T, Wilkerson F, Barber R (2002) One dimensional ecosystem model of the equatorial Pacific upwelling system. Part I : model development and silicon and nitrogen cycle. Deep-Sea Res II 49 (13–14): 2713–2745.
35. Gruber N, Frenzel H, Doney SC, Marchesiello P, McWilliams JC, et al. (2006) Eddy-resolving simulation of plankton ecosystem dynamics in the California Current System. Deep-Sea Res I 53 (9): 1483–1516.
36. Morozov AY (2010) Emergence of Holling type III zooplankton functional response: bringing together field evidence and mathematical modeling. Journal of theoretical biology 265 (1): 45–54.
37. Gentleman W, Leising A, Frost B, Strom S, Murray J (2003) Functional responses for zooplankton feeding on multiple resources : a review of assumptions and biological dynamics. Deep-Sea Res II 50 (22–26): 2847–2875.
38. Masumoto Y, Sasaki H, Kagimoto T, Komori N, Ishida A, et al. (2004) A fifty-year eddy-resolving simulation of the world ocean : Preliminary outcomes of OFES (OGCM for the Earth Simulator). Journal of Earth Simulator 1 (April): 35–56.
39. Sasaki H, Sasai Y, Kawahara M, Furuichi F, Araki A, et al. (2004) A series of eddy -resolving ocean simulations in the world ocean : OFES (OGCM for the Earth Simulator) project. OCEAN'04, pp. 1535–1541.
40. Sasaki H, Nonaka M, Masumoto Y, Sasai Y, Uehara H, et al. (2006) An eddy-resolving hindcast simulation of the quasi-global ocean from 1950 to 2003 on the Earth Simulator. High Resolution Numerical Modelling of the Atmosphere and Ocean, edited by W. Ohfuchi and K. Hamilton, springer ed., New York.
41. Chavez F (1989) Size distribution of phytoplankton in the central and eastern tropical Pacific. Global Biogeochem, Cycles 3 (1): 27–35.
42. Landry MR, Brown SL, Neveux J, Dupouy C, Blanchot J, et al. (2003) Phytoplankton growth and microzooplankton grazing in high-nutrient, low-chlorophyll waters of the equatorial Pacific: Community and taxon-specific rate assessments from pigment and flow cytometric analyses. J Geophys Res 108(C12): 8142.
43. Garcia HE, Locarnini RA, Boyer TP, Antonov JI (2006) World Ocean Atlas 2005. Vol. 4, Nutrients (phosphate, nitrate, silicate).
44. Capet X, McWilliams JC, Molemaker MJ, Shchepetkin AF (2008) Mesoscale to submesoscale transition in the California Current System. Part I: Flow structure, eddy flux, and observational tests. J Phys Oceanogr 38: 29–43.
45. Capet X, McWilliams JC, Molemaker MJ, Shchepetkin AF (2008) Mesoscale to submesoscale transition in the California Current System. Part II: Frontal processes. J Phys Oceanogr 38: 44–64.
46. Vignudelli S, Kostianoy AG, Cipollini P, Benveniste J (2011) Coastal Altimetry, Springer-Verlag, Berlin, 578 pp.

47. Cloern JE, Grenz C, Vidergar-Lucas L (1995) An empirical model of the phytoplankton chlorophyll : carbon ratio-the conversion factor between productivity and growth rate. Limnol Oceanogr 40 (7): 1313–1321.
48. Kruskal WH, Wallis WA (1972) Use of ranks in one-criterion variance analysis. J Am Stat Assoc 47: 583–621.
49. Philander SG, Yoon JH (1982) Eastern boundary currents and coastal upwelling. J Phys Oceanogr 12: 862–879.
50. Moriceau B, Gallinari M, Soetaert K (2007) Importance of particle formation to reconstructed water column biogenic silica fluxes. Global Biogeochem Cycles 21: GB3012.
51. Koné V, Machu E, Penven P, Andersen V, Garçon V, et al. (2005) Modeling the primary and secondary productions of the southern Benguela upwelling system: A comparative study through two biogeochemical models. Global Biogeochem Cycles 19: GB4021.
52. Gruber N, Lachkar Z, Frenzel H, Marchesiello P, Münnich M, et al. (2011) Eddy-induced reduction of biological production in eastern boundary upwelling systems, Nature Geosci 4, 11: 787–792.
53. Lathuilière C, Echevin V, Lévy M, Madec G (2010) On the role of mesoscale circulation on an idealized coastal upwelling ecosystem. J Geophys Res, 115: C09018.
54. Mackas DL, Peterson WT, Ohman MD, Lavaniegos BE (2006) Zooplankton anomalies in the California Current system before and during the warm ocean conditions of 2005. Geophys Res Lett 33: L22S07.

17

Climate and Land Use Controls on Soil Organic Carbon in the Loess Plateau Region of China

Yaai Dang[1,2,3], Wei Ren[2], Bo Tao[2], Guangsheng Chen[2], Chaoqun Lu[2], Jia Yang[2], Shufen Pan[2], Guodong Wang[3], Shiqing Li[1], Hanqin Tian[2]*

1 State Key Laboratory of Soil Erosion and Dryland Farming on the Loess Plateau, Institute of Soil and Water Conservation, Northwest A&F University, Yangling, Shaanxi, China, **2** International Center for Climate and Global Change Research, School of Forestry & Wildlife Sciences, Auburn University, Auburn, Alabama, United States of America, **3** College of Science, Northwest A&F University, Yangling, Shaanxi, China

Abstract

The Loess Plateau of China has the highest soil erosion rate in the world where billion tons of soil is annually washed into Yellow River. In recent decades this region has experienced significant climate change and policy-driven land conversion. However, it has not yet been well investigated how these changes in climate and land use have affected soil organic carbon (SOC) storage on the Loess Plateau. By using the Dynamic Land Ecosystem Model (DLEM), we quantified the effects of climate and land use on SOC storage on the Loess Plateau in the context of multiple environmental factors during the period of 1961–2005. Our results show that SOC storage increased by 0.27 Pg C on the Loess Plateau as a result of multiple environmental factors during the study period. About 55% (0.14 Pg C) of the SOC increase was caused by land conversion from cropland to grassland/forest owing to the government efforts to reduce soil erosion and improve the ecological conditions in the region. Historical climate change reduced SOC by 0.05 Pg C (approximately 19% of the total change) primarily due to a significant climate warming and a slight reduction in precipitation. Our results imply that the implementation of "Grain for Green" policy may effectively enhance regional soil carbon storage and hence starve off further soil erosion on the Loess Plateau.

Editor: Xiujun Wang, University of Maryland, United States of America

Funding: This study was supported by NASA Land Cover and Land Use Change Program (NNX08AL73G), NASA Interdisciplinary Science Program (NNG04GM39C), Chinese Universities Scientific Fund (z10921007), US National Science Foundation Grants (AGS-1243220, CNS-1059376), State Key Laboratory of Soil Erosion and Dryland Farming on the Loess Plateau Foundation (K318009902-1410), and Shaanxi Administration of Foreign Expert Affairs Science and Technology Activities Fundation (201327). Study design was supported by NASA Land Cover and Land Use Change Program (NNX08AL73G), NASA Interdisciplinary Science Program (NNG04GM39C), and Chinese Universities Scientific Fund (z10921007). Data collection and analysis was supported by NASA Land Cover and Land Use Change Program (NNX08AL73G), NASA Interdisciplinary Science Program (NNG04GM39C), and State Key Laboratory of Soil Erosion and Dryland Farming on the Loess Plateau Foundation (K318009902-1410). Preparation of the manuscript was supported by NASA Land Cover and Land Use Change Program (NNX08AL73G), US National Science Foundation Grants (AGS-1243220, CNS-1059376), and State Key Laboratory of Soil Erosion and Dryland Farming on the Loess Plateau Foundation (K318009902-1410).

Competing Interests: The authors have declared that no competing interests exist.

* E-mail: tianhan@auburn.edu

These authors contributed equally to this work.

Introduction

Soil organic carbon (SOC), the major component of soil organic matter, plays a key role in the terrestrial carbon cycle and thus has drawn great attention from scientific community. It is a dynamic component of terrestrial systems, affecting carbon exchange between terrestrial ecosystem and the atmosphere [1,2]. SOC storage is nearly three times as large as carbon storage in vegetation and twice as large as global atmospheric carbon storage [3]. Soil has higher potential to sequester more carbon (such as converting the type of land use) in the future [2,4], therefore, increasing soil carbon storage is one of the most economical and effective ways to alleviate the greenhouse effect, which has become a hot scientific and political issue during the past decades.

Changes in climate and land use, caused by both natural and anthropogenic processes, have greatly influenced the terrestrial carbon balance during the past decades [5,6,7,8]. It was reported that about one fourth of anthropogenic CO_2 emissions were due to land cover and land use change (LCLUC), especially deforestation [9]. Long-term experimental studies have confirmed that SOC is highly sensitive to land conversion from natural ecosystems, such as forest or grassland, to agricultural land, resulting in substantial SOC loss [6,10]. In addition, LCLUC may also cause carbon depletion by influencing soil respiration [11]. It was estimated that global carbon release from SOC mineralization owing to agricultural activities was approximately 0.80 Pg C/year (1 Pg = 10^{15} g) [12]. Globally, land use change resulted in a carbon release of (1.6±0.8) Pg C per year to the atmosphere during the period of 1990s [13]. However, the effects of conversions from cropland to grassland/forest on the SOC storage have not been fully understood and there still remains large uncertainty.

The Loess Plateau of China (Figure 1), located in the geographic center of China (33°43′N 100°54′E to 41°16′N 114°33′E), covers a total area of 628,000 km^2, which is about 6.5% of China's total land area. The Loess Plateau is characterized by highly erodible soils, steep slopes, being subjected to heavy rain, and low vegetation coverage due to excess exploitation of land resource and improper land use [4,14]. During the past decades, serious soil erosions caused by natural and anthropogenic disturbances (e.g., climate change, natural disasters, LCLUC etc.) occurred in the area of the Loess Plateau. Previous reports also indicated that the warming and drying climate in this region has significantly aggravated soil erosion [15]. As a result, the large amount of fine surface soil eroded from the loess area is transported into the Yellow River and acts as the main source of sediment of this river, which runs through the Loess Plateau and is considered to be the most turbid river in the world. Due to these disturbances, soil carbon storage on the Loess Plateau is much lower compared to other regions in China [16]. In general, adjusting the land use pattern so as to restore the degraded ecosystems and to modify the local rural income structure is regarded as the main measures to control soil and water erosion and to improve farmers' living conditions on the Loess Plateau. Since the 1950s, a series of conservation policies have been implemented in this region, such as extensive tree planting since the 1970s, integrated soil erosion controls on the watershed scale in the 1980s and the 1990s [17,18], and the government-funded project "Grain for Green" in 1999, aiming at transforming the low-yield slope cropland into grassland/forest. The implementation of these policies improved vegetation coverage, altered land use patterns, and changed the SOC storage. Although many field experiments have been performed to explore the impacts of both drying and warming climate and LCLUC on soil carbon storage on the Loess Plateau [19,20], little attention has been paid to the regional impacts of these factors and their interactions.

Over recent decades, many field observations and control experiments have been conducted to explore the effects of climate and land use change on the SOC in this region and make it possible to study the regional effects of climate change and LCLUC on SOC. In addition, many approaches, including eddy covariance flux tower, inventory, remote sensing techniques, forward and inversion models, have been used to examine the regional carbon budget on the Loess Plateau [21,22,23,24,25]. Among them, process-based ecosystem modeling is one of the most effective approaches to estimate regional SOC storage and fluxes in different terrestrial ecosystems driven by multiple global changes factors [7,11,24,26,27]. To address the effects of changes in climate, land use, and other environmental factors on SOC storage in this region, the Dynamic Land Ecosystem Model (DLEM), a highly integrated process-based model [28], was applied to evaluate the spatial and temporal patterns of SOC storage on the Loess Plateau during 1961–2005. The objectives of this study are: 1) to investigate the temporal and spatial patterns of SOC storage on the Loess Plateau; and 2) to identify the relative contribution of climate and land use changes to the SOC storage changes.

Methods

Model Description

The DLEM is a process-based terrestrial ecosystem model, which aims at simulating the impacts of natural and anthropogenic disturbances on the structure and functions of terrestrial ecosystems over the spatial and temporal contexts. The DLEM has been widely used to simulate the effects of climate variability and change, elevated atmospheric CO_2, tropospheric ozone pollution, land use change, and increasing nitrogen deposition, etc. on terrestrial carbon storage and fluxes in China and other regions across the globe [15,29,30,31,32,33].

In this study, DLEM simulates two kinds of LCLUC: land conversion from natural ecosystem to cropland, and cropland abandonment. In the DLEM model, the balance of soil organic matter depends on the transformation of litter (LIT) to soil organic matter, the fractions of conversion from gross primary production (GPP) to dissolved organic carbon (DOC), the returned organic matter from production decay (PRD) (e.g., manure), the growth of microbe, the methane production from dissolved organic carbon, and the carbon loss from soil organic matter (SOM) decomposition.

$$\frac{dC_{som}}{dt} = k_{tr}LITC_{loss} + k_{gppdoc}GPP + k_{prd}PRD_{docom} - k_{rh}SOMC_{docom} - k_{Lucc}C_{som} - DOC_{loss,methane}$$

where k_{tr} is the transfer rate of decomposed LIT to SOM; k_{gppdoc} is the fraction of GPP converted to soil DOC; k_{prd} is the returned rate of decomposed (or consumed) PRD to SOM pools as manure; k_{rh} is the fraction of decomposed SOM that is converted to CO_2 through heterotrophic respiration; k_{Lucc} is coefficient for quick carbon loss from SOM due to land use conversion; $DOC_{loss,methane}$ is DOC consumed for the growth of production of methane. More

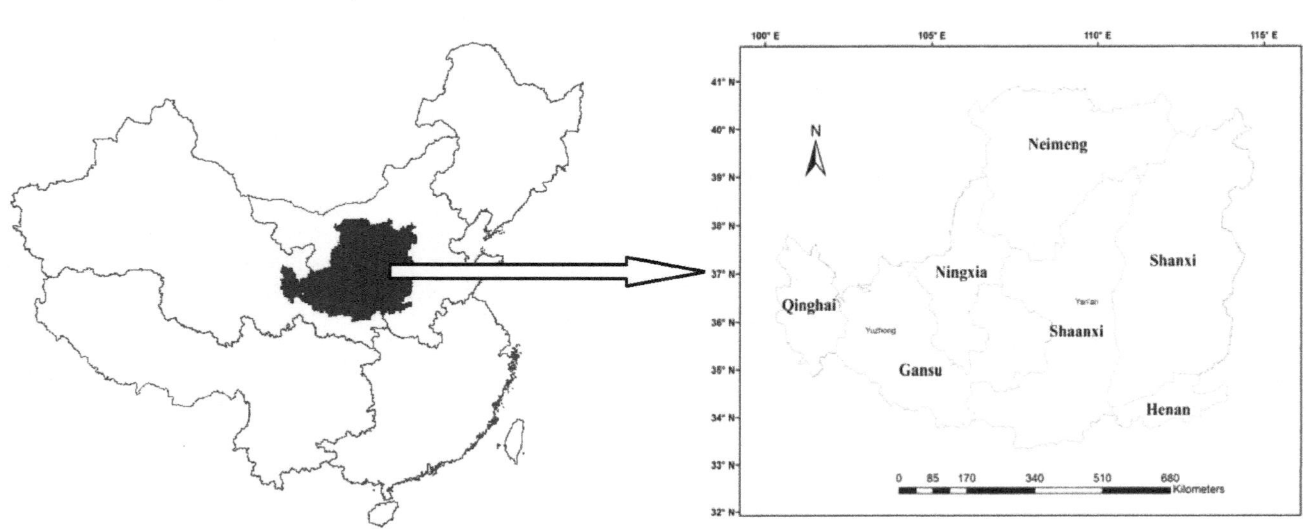

Figure 1. Location of the Loess Plateau, China.

detailed processes were described in our previous papers [15,29,30].

Input Data Description

The major input data in the DLEM include: (1) daily climatic data (i.e. maximum, minimum, average temperature, precipitation, relative humidity, and radiation) and atmospheric chemistry data (i.e. tropospheric O_3, atmospheric CO_2 and nitrogen deposition); (2) soil properties (including soil type, bulk density, depth, pH, soil texture) which are derived from the 1:1 million soil map based on the Second National Soil Survey of China [34,35,36]; (3) contemporary vegetation map for 2000 which was developed from Landsat Enhanced Thematic Mapper (ETM) imagery [37]; (4) long-term land use history which was developed on the basis of three recent (1990, 1995 and 2000) land cover maps and historical census datasets [38,39]. All the input datasets were developed at the spatial resolution of 10 km×0 km. Detailed information about other input data were described in our previous studies [15,29,30,40].

Climate change. Average air temperature on the Loess Plateau increased at a rate of 0.030°C/year from 1961 to 2005 (Figure 2a), higher than those reported for the entire China (about 0.029°C/year) and the global average level (about 0.010°C/year) [41] in the same period. The most rapid increase in temperature occurred during the 1990s. Air temperature increased from the north to the south of the Loess Plateau (Figure 2b), with the greatest increase in the northern Loess Plateau (e.g., the north of Shanxi and the northwest of Inner Mongolia).

The Loess Plateau can be mainly divided into three climate zones according to the precipitation: the northern Loess Plateau with precipitation below 400 mm, the central Loess Plateau with precipitation between 400 and 500 mm, and the southern Loess Plateau with precipitation above 500 mm [42,43]. The precipitation less than 550 mm/year occurred across most areas of the Loess Plateau. Over the past 45 years, the mean annual precipitation was approximately 423 mm, with the lowest of 288 mm in 1997 and the highest of 661 mm in 1964 (Figure 2c). A slightly decreasing trend at a rate of 1.27 mm/year in precipitation was found on the Loess Plateau from 1961 to 2005. This temporal trend of precipitation was consistent with a previous study based on the meteorological observations according to 99 stations for the period 1956–2005 on the Loess Plateau [44]. Figure 2d further indicated that decreases in precipitation occurred in the most areas of the central and southern Loess Plateau from 1961–1990 to 1991–2005. The regions with the most obvious drying trend were located in the central and southern Loess Plateau, especially in the north of Shaanxi and the center of Shanxi Province with a reduction of more than 80 mm in the recent 15 years (1991–2005) compared to the 1961–1990 average.

LCLUC. Expansion of cropland and pasture was driven by social-economic factors on the Loess Plateau during the past decades, though to some extent, the soil erosion and frequent drought events limited the massive expansion of cropland. Since the 1950s, many measures have been implemented to alleviate and control serious soil erosion on the Loess Plateau. For example, "Grain for Green" project, which was launched in 1999, recommended that cropland with slopes greater than 15° should be converted back to natural vegetation. Driven by such governmental policies, the Loess Plateau experienced a remarkable change in land use, characterizing by a large area of

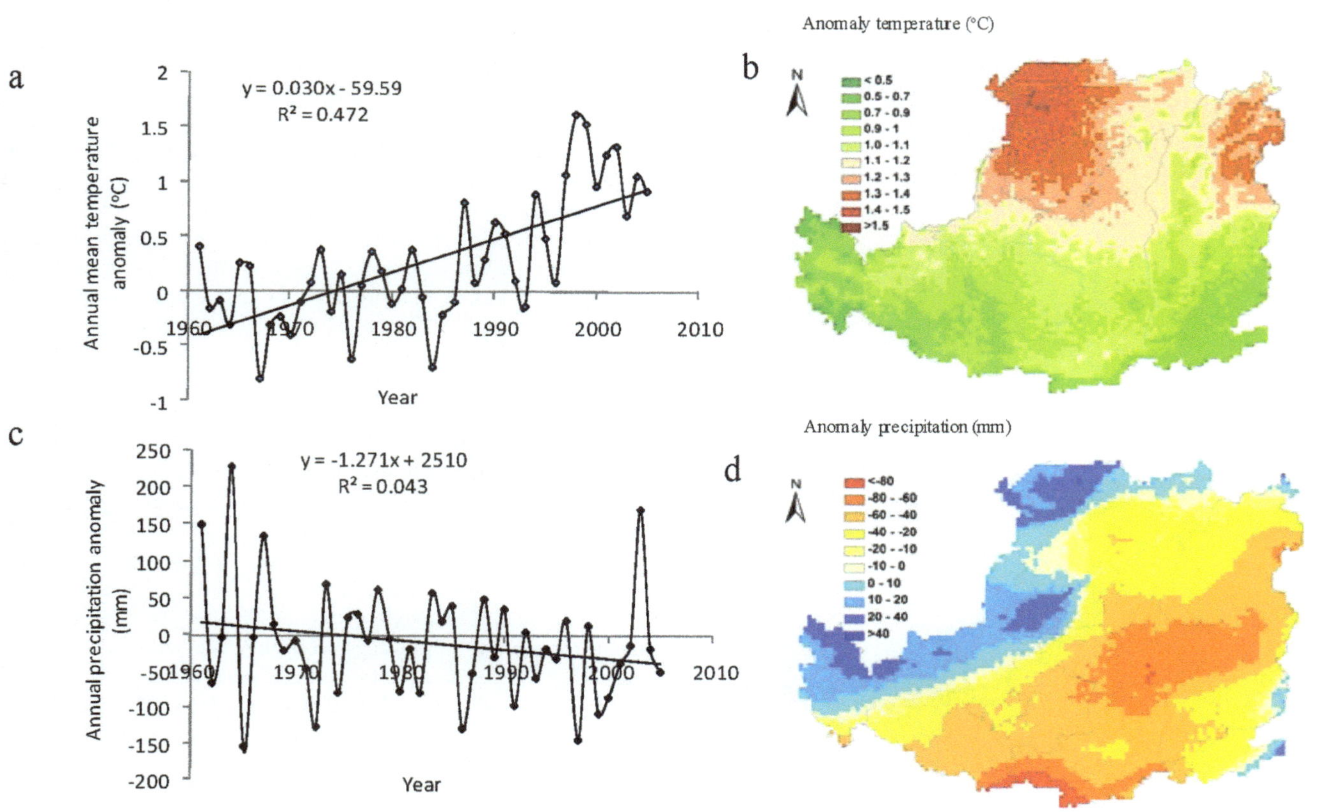

Figure 2. Anomaly of annual mean temperature (a), and precipitation (c) on the Loess Plateau from 1961 to 2005; and spatial distribution of temperature anomaly (b) and precipitation anomaly (d) during 1991–2005 (relative to 1961–1990 average);

conversion from cropland to grassland/forest. In order to well understand the LCLUC history on the Loess Plateau, we analyzed the spatial and temporal variations from 1961 to 2005 (Figure 3–4). Figure 3 showed that cropland area on the Loess Plateau decreased slowly until 1999, and followed by a sharp decrease. However, the grassland area showed an opposite changing trend in the same periods. Since 1961, the area of cropland decreased by 19.61%, while grassland and forest increased by 7.31% and 6.75%, respectively.

Land use patterns exhibited large spatial variations on the Loess Plateau (Figure 4). Grassland was the dominant vegetation type in the northern Loess Plateau. Cropland, grassland and shrubland were the main vegetation type in the middle Loess Plateau. In contrast, cropland and forest occupied large area of the southern Loess Plateau. Compared to 1960, the coverage of cropland decreased mainly due to land conversion from cropland to grassland, especially in the middle and southern part of Loess Plateau. During 1999–2005, cropland largely shrank in some areas of the middle Loess Plateau, especially in the northern Shaanxi and middle of Shanxi Province. These results were consistent with previous studies [18,45].

Model Parameterization and Evaluation

DLEM has been well calibrated and intensively validated against the site-level observed carbon fluxes and pool sizes from Chinese Ecosystem Research Network (CERN) and other previous studies [29,38,46]. In this study, we further compared our simulated results with field observation data and survey data from the Second National Soil Survey (1979–1983) to evaluate the model performance in simulating the SOC storage on the Loess Plateau. The simulated SOC storage as influenced by multiple environmental factors significantly correlated with the observed data ($R^2 = 0.738$, $p<0.01$, Figure 5), indicating that DLEM could capture the spatial and temporal patterns of SOC storage on the Loess Plateau. Several factors may contribute to the difference between model results and field observations. First, the input datasets were developed at a spatial resolution of 10 km×10 km for driving DLEM simulations. Each grid was assumed to have the uniform climate, land use type, and vegetation cover in the model. As a well-known climate-sensitive zone and a fragile ecological belt, some subtle changes in climate, land use or other factors on the Loess Plateau might cause large difference in the SOC storage even within one grid cell. This difference in the same grid will be reflected in the field experiments but neglected in the model simulations. Second, the shortage of field observation data may weaken the capability of model to realistically capture the magnitude and patterns in SOC changes, which has long been identified as one of the biases in the large-scale model development. In addition, model simplification and neglecting microbial biomass might be another potential reason [13,30].

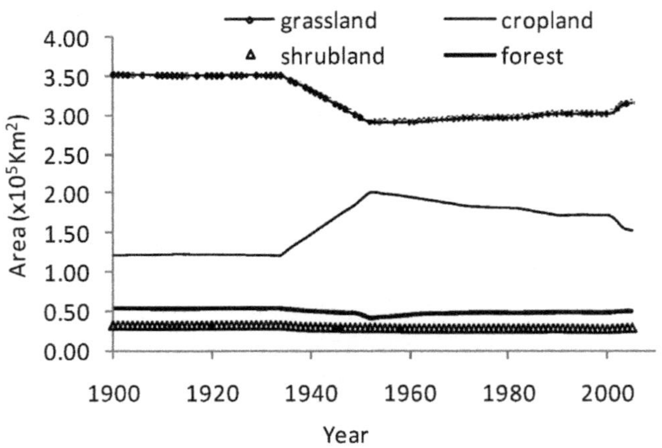

Figure 3. Area of major land use cover on the Loess Plateau during 1900–2005.

Simulation Experimental Design and Model Run

In this study, four main simulation experiments were designed to analyze the effects of climate change alone, LCLUC alone, the interaction of climate and LCLUC, and the combined effects of all environmental factors on the SOC storage on the Loess Plateau (Table 1). The experiment I was designed to provide a 'best-estimate' of spatial and temporal patterns of SOC driven by major environmental factors changes including climate variability, LCLUC, elevated CO_2, N fertilizer, N deposition, and O_3 pollution, etc. In experiment II and III, we simulated the contributions of climate variability alone and the LCLUC alone, respectively. In the experiment IV, we tried to understand the interactive effect between climate change and LCLUC on the SOC storage. The model simulations began with an equilibrium run to obtain the baseline carbon pools for each grid. A spin-up of about 100 years was applied if the climate change was included in the simulation experiments. Finally, the model was run in transient mode driven by the daily climate data and other time-variant or invariant input data.

Results and Analysis

Temporal Changes in SOC on the Loess Plateau

Model simulation indicated that the SOC storage over the entire Loess Plateau displayed substantial temporal and spatial variations in the context of multiple environmental factors changes (Figure 6a, b). As a whole, the combination of all these environmental factors considered in this study (e.g. climate, LCLUC and others environment factors) caused a net increase of about 0.27 Pg C in SOC storage from 1961 to 2005. The SOC storage kept relatively stable before the 1970s, and then gradually increased in the following two decades, and rapidly increased since 2000 (Figure 6a). Our further analysis found that the decrease of cropland area was relatively slower from the 1960s to 1999, and became rapid after then. Most of the abandoned cropland were replaced by grassland (Figure 3). Our results implied that temporal pattern in SOC change was partly related to the land use change on the Loess Plateau.

Spatial Variation in SOC on the Loess Plateau

We found that the spatial patterns of SOC storage were primarily controlled by the precipitation distribution. Spatially, the SOC storage increased gradually from the north to the south along an increasing precipitation gradient on the Loess Plateau. Large SOC increases were found in the southern and central Loess Plateau (south of Shaanxi and Shanxi Province in particular); and some other areas show a slight increase of SOC storage in the past decades. Due to changes in multiple environmental factors, the SOC storage increased throughout the majority of the Loess Plateau over the past 45 years (Figure 6b). Some areas showed a significant increase in SOC storage of more than 200 g C/m^2. However, a significant decrease occurred in the middle of central Loess Plateau (especially in the Northern Shaanxi Province), where experienced the most obvious warming and drying tendency in the past decades (Figure 2 b,d), releasing more than 100 g C/m^2 during the study period. We further found that the SOC increase

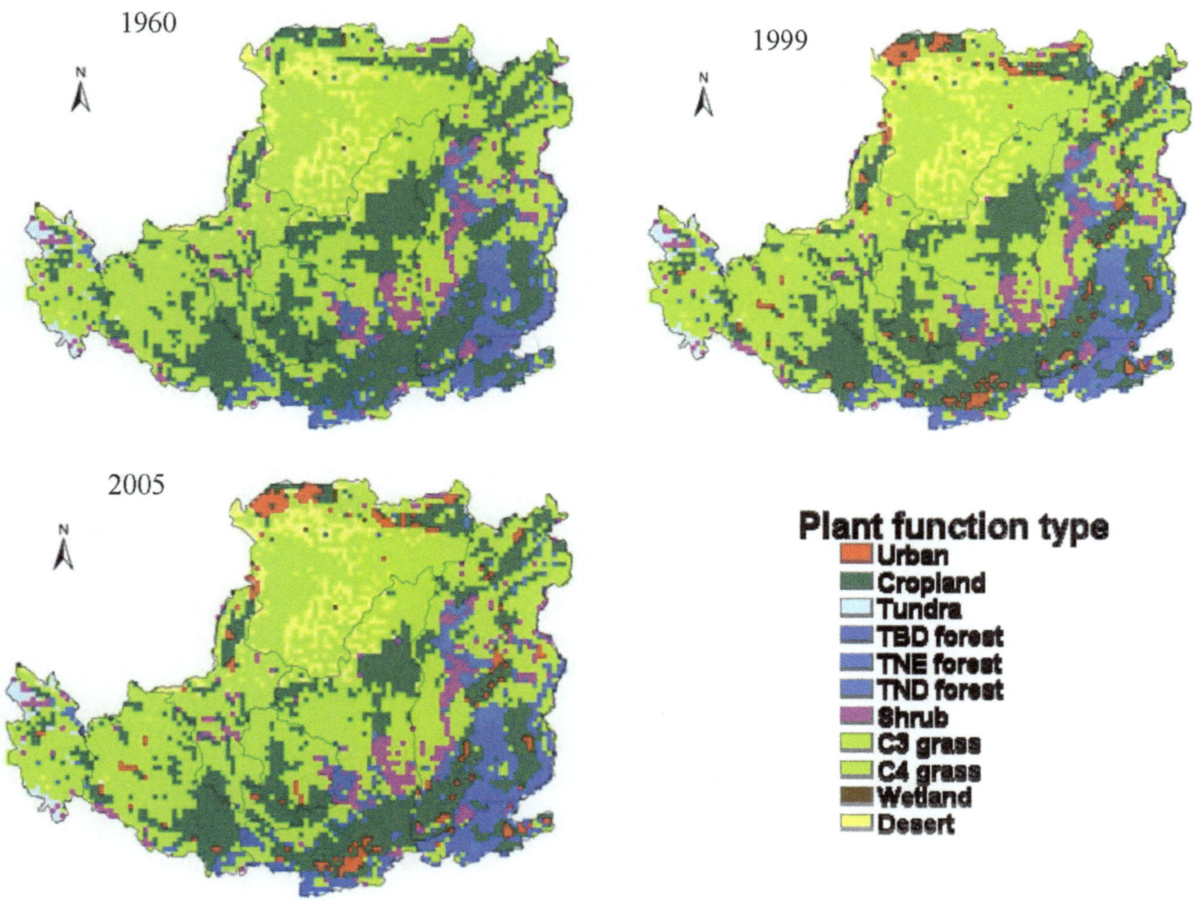

Figure 4. Spatial variations in LCLUC on the Loess Plateau in different year.

in the northern Loess Plateau was lower than that in the southern Loess Plateau during the past decades.

Individual Factorial Contributions to Changes in SOC Storage

To well understand the influence of climate on the SOC storage, we simulated the SOC storage on the Loess Plateau under the influence of climate change alone. DLEM simulated results

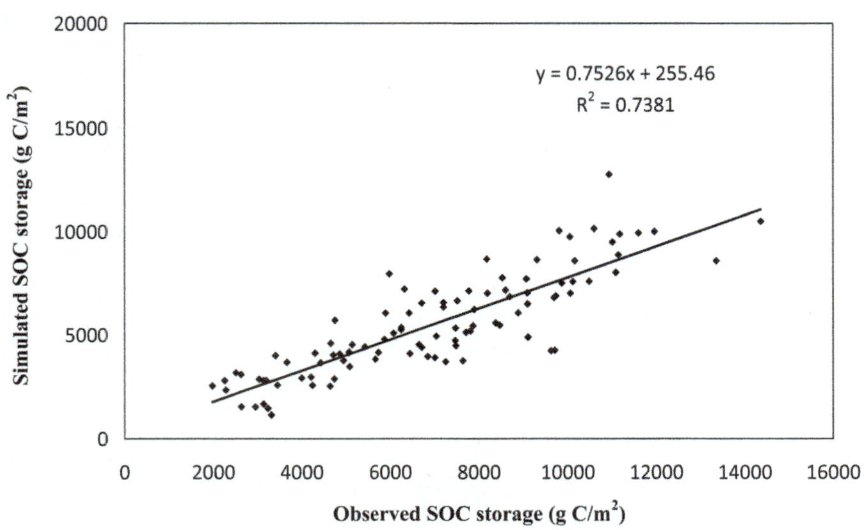

Figure 5. Correlation between simulated and observed SOC storage based on 91 soil samples.

Table 1. Simulation experiments.

Simulation Experiment	Environmental factors		
	Climate	LCLUC	Others
I. All	1960–2005	1960–2005	1960–2005
II. Climate only	1960–2005	1960	1960
III. LCLUC only	1960	1960–2005	1960
IV. Climate-LCLUC	1960–2005	1960–2005	1960

Notes: Climate-LCLUC means the combination effects of climate and LCLUC. Others include the effects of atmospheric CO_2, ozone pollution (AOT 40 index), nitrogen deposition, nitrogen fertilizer application, etc. All means the simulation experiment which includes all above environmental factors.

showed that SOC storage decreased by about 0.05 Pg C with substantial inter-annual fluctuations throughout the Loess Plateau from 1961 to 2005 (Figure 6c). SOC storage showed a slowly decreasing trend before the late 1990s, followed by a sharp decrease until the early 21st century. Since 1961, SOC storage decreased in most areas of the Loess Plateau under the influences of the climate, with a maximum carbon release of 200 g C/m^2 in some areas of the southern and the central Loess Plateau (Figure 6d). Figure 6d also indicated that SOC storage decreased in the southeastern humid monsoon climatic regions, and to a lesser extent in the continental dry climatic regions in the northern Loess Plateau during 1961–2005.

Considering the single effect of LCLUC, our results showed that the SOC storage increased by 0.14 Pg C during 1961–2005 (Figure 6e). We found that the most rapid increase in the SOC storage occurred after the late 1970s (Figure 6e). DLEM simulation results also showed that obvious increases of SOC storage were found in most areas of the central and the southern Loess Plateau (Figure 6f).

Relative Contributions of Climate, LCLUC, and their Interactions

Our simulated results indicated that with multiple environmental changes, SOC continuously increased from 1961 to 2005 on the Loess Plateau, resulting in a net increase in SOC storage by 0.27 Pg C (5.97 Tg C/year) (Figure 7). Among these factors, LCLUC was obviously the major factor affecting the magnitude of SOC change on the Loess Plateau, leading to a significant increase (0.14 Pg C) in SOC storage, accounting for 55% of the net increase in SOC in the past 45 years. However, in the same period, warming and drying climate greatly reduced SOC storage by 0.05 Pg C, approximately 19% of the total SOC storage change. The interaction between climate and LCLUC contributed to the net SOC increases by 3%.

Discussion

SOC Storage Change on the Loess Plateau

In this study, the combination of all environmental factors (climate, LCLUC and others environment factors) caused a net increase of about 0.27 Pg C in SOC storage, indicating that soil acted as a weak carbon sink on the Loess Plateau during the past 45 year. The temporal pattern of SOC storage was largely influenced by land use change and relevant land use policies.

Except the central Loess Plateau, the SOC storage increased throughout the entire region, with a smaller sink in the northern Loess Plateau and a larger sink in the southern Loess Plateau during 1961–2005, which is consistent with previous reports [43,47,48]. The less increase of SOC storage in the northern Loess Plateau might be due to more sandy soil, lower soil carbon input from plant biomass, and larger soil carbon loss from erosion. This is consistent with previous reports, indicating the soil in the northern Loess Plateau contained more sand which could accumulate less carbon than the soil with more clay [49,50,51,52]. On the other hand, the area of forest and shrubland, characterized by higher aboveground biomass and productivity than cropland and grassland [43,53], decreased from the south to north over the Loess Plateau (Figure 4). This also contributed to realtively less SOC increase in the northern Loess Plateau. In addition, the extensive soil erosions in the northern Loess Plateau was suggested to further explain the lower contents of SOC storage [2,17,54].

Comparisons with Field Observations

Our estimation of SOC storage and its change on the Loess Plateau were comparable to other studies. Based on the 0–100 cm SOC data collected from 382 sampling sites across the entire Loess Plateau in 2008, Liu et al. [55] indicated that mean SOC storage was 7.70 kg C/m^2. Using data gathered by the Second National Soil Survey of China (1979–1983), Xu et al. [56] estimated that SOC storage was 1.07 Pg in 0–20 cm soils layers on the Loess Plateau region (with a land area of 429,800 km^2). A soil survey conducted throughout the Loess Plateau region indicated that the SOC storage amounted to 1.23 Pg C in the 0–20 cm soil layers with land area of 592,900 km^2 during 1985–1988, [55]. Converting SOC storage from different soil depth to the same soil depth according to the summary of the vertical distribution of soil organic in the 0–100 cm soil [31], SOC storage were 6.22 kg C/m^2 and 5.19 kg C/m^2 in Xu et al.'s [57] and Liu et al.'s [55] reports (be mentioned but not published) respectively. In the same soil depth, DLEM-estimated average SOC storage was 6.45 kg C/m^2 in 2005, which falls in the range of previous studies.

Tian et al. [7] estimated an average SOC sink of 94 Tg C/year in the terrestrial ecosystem of China as influenced by multiple global change factors during 1961–2005. Our results showed that soil on the Loess Plateau acted as a weak sink of 5.97 Tg C/year under the combined experiment during the same period. Considering that the Loess Plateau accounts for approximately 6.5% of the country's land surface, implying that the SOC storage on the Loess Plateau was close to the country-level average. At national level, Wang et al. [35] suggested that SOC storage in China decreased by about 5.69 g C/m^2 per year during the 1960s–1980s, while our study showed that the SOC storage on the Loess Plateau significantly increased by 4.30 g C/m^2 per year during the same period. This indicated that successful implementation of land conservation measures was beneficial and effective to reduce soil erosion and improve soil properties, thus enhance the soil carbon sequestration over the Loess Plateau.

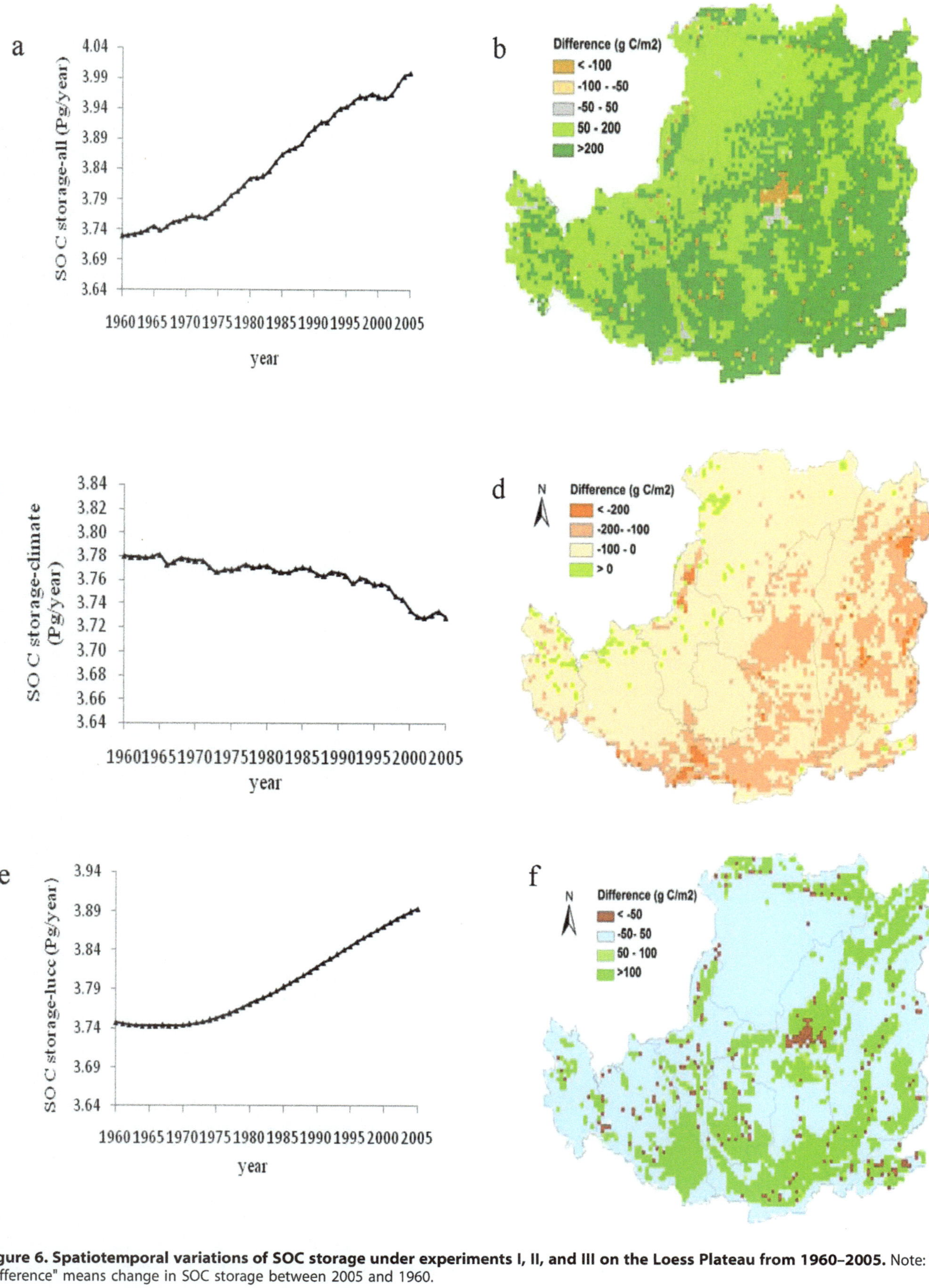

Figure 6. Spatiotemporal variations of SOC storage under experiments I, II, and III on the Loess Plateau from 1960–2005. Note: The "difference" means change in SOC storage between 2005 and 1960.

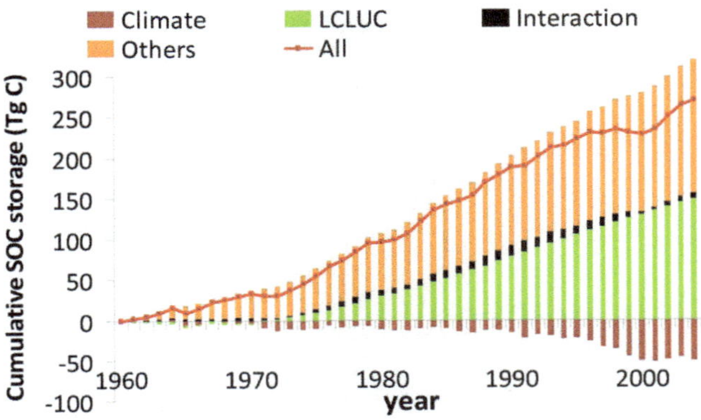

Figure 7. Contributions of multi-factor global changes to accumulative SOC change during 1961–2005. Note: "interaction" referrers to the interactive effects between climate and LCLUC.

Climate Controls on the Spatiotemporal Patterns of SOC Storage

Climatic factors, especially precipitation and temperature, play an important role in long-term variations of SOC due to their effects on the quantity and quality of organic residue inputs and on the rates of soil organic matter and litter decomposition [1,12,58]. DLEM-simulated results showed that SOC storage decreased by about 0.05 Pg C with a significant inter-annual fluctuations throughout the Loess Plateau from 1961 to 2005 as influenced by climate change alone (Figure 6c). As shown in Figure 2, precipitation decreased at a rate of 1.27 (mm/year) during 1961–2005, which was a notable factor that influenced the change of SOC on the Loess Plateau, especially after 1990. Meanwhile, increased air temperature (0.030°C/year) might accelerate the evapotranspiration and potentially aggravate the water deficiency, thus cause the formation of drying soil layer and suppress the growth of vegetation. Due to the warming and drying climate, dried soil layer was widely distributed in the hilly and gully areas of the Loess Plateau. The development of drying soil layer has been regarded as a key cause for the decrease of SOC storage in some areas on the Loess Plateau in the past decades [59].

Climate also had a substantial effect on the spatial distribution of SOC storage. Our results suggested that SOC storage decreased in most areas of the Loess Plateau under the influences of the climate change since 1961 (Figure 6d). This result could be explained partly by the distribution of temperature and precipitation, which is well-known to have a positive relationship with SOC decomposition [60]. During 1961–2005, both temperature and precipitation were higher in the central and southern Loess Plateau, compared to the northern Loess Plateau. Therefore, the higher loss of soil carbon from decomposition might be an important cause for lower SOC accumulation in the central and southern Loess Plateau. This result was supported by previous studies [43,50,57].

Land Use Controls on the Spatiotemporal Patterns of SOC Storage

During the past decades, the Loess Plateau has experienced a complex change in land use pattern (Figure 3). Considering the LCLUC alone, DLEM-simulated results showed that the SOC storage increased by 0.14 Pg C during 1961–2005 (Figure 6e), which was significantly larger than the influence of climate change alone. However, in the first few years, the SOC changed slowly, even had a decreasing tendency in the 1960s, and then increased gradually since the 1980s.

Since the 1950s, various soil and water conservation measures including afforestation, cropland abandonment, and terrace construction etc., have been implemented in this region [17,18]. All of these measures directly or indirectly affected the vegetation cover and further influenced the SOC storage on the Loess Plateau. Previous studies demonstrated that soil can lose up to 20–40 percent of organic carbon into the atmosphere when perennial vegetation land was converted into cultivation land [61,62]. On the contrary, conversion from cropland to perennial vegetation land or shrubland was found to accumulate SOC by increasing carbon derived from new vegetation and decreasing carbon loss from decomposition and erosion [11,43,51,63]. Liu et al. [47] reported that SOC in shrubland, which was converted from

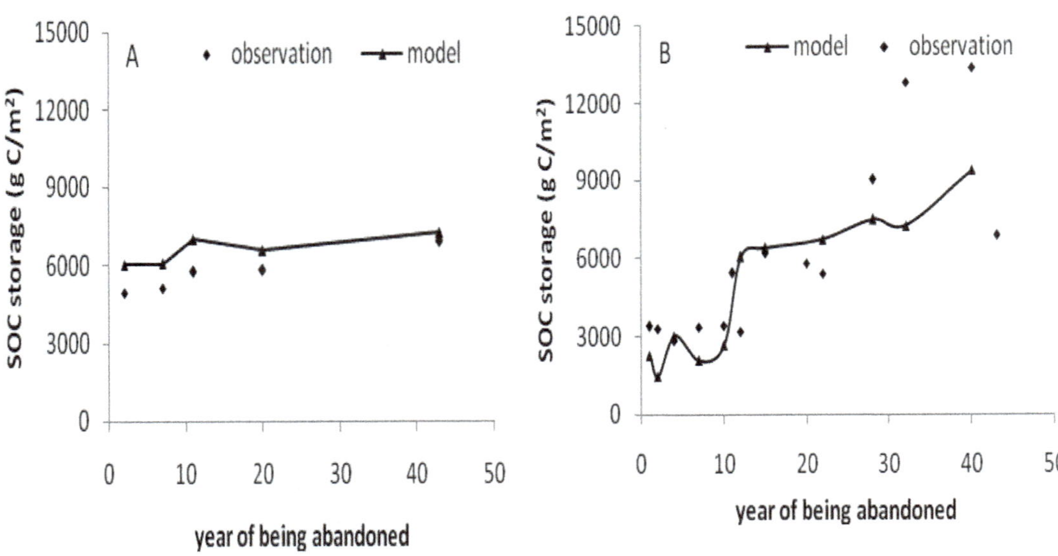

Figure 8. Comparison between simulated (DLEM) and observed SOC storage. Notes: A: cropland abandoned inYuzhong (Gansu Province), the observation data come from Jinping Jiang et al. [67]; B: cropland abandoned in Yan'an (Shaanxi Province), the observation data come from Junmin Wang et al. [2].

cropland in 1985, were 27.7%–34.8% higher than that of the cropland in 2010 on the Loess Plateau. Fu et al. [64] suggested that cropland abandonment significantly increased the density and stock of SOC in 0–100 cm soil profiles on the Loess Plateau. Feng et al. [48] found a total of 96.1 Tg of additional carbon had been sequestered on the Loess Plateau since China's "Grain for Green" program during 2000–2008, by using remote sensing techniques and ecosystem modelling. Our results are comparable with those previous findings.

LCLUC might induce an immediate change in vegetation coverage but a lagged effect on the change of SOC storage [65]. In this study, the SOC changed slowly, and even had a decreasing trend in the 1960s. We also found that the most rapid increase in the SOC storage occurred after the late 1970s, which might be partly due to extensive and conducted tree planting projects for alleviating soil erosion. The large-scale cropland abandonment in response to the recently implemented "Grain for Green" policy also contributed to the increase of SOC storage on the Loess Plateau. The plantation from "Grain for Green" Project would keep a large proportion of carbon in wood which need long time to return to soil. During the initial period of land conversion from cropland to forests, leaf biomass of trees were very low, so the litters on the ground decreased and resulted in a slight change (even decrease) in SOC. During the period of implementing "Grain for Green" project, litters on the ground accumulated gradually with the growth of planted trees, resulting in increasing SOC storage after a certain time period. Compared with simulated results in the context of multiple environmental factors, a similar temporal pattern for the SOC change was found when considering LCLUC alone from 1961 to 2005 (Figure 6a,e). This further implied that LCLUC was the dominant factor controlling temporal variations of SOC storage.

DLEM simulated results also showed obvious increases of SOC storage in most areas of the central and southern Loess Plateau (Figure 6f). Further analysis found that changes in SOC storage were smaller in the LCLUC alone simulation experiment than that with the combination of all environmental change (Figure 6b). However, SOC storage changed slightly in most areas of the northern Loess Plateau from 1961 to 2005, which were also suggested by other research results (e.g., [18,43]).

Interactive Effects of LCLUC and Climate on SOC Storage

Based on DLEM simulations, Tian et al. [7] found that LCLUC accounted for 17% of the net carbon increase, but climate change reduced it by 4% in China in the past decades. However, we found that LCLUC increased SOC storage by 0.14 Pg C, accounting for 55% of the net increase in SOC in the past 45 years over the Loess Plateau. In the same period, warming and drying climate greatly reduced SOC storage by 0.05 Pg C, approximately 19% of the total SOC storage change. These results implied that the SOC was more sensitive to LCLUC and climate change on the Loess Plateau comparing to other regions in China. In the past decades, LCLUC, particularly in the southern Loess Plateau, enhanced the vegetation coverage and reduced the anthropogenic disturbance, which further enhanced the SOC storage.

The interaction between climate and LCLUC contributed to the net SOC increases by 3%. Our results also showed that the interactive effects among environmental factors can't be neglected in attributing the changes of SOC storage in response to environmental factors on the Loess Plateau. Although the interaction among environmental factors has been recognized long before [66], most of the field experiments still overlook it. This study further demonstrated that the modeling approach may serve as one complementary tool for the field experiments in addressing interactive effects among multiple environmental factors.

Effects of Cropland Abandonment on SOC Storage

Cropland abandonment and natural vegetation recovery were important implementations to mitigate soil loss on the Loess Plateau. Previous studies demonstrated that former land use types, soil property, climate change, and soil management practices were crucial to changes of SOC storage in the establishment of perennial vegetation type [49,50]. Other factors, such as the abandonment age, have also been found to play a significant role in SOC accumulation and should not be ignored [4,50].

In this study, we chose two sites (Yuzhong City and Yan'an City) (Marked in Figure 1) located on the Loess Plateau to explore SOC storage pattern at different abandonment stages (or restoration age) and to further evaluate the DLEM simulated results against the observations (Figure 8). It showed that DLEM results could well capture the distribution characteristics of the SOC storage in different abandonment stages. Both model results and field observations demonstrated that the restoration age played a key role in SOC accumulation. Jiang et al. [67] indicated that SOC storage changed little in early abandonment stage in a semiarid hilly area of Yuzhong City, followed by an obvious increase after 9–12 years, and then increased stably (Figure 8A). Wang et al. [2] found the similar temporal pattern through studying the change of SOC storage during different successional stages of rehabilitated grassland in Yan'an City. Compared to cropland, the rehabilitated grassland had a lower SOC storage at the early stage (1–12 years). However, SOC storage was higher than that in cropland after 15 years and then increased steadily (Figure 8B). Generally, long-term abandonment (>10 years) and the following colonization of natural vegetation could lead to substantial increase of SOC storage. The difference of the SOC storage between two sites over the early abandonment stage might be attributed to difference in environmental factors such as climate and soil property. Yan'an City is located in a semi-arid and warm temperate zone with less precipitation and more obvious drying and warming tendency, compared with Yuzhong City during the past years. This also partly explained the lower SOC storage in cropland in this climate zone than others over the Loess Plateau.

Our results indicated that cropland abandonment could increase SOC storage, improve soil quality and promote ecosystem restoration, especially in the warm and dry climate zone. However, these effects will emerge after a relative longer time period (e.g., >10 years) under local environmental conditions.

Uncertainties

This study examined temporal and spatial patterns of SOC storage and attributed these patterns to multiple environmental factors on the Loess Plateau during 1961–2005. There were several uncertainties which need to be addressed in our future work. First, impacts of some ecological processes, such as soil erosion, were not separated from other environmental factors in this study. The soil erosion area could be as high as 45.4×10^4 km^2 (72.3% of the land area) over the Loess Plateau. Soil erosion has long been identified as one of key factors controlling the change of SOC storage [47]. Further efforts should be put on interactions among soil erosion and other environmental factors. Second, the input datasets were developed at the spatial resolution of 10 km×10 km which was the finest dataset for the Loess Plateau. However, it is still difficult to capture subtle change of SOC storage due to high heterogeneity in land surface processes. In the long run, finer gidded datasets would be greatly helpful for further quantifying temporal and spatial changes in SOC storage on the Loess Plateau. In addition, the uncertainties from other input data, model structure, and parameterization need to be further specified in the furture efforts.

Conclusions

This study examined effects of climate and land use changes on SOC storage on the Loess Plateau of China in the context of multiple global changes by using an integrated ecosystem model. The results showed that temperature on the Loess Plateau has significantly increased, while precipitation slightly decreased during 1961–2005. Meanwhile, this region experienced a remarkable change in land cover and land use, characterized by conversions from cropland to grassland/forest owing to the government policies to alleviate soil and water losses during the past decades. The overall change in SOC storage due to multiple environmental factors was estimated to be a net increase of SOC storage by 0.27 Pg C during 1961–2005, indicating that soil on the Loess Plateau acted as a carbon sink in this period. Among multiple factors, LCLUC led to a significant increase in SOC storage of 0.14 Pg C, accounting for 55% of the net increase in SOC. In contrast, climate change reduced SOC storage by 0.05 Pg C (approximately 19% of the total SOC change). The interaction of climate and LCLUC accounted for 3% of the net increase in SOC. Our results were consistent with field observation data and both of them suggested that SOC storage could be enhanced significantly by the conversion of cropland to grassland along with the increasing abandonment age on the Loess Plateau. However, the magnitude could be influenced by local environmental conditions.

This study provides the first attempt to quantify relative effects of multiple environmental factors (climate and LCLUC in particular) on regional SOC storage on the Loess Plateau over the past decades. The results drawn from this study provide insight for land management as well as policy-making to enhance carbon sequestration and alleviate the serious soil erosion conditions on the Loess Plateau. To reduce uncertainties in estimating effects of climate and land use changes on SOC storage, it is needed to put further efforts in developing more reliable and fine-resolution input data and improve model representation of some other processes relevant to SOC, such as soil erosion.

Acknowledgments

This study was supported by NASA Land Cover and Land Use Change Program (NNX08AL73G), NASA Interdisciplinary Science Program (NNG04GM39C), Chinese Universities Scientific Fund (z10921007), US National Science Foundation Grants (AGS-1243220, CNS-1059376), State Key Laboratory of Soil Erosion and Dryland Farming on the Loess Plateau Foundation (K318009902-1410), and Shaanxi Administration of Foreign Expert Affairs Science and Technology Activities Fundation (201327). Thank Mingliang Liu, Xiaofeng Xu, and Jiyuan Liu for their contributions in the development of spatial data and the DLEM model.

Author Contributions

Conceived and designed the experiments: HT YD WR. Performed the experiments: YD WR. Analyzed the data: YD WR BT GC. Contributed reagents/materials/analysis tools: GC BT JY SP. Wrote the paper: YD WR HT BT GC CL JY SP GW SL.

References

1. Lal R (2004) Soil carbon sequestration impacts on global climate change and food security. Science 304: 1623–1627.
2. Wang Y, Fu B, Lü Y, Chen L (2011) Effects of vegetation restoration on soil organic carbon sequestration at multiple scales in semi-arid Loess Plateau, China. Catena 85: 58–66.
3. Post WM, Emanuel WR, Zinke PJ, Stangenberger AG (1982) Soil carbon pools and world life zones.
4. Zhang K, Dang H, Tan S, Cheng X, Zhang Q (2010) Change in soil organic carbon following the 'Grain-for-Green' programme in China. Land Degradation & Development 21: 13–23.
5. Eglin T, Ciais P, Piao S, Barre P, Bellassen V, et al. (2010) Historical and future perspectives of global soil carbon response to climate and land-use changes. Tellus B 62: 700–718.
6. Martin D, Lal T, Sachdev C, Sharma J (2010) Soil organic carbon storage changes with climate change, landform and land use conditions in Garhwal hills of the Indian Himalayan mountains. Agriculture, Ecosystems & Environment 138: 64–73.
7. Tian H, Melillo J, Lu C, Kicklighter D, Liu M, et al. (2011) China's terrestrial carbon balance: Contributions from multiple global change factors. Global Biogeochemical Cycles 25.
8. Wang X, Feng Z, Ouyang Z (2001) The impact of human disturbance on vegetative carbon storage in forest ecosystems in China. Forest ecology and management 148: 117–123.
9. Barnett TP, Adam JC, Lettenmaier DP (2005) Potential impacts of a warming climate on water availability in snow-dominated regions. Nature 438: 303–309.
10. Paul EA (1997) Soil organic matter in temperate agroecosystems: long term experiments in North America: CRC PressI Llc.
11. Post WM, Kwon KC (2000) Soil carbon sequestration and land-use change: processes and potential. Global change biology 6: 317–327.
12. Lal R, Bruce J (1999) The potential of world cropland soils to sequester C and mitigate the greenhouse effect. Environmental Science & Policy 2: 177–185.
13. Schimel J (2001) 1.13-Biogeochemical Models: Implicit versus Explicit Microbiology. Global Biogeochemical Cycles in the Climate System: 177–183.
14. Liu S, Bliss N, Sundquist E, Huntington TG (2003) Modeling carbon dynamics in vegetation and soil under the impact of soil erosion and deposition. Global Biogeochemical Cycles 17.
15. Zhang C, Tian H, Pan S, Liu M, Lockaby G, et al. (2008) Effects of forest regrowth and urbanization on ecosystem carbon storage in a rural–urban gradient in the southeastern United States. Ecosystems 11: 1211–1222.
16. Li Z, Liu W-z, Zhang X-c, Zheng F-l (2009) Impacts of land use change and climate variability on hydrology in an agricultural catchment on the Loess Plateau of China. Journal of hydrology 377: 35–42.
17. Chen L, Gong J, Fu B, Huang Z, Huang Y, et al. (2007) Effect of land use conversion on soil organic carbon sequestration in the loess hilly area, loess plateau of China. Ecological Research 22: 641–648.
18. Xin Z, Xu J, Zheng W (2008) Spatiotemporal variations of vegetation cover on the Chinese Loess Plateau (1981–2006): impacts of climate changes and human activities. Science in China Series D: Earth Sciences 51: 67–78.
19. Fu B-J, Wang Y-F, Lu Y-H, He C-S, Chen L-D, et al. (2009) The effects of land-use combinations on soil erosion: a case study in the Loess Plateau of China. Progress in Physical Geography 33: 793–804.
20. Wang G, Huang J, Guo W, Zuo J, Wang J, et al. (2010) Observation analysis of land-atmosphere interactions over the Loess Plateau of northwest China. Journal of Geophysical Research: Atmospheres (1984–2012) 115.
21. Zhao M, Heinsch FA, Nemani RR, Running SW (2005) Improvements of the MODIS terrestrial gross and net primary production global data set. Remote Sensing of Environment 95: 164–176.
22. Fang J, Guo Z, Piao S, Chen A (2007) Terrestrial vegetation carbon sinks in China, 1981–2000. Science in China Series D: Earth Sciences 50: 1341–1350.
23. Peters W, Jacobson AR, Sweeney C, Andrews AE, Conway TJ, et al. (2007) An atmospheric perspective on North American carbon dioxide exchange: CarbonTracker. Proceedings of the National Academy of Sciences 104: 18925–18930.
24. Piao S, Ciais P, Friedlingstein P, de Noblet-Ducoudré N, Cadule P, et al. (2009) Spatiotemporal patterns of terrestrial carbon cycle during the 20th century. Global Biogeochemical Cycles 23.
25. Li K, Wang S, Cao M (2004) Vegetation and soil carbon storage in China. SCIENCE IN CHINA SERIES D EARTH SCIENCES-ENGLISH EDITION- 47: 49–57.
26. King J, Bradley R, Harrison R, Carter A (2004) Carbon sequestration and saving potential associated with changes to the management of agricultural soils in England. Soil Use and Management 20: 394–402.
27. Yan H, Cao M, Liu J, Tao B (2007) Potential and sustainability for carbon sequestration with improved soil management in agricultural soils of China. Agriculture, Ecosystems & Environment 121: 325–335.
28. Tian H, Chen G, Liu M, Zhang C, Sun G, et al. (2010) Model estimates of net primary productivity, evapotranspiration, and water use efficiency in the terrestrial ecosystems of the southern United States during 1895–2007. Forest ecology and management 259: 1311–1327.
29. Ren W, Tian H, Liu M, Zhang C, Chen G, et al. (2007) Effects of tropospheric ozone pollution on net primary productivity and carbon storage in terrestrial ecosystems of China. Journal of Geophysical Research: Atmospheres (1984–2012) 112.

30. Xu X, Tian H, Zhang C, Liu M, Ren W, et al. (2010) Attribution of spatial and temporal variations in terrestrial methane flux over North America. Biogeosciences Discussions 7: 5383–5428.
31. Tian H, Xu X, Lu C, Liu M, Ren W, et al. (2011) Net exchanges of CO2, CH4, and N2O between China's terrestrial ecosystems and the atmosphere and their contributions to global climate warming. Journal of Geophysical Research: Biogeosciences (2005–2012) 116.
32. Ren W, Tian H, Tao B, Huang Y, Pan S (2012) China's crop productivity and soil carbon storage as influenced by multifactor global change. Global Change Biology 18: 2945–2957.
33. Tian H, Chen G, Zhang C, Liu M, Sun G, et al. (2012) Century-scale responses of ecosystem carbon storage and flux to multiple environmental changes in the southern United States. Ecosystems 15: 674–694.
34. Shi H, Shao M (2000) Soil and water loss from the Loess Plateau in China. Journal of Arid Environments 45: 9–20.
35. Wang S, Tian H, Liu J, Pan S (2003) Pattern and change of soil organic carbon storage in China: 1960s–1980s. Tellus B 55: 416–427.
36. Zhang XP, Zhang L, McVicar TR, Van Niel TG, Li LT, et al. (2008) Modelling the impact of afforestation on average annual streamflow in the Loess Plateau, China. Hydrological Processes 22: 1996–2004.
37. Liu J, Tian H, Liu M, Zhuang D, Melillo JM, et al. (2005) China's changing landscape during the 1990s: Large-scale land transformations estimated with satellite data. Geophysical Research Letters 32.
38. Liu M, Tian H (2010) China's land cover and land use change from 1700 to 2005: Estimations from high-resolution satellite data and historical archives. Global Biogeochemical Cycles 24.
39. Tian H, Chen G, Zhang C, Melillo JM, Hall CA (2010) Pattern and variation of C: N: P ratios in China's soils: a synthesis of observational data. Biogeochemistry 98: 139–151.
40. Chen H, Tian H, Liu M, Melillo J, Pan S, et al. (2006) Effect of land-cover change on terrestrial carbon dynamics in the southern United States. Journal of environmental quality 35: 1533–1547.
41. Trenberth K, Jones P, Ambenje P, Bojariu R, Easterling D, et al. (2007) Observations: Surface and Atmospheric Climate Change, chap. 3 of Climate Change 2007: The Physical Science Basis. Contribution of Working Group I to the Fourth Assessment Report of the Intergovernmental Panel on Climate Change [Solomon, S., Qin, D., Manning, M., Marquis, M., Averyt, KB, Tignor, M., Miller, HL and Chen, Z.(eds.)]., 235–336. Cambridge University Press, Cambridge, UK and New York, NY, USA.
42. Li R, Yang W, Li B (2008) Research and future prospects for the Loess Plateau of China. SciencePress, Beijing (in Chinese).
43. Chang R, Fu B, Liu G, Liu S (2011) Soil carbon sequestration potential for "Grain for Green" project in Loess Plateau, China. Environmental management 48: 1158–1172.
44. Xin Z, Xu J, Ma Y (2009) Spatio-temporal variation of erosive precipitation in Loess Plateau during past 50 years. Scientia Geographica Sinica 29: 89–104.
45. Song Y, Ma M-g (2007) Study on vegetation cover change in Northwest China based on SPOT VEGETATION data. Journal of Desert Research 27: 89–93.
46. Ren W, Tian H, Xu X, Liu M, Lu C, et al. (2011) Spatial and temporal patterns of CO2 and CH4 fluxes in China's croplands in response to multifactor environmental changes. Tellus B 63: 222–240.
47. Liu X, Li F-M, Liu D-Q, Sun G-J (2010) Soil organic carbon, carbon fractions and nutrients as affected by land use in semi-arid region of Loess Plateau of China. Pedosphere 20: 146–152.
48. Feng X, Fu B, Lu N, Zeng Y, Wu B (2013) How ecological restoration alters ecosystem services: an analysis of carbon sequestration in China's Loess Plateau. Scientific reports 3.
49. Guo Z, Ruddiman WF, Hao Q, Wu H, Qiao Y, et al. (2002) Onset of Asian desertification by 22 Myr ago inferred from loess deposits in China. Nature 416: 159–163.
50. Paul K, Polglase P, Nyakuengama J, Khanna P (2002) Change in soil carbon following afforestation. Forest ecology and management 168: 241–257.
51. LAGANIÈRE J, ANGERS DA, PARÉ D (2010) Carbon accumulation in agricultural soils after afforestation: a meta-analysis. Global Change Biology 16: 439–453.
52. Cheng L, Zhao X (2011) Soil mineralized nutrients changes and soil conservation benefit evaluation on 'green project grain' in ecologically fragile areas in the south of Yulin city, Loess Plateau. African Journal of Biotechnology 10: 2230–2237.
53. Xiao Y (1990) Comparative studies on biomass and productivity of Pinus tabulaeformis plantations in different climatic zones in Shaanxi Province [China]. Acta Phytoecologica et Geobotanica Sinica 14.
54. Fu X, Shao M, Wei X, Horton R (2010) Soil organic carbon and total nitrogen as affected by vegetation types in Northern Loess Plateau of China. Geoderma 155: 31–35.
55. Liu Z, Shao Ma, Wang Y (2011) Effect of environmental factors on regional soil organic carbon stocks across the Loess Plateau region, China. Agriculture, Ecosystems & Environment 142: 184–194.
56. Xu X, Zhang K, Peng W (2003) Spatial distribution and estimating of soil organic carbon on Loess Plateau. J Soil Water Conserv 17: 13–15.
57. Xu X, Liu W, Kiely G (2011) Modeling the change in soil organic carbon of grassland in response to climate change: effects of measured versus modelled carbon pools for initializing the Rothamsted Carbon model. Agriculture, Ecosystems & Environment 140: 372–381.
58. Schlesinger WH (1990) Evidence from chronosequence studies for a low carbon-storage potential of soils. Nature 348: 232–234.
59. Han F, Hu W, Zheng J, Du F, Zhang X (2010) Estimating soil organic carbon storage and distribution in a catchment of Loess Plateau, China. Geoderma 154: 261–266.
60. Lehmann J, Skjemstad J, Sohi S, Carter J, Barson M, et al. (2008) Australian climate–carbon cycle feedback reduced by soil black carbon. Nature Geoscience 1: 832–835.
61. Houghton R, Hackler J, Lawrence K (1999) The US carbon budget: contributions from land-use change. Science 285: 574–578.
62. van der Werf GR, Morton DC, DeFries RS, Olivier JG, Kasibhatla PS, et al. (2009) CO2 emissions from forest loss. Nature Geoscience 2: 737–738.
63. Richter DD, Markewitz D, Trumbore SE, Wells CG (1999) Rapid accumulation and turnover of soil carbon in a re-establishing forest. Nature 400: 56–58.
64. Fu B-J, Zhang Q-J, Chen L-D, Zhao W-W, Gulinck H, et al. (2006) Temporal change in land use and its relationship to slope degree and soil type in a small catchment on the Loess Plateau of China. Catena 65: 41–48.
65. Kuzyakov Y, Gavrichkova O (2010) Review: Time lag between photosynthesis and carbon dioxide efflux from soil: a review of mechanisms and controls. Global Change Biology 16: 3386–3406.
66. Dermody O (2006) Mucking through multifactor experiments; design and analysis of multifactor studies in global change research. New Phytologist 172: 598–600.
67. Jiang J-P, Xiong Y-C, Jiang H-M, Ye D-Y, Song Y-J, et al. (2009) Soil microbial activity during secondary vegetation succession in semiarid abandoned lands of Loess Plateau. Pedosphere 19: 735–747.

Global Validation of a Process-Based Model on Vegetation Gross Primary Production Using Eddy Covariance Observations

Dan Liu[1], Wenwen Cai[1], Jiangzhou Xia[1], Wenjie Dong[1], Guangsheng Zhou[2,3], Yang Chen[1], Haicheng Zhang[1], Wenping Yuan[1]*

1 State Key Laboratory of Earth Surface Processes and Resource Ecology, Beijing Normal University, Beijing, China, 2 Chinese Academy of Sciences, Institute of Botany, State Key Laboratory of Vegetation and Environmental Change, Beijing, China, 3 Chinese Academy of Meteorological Sciences, Beijing, China

Abstract

Gross Primary Production (GPP) is the largest flux in the global carbon cycle. However, large uncertainties in current global estimations persist. In this study, we examined the performance of a process-based model (Integrated BIosphere Simulator, IBIS) at 62 eddy covariance sites around the world. Our results indicated that the IBIS model explained 60% of the observed variation in daily GPP at all validation sites. Comparison with a satellite-based vegetation model (Eddy Covariance-Light Use Efficiency, EC-LUE) revealed that the IBIS simulations yielded comparable GPP results as the EC-LUE model. Global mean GPP estimated by the IBIS model was 107.50 ± 1.37 Pg C $year^{-1}$ (mean value $\pm$ standard deviation) across the vegetated area for the period 2000–2006, consistent with the results of the EC-LUE model (109.39 ± 1.48 Pg C $year^{-1}$). To evaluate the uncertainty introduced by the parameter V_{cmax}, which represents the maximum photosynthetic capacity, we inversed V_{cmax} using Markov Chain-Monte Carlo (MCMC) procedures. Using the inversed V_{cmax} values, the simulated global GPP increased by 16.5 Pg C $year^{-1}$, indicating that IBIS model is sensitive to V_{cmax}, and large uncertainty exists in model parameterization.

Editor: Christopher Carcaillet, Ecole Pratique des Hautes Etudes, France

Funding: This study was supported by the National Science Foundation for Excellent Young Scholars of China (41322005), the National High Technology Research and Development Program of China (863 Program) (2013AA122003), National Natural Science Foundation of China (41201078), Program for New Century Excellent Talents in University (NCET-322 12-0060) and the Fundamental Research Funds for the Central Universities. The funders had no role in study design, data collection and analysis, decision to publish, or preparation of the manuscript.

Competing Interests: The authors have declared that no competing interests exist.

* Email: wenpingyuancn@yahoo.com

Introduction

Terrestrial gross primary production (GPP) is the largest carbon flux in terrestrial ecosystems, and it is approximately 20 times larger than the amount of carbon introduced from anthropogenic sources [1]. Thus, even small fluctuations in GPP can cause large changes in the airborne fraction of carbon and subsequently influence future climate change [2]. Vegetation also contributes to human welfare by providing food, fiber and energy [3–4]. Therefore, regular monitoring and reliable estimation of global terrestrial GPP is important for improving our understanding of the global carbon cycle, accurately predicting future climate, and ensuring the long-term sustainability of terrestrial ecosystem services.

Ecosystem models serve as a backbone for evaluating large-scale and global GPP. Two categories of ecosystem models are widely used: process-based and satellite-based. Satellite-based models are driven by remotely sensed data and provide simple means of estimating GPP [5]; however, they are limited in their ability to model mechanisms. Process-based models typically exhibit detailed expressions of terrestrial processes, such as photosynthesis, respiration, phenology, and hydrological cycle. Therefore, process-based models play important roles in investigating the mechanisms underlying current biases in estimated ecosystem production [6], predicting the future conditions of the terrestrial carbon cycle, and exploring its feedback to climate change [7].

Numerous attempts have been made to develop and improve process-based models. However, a recent study from the North American Carbon Project (NACP) showed that current models perform poorly and difference between observations and simulations far exceed the observational uncertainty [8]. Model parameter uncertainty is a key source limiting the accuracy of process-based models. Knorr and Heimann analyzed the uncertainties of process-based models [9]. They found that parameter uncertainties could explain much of the large variance among models and that the largest uncertainties arose from plant photosynthesis, respiration and soil water storage.

The maximum rate of carboxylation by the enzyme Rubisco (V_{cmax}) is fundamental in modeling photosynthesis [10]. Sensitivity analysis shows that the projections of ecosystem production are particularly sensitive to the fixed parameters associated with V_{cmax} [11]. Therefore, the parameterization scheme of V_{cmax} is essential for GPP simulation, and its impacts on model performance need to be tested.

Eddy covariance measurements recorded by the increasing number of eddy covariance (EC) towers provide a great opportunity for model validation and improvement. Concurrent measurements include carbon fluxes, latent heat, and sensible

Table 1. The values of V_{cmax} set in the IBIS model and the initial range of MCMC inversion.

Plant Functional Type (PFT)	V_{cmax} (10^{-6} mol CO_2 m^{-2} s^{-1})	V_{cmax} range (10^{-6} mol CO_2 m^{-2} s^{-1})
Tropical broadleaf trees	65	1–300
Warm-temperate broadleaf trees	40	1–300
Temperate broadleaf trees	30	1–300
Boreal broadleaf trees	30	1–300
Temperate conifer trees	30	1–300
Boreal conifer trees	20	1–300
Shrub	27.5	1–300
C3 herbaceous	25	1–300
C4 herbaceous	15	1–300

V_{cmax} set according to Kucharik et al. [47].

heat, as well as meteorological conditions such as air temperature and relative humidity, and provide unprecedented datasets for model validation and evaluations of parameter constraints [12]. The current network of EC sites covers a wide range of ecosystem types, thus it has the potential to significantly improve our understanding of the variation in GPP across time, space and biomes [13].

The goal of this study was to validate a process-based ecosystem model (Integrated BIosphere Simulator, IBIS) based on measurements from 62 EC sites [14]. The specific objectives were to (1) examine the performance of IBIS over several ecosystem types, (2) compare IBIS model performance with a satellite-based model (i.e., Eddy Covariance-Light Use Efficiency, EC-LUE), and (3) investigate the impacts of the parameter V_{cmax} on model performance.

Data and Methods

2.1 IBIS model and parameter inversion

The Integrated BIosphere Simulator (IBIS) is designed to integrate a variety of terrestrial ecosystem processes within a single, physically consistent modeling framework. It represents land surface processes, canopy physiology, vegetation phenology, long-term vegetation dynamics, and carbon and water cycling [14]. The photosynthesis module of the IBIS model is provided by the formulations of Farquhar [15].

C_3 photosynthesis and C_4 photosynthesis are expressed separately in the IBIS model. For C_3 plants, the gross photosynthesis rate per unit leaf area, A_g(mol CO_2 m^{-2} s^{-1}) is expressed as

$$A_g = \min(J_e, J_c, J_s) \tag{1}$$

where J_e is light-limited rate of photosynthesis, J_c represents the Rubisco-limited rate of photosynthesis, and J_s is the photosynthesis limited by the inadequate rate of utilization of triose phosphate.

The light-limited rate of photosynthesis is given as

$$J_e = \alpha_3 Q_p \frac{C_i - \Gamma_*}{C_i + 2\Gamma_*} \tag{2}$$

where α_3 is the intrinsic quantum efficiency of CO_2 uptake in C_3 plants (mol CO_2 mol^{-1} quanta), Q_p is the flux density of photosynthetically active radiation absorbed by leaf (mol quanta m^{-2} s^{-1}), C_i is the concentration of CO_2 in the intercellular air spaces of the leaf (mol mol^{-1}), and Γ_* is the compensation point for gross photosynthesis (mol mol^{-1}).

The Rubisco-limited rate of photosynthesis is calculated as

$$J_c = \frac{V_m(C_i - \Gamma_*)}{C_i + K_C(1 + \frac{[O_2]}{K_O})} \tag{3}$$

where V_m is the maximum carboxylase capacity of Rubisco (mol CO_2 m^{-2} s^{-1}) and K_C and K_O are the Michaelis-Menten coefficients (mol mol^{-1}) for CO_2 and O_2, respectively.

Under conditions of high intercellular CO_2 concentrations and high irradiance, photosynthesis is limited by the inadequate rate of utilization of triose phosphate. This limitation is expressed as

$$J_s = 3T(1 - \frac{\Gamma_*}{C_i}) + \frac{J_p \Gamma_*}{C_i} \tag{4}$$

where T is the rate of triose phosphate utilization.

Photosynthesis in C_4 plants is similarly modeled as the minimum of three potential capacities to fix carbon [16]. The gross photosynthesis rate is given by

$$A_g = \min(J_i, J_e, J_c) \tag{5}$$

where $J_i = \alpha_4 Q_p$ is the light-limited rate of photosynthesis, $J_e = V_m$ is the Rubisco-limited rate of photosynthesis and $J_c = k$ is the CO_2-limited rate of photosynthesis at low CO_2 concentrations.

The parameter V_{cmax} (mol CO_2 m^{-2} s^{-1}) is very important for simulating the photosynthesis process. In the IBIS model, it is established as a constant that differs among plant functional types (PFTs) (Table 1). To validate the IBIS model and investigate the impact of the V_{cmax} parameter scheme on model performance, we conducted two simulations (IBIS and IBIS-Type) for each site. IBIS simulation used the default values of V_{cmax} (Table 1). For the IBIS-Type simulation, the vegetation-specific V_{cmax} values were inversed for each PFT. The Markov Chain-Monte Carlo (MCMC) procedure was used for the parameter inversion, and the Metropolis-Hastings (M-H) algorithm was used as the MCMC sampler [17–18] (see Xu et al. [19] and Yuan et al. [20] for detailed descriptions of the MCMC procedure). We conducted 10000 samples for each site and assigned the V_{cmax} value with the highest frequency as the optimal value of V_{cmax}. Finally, we ran the model using the inversed V_{cmax} for each PFT as the IBIS-Type simulation.

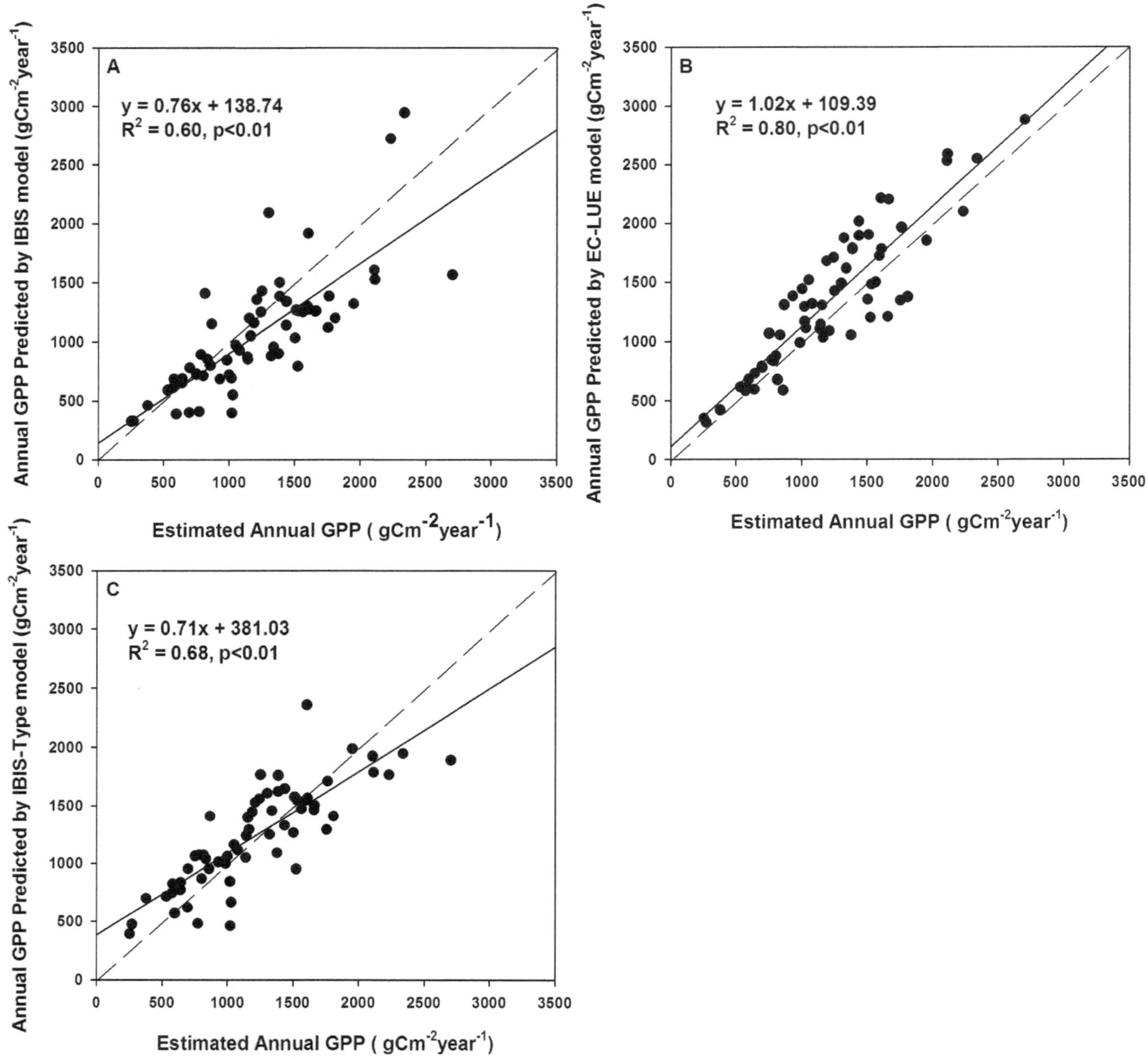

Figure 1. Comparison of predicted and measured GPP. Comparison between gross primary production (GPP) estimated from Eddy Covariance (EC) measurements and GPP predicted from different model simulations: (a) IBIS, (b) EC-LUE and (c) IBIS-Type. The solid lines are the linear regression lines and the short dashed lines are the 1:1 lines.

2.2 EC-LUE model

A satellite-based model (i.e. EC-LUE) [5,21–23] was used to compare the local and global GPP simulations with those of the IBIS model. The EC-LUE model is based on two assumptions: (1) ecosystem GPP has a direct relationship with the absorbed photosynthetically active radiation (APAR) via light use efficiency (LUE), where LUE is defined as the amount of carbon produced per unit of APAR; and (2) realized LUE may be reduced below its theoretical potential value by environmental stressors, such as low temperatures or water shortages. The EC-LUE model is driven by four variables: the normalized difference vegetation index (NDVI), photosynthetically active radiation (PAR), air temperature, and the ratio of sensible to latent heat flux (Bowen ratio).

2.3 Data

We used the eddy covariance (EC) data to validate the IBIS model. We used data obtained from the LaThuile dataset (http://www.fluxdata.org). The daily GPP values were estimated from the eddy covariance measurements using a community standard method [24]. Briefly, GPP was estimated from the equation:

$$GPP = R_{day} - NEE_{day} \tag{6}$$

where NEE_{day} is daytime NEE. Daytime ecosystem respiration R_{day} was estimated using daytime temperature and an equation describing the temperature dependence of respiration, which was formulated from nighttime NEE measurements. For further details on the algorithm, see Reichstein et al. (2005) [24]. The gap filling

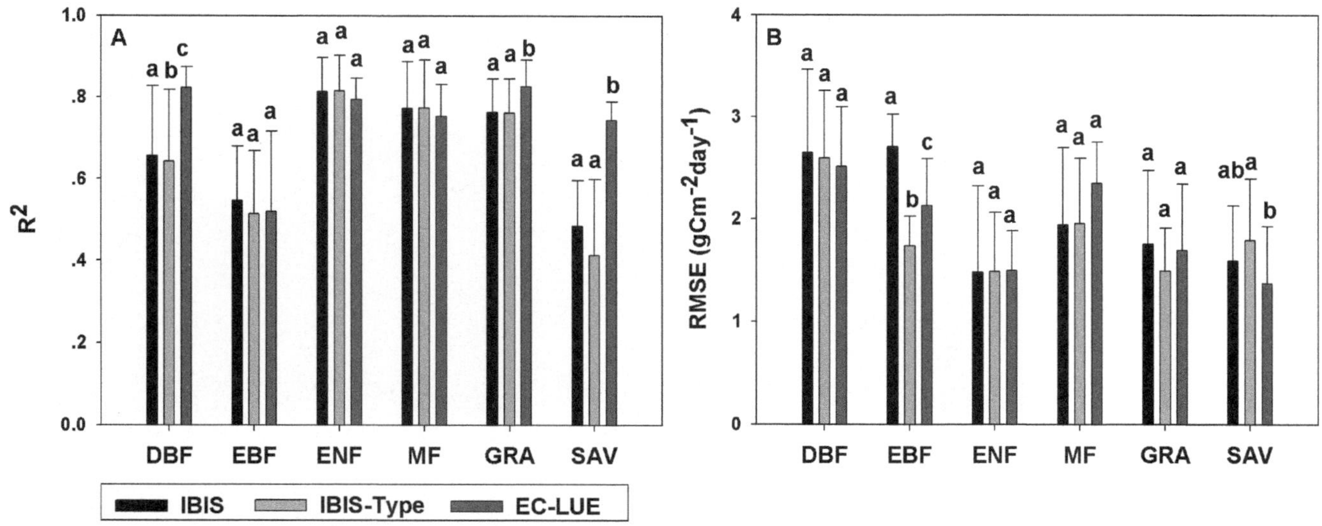

Figure 2. Comparison among models for different PFTs. Comparisons among IBIS, IBIS-Type and EC-LUE models for each plant functional type (PFT), where (a) and (b) are the results of R^2 and RMSE, respectively. Lowercase letters above the bars indicate significant differences among models.

and quality control of this dataset are conducted according to standard criteria [25–26], and the uncertainty in annual GPP can be controlled to some extent below 100 g C m^{-2} $year^{-1}$ [25]. This dataset is widely used in model studies. In the present study, we selected 62 EC sites for model validation (Table S1). The selected data covered six major terrestrial biomes: evergreen needleleaf forest (ENF), deciduous broadleaf forest (DBF), evergreen broad-leaf forest (EBF), mixed forest (MF), grassland (GRA) and savanna (SAV). Additional information on the vegetation, climate and soil characteristics of each site was collected from the associated metadata of the LaThuile dataset. Daily average, maximum and minimum temperature, relative humidity, precipitation, cloud fraction, photosynthetically active radiation, latent heat and sensible heat were used to drive the IBIS, and GPP was used to evaluate its performance.

To decrease model uncertainty, we used the satellite-based leaf area index (LAI) from the Moderate Resolution Imaging Spectro-radiometer (MODIS) as a model input. The Normalized Difference Vegetation Index (NDVI) data derived from MODIS were used to drive the EC-LUE model. The 8-day MODIS-NDVI data (MOD13) and the MODIS-LAI (MOD15) data with 1-km spatial resolution were used for model verification at the EC sites. Quality control (QC) flags, which signal cloud contamination in each pixel, were examined to filter out NDVI and LAI data of insufficient quality. We temporally filled missing or unreliable values for each 1-km MODIS pixel based on their corresponding quality assessment data fields, as proposed by Zhao et al. [27]. In addition, data on soil properties, including soil texture, organic carbon content and nitrogen content, were required to input soil information into the model, and were therefore collected from the sites where EC towers are established.

For global simulation, we used meteorological datasets from the Modern Era Retrospective Analysis for Research and Applications (MERRA) archive for 2000–2006 to drive the IBIS and EC-LUE models. MERRA is a NASA reanalysis dataset for the satellite era which uses a new version of the Goddard Earth Observing System Data Assimilation System Version 5 (GEO-5). We used climate conditions at 10 meters above the land surface and at a resolution of 0.5° latitude by 0.6° longitude. The Global Gridded Surfaces of Selected Soil Characteristics datasets were used to supply soil properties for the IBIS model; detailed information is available from the website (http://ww.isric.org). The global distribution of plant functional types (PFTs) was derived by overlapping the MODIS land-cover type product with the Köppen-Geiger climate classification map, with the land cover classifications aggregated into nine PFTs (Table 1). The IBIS model was unable to simulate carbon cycle processes in cropland or wetland; therefore, these two vegetation types were replaced with C_3 grassland in this study.

2.4 Statistical analysis

Three metrics were used to evaluate model performance:

(1) the coefficient of determination, R^2, which represents how much variation in the observations is explained by the model simulations;

(2) the root mean square error (RMSE), which represents the total difference between the simulated and measured values;

(3) the relative predictive error (RPE), which represents the ratio of error to observation. It is computed as

$$RPE = \frac{\overline{P} - \overline{O}}{\overline{O}} \times 100 \qquad (7)$$

where $\overline{P}$ and $\overline{O}$ are the mean simulated and measured values, respectively.

One-way ANOVA was employed using SPSS software to test the significance of the differences in optimal V_{cmax} for each biome in the IBIS-Type scheme, and the paired-samples t-test was used to test the significance of the difference in statistical metrics between different models.

Results

3.1 Model validation at EC towers

The overall comparison of the estimated GPP with the EC measurements showed that the IBIS model performed well in capturing the variability in GPP. Across all study sites, the IBIS model explained approximately 60% of the variation in site-averaged GPP (Fig. 1). The coefficient of determination (R^2) varied from 0.11 at the ES-LMa site to 0.94 at the CA-Man site, with a mean value of 0.71 across all EC sites. The mean R^2 values

Table 2. The performance of the IBIS model and the EC-LUE model.

Site	Vegetation Type	IBIS			EC-LUE		
		R^2	RMSE	RPE	R^2	RMSE	RPE
CA-Oas	DBF	0.75	2.57	−32.34	0.87	2.26	26.75
DE-Hai	DBF	0.84	2.26	−17.92	0.88	1.62	−3.42
FR-Fon	DBF	0.68	2.98	−24.83	0.77	3.23	22.55
FR-Hes	DBF	0.74	3.07	−20.36	0.80	2.72	9.49
IT-Col	DBF	0.75	2.62	−18.94	0.75	2.88	20.36
IT-Ro1	DBF	0.61	2.03	6.49	0.77	2.33	32.39
IT-Ro2	DBF	0.24	4.05	−25.93	0.83	2.85	30.68
US-Bar	DBF	0.80	1.80	1.10	0.83	2.82	37.93
US-Bn2	DBF	0.72	1.12	10.49	0.86	1.04	14.97
US-Dk2	DBF	0.68	3.74	−23.88	0.81	2.86	19.87
US-Ha1	DBF	0.76	2.82	−19.57	0.82	2.38	11.84
US-MOz	DBF	0.32	3.62	−24.46	0.85	3.02	34.81
US-UMB	DBF	0.64	2.57	−2.44	0.92	2.47	41.20
US-Wi8	DBF	0.65	1.85	19.52	0.76	2.72	37.90
AU-Tum	EBF	0.69	2.80	22.60	0.75	2.27	−5.86
AU-Wac	EBF	0.43	2.92	24.27	0.39	2.66	5.82
FR-Pue	EBF	0.63	2.89	59.41	0.62	2.03	15.37
PT-Mi1	EBF	0.44	2.24	72.49	0.33	1.55	−17.23
CA-Ca3	ENF	0.74	1.38	14.26	0.76	1.87	14.11
CA-Man	ENF	0.94	0.61	6.83	0.80	1.19	13.54
CA-Obs	ENF	0.91	0.80	−11.11	0.86	1.16	9.49
CA-Ojp	ENF	0.88	0.77	17.75	0.78	0.99	5.13
CA-Qfo	ENF	0.84	0.91	4.25	0.80	1.19	9.92
CA-SJ3	ENF	0.79	1.24	33.48	0.75	2.12	51.35
FI-Hyy	ENF	0.87	1.43	−22.85	0.89	1.20	−2.87
FI-Sod	ENF	0.85	0.89	2.78	0.77	1.06	−7.73
IT-Ren	ENF	0.73	1.49	−7.29	0.69	1.74	−12.11
NL-Loo	ENF	0.81	2.28	−32.99	0.78	1.95	−21.84
RU-Fyo	ENF	0.87	2.30	−34.64	0.85	1.82	−23.40
RU-Zot	ENF	0.86	1.05	−14.51	0.85	1.00	0.63
SE-Fla	ENF	0.86	0.98	13.78	0.76	1.46	6.35
US-Bn1	ENF	0.75	0.99	6.62	0.76	1.10	1.76
US-Fmf	ENF	0.73	1.09	−14.31	0.73	1.50	21.17
US-Ha2	ENF	0.87	1.61	−21.33	0.85	1.96	11.45

Table 2. Cont.

Site	Vegetation Type	IBIS			EC-LUE		
		R^2	RMSE	RPE	R^2	RMSE	RPE
US-Ho1	ENF	0.92	2.11	−31.29	0.84	1.69	−9.90
US-Me2	ENF	0.61	2.73	−47.52	0.77	1.63	−20.61
US-Me3	ENF	0.75	0.89	−6.74	0.76	1.15	−31.66
US-NC2	ENF	0.71	4.05	−42.25	0.83	2.08	6.32
CA-Let	GRA	0.84	1.68	−34.70	0.90	1.10	13.64
CN-HaM	GRA	0.82	1.68	−42.69	0.83	1.47	12.36
CN-Xi2	GRA	0.67	0.77	21.88	0.70	0.82	10.69
DE-Meh	GRA	0.84	1.90	−25.53	0.85	1.49	0.20
HU-Bug	GRA	0.70	1.92	−30.21	0.85	2.30	46.43
IE-Dri	GRA	0.65	2.80	−33.36	0.75	2.25	−6.00
IT-MBo	GRA	0.84	2.32	−28.74	0.83	2.49	20.46
PT-Mi2	GRA	0.76	1.43	−24.61	0.77	2.14	52.77
RU-Ha1	GRA	0.79	0.86	−3.50	0.82	1.77	42.51
US-ARc	GRA	0.64	2.91	−33.45	0.90	2.21	41.67
US-Aud	GRA	0.84	1.03	16.20	0.89	0.57	10.06
BE-Bra	MF	0.81	1.31	12.90	0.71	1.91	0.95
BE-Jal	MF	0.90	1.48	−19.78	0.84	1.99	−4.06
BE-Vie	MF	0.82	1.88	−25.86	0.78	2.10	−27.43
CA-Gro	MF	0.75	1.72	−7.45	0.75	3.07	44.54
JP-Tef	MF	0.54	2.29	0.07	0.61	2.33	9.86
JP-Tom	MF	0.78	3.48	−34.21	0.83	2.55	−21.88
US-PFa	MF	0.81	1.44	−0.10	0.76	2.50	28.45
ES-LMa	SAV	0.59	1.55	−39.30	0.77	1.27	8.16
ZA-Kru	SAV	0.32	2.10	−51.83	0.74	1.35	22.69
BW-Ma1	SAV	0.42	1.58	−48.15	0.74	0.90	9.73
US-FR2	SAV	0.11	2.17	−46.41	0.77	1.27	8.16
US-Ton	SAV	0.50	1.48	−20.37	0.79	2.14	31.26
US-SRM	SAV	0.59	1.49	4.82	0.66	1.88	25.23

DBF: deciduous broadleaf forest; EBF: evergreen broadleaf forest; ENF: evergreen needleleaf forest; GRA: grassland; MF: mixed forest; SAV: savanna.

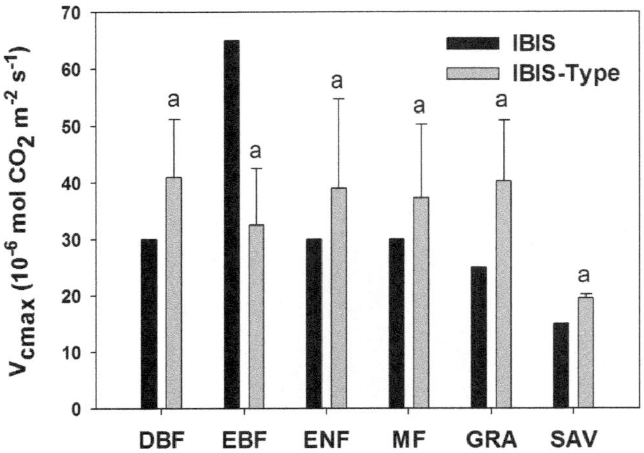

Figure 3. Comparison of V_{cmax} from the original scheme and the inversed V_{cmax} values. Comparison of model V_{cmax} values with the inversed values of the IBIS-Type. Same letter on the top of the bar indicates no significant difference among biomes.

were 0.66, 0.55, 0.81, 0.77, 0.76 and 0.48 for deciduous broadleaf forest, evergreen broadleaf forest, evergreen needleleaf forest, mixed forest, grassland and savanna, respectively (Fig. 2). The root mean square error (RMSE) varied from 1.48 to 2.71 g C m^{-2} day^{-1} among the six vegetation types (Fig. 2). On average, the mean relative predictive error (RPE) of all sites was −9.68%, and the RPE at most sites was less than 20% (Table 2).

Although the IBIS model explained most of the GPP variability at individual sites, large differences between the predicted and estimated GPP values from the EC measurements were apparent at some sites and for some vegetation types. The IBIS model underestimated GPP for the majority of PFTs and overestimated GPP in evergreen broadleaf forest. Specifically, over 40 of the 62 sites had negative RPE values, and the RPEs of 26 sites were below −20%. The largest underestimation occurred at two savanna sites, ZA-Kru and BW-Ma1, with PREs of −51.83% and −48.15%, respectively. The other 24 sites with RPEs below −20% were predominantly deciduous broadleaf forest (6 sites), evergreen needleleaf forest (7 sites) and grassland (8 sites). In addition, the model overestimated GPP at 6 sites with RPEs greater than 20%, four of which were evergreen broadleaf forest. Extreme overestimation occurred at two sites of evergreen broadleaf forest, FR-Pue and PT-Mi1, with RPEs of 59.41% and 72.49%, respectively.

3.2 Comparison of IBIS and EC-LUE

Compared to the satellite-based EC-LUE model, the IBIS model performed comparably at most sites, according to the R^2, RMSE and RPE values (Table 2). The IBIS mean R^2 values for evergreen broadleaf forest, evergreen needleleaf forest, mixed forest and grassland were similar to those of the EC-LUE model, which were 0.52, 0.79, 0.75, 0.83, respectively; no significant differences were found for most PFTs. Comparable results were also found for the RMSE, and no significant differences in RMSE were detected for any PFTs except for evergreen broadleaf forest.

However, the EC-LUE model had some advantages for some PFTs (Figure 2). For broadleaf forest, grassland and savanna, the R^2 of the EC-LUE was significantly higher than that of the IBIS, particularly for savanna, where the mean was 76% higher than that of the IBIS (0.74 for EC-LUE and 0.42 for IBIS). In addition, the EC-LUE had significantly low model error for evergreen broadleaf forest, with a mean RMSE 22% lower than that of the IBIS (2.12 g C m^{-2} day^{-1} for EC-LUE and 2.71 g C m^{-2} day^{-1} for IBIS).

3.3 IBIS Performance with inversed V_{cmax}

Model parameterization did not significantly improve the performance of the IBIS, as indicated by the R^2 and RMSE. The mean R^2 values of the IBIS-Type were 0.64, 0.51, 0.81, 0.77, 0.76, 0.41 for deciduous broadleaf forest, evergreen broadleaf forest, evergreen needleleaf forest, mixed forest, grassland and savanna, respectively. These values were very similar to those of the IBIS model; only the R^2 of the deciduous broadleaf forest differed significantly, being higher in the IBIS-Type model. The overall R^2 increased from 0.60 to 0.68 after revising V_{cmax} (Fig. 1). The RMSEs of the IBIS and the IBIS-Type models were also comparable. The mean RMSE of the IBIS-Type varied from 1.49 g C m^{-2} day^{-1} for evergreen needleleaf forest to 2.60 g C m^{-2} day^{-1} for deciduous broadleaf forest. Evergreen broadleaf forest was the only vegetation type yielding a significant difference in RMSE between the two models.

The IBIS employs a set of parameter values for a given PFT (Table 1). In the IBIS-Type scheme, the V_{cmax} value, which was set to as the mean value of V_{cmax} inversed from each site within a given PFT, was largely differentiated from the original parameter values. V_{cmax} of the IBIS-Type was 36.67%, 30.00%, 24.29%, 61.09% and 30.00% higher than that of the IBIS for deciduous broadleaf forest, evergreen needleleaf forest, mixed forest, grassland and savanna, respectively; it was 50% lower for evergreen broadleaf forest. In addition, the V_{cmax} value inversed at each site did not indicate significant differences among different PFTs (Fig. 3).

3.4 Temporal and spatial patterns in global averaged GPP

The spatial pattern in average annual GPP estimated using the original IBIS, the IBIS-Type and the EC-LUE models from 2000 to 2006 were generally consistent (Fig. 4). The highest value was from the humid tropics (Amazonia, Central Africa and Southeast Asia), with an annual GPP over 2000 g C m^{-2} $year^{-1}$. Temperate regions had intermediate levels of GPP, and the lowest GPP was found in both cold and arid regions.

The magnitude of GPP estimated by the IBIS and EC-LUE models were comparable, reaching 107.50±1.37 Pg C $year^{-1}$ and 109.39±1.48 Pg C $year^{-1}$ (mean value ± standard deviation) globally, respectively (Fig. 5). Two-model comparisons revealed consistent GPP estimations for the various PFTs (Table 3) with the exception of savanna, for which the IBIS model greatly underestimated GPP. The GPP estimate of the IBIS-Type scheme was much higher than those of the other two models, with a global value of 123.97±1.76 Pg C $year^{-1}$ (Fig. 5). Larger GPP estimations using the IBIS-Type simulation were found for most biomes (Table 3).

Discussion

Process-based ecosystem models are one of the most important components of earth system models used to predict future climate change [1]. The IPCC AR5 requested that all earth system models integrate the global carbon cycle module [7]. Previous studies have shown that the uncertainty in carbon cycle models can produce 40% differences in the predicted temperature by 2100 [28]. The GPP is the total photosynthetic uptake or carbon assimilation by plants, and it is a key component of terrestrial carbon balance. Any errors in the GPP simulations will propagate through the model, introducing errors into the simulated biomass and net ecosystem

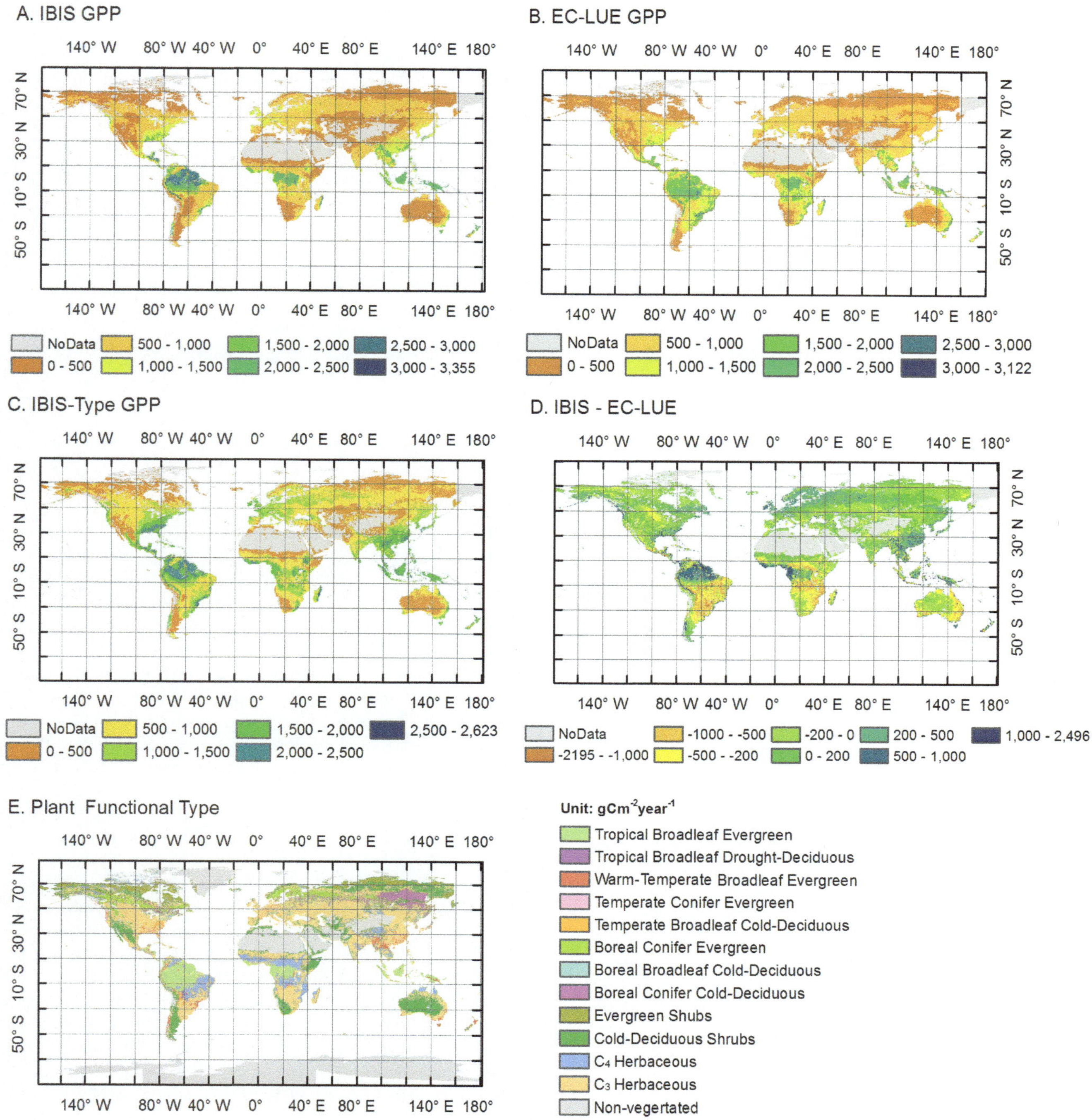

Figure 4. The global pattern of GPP. The global pattern of annual vegetation gross primary production (GPP) from 2000 to 2006. (a) estimated GPP using the IBIS model, (b) estimated GPP using the EC-LUE model, (c) estimated GPP using the IBIS-Type model, (d) the difference between the IBIS and EC-LUE models and (e) the spatial distribution of plant functional types (PFTs).

fluxes. If the simulated GPP is too low or too high, predicted leaf area index, wood biomass, crop yield, and soil biomass may also be too low or too high [29].

In this study, we examined the performance of the IBIS, which has been widely used to evaluate the regional and global terrestrial ecosystem carbon balance and has been integrated into earth system models (e.g., Brazilian Earth System) [30]. Our results indicate that the IBIS model is a good candidate for simulating GPP at regional-to-global scales, and its performance was comparable to that of the satellite-based EC-LUE model based on EC site validation and comparison. The magnitude of global GPP estimated by the IBIS is also consistent with results of previous studies. Our estimate of annual global mean GPP was 107.50 ± 1.37 Pg C year^{-1}(mean value ± standard deviation). Beer et al. estimated global GPP as 123.8 Pg C year^{-1} [31]. Two satellite-based light-use efficiency models revealed similar estimates of global GPP: 109.29 Pg C year^{-1} by the MODIS algorithm [27] and 111 Pg C year^{-1} by the EC-LUE [21]. Interestingly, we found that the IBIS was consistent with the EC-LUE model across different PFTs (Table 3), despite the large

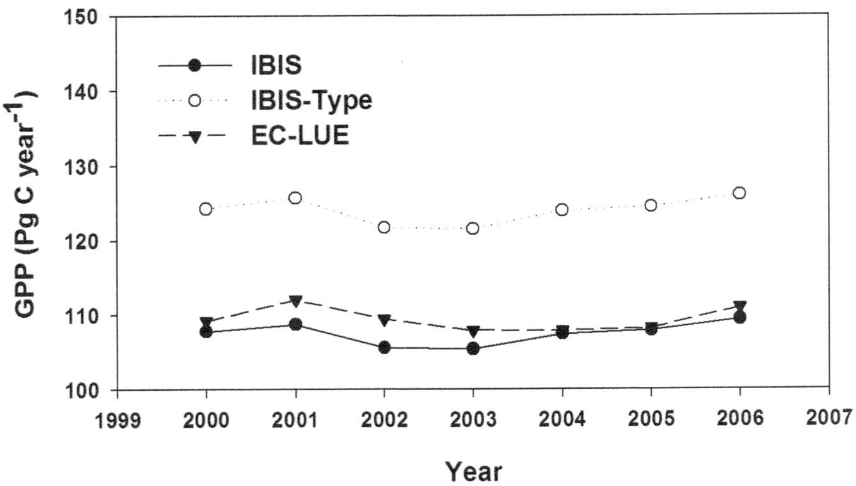

Figure 5. Interannual variability in GPP derived from different models. Interannual variability in global mean gross primary production (GPP) derived from the IBIS, IBIS-Type and EC-LUE models.

differences between the two approaches (i.e., satellite-based vs. process-based).

However, the IBIS did not perform well for two PFTs: evergreen broadleaf forest and savanna. Among the four evergreen broadleaf forest sites, two (PT-Mi1 and FR-Pue) have subtropical Mediterranean climate with dry summers and wet winters [32–33], and the AU-Tum site also suffers drought in summer [34]. At these sites, rainfall is the key driver of water and carbon fluxes [35]. Leuning analyzed the CO_2 and H_2O fluxes at the AU-Tum site [34], and found that carbon uptake was more strongly constrained by water stress than by temperature, and strongly affected by soil water availability. Moreover, savannas are characterized by climate with distinct wet and dry seasons, and this climate forcing causes savanna to form open, heterogeneous woodland canopies with grass understories [36].

These particular ecosystem properties result in greater complexity of modeling fluxes [37]. Process-based models need to simulate variation in soil moisture and plant phenology. However, previous studies have identified significant biases when simulating soil moisture [38]. We examined the performance of the IBIS on soil water at the savanna sites and also found obvious differences between the simulated values and the EC measurements (Fig. 6). Moreover, the IBIS integrated temperature-dominated phenology algorithms developed by Botta et al. [39]. However, field studies suggest that for many drought-deciduous species, the first large precipitation event at the start of the rainy season initiates rapid leaf flush [40–41], and leaf senescence is closely related to soil water availability in the dry season [41–42]. This relationship may explain why the IBIS model did not effectively capture the variance in GPP at savanna sites.

Parameterization is another large source of uncertainty in process-based models. V_{cmax} is the key parameter of the photosynthesis process [11], and a study by Bonan et al. [43] suggests that uncertainty in this parameter could account for 30 Pg C year^{-1} variations in model estimation of global GPP. In the present study, the differences in setting V_{cmax} values between the IBIS and IBIS-Type schemes caused an increase in global GPP of over 16.5 Pg C year^{-1}, which also indicated that V_{cmax} is a salient parameter for simulating GPP. Unfortunately, the determination of V_{cmax} in current models contains large uncertainty. Rogers surveyed V_{cmax} in current state-of-the-art models [44] and found that V_{cmax} varied within a wide range of −46 to +77% of the PFT mean. Thus, the determination of V_{cmax} and the reduction of uncertainty in this parameter are important issues for model development.

Many parameters in process-based models are established by PFTs, which are based on the assumption that the same type of vegetation responds similarly to the environment. However, a current study found that model parameters were more variable than previously assumed within the given PFTs [45] and that categorization of vegetation into less than eight PFTs may result in artificial multiple steady-states in a model of the Earth's climate-vegetation system depending on the number of PFTs used [46]. In the present study, the variation analysis showed that V_{cmax} did not significantly differ among PFTs. The predetermined parameterization scheme that sets the V_{cmax} constant values for each PFT in the IBIS model may cause systematic error.

We attempted to test the impact of V_{cmax} on model performance. However, variations in V_{cmax} cannot explain the overall uncertainty of the IBIS model. Validation of the IBIS on three flux sites demonstrated that parameterization and formula-

Table 3. The magnitude of gross primary production (GPP) in each plant functional type (PFT).

GPP (kgCm^{-2}year^{-1})	DBF	EBF	ENF	MF	GRA	SAV
IBIS	0.89	2.00	1.22	1.19	0.68	0.63
IBIS-TYPE	1.05	1.68	1.51	1.53	0.89	0.74
EC-LUE	0.78	1.76	0.96	1.16	0.73	0.92

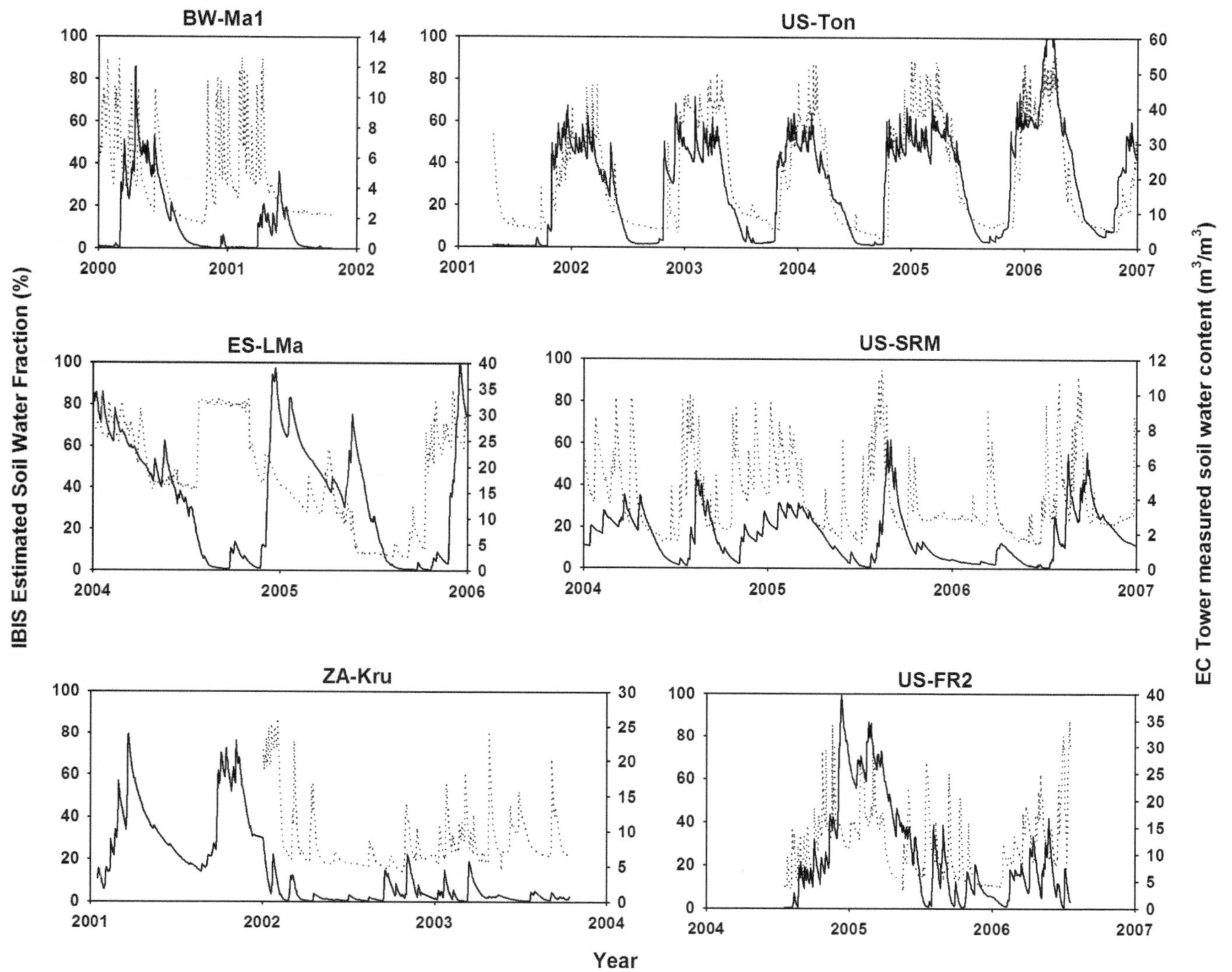

Figure 6. Comparison of estimated and observed soil water. Daily variation in the estimated soil water fraction of the IBIS model (i.e., fraction of soil pore space containing liquid water) and in the observed soil water content at EC sites. The solid lines represent simulation data, and the dotted lines represent the observed data.

tions of phenology also limit the model's ability to capture seasonal fluctuations in carbon and water exchange [38]. Particularly in regions with summer drought, phenology is primarily controlled by water supply rather than temperature [34]. This relationship hinders model simulation because biases in phenology and the dynamics of the leaf area index can affect the simulation of evapotranspiration. Such errors can pass to simulations of soil water content and other variables associated with the water and carbon cycles [38]. Therefore, revisions of not only the parameterization of V_{cmax} but also other parameters and formulations are needed for model improvement.

Summary

Process-based models are important tools for carbon cycle research, but current models incorporate substantial uncertainty. This study examines the performance of the IBIS model at global EC sites. Our results showed that the IBIS model explained 60% of the variation in GPP at all EC sites and performed comparably to the EC-LUE model, which explained 80% of the variation in observed GPP. At the global scale, the magnitudes of GPP estimated by the IBIS and EC-LUE models were comparable, being 107.50±1.37 Pg C year^{-1} and 109.39±1.48 Pg C year^{-1}, respectively. The parameter V_{cmax} is a key parameter in the photosynthesis model. In the IBIS model, V_{cmax} was set as a constant for each PFT. The inversed V_{cmax} value was largely differentiated from the original setting, and no significant differences were detected among PFTs.

Acknowledgments

This work used eddy covariance data acquired by the FLUXNET community; in particular, by the following networks: AmeriFlux (U.S. Department of Energy, Biological and Environmental Research, Terrestrial Carbon Program (DE-FG02-04ER63917 and DE-FG02-04ER63911), AfriFlux, AsiaFlux, CarboAfrica, CarboEuropeIP, CarboItaly, CarboMont, ChinaFlux, Fluxnet-Canada (supported by CFCAS, NSERC, BIOCAP, Environment Canada, and NRCan), GreenGrass, KoFlux,

LBA, NECC, OzFlux, TCOS-Siberia, USCCC. We acknowledge the financial support of the eddy covariance data harmonization provided by CarboEuropeIP, FAO-GTOS-TCO, iLEAPS, Max Planck Institute for Biogeochemistry, National Science Foundation, University of Tuscia, Université Laval, Environment Canada and the US Department of Energy. We also acknowledge the database development and technical support from Berkeley Water Center, Lawrence Berkeley National Laboratory, Microsoft Research Science, Oak Ridge National Laboratory, University of California–Berkeley and the University of Virginia.

Author Contributions

Conceived and designed the experiments: WY. Performed the experiments: DL. Analyzed the data: DL. Wrote the paper: DL WC JX WD GZ YC HZ WY.

References

1. IPCC (2007) Climate Change 2007: The Physical Science Basis. Contribution of Working Group I to the Fourth Assessment Report of the Intergovernmental Panel on Climate Change [Solomon S, Qin D, Manning M, Chen Z, Marquis M, et al. (eds).]. Cambridge University Press, Cambridge, United Kingdom and New York, NY, USA.
2. Raupach MR, Canadell JG, Quere CL (2008) Anthropogenic and biophysical contributions to increasing atmospheric CO_2 growth rate and airborne fraction. Biogeosciences 5(6): 1601–1613.
3. Rojstaczer S, Sterling SM, Moore NJ (2001) Human appropriation of photosynthesis products. Science 294(5551): 2549–2552.
4. Imhoff ML, Bounoua L, Ricketts T, Loucks C, Harriss R, et al. (2004) Global patterns in human consumption of net primary production. Nature 429(6994): 870–873.
5. Yuan WP, Liu SG, Zhou GS, Zhou GY, Tieszen LL, et al. (2007) Deriving a light use efficiency model from eddy covariance flux data for predicting daily gross primary production across biomes. Agricultural and Forest Meteorology 143(3–4): 189–207.
6. Zhang FM, Chen JM, Chen JQ, Gough CM, Martin TA, et al. (2012) Evaluating spatial and temporal patterns of MODIS GPP over the conterminous US against flux measurements and a process model. Remote Sensing of Environment 124: 717–729.
7. Friedlingstein P, Cox P, Betts R, Bopp L, von Bloh W, et al. (2006) Climate-carbon cycle feedback analysis: Results from the C4MIP model intercomparison. Journal of Climate 19(14): 3337–3353.
8. Schwalm CR, Williams CA, Schaefer K, Anderson R, Arain MA, et al. (2010) A model-data intercomparison of CO_2 exchange across North America: Results from the North American Carbon Program site synthesis. Journal of Geophysical Research: Biogeosciences 153(G3): G00H05.
9. Knorr W, Heimann M (2001) Uncertainties in global terrestrial biosphere modeling 1. A comprehensive sensitivity analysis with a new photosynthesis and energy balance scheme. Global Biogeochemical Cycles 15(1): 207–225.
10. Wolf A, Akshalov K, Saliendra N, Johnson DA, Laca EA (2006) Inverse estimation of V_{cmax}, leaf area index, and the Ball-Berry parameter from carbon and energy fluxes. Journal of Geophysical Research: Atmosphere 111(D8) DOI: 10.1029/2005JD005927.
11. Friend AD (2010) Terrestrial plant production and climate change. Journal of Experimental Botany 61(5): 1293–1309.
12. Friend AD, Arneth A, Kiang NY, Lomas M, Ogee J, et al. (2007) FLUXNET and modeling the global carbon cycle. Global Change Biology 13(3): 610–633.
13. Baldocchi D, Falge E, Gu LH, Olson R, Hollinger D, et al. (2001) FLUXNET: A new tool to study the temporal and spatial variability of ecosystem-scale carbon dioxide, water vapor, and energy flux densities. Bulletin of the American Meteorological Society 82(11): 2415–2434.
14. Foley JA, Prentice IC, Ramankutty N, Levis S, Pollard D, et al. (1996) An integrated biosphere model of land surface processes, terrestrial carbon balance, and vegetation dynamics. Global Biogeochemical Cycles 10(4): 603–628.
15. Farquhar GD, Caemmerer SV, Berry JA (1980) A biochemical-model of photosynthetic CO_2 assimilation in leave of C-3 species. Planta 149(1): 79–90.
16. Collatz GJ, Ball JT, Grivet C, Berry JA (1991) Physiological and environmental-regulation of stomata conductance, photosynthesis and transpiration - a model that includes a laminar boundary-layer. Agricultural and Forest Meteorology 54(2–4): 107–136.
17. Metropolis N, Rosenbluth AW, Rosenbluth MN, Teller AH, Teller E (1953) Equation of state calculations by fast computing machimes. Journal of Chemical Physics 21(6): 1087–1092.
18. Hastings WK (1970) Monte-Carlo sampling methods using Markov Chains and their applications. Biometrika 57(1): 97.
19. Xu T, White L, Hui DF, Luo YQ (2006) Probabilistic inversion of a terrestrial ecosystem model: Analysis of uncertainty in parameter estimation and model prediction, Global Biogeochemical Cycles 20(2) DOI: 10.1029/2005GB002468.
20. Yuan WP, Liang SL, Liu SG, Weng ES, Luo YQ, et al. (2012) Improving model parameter estimation using coupling relationships between vegetation production and ecosystem respiration. Ecological Modelling 240: 29–40.
21. Yuan WP, Liu SG, Yu GR, Bonnefond JM, Chen JQ, et al. (2010) Global estimates of evapotranspiration and gross primary production based on MODIS and global meteorology data. Remote Sensing of Environment 114(7): 1416–1431.
22. Yuan W, Luo Y, Li X, Liu S, Yu G, et al. (2011) Redefinition and global estimation of basal ecosystem respiration rate. Global Biogeochemical Cycles 25(4) DOI: 10.1029/2011GB004150.
23. Li X, Liang S, Yu G, Yuan W, Cheng X, et al. (2013) Estimation of gross primary production over the terrestrial ecosystems in China. Ecological Modelling 261–262(0): 80–92.
24. Reichstein M, Falge E, Baldocchi D, Papale D, Aubinet M, et al. (2005) On the separation of net 20 ecosystem exchange into assimilation and ecosystem respiration: review and improved algorithm. Global Change Biology 11: 1424–1439.
25. Papale D, Reichstein M, Aubinet M, Canfora E, Bernhofer C, et al. (2006) Towards a standardized processing of Net Ecosystem Exchange measured with eddy covariance technique: algorithms and uncertainty estimation. Biogeosciences 3: 571–583.
26. Moffat AM, Papale D, Reichstein M, Hollinger DY, Richardson AD, et al. (2007) Comprehensive comparison of gap-filling techniques for eddy covariance net carbon fluxes. Agricultural and Forest Meteorology 147(3–4): 209–232.
27. Zhao M, Heinsch FA, Nemani RR, Runing SW (2005) Improvements of the MODIS terrestrial gross and net primary production global data set. Remote Sensing of Environment 95(2): 164–176.
28. Huntingford C, Lowe JA, Booth BBB, Jones CD, Harris GR, et al. (2009) Contributions of carbon cycle uncertainty to future climate projection spread. Tellus B 61(2): 355–360.
29. Schaefer K, Collatz JG, Tans P, Denning SA, Baker I, et al. (2008) Combined Simple Biosphere/Carnegie-Ames-Stanford Approach terrestrial carbon cycle model. Journal of Geophysical Research-Biogeosciences 113(G3) DOI: 10.1029/2007JG000603.
30. Nobre P, Siqueira LSP, de Almeida RAF, Malaguitti M, Giarolla E, et al. (2013) Climate simulation and change in the Brazilian Climate Model. Journal of Climate 26(17): 6716–6732.
31. Beer C, Reichstein M, Tomelleri E, Ciais P, Jung M, et al. (2010) Terrestrial gross carbon dioxide uptake: Global distribution and covariation with climate. Science 329(5993): 834–838.
32. Reichstein M, Tenhunen J, Roupsard O, Ourcival JM, Rambal S, et al. (2003) Inverse modeling of seasonal drought effects on canopy CO_2/H_2O exchange in three Mediterranean ecosystems. Journal of Geophysical Research Atmospheres 108(D23): 4726.
33. David TS, Ferreira MI, Cohen S, Pereira JS, David JS (2004) Constraints on transpiration from an evergreen oak tree in southern Portugal. Agricultural and Forest Meteorology 122(3–4): 193–205.
34. Leuning R, Cleugh HA, Zegelin SJ, Hughes D (2005) Carbon and water fluxes over a temperate Eucalyptus forest and a tropical wet/dry savanna in Australia: measurements and comparison with MODIS remote sensing estimates. Agricultural and Forest Meteorology 129(3–4): 151–173.
35. Yuan W, Luo Y, Richardson AD, Oren R, Luyssaert S, et al. (2009) Latitudinal patterns of magnitude and interannual variability in net ecosystem exchange regulated by biological and environmental variables. Global Change Biology 15(12): 2905–2920.
36. Eamus D, Prior L (2001) Ecophysiology of trees of seasonally dry tropics: Comparisons among phenologies. Advances in Ecological Research 32: 113–197.
37. Baldocchi D, Xu L, Kiang N (2004) How plant functional-type, weather, seasonal drought, and soil physical properties alter water energy fluxes for an oak-grass savanna and an annual grassland. Agricultural and Forest Meteorology 123(1–2): 13–39.
38. Kucharik CJ, Barford CC, Maayar ME, Wofsy SC, Monson RK, et al. (2006) A multiyear evaluation of a Dynamic Global Vegetation Model at three AmeriFlux forest sites: Vegetation structure, phenology, soil temperature, and CO_2 and H_2O vapor exchange. Ecological Modelling 196(1–2): 1–31.
39. Botta A, Viovy N, Ciais P, Friedlingstein P, Monfray P (2000) A global prognostic scheme of leaf onset using satellite data. Global Change Biology 6(7): 709–725.
40. Monasterio M, Sarmiento G (1976) Phenological strategies of plant species in the tropical savanna and the semi-deciduous forest of the Venezuelan Llanos. Journal of Biogeography 3(4): 325–355.
41. Borchert R (1994) Soil and stem water storage determine phenology and distribution of tropical dry forest trees. Ecology 75(5): 1437–1449.
42. Childes SL (1988) Phenology of nine common woody species in semi-arid, deciduous Kalahari Sand vegetation. Vegetatio 79(3): 151–163.
43. Bonan GB, Lawrence PJ, Oleson KW, Levis S, Jung M, Reichstein M, et al. (2011) Improving canopy processes in the Community Land Model version 4 (CLM4) using global flux fields empirically inferred from FLUXNET data. Journal of Geophysical Research-Biogeosciences 116(G2) DOI: 10.1029/2010JG001593.

44. Rogers A (2013) The use and misuse of V_{cmax} in Earth System Models. Photosynthesis Research 1–15.
45. Groenendijk M, Dolman AJ, van der Molen MK, Leuning R, Arneth A, et al. (2011) Assessing parameter variability in a photosynthesis model within and between plant functional types using global Fluxnet eddy covariance data. Agricultural and Forest Meteorology 151(1): 22–38.
46. Kleidon A, Fraedrich K, Low C (2007) Multiple steady-states in the terrestrial atmosphere-biosphere system: a result of a discrete vegetation classification. Biogeosciences 4(5): 707–714.
47. Kucharik CJ, Foley JA, Delire C, Fisher VA, Coe MT, et al. (2000) Testing the performance of a Dynamic Global Ecosystem Model: Water balance, carbon balance and vegetation structure. Global Biogeochemical Cycles 14(3): 795–825.

19

Inorganic Nitrogen Supply and Dissolved Organic Nitrogen Abundance across the US Great Plains

Megan L. Mobley[1,2]*, Matthew J. Cleary[1], Ingrid C. Burke[1,2,3]

1 Department of Botany, University of Wyoming, Laramie, Wyoming, United States of America, **2** Helga Otto Haub School of Environment and Natural Resources, University of Wyoming, Laramie, Wyoming, United States of America, **3** Program in Ecology, University of Wyoming, Laramie, Wyoming, United States of America

Abstract

Across US Great Plains grasslands, a gradient of increasing mean annual precipitation from west to east corresponds to increasing aboveground net primary productivity (ANPP) and increasing N-limitation. Previous work has shown that there is no increase in net N mineralization rates across this gradient, leading to the question of where eastern prairie grasses obtain the nitrogen to support production. One as-yet unexamined source is soil organic N, despite abundant literature from other ecosystems showing that plants take up dissolved soil organic N. This study measured KCl-extractable dissolved organic N (DON) in surface soils across the grassland productivity gradient. We found that KCl-extractable DON pools increased from west to east. If available to and used by plants, this DON may help explain the high ANPP in the eastern Great Plains. These results suggest a need for future research to determine whether, in what quantities, and in what forms prairie grasses use organic N to support primary production.

Editor: Hormoz BassiriRad, University of Illinois at Chicago, United States of America

Funding: This research was funded by the University of Wyoming Excellence Funds. The funders had no role in study design, data collection and analysis, decision to publish, or preparation of the manuscript.

Competing Interests: The authors have declared that no competing interests exist.

* Email: meg.mobley@alumni.duke.edu

Introduction

It has in the past been widely accepted that nitrogen compounds must undergo mineralization – that is, transformation from organic to inorganic forms - to become available to plants [1]. Yet, in a variety of ecosystems, plants have been shown to take up dissolved organic nitrogen (DON) directly from the soil, as reviewed by [2,3]. These plant-accessible organic N compounds include amino acids, amino sugars, amino-polysaccharides, nucleic acids, and various purine and pyrimidine compounds [2]. A new paradigm concerning N cycling has gained acceptance [2], in which the rate of depolymerization of organic N polymers into monomers by microbial enzymes, i.e., the production of DON, regulates ecosystem N cycling rather than the rate of N mineralization.

The ecological importance of DON acquisition is still unclear in many ecosystems [4], including temperate grasslands. Jones et al. [5,6] suggest that temperate grassland species can take up DON directly from the soil, although it is unclear how important DON is relative to other sources. Knowledge about the magnitude and importance of plant uptake of DON is important for our understanding and predictions of N cycling using models, and for clarifying the dynamics of plant-microbial competition. Understanding the potential for grassland plant use of DON could lead to management strategies that reduce fertilizer use and associated nutrient runoff and pollution while still supporting productive and N-retentive cereal croplands and biofuel-producing grasslands.

For grasslands of the United States, the combined results of previous research do not frame a coherent understanding of grassland N cycling. A number of studies have shown grassland plant production to respond to experimental N additions, suggesting that these ecosystems are N–limited [7–10]; a literature review concluded that although water was an important co-limiter of the semi-arid western grasslands, N-limitation was the rule across the Great Plains [11]. Aboveground net primary production (ANPP) increases with mean annual precipitation from semiarid shortgrass steppe in the west through mixed grass prairie to the subhumid tallgrass prairie in the east [12]. Both the concentration of N in biomass and the litter decomposition rate decrease, and nutrient use efficiency increases, along this precipitation gradient [13]. Nevertheless, plant N uptake, as indicated by the quantity of N in aboveground production (ANPP-N), increases along this precipitation and production gradient [14,15]. These lines of evidence suggest that the severity of N-limitation increases from west to east.

To explain observed patterns of ANPP, simulations of the Century model predict a linear increase of net N mineralization rates with precipitation [16], yet field studies have reported no west-to-east trend in net N mineralization [14,15]. Total soil N, on the other hand, has been shown to be higher in tallgrass than shortgrass prairie [13–15,17], suggesting that organic N may play a role in supplying the additional N needed to support eastern tallgrass prairie production. This could explain the discrepancy between model predictions and field studies, as commonly used ecosystem models do not consider the rate of organic N depolymerization or plant uptake of organic N, focusing only on inorganic N mineralization [18–21].

This exploratory study examined the spatial pattern in abundances and rates of production of KCl-extractable dissolved inorganic N (DIN) and DON along the west to east precipitation gradient across US grasslands.

Materials and Methods

Site Description

The Nature Conservancy, Fort Hays State University, USDA-ARS, and the Long-Term Ecological Research programs at Konza Prairie and Shortgrass Steppe provided access to field sites spanning from eastern Colorado to eastern Kansas [14,15,17]. The entire gradient spans precipitation and temperature ranges of 500 mm and 4.4°C, respectively. The two westernmost sites, Shortgrass Steppe (SGS) Long-Term Ecological Research (LTER) and Arikaree River (ARI) Nature Conservancy site, are located in the drier shortgrass steppe. Two sites, Smokey River (SVR) Nature Conservancy site and a property of Ft. Hays State University (HAYS), were located in the northern mixed grass prairie. The easternmost site was located in the wetter tallgrass prairie, Konza Prairie LTER (KON) (Table 1). Each sampling location was located on native upland grasslands with moderate grazing by cattle in the summer and minimal disturbance. The soils at all sites ranged from loam to sandy loam. The vegetation of the shortgrass steppe was dominated by *Bouteloua gracilis* (Willd. ex Kunth) Lag. ex Griffiths and *Buchloë dactyloides* (Nutt.) J.T. Columbus, whereas *Andropogon gerardii* Vitman and *Schizachyrium scoparium* (Michx.) Nash dominated the tallgrass prairie. The mixed-grass prairie comprised a combination of all of the species above.

Field Sampling

At each site, we set up three 25 m long transects within a 100 m radius of each other in undisturbed areas representative of the surrounding vegetation. From June 6–15, 2011, we collected soil cores (10 cm length ×5 cm diameter) from two microsites (under plant canopies and interspaces between plants) and at two depths (0–5 cm and 5–10 cm) at three random locations on each transect, for a total of 180 soil samples (2 microsites ×2 depths ×3 locations ×3 transects ×5 sites). Soils were immediately sieved through a 2 mm mesh and homogenized in a plastic bag.

We extracted soils immediately for dissolved N pools, in order to minimize disturbance-induced N transformations associated with delays in processing, and to obtain the most realistic estimates of DON and DIN concentrations possible [22]. We extracted dissolved pools of N following the procedure outlined by Robertson et al. [23] and Bardgett et al. [24]. A 10 g subsample of field-fresh, sieved, homogenized soil was placed in an extraction cup with 100 mL 2M KCl solution and shaken for 10 minutes. The extraction cups were returned to the laboratory at the University of Wyoming, Laramie, WY, to be filtered and frozen within two days of sampling. The remaining soil was air-dried and stored in paper bags until further analysis.

We installed one pair of Plant Root Simulator (PRS) resin probes (Western Ag Innovations Inc., Saskatoon, Saskatchewan) per microsite at 10 random locations along each transect (2 microsites ×10 locations ×3 transects ×5 sites = 300 pairs), to estimate inorganic N supply rate. Each pair consisted of one anion and one cation probe to collect mobile NO_3^- and NH_4^+, respectively. Probes were inserted into the soil with the adsorption membrane extending from the surface to 5.5 cm depth, and were buried for 30 days from early May to early June 2011. At each site, the probes were removed on the same day as the soil sampling. Probes were sorted into two groups of five pairs of probes from each microsite type. This resulted in two replicate sets of probes per each microsite type per each transect for a total of 60 (2 replicates ×2 microsites ×3 transects ×5 sites) probe sets. Probes were washed in the laboratory with DI water within two days of unearthing and then sent to Western Ag Innovations, Inc. (Saskatoon, Saskatchewan) for inorganic N analysis.

Laboratory Analyses

The eluate from the extraction cups was shaken for 1 hour and filtered in the laboratory using Whatman #1 filter paper and frozen immediately thereafter. For measurement of total KCl-extractable dissolved total N and inorganic N, we thawed the eluate and diluted a 2.5 mL subsample with 17.5 mL DI water. We analyzed total dissolved N on a Shimadzu TOC-V Analyzer (Shimadzu Scientific Instruments, Wood Dale, IL, USA). Dissolved inorganic N was analyzed at the Natural Resource Ecology Laboratory (Colorado State University, Fort Collins, CO) using an Alpkem Flow Solution IV Automated wet chemistry system (O.I. Analytical, College

Table 1. Locations and attributes of grassland sites.

	SGS:	ARI:	SVR:	HAYS:	KON:
	Shortgrass Steppe LTER	**Arikaree River TNC**	**Smokey River TNC**	**Ft. Hays State Univ.**	**Konza Prairie LTER**
Latitude, Longitude	40°48.8′N, 104°42.8′W	39°45.4′N, 102°26.9′W	38°53.6′N, 100°58.7′W	38°52.4′N, 99°23.3′W	39°4.8′N, 96°33′W
MAP (mm)	320	450	502	575	835
MAT (°C)	8.6	9.7	10.8	11.9	13.0
Grassland Type	Shortgrass Steppe	Shortgrass Steppe	Mixed Grass Prairie	Mixed Grass Prairie	Tallgrass Prairie
ANPP-N [15] (g N m^{-2} y^{-1})	1.1	1.7	1.5	1.6	2.4
ANPP-C [15] (g C m^{-2} y^{-1})	143	123	221	211	337
Net N min. [14] (g N m^{-2} y^{-1})	1.5 (0.49)	3.0 (1.04)	1.5 (0.53)	3.0 (0.63)	2.4 (0.26)
Soil C (g C m^{-2})	1367 (67.9)	1440 (68.6)	1851 (52.9)	5204 (155.4)	2373 (151.2)
Soil N (g N m^{-2})	287.2 (13.1)	274.2 (13.0)	214.6 (13.3)	285.1 (5.9)	210.7 (11.0)
Soil C:N	4.91 (0.31)	5.33 (0.25)	9.05 (0.50)	18.3 (0.44)	11.2 (0.32)
Volumetric Soil Moisture (%)	13.8 (0.40)	7.3 (0.53)	3.7 (0.17)	3.4 (0.36)	6.3 (0.30)

MAP = mean annual precipitation (mm), MAT = mean annual temperature (°C), TNC = The Nature Conservancy, LTER = National Science Foundation Long-Term Ecological Research site. Data in lower panel is from this study; values in parentheses represent one standard error of the mean.

Station, TX, USA). We estimated KCl-extractable DON as the difference between total dissolved N and dissolved inorganic N.

We mixed and subsampled remaining air-dried soils for analysis of total C and N. Subsamples were ground on a rotary soil grinder (Planetary Micro Mill Pulverisette 7, Fritsch GmbH, Idar-Oberstein, Germany), and 2 mg of ground soil were analyzed on a Perkin Elmer 2400 Elemental Analyzer (Perkin Elmer, Inc., Waltham, MA, USA) for total C and N. Volumetric water content was estimated after drying a 100 g subsample of field moist soil at 105°C for one day [23].

Statistical Analyses

We performed all analyses in the statistical program R, version 2.14.2 [25]. Extractable DIN and DON concentrations were multiplied by the soil sample depth (5 cm) and bulk density (mean $= 0.877$ g cm^{-3}) to obtain DIN and DON pools on a per-area basis. Data for the two soil depth increments (0–5 cm, 5–10 cm) were aggregated such that all analyses were of soil properties and constituents to 10 cm depth, with the exception of PRS probe data that characterized the top 5 cm of soil. We employed two-way ANOVAs coupled to Tukey multiple comparison tests [26] to test differences in the dependent variables total soil C (g C m^{-2}), N (g N m^{-2}), C:N (mass C:N), and 0–5 cm soil inorganic N supply rate (µg N 10 cm^{-2} $month^{-1}$) among sites and between microsites. Two-way ANCOVAs tested differences in the dependent variables total soil KCl-extractable DIN (g DIN m^{-2}) and KCl-extractable DON (g DON m^{-2}) among sites and between microsites with volumetric soil moisture as a continuous independent variable. One extreme outlier value for DON with strong influence on the ANCOVA (DON>17 g N m^{-2}; outlier test Bonferonni $p = 4.4 \times 10^{-26}$) was excluded from DON analyses. ANOVAs were called using aov(), with drop1(model, test = "F") to extract type III sums of squares. Multiple comparisons were performed using TukeyHSD(). We performed linear regressions using lm() to assess the effect of soil moisture on DIN and DON measurements, and to test for a relationship between KCl-extractable DON pools and MAP. A paired t-test tested differences in DON and DIN abundance (g N m^{-2}) across all sites and microsites. Significance was assessed at $\alpha = 0.05$. Data for this study can be found in the Supporting Information.

Results

Soil C differed significantly among sites (among-sites $F = 212$, d.f. = 4, $p<0.0001$) and was significantly higher in the tallgrass prairie than shortgrass steppe ($p<0.0001$, Table 1). Total soil N (g N m^{-2}) decreased from west to east across the grassland gradient (among-sites $F = 10.7$, d.f. = 4, $p<0.0001$, Table 1). Soil C:N increased significantly from west to east across the gradient (among-sites $F = 211$, d.f. = 4, $p<0.0001$). There were no significant differences in soil C ($p = 0.439$), N ($p = 0.537$), or C:N between under-plant and plant interspace microsites ($p = 0.101$).

The 30-day inorganic N supply rate estimated with PRS probes was highly variable and showed a slight decline across the gradient, with the supply rate at the easternmost site significantly lower than that of the westernmost site (among-sites $F = 8.07$, d.f. = 4, $p<0.0001$; Figure 1a). There were no significant differences in inorganic N supply rates between microsites.

The pool of KCl-extractable dissolved inorganic nitrogen (DIN) was greater in the eastern tallgrass prairie compared to the western shortgrass steppe (among-sites $F = 44.9$, d.f. = 4, $p<0.0001$; Figure 1b). KCl-extractable dissolved organic N was significantly more abundant in the tallgrass region compared to the shortgrass steppe (among-sites $F = 10.6$, d.f. = 4, $p<0.0001$; Figure 1b).

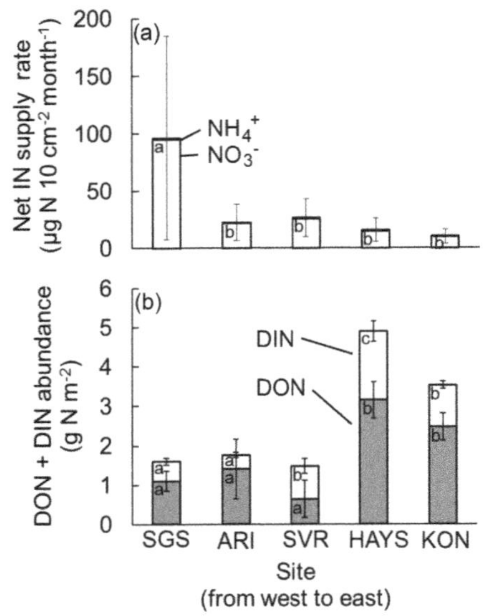

Figure 1. Inorganic N supply rates and extractable dissolved N pools at grassland sites. a) 0–5 cm soil net inorganic N supply rates (NH_4^+ in black bars, NO_3^- in white bars) at five grasslands of the US Great Plains, arranged from west to east. Error bars represent 95% confidence intervals on total ($NH_4^+ + NO_3^-$) inorganic N supply rate. Lowercase letters represent significant site differences at $\alpha = 0.05$. b) Total extractable dissolved soil N (DIN+DON) to 10 cm soil depth across the US Great Plains. Lowercase letters in the gray columns represent significant differences in DON among sites; lowercase letters in the white columns indicate significant differences in DIN among sites. Multiple comparisons were performed on output of ANOVAs that did not include soil moisture as a covariate. Error bars represent 95% confidence intervals. Significance assessed at $\alpha = 0.05$. One extreme outlier (DON>17 g N m^{-2} at SVR) was excluded from the analysis.

There were no significant differences in DIN ($p = 0.128$) or DON ($p = 0.461$) between microsites. Across all sites, abundances of KCl-extractable DON were significantly higher than DIN (paired $t = 6.648$, d.f. = 82, $p<0.0001$). Both DIN (F = 10.4, d.f. = 79, $R^2 = 0.106$, p = 0.0018) and DON (F = 21.1, d.f. = 78, $R^2 = 0.202$, p<0.0001) increased with soil moisture (Figure 2), but soil moisture was not significant as a covariate of DIN ($p = 0.277$) or DON ($p = 0.241$) in the ANCOVAs. There was a positive relationship between DON measurements and site-level ANPP (F = 6.257, d.f. = 42, $R^2 = 0.130$, $p = 0.0164$; Figure S1).

Discussion

This study tested the hypothesis that extractable dissolved organic N pools would increase along a precipitation and production gradient from west to east across US grasslands. Unlike two previous studies at the same sites [14,15], we did not find a west to east increase in total soil N; soil C and C:N did increase from west to east as in previous studies. Also like previous studies, we found no increase in inorganic N supply rates from west to east, even though plant uptake does increase (ANPP-N, Table 1). This supports previous conclusions that N becomes more limiting, and the N cycle grows tighter, from west to east. We did observe, however, that pools of KCl-extractable dissolved inorganic and organic N, representing exchangeable + free soil N [27], increased from west to east.

These data must be considered as suggestive, rather than conclusive, given a number of limitations to this small study. Our one-time measurements of KCl-extractable DIN and DON and

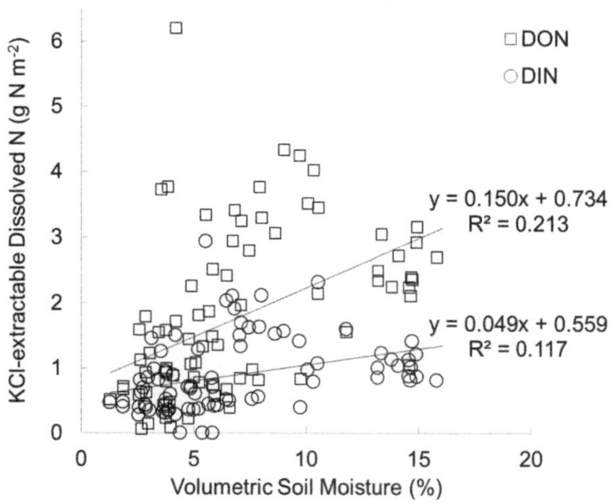

Figure 2. Simple linear regressions of KCl-extractable soil dissolved inorganic (DIN; circles) and organic nitrogen (DON; squares) against volumetric soil moisture for 0–10 cm soil samples from five sites across the Great Plains.

our one resin probe incubation give an incomplete characterization of the long-term mean pools and rates for these grassland sites. Pool size is not necessarily indicative of cycling rate or ecological importance, as a large pool might result from high production or low demand. We did not characterize the chemical composition or molecular weight distribution of DON, thus we cannot speculate on how much of the DON is potentially plant-available. There are artifacts associated with each of the field methods that could reduce the sensitivity of N pool and production estimates. The soil and root disturbance associated with soil sampling may release DON that would not naturally be available in the soil [4], and use of KCl as an extractant may result in a pool of extractable dissolved N that is quite different in composition, quantity, and quality than that found under natural conditions [28,29]. As evidenced by the significant relationships between KCl-extractable N pools and soil moisture (Figure 2), the KCl-extraction likely overestimates what is actually plant-accessible under normal dry soil conditions. Because less nitrate is likely to reach the resin probes when soil moisture is limiting, regardless of actual mineralization rates, we consider resin probes to be better indicators than extractions of the available N that is accessible by plants [30]. These methodological artifacts should not affect comparisons among sites within a given study if consistent methods are used at all sites, as was the case in this study and previous studies at these sites [14,15].

Alternative or additional sources of ANPP-N may include biological N fixation and atmospheric N deposition. N fixation, largely by free-living organisms, has been estimated at <0.1 g N m^{-2} y^{-1} in shortgrass steppe [31] and ~0.8 g N m^{-2} y^{-1} in tallgrass prairie [32], although it is unclear how much of that fixed N enters the plant-soil N cycle versus that volatilized or denitrified [33]. Total wet+dry inorganic N deposition increases from approximately 0.35 g N m^{-2} y^{-1} at SGS to approximately 0.71 g N m^{-2} y^{-1} at KON [34,35], and so could also provide a portion of the additional N required to support ANPP.

This study provides one piece of support for the broader hypothesis that organic N may serve as a nitrogen source for grassland production—a source that would explain how plant N uptake (ANPP-N) is able to increase along the precipitation gradient without a concomitant increase in N mineralization rates. Our results show that a sizeable pool of KCl-extractable dissolved organic N resides in these grassland soils, and that this N pool increases from west to east across the Great Plains precipitation gradient. The larger KCl-extractable DON pools at the eastern end of the precipitation gradient open the possibility that DON as a source of plant N may help resolve discrepancies in between model predictions and measurements of available N. If so, ecosystem models such as Biome-BGC, Century, PnET, and TEM may be missing a source of N for grasslands. These models represent the N required for primary productivity as "mineral N" generated by N mineralization [18–21]. If organic N, which these results indicate is abundant at these sites, is important for plant growth, then an important nutrient source is being left out of model simulations.

Conclusions

We found that the abundance of DON increased across the west-to-east along the precipitation and productivity gradient of US Great Plains grasslands, suggesting that it could be a significant source of N if plants are able to utilize it. If organic N is an important component of available N in US grasslands, it may explain why ecosystem models that do not include plant use of organic N are unable to predict N supply rates across the US grasslands. Additional studies must be conducted to determine if grass species dominating the US Great Plains assimilate biologically-relevant quantities of organic N.

Supporting Information

Figure S1 The relationship between KCl-extractable soil DON and ANPP. Simple linear regression of KCl-extractable soil dissolved organic nitrogen (DON; microsites and depths for each sample location mathematically combined to yield $n = 9$ replicate measurements per site) against site-level mean aboveground net primary production (ANPP; [15]) for five grassland sites across the US Great Plains.

Acknowledgments

We wish to thank The Nature Conservancy, Fort Hays State University, USDA-ARS, and the Long-Term Ecological Research Sites at Konza Prairie and Shortgrass Steppe for access to field sites. We thank Colin Pinney (EcoCore Analytical Services, a Colorado State University Core Facility) and Leann Naughton for assistance with nitrogen analysis. This manuscript was improved by comments from William Lauenroth, Urszula Norton, David Bell, Daniel Schlaepfer, Otgonsuren Avirmed, Ariuna Lkhagva, Lynn Moore, Cait Rottler, and four anonymous reviewers.

Author Contributions

Conceived and designed the experiments: MJC ICB. Performed the experiments: MJC. Analyzed the data: MJC MLM. Contributed reagents/materials/analysis tools: ICB. Wrote the paper: MJC MLM ICB.

References

1. Stanford G, Smith S (1972) Nitrogen mineralization potentials of soils. Soil Science Society of America Journal 36: 465–472.
2. Schimel JP, Bennett J (2004) Nitrogen mineralization: challenges of a changing paradigm. Ecology 85: 591–602.
3. Neff JC, Chapin FS, Vitousek PM (2003) Breaks in the cycle: dissolved organic nitrogen in terrestrial ecosystems. Frontiers in Ecology and the Environment 1: 205–211.
4. Hobbie JE, Hobbie EA (2013) Microbes in nature are limited by carbon and energy: the starving-survival lifestyle in soil and consequences for estimating microbial rates. Frontiers in Microbiology 4: Article 324.
5. Jones DL, Healey JR, Willett VB, Farrar JF, Hodge A (2005) Dissolved organic nitrogen uptake by plants–an important N uptake pathway? Soil Biology and Biochemistry 37: 413–423.
6. Jones DL, Shannon D, Murphy DV, Farrar J (2004) Role of dissolved organic nitrogen (DON) in soil N cycling in grassland soils. Soil Biology and Biochemistry 36: 749–756.
7. Dodd JL, Lauenroth WK (1979) Analysis of the Response of a Grassland Ecosystem to Stress. In: French NR, editor.Perspectives in Grassland Ecology: Springer New York.pp. 43–58.
8. Lauenroth WK, Dodd JL, Sims PL (1978) The effects of water- and nitrogen-induced stresses on plant community structure in a semiarid grassland. Oecologia 36: 211–222.
9. Risser PG, Parton WJ (1982) Ecosystem Analysis of the Tallgrass Prairie: Nitrogen Cycle. Ecology 63: 1342–1351.
10. Clenton EO, Hyde RM, Anderson KL (1970) Effects of Clipping and Supplemental Nitrogen and Water on Loamy Upland Bluestem Range. Journal of Range Management 23: 341–346.
11. Hooper D, Johnson L (1999) Nitrogen limitation in dryland ecosystems: Responses to geographical and temporal variation in precipitation. Biogeochemistry 46: 247–293.
12. Sala OE, Parton W, Joyce L, Lauenroth W (1988) Primary production of the central grassland region of the United States. Ecology 69: 40–45.
13. Murphy KL, Burke IC, Vinton MA, Lauenroth WK, Aguiar MR, et al. (2002) Regional analysis of litter quality in the central grassland region of North America. Journal of Vegetation Science 13: 395–402.
14. McCulley RL, Burke IC, Lauenroth WK (2009) Conservation of nitrogen increases with precipitation across a major grassland gradient in the Central Great Plains of North America. Oecologia 159: 571–581.
15. Barrett J, McCulley R, Lane D, Burke I, Lauenroth W (2002) Influence of climate variability on plant production and N-mineralization in Central US grasslands. Journal of Vegetation Science 13: 383–394.
16. Burke IC, Lauenroth WK, Parton WJ (1997) Regional and temporal variation in net primary production and nitrogen mineralization in grasslands. Ecology 78: 1330–1340.
17. Vinton MA, Burke IC (1997) Contingent effects of plant species on soils along a regional moisture gradient in the Great Plains. Oecologia 110: 393–402.
18. Parton W, Schimel DS, Cole C, Ojima D (1987) Analysis of factors controlling soil organic matter levels in Great Plains grasslands. Soil Science Society of America Journal 51: 1173–1179.
19. McGuire AD, Melillo JM, Kicklighter DW, Joyce LA (1995) Equilibrium responses of soil carbon to climate change: empirical and process-based estimates. Journal of Biogeography 22: 785–796.
20. Running SW, Hunt ER Jr (1993) Generalization of a forest ecosystem process model for other biomes, BIOME-BGC, and an application for global-scale models. In: Ehleringer JR, Field CB, editors.Scaling Physiological Processes: Leaf to Globe.New York: Academic Press, Inc. pp. 141–158.
21. Aber JD, Ollinger SV, Driscoll CT (1997) Modeling nitrogen saturation in forest ecosystems in response to land use and atmospheric deposition. Ecological Modelling 101: 61–78.
22. Van Miegroet H (1995) Inorganic nitrogen determined by laboratory and field extractions of two forest soils. Soil Science Society of America Journal 59: 549–553.
23. Robertson GP, Wedin DA, Groffman PM, Blair JM, Holland EA, et al. (1999) Soil carbon and nitrogen availability. Nitrogen mineralization, nitrification and soil respiration potentials. In: Robertson GP, Coleman DC, Bledsoe CS, Sollins P, editors.Standard Soil Methods for Long-Term Ecological Research.New York: Oxford University Press. pp. 258–271.
24. Bardgett RD, Streeter TC, Bol R (2003) Soil microbes compete effectively with plants for organic-nitrogen inputs to temperate grasslands. Ecology 84: 1277–1287.
25. R Development Core Team (2012) R: A Language and Environment for Statistical Computing. 2.14.2 ed. Vienna, Austria: R Foundation for Statistical Computing.
26. Hsu JC (1996) Multiple comparisons: theory and methods. London: Chapman & Hall. xiv, 277 p., 274 p. of plates p.
27. Näsholm T, Kielland K, Ganeteg U (2009) Uptake of organic nitrogen by plants. New Phytologist 182: 31–48.
28. Inselsbacher E, Näsholm T (2012) The below-ground perspective of forest plants: soil provides mainly organic nitrogen for plants and mycorrhizal fungi. New Phytologist 195: 329–334.
29. Jones DL, Willett VB (2006) Experimental evaluation of methods to quantify dissolved organic nitrogen (DON) and dissolved organic carbon (DOC) in soil. Soil Biology and Biochemistry 38: 991–999.
30. Hook PB, Burke IC (1995) Evaluation of methods for estimating net nitrogen mineralization in a semiarid grassland. Soil Science Society of America Journal 59: 831–837.
31. Copley PW (1972) Evaluation of biological N_2 fixation in a grassland ecosystem. Fort Collins, CO: Colorado State University. 90 p.
32. DuBois JD, Kapustka LA (1983) Biological Nitrogen Influx in an Ohio Relict Prairie. American Journal of Botany 70: 8–16.
33. West NE, Skujins J (1977) The nitrogen cycle in North American cold-winter semi-desert ecosystems. Oecologia plantarum 12: 45–53.
34. National Atmospheric Deposition Program (NRSP-3) (2011) NADP Program Office, Illinois State Water Survey, 2204 Griffith Dr., Champaign, IL 61820.
35. U.S. Environmental Protection Agency Clean Air Markets Division (2011) Clean Air Status and Trends Network (CASTNET) Annual Total Deposition.

Non-Deterministic Modelling of Food-Web Dynamics

Benjamin Planque[1]*, Ulf Lindstrøm[1], Sam Subbey[2]

1 Institute of Marine Research, Tromsø, Norway, **2** Institute of Marine Research, Bergen, Norway

Abstract

A novel approach to model food-web dynamics, based on a combination of chance (randomness) and necessity (system constraints), was presented by Mullon et al. in 2009. Based on simulations for the Benguela ecosystem, they concluded that *observed patterns of ecosystem variability may simply result from basic structural constraints within which the ecosystem functions*. To date, and despite the importance of these conclusions, this work has received little attention. The objective of the present paper is to replicate this original model and evaluate the conclusions that were derived from its simulations. For this purpose, we revisit the equations and input parameters that form the structure of the original model and implement a comparable simulation model. We restate the model principles and provide a detailed account of the model structure, equations, and parameters. Our model can reproduce several ecosystem dynamic patterns: pseudo-cycles, variation and volatility, diet, stock-recruitment relationships, and correlations between species biomass series. The original conclusions are supported to a large extent by the current replication of the model. Model parameterisation and computational aspects remain difficult and these need to be investigated further. Hopefully, the present contribution will make this approach available to a larger research community and will promote the use of non-deterministic-network-dynamics models as 'null models of food-webs' as originally advocated.

Editor: André Chiaradia, Phillip Island Nature Parks, Australia

Funding: This work was supported by the Norwegian Research Council project 200796/E10 BarEcoRe: Barents Sea Ecosystem Resilience under global environmental change. The Norwegian Research Council had no role in study design, data collection and analysis, decision to publish, or preparation of the manuscript.

Competing Interests: The authors have declared that no competing interests exist.

* Email: benjamin.planque@imr.no

Introduction

Natural living systems are characterised by a high level of complexity, which results from the diversity of biological components at many levels of organisation (molecules, cells, organs, individuals, species, communities) and from the diversity of possible interaction types (physical, chemical, trophic, behavioural, cognitive). In addition, many biological interactions are non-linear, include feedback loops, and biological systems display a remarkable ability to constantly adapt and reconfigure themselves. Such systems, which display high complexity, non-linearity, and adaptability have been termed *Complex Adaptive Systems* (CAS, [1]).

The terms *chance*, *randomness* or *stochasticity* are different expressions related to the unpredictability of some events. Whether chance is a true feature in nature, as suggested by Prigogine [2], or simply the result of our inability to accurately observe and model natural phenomena is a matter of debate. However, the existence of *apparent* stochastic phenomena is undisputed. Throwing a dice or playing roulette are ways to produce a random outcome (otherwise there would be no game), despite the fact that these processes are believed to be ruled by the deterministic laws of Newtonian physics. In biology, the two pillars of the theory of evolution are selection and variation. The latter assumes randomness in the way DNA mutations and recombination take place. At a high level of biological organisation, the exact timing, location, and amplitude of extreme events such as pest outbreaks cannot be precisely predicted, although after they have occurred, their space-time evolution may be modelled statistically. These examples point to the central role of stochastic phenomena in real world physical and biological systems, and to the importance of chance in shaping the dynamics of such systems. Natural systems are therefore complex and adaptive systems partially controlled by stochastic phenomena, which makes them difficult to analyse and even harder to predict.

In his seminal work on ecosystem resilience, Holling [3] pointed to the incapability of conventional deterministic models to represent real world living systems because of their inability to integrate complexity, non-linearities, and stochasticity in an appropriate manner. Since Holling's contribution, developments in biological and ecological modelling have explicitly incorporated stochastic processes, although these have generally been built on deterministic *skeletons* [4]. The advent of individual based modelling has also led to population models directly built on stochastic processes occurring at the individual level [5]. Both types of approaches have shown the importance of stochastic processes on population dynamics, confirming the original insight of Holling. The theoretical and mathematical developments of stochastic models (or model components) of animal populations have greatly progressed, but this is not the case for ecosystem models in general and food-web dynamics models in particular. These still depend primarily on deterministic equations that relate predator species and their prey, though stochastic components are sometimes considered in addition to deterministic skeletons.

Apparent stochasticity in ecosystems may be the rule, but this does not mean that ecosystem dynamics are totally random. This is because *Nature*'s configurations are constrained by an ensemble of physical laws (e.g. gravitation, conservation of mass and energy) and evolutionary contingencies (e.g. pool of existing species, rate of genetic mutations). As a result, and as Cury et al. [6] point out, '*Nature may not be predictable but it is not totally unpredictable either*' and it is the combination of stochasticity and constraints that drives the spatial and temporal dynamics of ecosystems. Given the importance of chance (stochasticity) and necessity (constraints), these two elements should have a central place in the development of ecosystem models.

A central issue in the modelling of food-web dynamics has been the use of functional responses that describe trophic functional relationships. The debate on the theoretical foundations for particular functional responses is still open [7,8]. Deriving functional responses from empirical observations has proven difficult because trophic data that can be raised to the population level are parsimonious and empirical relationships are usually masked [9]. In addition, models of food-web dynamics are known to be highly sensitive to small variations in the shape of the functional responses [3,10]. Even when stochastic predator-prey systems have been modelled, these rely on deterministic skeletons and therefore assume that the underlying relationship between prey and predators can be defined deterministically [11].

To our knowledge, only one modelling approach has escaped the deterministic formulation of functional responses. This work was published by Mullon et al. [12] and constitutes a novel alternative to existing food-web dynamics models. A major innovation in this work is the combination of the two fundamental ingredients: randomness (chance) and constraints (necessity). More precisely, in this model, trophic flows (the amount of prey eaten by a predator) are randomly chosen within a set of possible values that fulfil specific physical and biological constraints. The model structure is that of a Non-Deterministic Network Dynamics (NDND) model. Its design is general and can, in principle, be used to simulate the dynamics of any food-web or other similar networks [13]. Mullon et al. present a specific application for the Benguela ecosystem on the basis of which they conclude that "*this model reproduces in a robust manner observed patterns of variability and can be used to question the relevance of other modelling approaches of ecosystem dynamics with regard to determinism, constraints and stochasticity. Referring to a non-deterministic model without any functional relationships and environmental or anthropogenic forcing can help in avoiding misleading advice based on the belief that we can explain the causes of observed patterns, which may simply result from basic structural constraints within which the ecosystem functions*". Despite the important implications of such conclusion for the research community working on ecosystem models, the work of Mullon et al. has received very little attention. Five years after it was published, the article by Mullon et al. has been cited once, and to date, no application of this model has been published for other areas. Does the model really work? Are the conclusions robust?

The objective of the present paper is to evaluate the replicability of the model presented by Mullon et al. For this purpose, we revisit the equations and input parameters that form the structure of the original model, which we term 'MMM' for 'Mullon's Minimal Model' and implement a comparable simulation model. In this contribution, we restate the model principles and provide a detailed account of the model structure, equations, and parameters following a model description protocol known as ODD (Overview, Design concepts, and Details, [14,15]). We use this platform to simulate the dynamics of the Benguela food-web, the original case study of the MMM. We use these simulations to evaluate the replicability of the MMM and the conclusions reached by Mullon et al. in their original contribution.

Model Formulation

The formulation of the non-deterministic network dynamics model for food-web is given below, following the ODD protocol [14,15]. The ODD protocol was designed to provide detailed information about simulation models so that these can be made easier to understand and to duplicate. The protocol was originally designed for individual based models (IBMs), and several components of this framework are specific to IBMs. However, we found many of components of the ODD protocol well suited for the description of the NDND model and the sections below follow the suggested headings of the original ODD protocol. The present model sometimes departs from the original formulation of the MMM. When this is the case, we have highlighted the difference between the two models and presented the justification for such departure.

Model purpose and principle

The purpose of the NDND model is to simulate food-web dynamics, i.e. the interannual fluctuations of the biomass of species and the trophic flows between them. The fundamental principle is that the flows of biomass between predators and prey are not deterministic, but are instead drawn randomly, given that they satisfy an ensemble of physical, physiological, and life-history constraints. Physical constraints are set by the law of conservation of mass, i.e. the total biomass in the food-web is maintained constant if the system is isolated (i.e. when there is no import, loss or export of biomass). Thus, fluctuations of total biomass in the food-web are solely the result of the balance between import of biomass into the food-web (e.g. new production) and export of biomass outside the food-web (e.g. fishing, egestion and metabolic losses). This is a so-called mass-balanced model and, as such, it shares similarities with other mass-balanced models like Ecopath with Ecosim (EwE, [16,17]). Physiological constraints are set by the maximum rates of ingestion by individual organisms of a given species, where this upper limit (i.e., the maximum *consumption over biomass* ratio) is termed *satiation*. In addition, the maximum growth and mortality rates of a population are limited as a function of the species' lifespan, where populations of short-lived species fluctuate at potentially higher rates than populations of long-lived species. The relationship between life span and population growth and mortality rates has both theoretical and empirical support [18,19].

The result is a mass balance model in which 1) trophic flows are drawn randomly for all species, 2) ingestion never exceeds satiation, and 3) the rate of biomass variation is bounded. These principles are identical to those of the MMM. We complemented the principle that the minimum biomass attainable by a population is greater than zero and corresponds to a 'refuge' biomass below which the species is no longer accessible to predators.

We present below the mathematical formulation of the model dynamic equation and constraints, as well as the definition of the model input parameters and how these can be related to ecologically meaningful quantities.

Entities, state variables, and scales

The structuring elements of the model are trophospecies and trophic interactions, which together constitute a food-web topology. A trophospecies represents an ensemble of organisms that share the same set of trophic interactions. All individuals in a trophospecies may not belong to the same taxonomic species; in the following text, we use the word *species* to mean trophospecies. Trophic interactions indicate the possible transfer of mass between two species. Contrary to food-web structure models, which are concerned with how food-webs assemble and how their structure can evolve with speciation and extinctions, the food-web topology of the NDND is fixed and provided as input data. The indices, variables, parameters, constraints, and master equation of the NDND are summarised in Appendix S1, alongside the original formulation of the MMM and their equivalence in the Ecopath with Ecosim framework (EwE, [17]).

State variables. The following variables are required to fully define the state of the system at any time: biomass of individual species i (B_i), trophic fluxes, i.e. biomass flux from species i to species j (F_{ij}), and import (I_i) and export (E_i) of biomass of individual species i in or out of the system.

Input parameters consist of the following:

- assimilation efficiency (γ_i): the proportion of biomass ingested by species i that can contribute to growth, reproduction, and maintenance ($0<\gamma_i<1$),
- satiation (σ_i): the maximum consumption rate by species i, expressed as a proportion of the current biomass of species i ($\sigma_i>0$),
- inertia (ρ_i): the minimum and maximum rates of biomass change are set respectively to $-\rho_i B_i$ and $+\rho_i B_i$ ($\rho_i>0$),
- metabolic and other losses (μ_i): The quantity μ_i represents the rate of biomass loss through metabolism, e.g. locomotion and maintenance, and 'other' mortality (mortality other than the predation explicitly represented in the model) ($\mu_i>0$),
- Refuge biomass (β_i): the irreducible biomass of a species ($\beta_i>0$).

Scales

The temporal resolution of the model is annual, i.e. each time-step in a simulation is one year. The spatial scale is a large 'self contained' ecosystem. The ecological scale is a 'food-web', typically from primary producers to top predators. In the present study, we have applied the model to the simplified Benguela ecosystem as defined in Mullon et al. [12].

Process overview and scheduling

The NDND model represents variations in the biomass of individual species as the result of import, export, and trophic interaction terms. The food-web dynamics is fully defined by the equation that describes variation in the biomass of individual species. This master equation, in continuous time, is:

$$\frac{dB_i}{dt} = \gamma_i \sum_j F_{ji} + I_i - \sum_j F_{ij} - E_i - \mu_i B_i \tag{1}$$

For simulation purpose, the NDND operates in discrete time steps, with one-year time intervals. Assuming that trophic flows, imports, and exports are constant during the integration period, the discretised form of the master equation can be derived (details of this derivation are presented in Appendix S2):

$$B_{i,t+1} = e^{(-\mu_i)} B_{i,t} + \frac{(1-e^{(-\mu_i)})}{\mu_i}\left[\gamma i \sum_j F_{ji,t} + I_i - \sum_j F_{ij,t} - E_i\right] \tag{2}$$

In equation 2, the term $\sum_j F_{ji,t}$ represents the total biomass of prey species consumed by species i between two consecutive time steps; the term $\sum_j F_{ij,t}$ represents the total biomass of species i consumed by its predators during the same period.

Contrary to standard food-web models, the trophic flows are not defined by a deterministic equation. Instead, they are drawn from the set of possible flows which satisfy the following constraints:

-flows are possible (i.e. species j is a predator of species i),

-flows are positive:

$$F_{ij} > 0, \tag{3}$$

-resulting biomasses are not below 'refuge level':

$$B_i \geq \beta_i, \tag{4}$$

-change in biomass is constrained by inertia:

$$e^{-\rho_i} B_{i,t} \leq B_{i,t+1} - I_{i,t} + E_{i,t} \leq e^{\rho_i} B_{i,t}, \tag{5}$$

-total food intake is limited by satiation:

$$\sum_j F_{ji} \leq \sigma_i B_i \tag{6}$$

At each time step, a vector of random flows ($\vec{F}$), which satisfies all the above constraints, is drawn. Solving the multiple constraints equation is a complex computational problem and the solution employed here is presented in details in Subbey et al. (Subbey S, Planque B, Lindstrøm U, submitted. Exploring stochasticity and imprecise knowledge based on linear inequality constraints. SIAM Journal on Scientific Computing).

Design concepts

The main design concept of the model is a food-web topology (species and their trophic links) dynamically modelled by the combination of chance and necessity; chance being modelled by a stochastic process (the random drawing of trophic flows) and

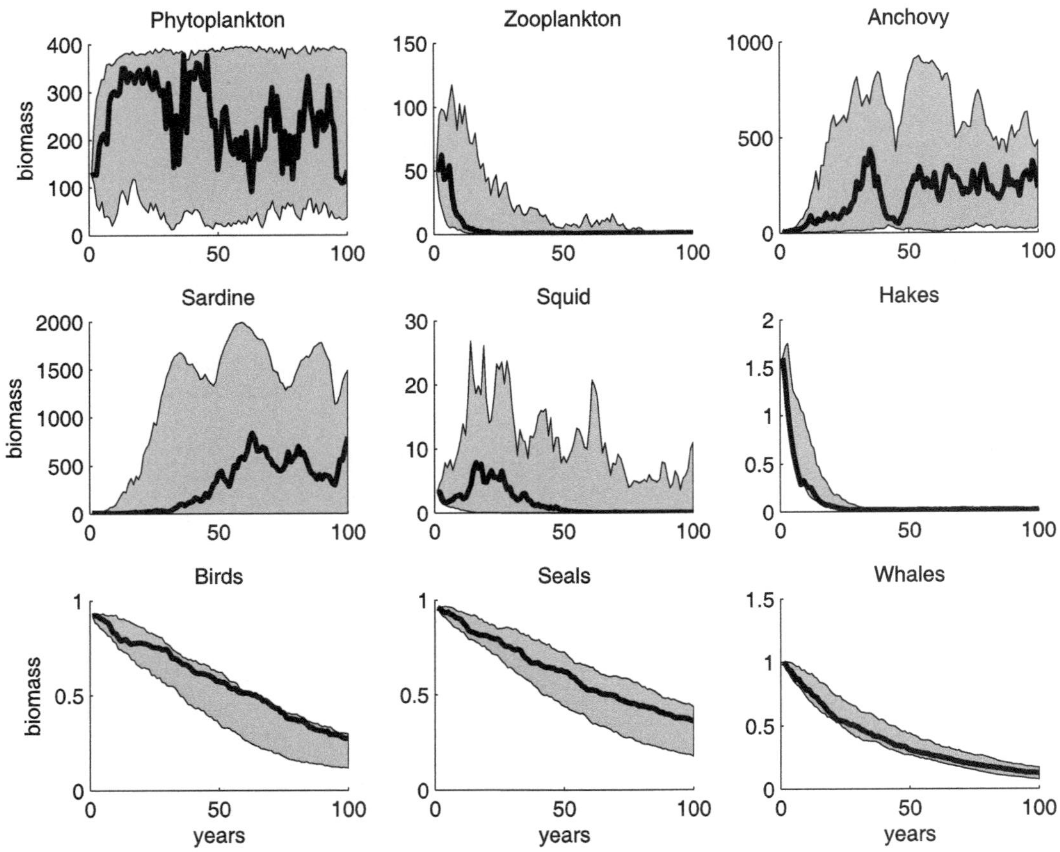

Figure 1. Fifty simulations of the NDND for the Benguela ecosystem, using initial biomass, input parameters and food-web topology as in the original MMM paper. The black thick lines show biomass trajectories for an individual simulation. The grey areas indicate the ranges of biomass values covered by the 50 simulations.

necessity being expressed by the limited set of physical, physiological and life-history constraints outlined above.

Emergence

There are several emerging properties that can be expected from the model, several of which were explored with the MMM. These include:

- Temporal dynamics of individual species, e.g. temporal autocorrelation, quasi-cycles, abrupt shifts,
- Temporal dynamics of integrated food-web properties, e.g. total biomass, total trophic flows, total assimilation efficiency, mean trophic level,
- Diet fraction, i.e. the percentage of various prey consumed by a predator,
- Stock-recruitment relationships,
- Trophic regulations, e.g. top-down *vs* bottom-up controls measured as the temporal correlation between biomass of predators and prey,
- Predator-prey functional responses.

Many of the above quantities can be derived from field or experimental data and therefore constitute an ensemble of criteria against which model results can be evaluated. The above list is not exhaustive and creative researchers will surely find other ways to describe additional emerging properties of the system.

Adaptation, individual level properties, and collectives

Adaptation is not included in the model design. The topology of the food-web and model parameters are set and the model does not include adaptive mechanism by which these might change. The structuring elements of the model are trophospecies and properties at the level of individual organisms are irrelevant, as are collectives.

Interactions

The only interactions in this model are trophic interactions, i.e. flow of biomass between predator and prey species.

Stochasticity

Randomness is central to the NDND model since the key elements, i.e. the trophic flows between species, are drawn randomly from a set of possible flows. In essence the model is tychastic [20], i.e. it is concerned with the subset of all possible transitions from one time step to the next. In practice, the simulations are stochastic, i.e. only one combination of trophic flows is drawn from the set of possible ones.

Initialisation and input parameter values

The structure of the food-web (i.e. the food-web topology defined by the list of species and possible links between them) must be set. In addition, NDND requires initial values for individual species biomass ($B_{i,t0}$), and the five species-specific parameters (γ_i, σ_i, ρ_i, μ_i and β_i).

Table 1. Input parameters of the Benguela pelagic ecosystem model.

Species	Import (t.km^{-2})	Initial Biomass (t.km^{-2})	Assimilation efficiency (γ)	Other losses (μ)	Inertia (ρ)	Satiation (σ)	Refuge Biomass (β t.km^{-2})
Phytoplankton	700	25.00	-	15	0.683	-	0.250
Zooplankton	-	2.00	0.45	12	0.615	80.0	0.200
Anchovy	-	10.21	0.85	1	0.470	15.1	0.102
Sardine	-	10.59	0.50	1	0.405	6.0	0.105
Squid	-	0.70	0.85	1	0.531	4.7	0.001
Hake	-	2.00	0.85	1	0.300	4.7	0.002
Birds	-	0.20	0.85	10	0.113	15.3	0.001
Seals	-	0.53	0.85	8	0.095	12.2	0.001
Whales	-	0.10	0.85	9	0.049	13.8	0.001

Input data

Species-specific data is required for import (I_i) and export (E_i) to express flows to and from the model domain. These can be invariant over time or variable at each time step.

Departure from the original MMM model

We have tried to keep the formulation of the NDND food-web model as close as possible to the original formulation of the MMM. There are however some differences. These are as follows:

- *Discrete formulation of the master equation:* In the MMM, the main equation was presented for populations at equilibrium (eq 2 in Mullon et al.) and can be used to express the dynamics of the system in continuous time. The continuous form of the master equation of the NDND model is presented above (eq. 1) and is very close to the one presented in the MMM (Appendix S1, but see below the difference concerning the import term). In addition, we provide the discrete form of the master equation, which is necessary for iterative computation of the model (eq. 2). The conversion of the continuous equation (eq. 1) into its discrete analogue (eq. 2) is presented in Appendix S2.
- *Refuge biomass* (β_i). This was not a feature of the MMM, in which biomasses were only constrained to remain positive. We found that this could lead to situations where a species would reach extremely small biomass and therefore flows to and from this species would be very small in comparison with other flows. This could result in computational problems due to scaling difficulties. Modelled biomass levels could fall below the weight of a single individual, which is biologically implausible. The combination of very low biomass with bounded growth rates could also lead to unrealistically long recovery times. The introduction of the additional parameter β for refuge biomass in the NDND solves these problems. It also introduces the possibility to model actual refuge strategies for species which can become inaccessible to predation at low densities. For the current simulations, the refuge biomass levels were arbitrarily set to 1% of the starting biomass levels.
- *Inertia*: In the MMM, inertia was expressed in such way that the maximum potential increase and decrease in species

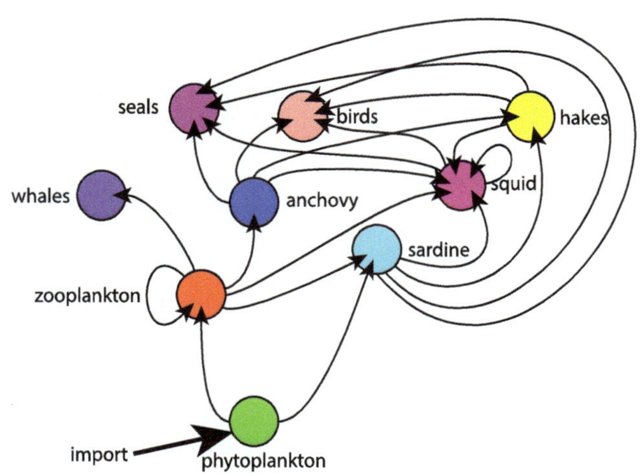

Figure 2. The food-web topology of the modelled Benguela ecosystem. Arrows indicate a trophic relationship and point towards predators. The primary production is set by a fixed annual import of phytoplankton.

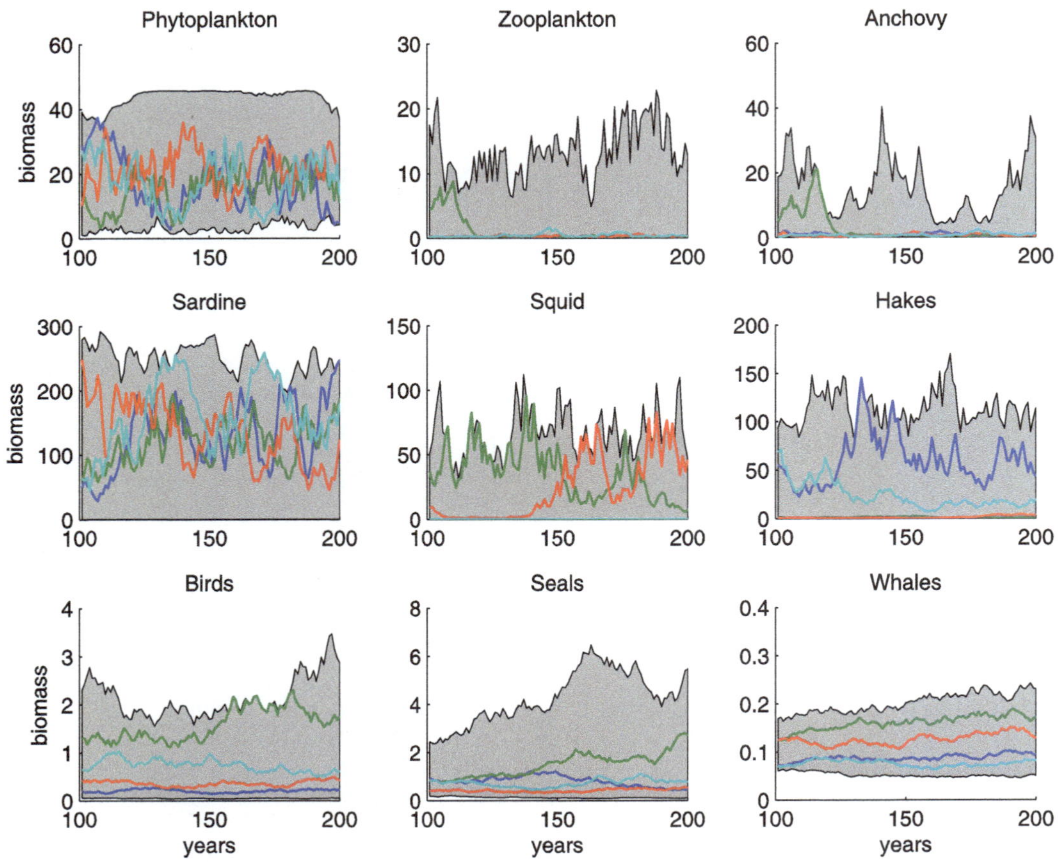

Figure 3. Fifty simulations of the NDND for the Benguela ecosystem, using revised initial biomass, input parameters, and food-web topology. The coloured lines show biomass trajectories for four individual simulations. The grey areas indicate the ranges of biomass values covered by the 50 simulations.

biomasses were equal. As a result species biomass could decline at much faster rates than they could recover. For example, in the MMM, zooplankton biomass could vary by up to 85%. This implied that the population could decline by 85% in one year, to reach 15% of its original biomass. However, to grow back to its original biomass at a maximum increase of 85% every year would require about 3 years (reaching 28% of the original biomass in year 1, 51% in year 2 and ≈100% in year 3). We used an alternative formulation for inertia to avoid such asymmetry between the rate of biomass decline and recovery. Instead of setting maximum biomass increase and decrease to be equal, it is the minimum and maximum rates of biomass change that are equal in the NDND (i.e. respectively $-\rho\beta$ and $+\rho\beta$). This is presented in more details in Appendix S3.

- *Import and assimilation terms*: In the MMM, the import term I strictly referred to inflow of nutrients to autotroph species (e.g. phytoplankton). For those species, the term γ did not refer to assimilation efficiency (as it did for other species), but to conversion efficiency from nutrient to biomass. This was confusing because the term γ could refer to two different biological processes: 1) conversion of nutrients into species biomass through photosynthesis, when it is applied to I, and 2) assimilation efficiency of ingested food otherwise. We also found this notation restrictive because it is only possible in the MMM to use the term I for import to autotroph species, but not to heterotroph or mixotroph species. In the NDND model, I represents the direct import of biomass, either through new production (autotrophs, mixotrophs) or through migration and transport. In the case of production, I represents the converted biomass and there is no longer need for the conversion efficiency. In the present model, the term γ is only used to describe assimilation efficiency.
- *Sampling*: In the MMM, at each time step, the biomasses of all species were randomly drawn from the set of possible solutions that satisfied the model constraints using linear programming. This approach can be viewed as the inverse problem of determining a set of flows, which result in the drawn biomasses. The approach in NDND, however, involves solving the forward problem of determining an ensemble of vectors of flows that satisfies a set of constraints and the use of flows to determine future biomasses. Viewed in terms of probabilities, the MMM approach considers all possible future states (biomass vectors) to be equiprobable, while in the current model, all possible transitions (flow vectors) have equal probability.

We have provided above an update of the original formulation of the MMM and a detailed description of the model state variables, input parameters, and constraints. We provide in the supplementary material the correspondence between state variables and input parameters in the NDND model and their formulation in the MMM as well as in the EwE context (Appendix S1). The implementation of the model was done in Matlab.

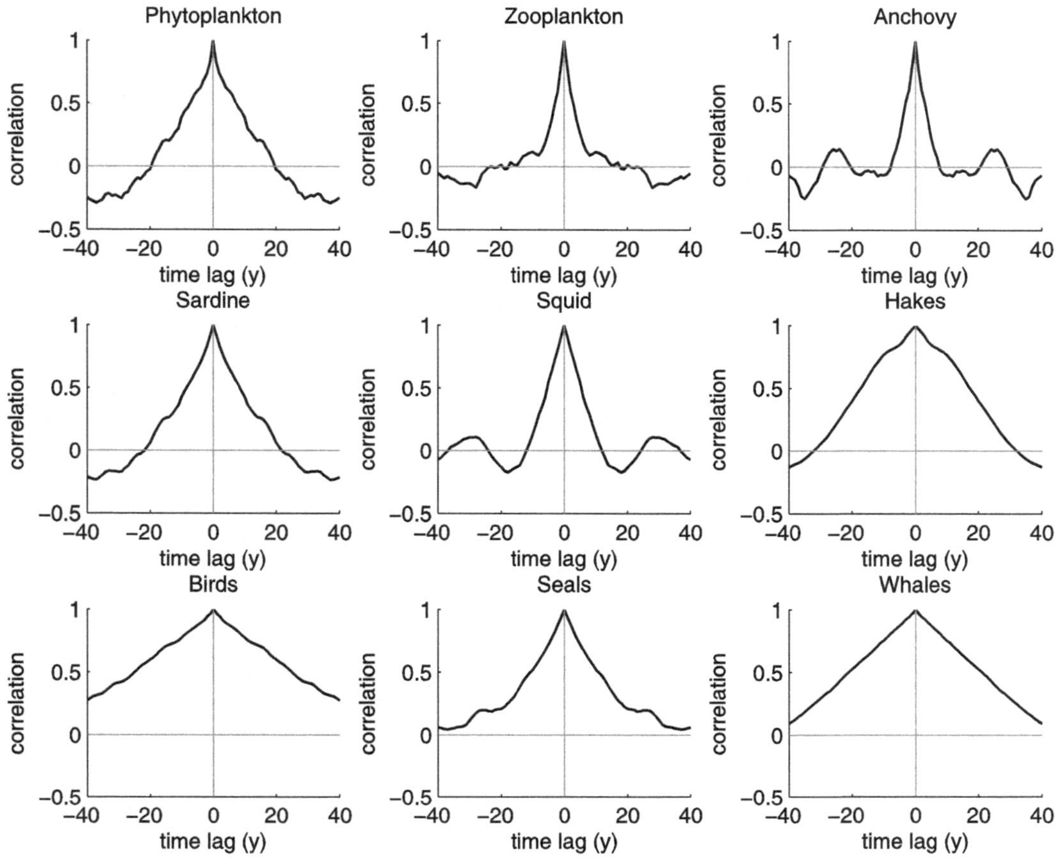

Figure 4. Autocorrelation series for a given run (time steps of the simulation on the X-axis; correlation coefficient on the Y-axis). An estimation of the cycle duration is provided by the lag associated with positive peaks, or twice the lag with negative peaks.

Results

Reproduction of the MMM simulations

We ran 50 simulations of the Benguela ecosystem dynamics using the same set of parameters as in the MMM over a period of 100 years. The simulations produced time series of individual species, but failed to convincingly reproduce the ecosystem dynamics presented in the original study. In particular, the modelled populations of zooplankton, hakes, birds, seals, and whales declined to low or very low levels, whilst anchovy and sardine populations reached levels about one to two orders of magnitude greater than those simulated in the MMM or observed in the wild. These features were consistently observed in the 50 simulations (Figure 1).

Parametrisation of the Benguela ecosystem model

Since the initial model configuration did not convincingly produce patterns of variability similar to those of the MMM, we explored if other model parameterisations could produce realistic patterns similar to those presented in the original study. This was done by trial and error, on the basis of the original food-web modelling study by Shannon et al. [21]. We kept the original values for the 'other' mortality coefficient (μ and derived the inertia coefficient (ρ) from the original values, following the equations provided in appendices 1 and 3. The initial biomasses were set by trial and error and the refuge biomasses (β) were set to 1% of the starting biomass values. The derivation of assimilation efficiency coefficients in the MMM was unclear. Assimilation efficiency is likely to vary within and between species because it depends on food quality; generally, carnivores have higher assimilation efficiencies than herbivores. This is now reflected in the use of assimilation efficiency (γ) derived from the work of Yodzis and Innes [22]. The satiation coefficient was derived from μ and γ to ensure that maximum feeding rates were greater than the requirements for species maintenance under absence of predation. These coefficients were then adjusted by trial-and-error. Model parameters in the revised configuration are given in Table 1. The topology of the food-web remained unchanged except for anchovy, which was changed to only feed on zooplankton (Figure 2), while it also fed on phytoplankton in the MMM. This seemed unrealistic given that bite-feeding, rather than filter-feeding, is the dominant or exclusive pattern of anchovy feeding [23,24]. Despite the corrections above, the model structure (topology) and parametrisation was not satisfactory because some key components of the ecosystem were missing in comparison with the original study of Shannon et al. For example, the MMM does not include meso-pelagic or benthic species that play an important role in the energy transfer of the Benguela system. In such a situation, one should not expect that the food-web model could represent realistic dynamics and biomass levels for all species simultaneously. As in the original study, we ran 50 simulations of the model with the revised structure and parameters. The outputs are presented for years of simulation 101 to 200, to avoid patterns eventually driven by initial biomass conditions.

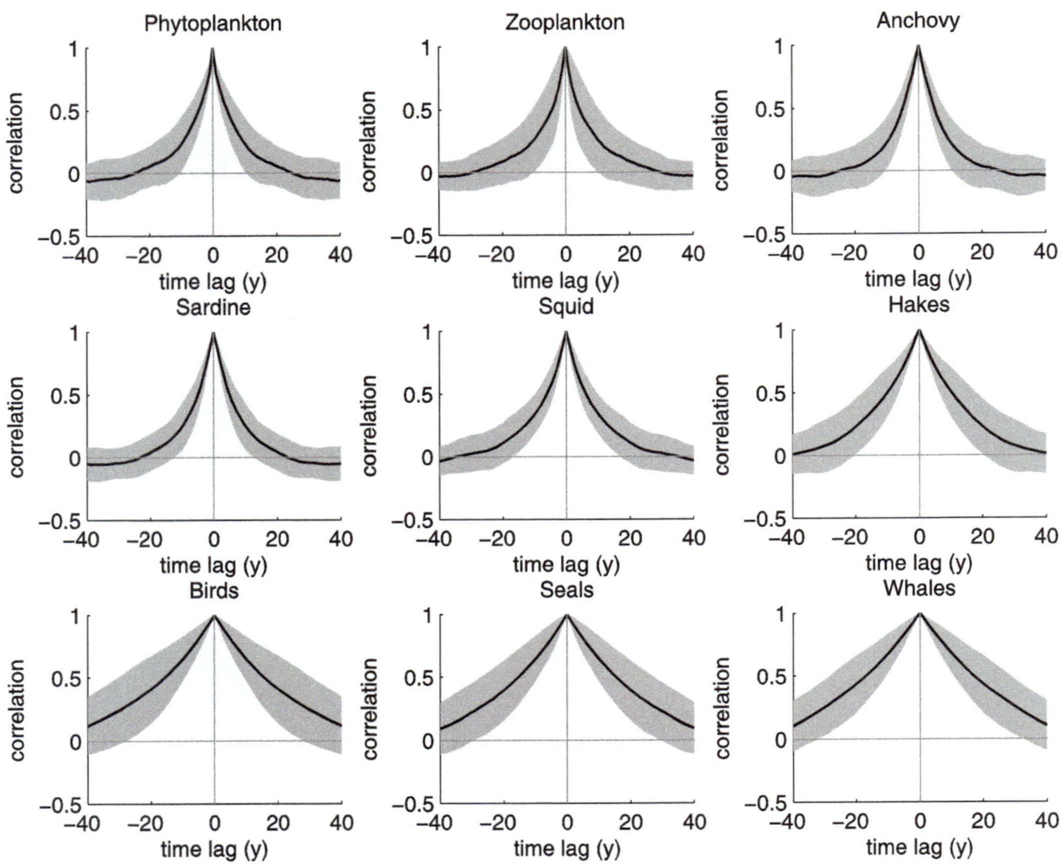

Figure 5. Mean (central line) and SD (shadow area, mean ± SD) on each side of the central line of autocorrelations for a model experiment of 50 runs (same axis as in Fig. 4).

Reproduction of pseudo-cycles

Individual simulations were uncorrelated, as expected, given the stochastic nature of the model (Figure 3). For all species, the simulated biomasses covered a large range, sometimes spanning several orders of magnitude. Year-to-year and decadal variations were evident from the simulations. As in the MMM, the model appeared to produce series with pseudo-cycles. It was not as clear in our simulations that these were predominantly seen at intermediate trophic levels. We also performed an analysis of the autocorrelation of the simulated abundance series (Figure 4). We found that the autocorrelation functions could vary considerably and that the cyclic patterns and trends could substantially differ between the different simulations. The values of autocorrelation coefficients averaged over the 50 runs indicated an increase in the length of pseudo-cycles (or a dominance of trends) with increasing life span, but not necessarily with trophic level (Figure 5). Like Mullon et al., we conclude that, for a given species, the length of its cycle is highly variable and that the pattern of pseudocycles of irregular lengths can be related to the species life span.

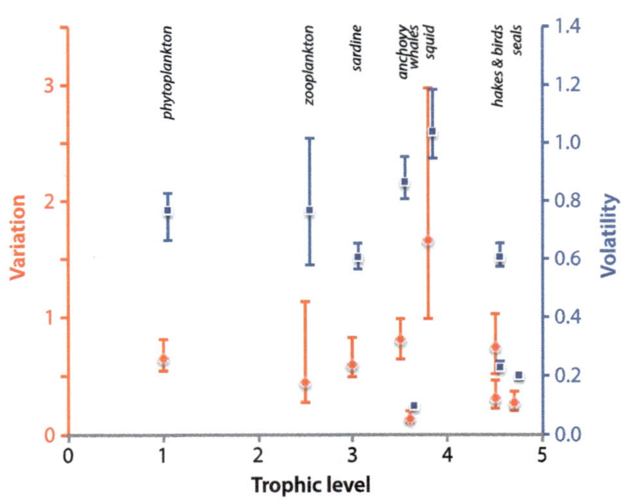

Figure 6. Variability patterns in the simulated ecosystem: values of variation (red circles) and volatility (blue squares) as a function of trophic level. Central points indicate median value and error bars indicate the interquartile ranges for the 50 simulations.

Reproduction of other patterns of variability

In the MMM paper, other patterns of variability were quantified by considering the variation and volatility of abundance series. We followed the same approach. Variation was measured as the ratio of interquartile range to the median of the series and volatility was measured as the ratio between the range of observed values in a given period and the value central to this period, which represents the ratio of short- to long-term variation. We found that variation and volatility had highest values for squid, sardine, anchovy, and zooplankton (Figure 6). In our simulations, the relationship between variation, volatility, and trophic levels was not clearly apparent, although top predators (hakes, birds, seals,

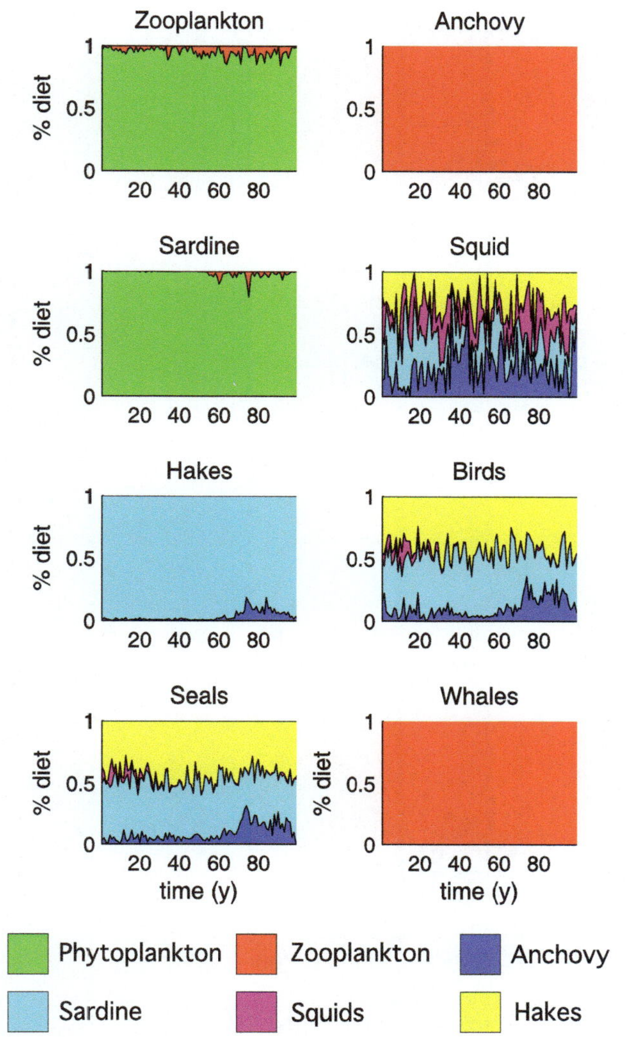

Figure 7. Diet composition dynamics (one single run of 100 years) for all model compartments except phytoplankton. This figure also shows the connectivity of the network and its variability in time.

and whales) displayed the lowest variability. Unlike Mullon et al., we cannot conclude that the indices of variation and volatility indicated a dome-shaped pattern, with highest values at intermediate trophic levels, which was interpreted as a simple illustration of a wasp-waisted system, *sensu* Bakun [25].

Reproduction of diet patterns

The list of prey species that determine the diet of a predator was defined by the food-web topology (Figure 2). However, how the proportions of various prey vary in time was determined by the relative abundances of all prey and predators in the food-web and by the stochastic process from which random trophic flows were drawn at each time step. For some species (e.g. squid), diet composition varied substantially from year-to-year, whilst for others (e.g. seals), it remained relatively stable (Figure 7). Interestingly, the diet of squid presented in the MMM was dominated by zooplankton, with minor predation on sardine and anchovy, but in our simulation, zooplankton was virtually absent from squid diet, which was mainly composed of anchovy, sardine, squid, and hakes in highly variable proportions. This variation in diet between the two models illustrated how model configuration and stochastic simulations can lead to a wide range of system configurations.

Trophic functional relationships, which are formally excluded from the model structure, can be investigated as emerging properties. From the single simulation represented in Figure 8, we observed three types of configurations: positive relationship (e.g. hake feeding on anchovy), negative relationship (e.g. sardine feeding on phytoplankton), and absence of relationships (e.g. birds feeding on sardine). Even in the case of apparent relationships, there was a large scattering of the simulated data, which reflected diet variations that have been observed in the field through the analysis of stomach contents of predators in relation to prey abundance [26,27].

Reproduction of stock-recruitment relationships

We defined a proxy for recruitment, expressed as the variation in population biomass corrected for losses due to metabolic activities and other losses: $R_i = B_{i,t+1} - e^{-\mu} B_{i,t}$ (note that this recruitment equation is different from that used in the MMM paper). The resulting stock-recruitment plots (Figure 9) displayed noisy linear positive relationships. Here, the stock-recruitment relationship emerged as a consequence of ecosystem functioning, not as a causal principle. The simulated patterns mimicked published data, although we saw no clear sign of density dependence, unlike in the MMM and as is generally assumed in fisheries stock recruitment models (e.g. Beverton–Holt and Ricker funtions) [28].

Reproduction of interdecadal variations in trophic interactions

Interactions between prey and predator in marine systems can lead to apparent correlations between biomass time series. When the correlation is negative, the relationship is generally described as being top-down controlled (the predator controlling the abundance of the prey), and when it is positive, the prey-predator system is said to be bottom-up controlled (the prey controlling the abundance of the predator) [29]. Although simplistic, this description of the trophic controls between predator and prey is easy to construct from field data. These trophic controls have been shown to fluctuate at interdecadal time scales (see e.g. [30]). The modelled dynamics highlighted the strong negative correlation between sardine and phytoplankton over the 100 y simulation period, indicating that the standing stock of phytoplankton was controlled by grazing from sardine (Figure 10). The relationship between hake and anchovy was also negative, but large fluctuations appeared on a decadal time-scale, which occasionally resulted in positive correlations) thereby mimicking interdecadal fluctuations observed in real systems.

Discussion

Replicability of the original model

The original study of Mullon et al. used a novel non-deterministic modelling approach to simulate the dynamics of marine food-webs. In this way, the 'null model' could serve as a reference for other deterministic models. Based on this approach, Mullon et al. reached important conclusions regarding the dynamics of food-webs when modelled with only few assumptions and without explicit deterministic formulations of trophic functional relationships. Here, we have tried to reproduce the original model structure and dynamics and test whether the conclusions reached in the original study were still valid in our replication.

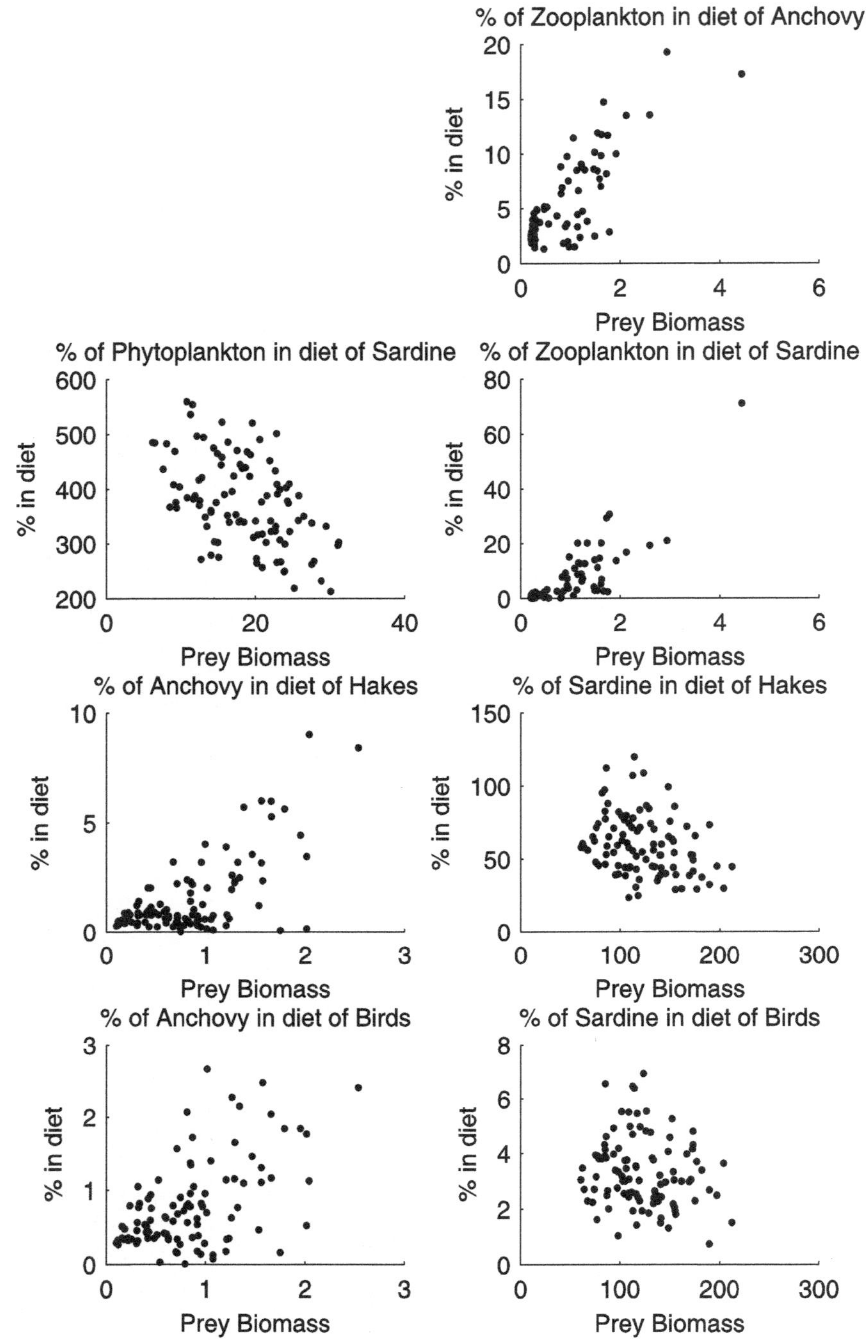

Figure 8. Simulated diet/abundance relationships. Captions indicate the name of predator species followed by the name of one of the major prey species for a predator. The **Y**-axis represents the proportion of prey in the predator diet and the **X**-axis represents the abundance of the prey species.

We found that when we used the input parameters from the original model, we could not reproduce the food-web dynamics presented in the original study. This is surprising given that the equations of the two models are identical except for the constraints on inertia and refuge biomass. Our revision of these equations should have led to a more stable system with a lower probability of rapid population collapse. It is not clear how input parameters and starting biomasses presented in the MMM paper were derived from the original modelling study of Shannon et al. [21], so we went back to the original study to revise these values, using input from other sources to document assimilation efficiencies [22]. Even with these revisions, we had to perform trial-and-error runs until we could produce a 'realistic' set of simulations for the Benguela pelagic food-web dynamics.

The major conclusions from the MMM paper concern the reproduction of several ecosystem dynamic patterns: pseudo-cycles, variation and volatility, diet, stock-recruitment relationships, and correlations between species biomass series. For all of these aspects, we reached similar conclusions, although these were often not as strongly supported as suggested in the original study.

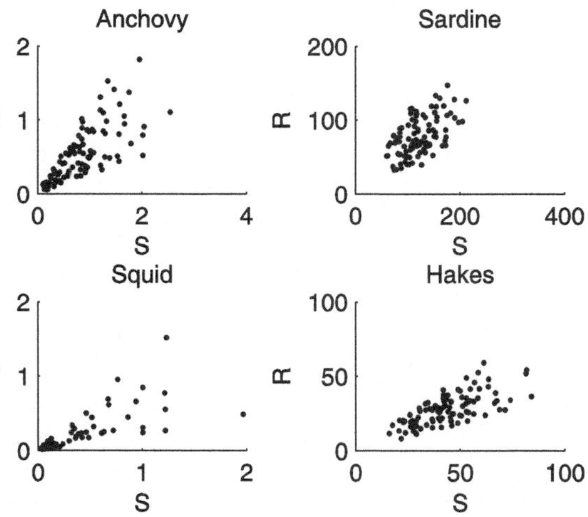

Figure 9. Examples of stock-recruitment (S, R) relationships using the variation from population biomass minus annual metabolic losses to new population biomass as a proxy for recruitment: $R_{i,t} = B_{i,t+1} - e^{-\mu} B_{i,t}$.

We found pseudo-cycles of various periods for the different species in the food-web and that variation and volatility varied between species and were lower for higher trophic levels. Noisy trophic functional relationships emerged from our model simulations and large year-to-year fluctuations in diet composition could happen for intermediate predators with a range of potential prey. Simulations also indicated that noisy stock-recruitment relationships were possible. We also found correlations between species biomass series, which could either be stable or vary at interdecadal time-scales. Despite differences between the original MMM and the present model, we show that a non-deterministic model *with a minimum set of constraints* can recreate important ecosystem dynamical patterns. This result is crucial, in particular, because deterministic models that require a large amount of assumptions and input parameters generally fail to reproduce most of the patterns described above. For example, EwE models, which include explicit trophic functional responses and are fitted to historical data, usually fail to produce realistic year-to-year variability in population biomass when projected forward, but instead generate smooth patterns over longer time scales (e.g., [31]). In addition, most food-web models, including EwE, are highly sensitive to assumptions and data uncertainties regarding trophic functional relationships [10,17,32], a problem that is absent from the current modelling approach, in which such relationships are not an input to the model but emerging properties of the system.

Ecosystem variability

In several instances during our trial runs, the food-web state could be trapped, i.e. it was not possible to find a combination of trophic flows that would satisfy all constraints. Our simulations also appeared more variable than in the original study. Since the NDND has an additional constraint (refuge biomass) and the constraint on inertia is stronger (see Appendix S3), the most plausible explanation for the increased variability lies in the method used to generate the random transition from one food-web state to the next. There are fundamental differences in the two approaches. The inverse problem of determining flows given biomasses should, in principle, not result in a unique solution (there may be an infinite combination of flow configurations for the same observed biomass). The linear programming approach adopted in MMM limits the solution space to flow values at the vertices of the polytope defining the constraints. In the NDND approach, however, the vertices are a subset of the solution space, which also includes the interior points of the polytope. Hence, sampling in the case of NDND is expected to show larger variability than with the MMM. Further, by the nature of the sampling procedure, it is possible to generate realisations, which though mathematically right (i.e. satisfying the constraints on the flow functions) are biologically implausible. Variability in the model may exceed what is observed in the wild. If this is the case, it would mean that the set of constraints currently used is insufficient to restrict model variability within observable limits

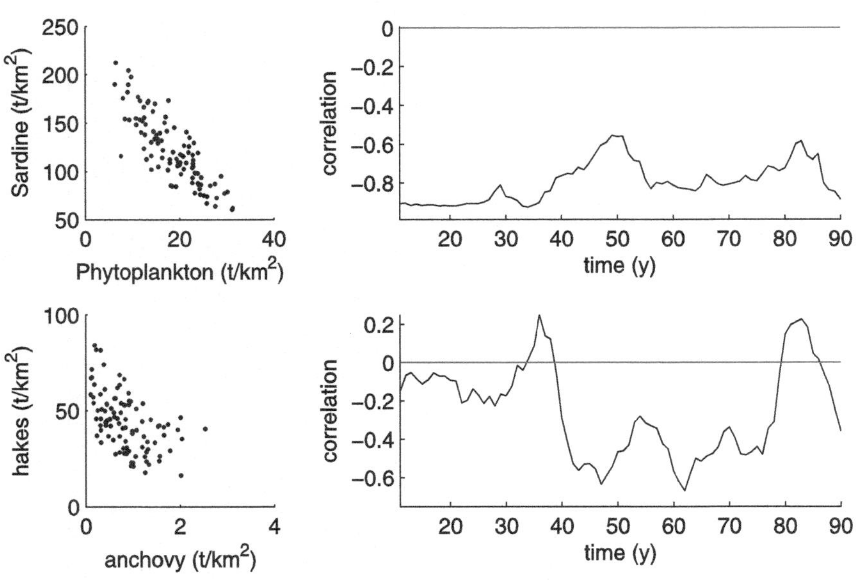

Figure 10. Left: Biomass of predator versus biomass of prey over 100 y for sardine-phytoplankton (top) and hakes-anchovy (bottom). Right: sliding correlations between predator and prey biomass. The correlation is calculated over a 20 y window. These highlight the general relationship between prey and predator (top-down vs. bottom-up) and the interdecadal variations in trophic controls.

and additional constraints may be necessary. This would require further investigation and comparisons with real systems.

Model complexity

The model configuration for the Benguela is a simple one, with only nine trophospecies and 20 trophic links, so the computational problem is both relatively simple (a polytope of 20 dimensions to explore) and fast (<0.1 s per year of simulation). However, more complex food-web models, such as those constructed in EwE containing >50 species and several hundred links will likely lead to computational challenges. For instance, the simplest case of uniform sampling of a convex polytope in high dimensions leads theoretically to what is termed an NP-hard problem (i.e., belonging to the class of problems that are, informally, "at least as hard as the hardest problems in Non-deterministic Polynomial-time", see e.g. [33]). Since the number of constraints is finite, a complete characterization of the polytope could be argued to be known. The problem therefore reduces to the task of finding fast algorithms (usually deterministic) that perform exhaustive enumeration on a convex polytope. Unfortunately, deterministic algorithms that perform can be infeasible in high dimensions since the number of such vertices could scale exponentially with the dimension [34]. An alternative approach to the deterministic approach is to use stochastic algorithms (Monte Carlo based with acceptance-rejection rules) to sample the polytope. However, even the fastest of such algorithms is known to suffer scaling problems (with respect to mixing time, see e.g. [35]) in high dimensions, especially when the polytope is highly heterogeneous, as in the case of our food-web model. Fortunately, to investigate the dynamics of whole food-webs (i.e. rather than the dynamics of individual taxa), such levels of food-web complexity are often not neccessary and an approach based on the use of few, well defined trophospecies groups might capture most the food-web properties [36].

Model improvements

The NDND framework is a potentially powerful framework to investigate food-web dynamical patterns that are driven by a few sets of constraints and can therefore be used as a reference model against which more complicated models can be evaluated, as originally suggested by Mullon et al. However, several developments appear necessary for such a model to become a general and powerful tool in the study of ecosystem dynamics. Setting the values of the model input parameters is a difficult task for which there is yet no objective and transparent methodology. This is no surprise, given the complexity of the task. It took nearly 20 years before such method became available for Ecopath models [37]. Hopefully methods used for parameter optimization in Ecopath can be reformulated for the NDND models, so that setting model parameters will be less based on trial and error and more on ecological theory and available field data. Using metabolic theory of ecology [38] to estimate metabolic losses and maximum consumption rates (satiation) and life-history theory to derive growth rate (inertia) estimates will also make the model more general and easier to parameterise.

The evaluation of model performance was not properly addressed in the MMM and this limitation remains in the current model. Since these are stochastic models which are not required to fit data in the conventional way (i.e. by fitting time series of biomass for example) there is, as of yet, no simple and accessible methodology to evaluate model performance or to perform sensitivity tests. Approaches to this problem may be rather different from the conventional techniques favoured by ecological modellers today (e.g., [39,40]) and would likely involve a pattern oriented approach as advocated by Grimm and colleagues [41–43] combined with dedicated statistical inference for stochastic models [44] and hierarchical model evaluation techniques [45].

The original form of the MMM and the current form of the NDND model are prototypes and clearly, these models do not benefit from the experience of other species-based or size-based ecosystem models [32]. In the present contribution and the associated appendices, we have detailed as much as possible the hypotheses, equations, and computational aspects of the NDND to allow other researchers to test this approach and contribute to the model development. This should allow for the use of the NDND as a null model for comparison with other deterministic modelling approaches.

In its current form, the NDND simulates the dynamics of simple food-webs over annual time steps and in a single area. However, the mathematical formulation and computation can readily allow inclusion of spatialised food-webs (in a way similar to Ecospace or GADGET models [46,47]), shorter simulation time-steps (seasons, months, days), or the modelling of age-structured populations.

Conclusions

While randomness is generally considered a source of uncertainty for deterministic models, the current study supports the original conclusions of Mullon et al. that stochasticity can play a structural and central role in shaping key features of food-webs. In this, the NDND model shares similarities with models issued from the ecological neutral theory, which can also reproduce ecological patterns on the basis of few assumptions combined with stochasticity [48]. We do not claim that real systems resemble stochastic food-webs, but rather that the NDND approach can improve our understanding by making simplifying assumptions about complex systems.

The approach proposed by Mullon et al. [12], which combines chance (randomness) and necessity (constraints), is unique in the field of ecosystem modelling, although these two elements have long been recognised as shaping biological systems [49]. Their original conclusion, i.e. that *observed patterns of ecosystem variability may simply result from basic structural constraints within which the ecosystem functions*, is of great importance to ecosystem modellers and those who may use model outputs as a support for management decisions. These conclusions are supported, to a large extent, by the current replication of the model. However, model parameterisation and computational aspects remain difficult and these need to be investigated further. Hopefully, the present contribution will make this approach available to a larger research community and will promote the use of NDND as 'null models of food-webs' as originally advocated.

Supporting Information

Appendix S1 Table presenting the main elements of the MMM, the NDND and the Ecopath with Ecosim models. The table provides a comparative description of the indices, state variables, input parameters, constraints and equations for the three modeling approaches.

Appendix S2 Integration of the model master equation in continuous form (eq. 1) into its discrete form, for time step *t* to time step *t+1* (eq. 2).

Appendix S3 Comparison of the variance in rate of biomass change in the MMM and NDND models as a result of distinct formulation of the inertia constraint (ρ).

Acknowledgments

The authors wish to thank Christian Mullon for inspiring discussions and advice on the design, structure, and programming of the first version of this model, and Jennifer Devine for reviewing the English.

Author Contributions

Analyzed the data: BP UL. Wrote the paper: BP UL SS. Conception and design of the model: BP UL SS. Coding of the model: BP SS.

References

1. Levin SA (2005) Self-organization and the emergence of complexity in ecological systems. BioScience 55: 1075–1079.
2. Prigogine I (1996) La fin des certitudes. Paris: Odile Jacob.
3. Holling CS (1973) Resilience and stability in ecological systems. Annual Review of Ecology and Systematics 4: 1–24.
4. Coulson T, Rohani P, Pascual M (2004) Skeletons, noise and population growth: the end of an old debate? Trends in Ecology and Evolution 19: 359–364.
5. Black A, McKane AJ (2012) Stochastic formulation of ecological models and their applications. Trends in Ecology and Evolution 27: 337–345.
6. Cury PM, Mullon C, Garcia SM, Shannon LJ (2005) Viability theory for an ecosystem approach to fisheries. ICES Journal of Marine Science 62: 577–584.
7. Abrams PA, Ginzburg LR (2000) The nature of predation: prey dependent, ratio dependent or neither? Trends in Ecology & Evolution 15: 337–341.
8. Koen-Alonso M (2007) A process-oriented approach to the multi-species functional response In: Rooney N, McCann KS, Doakes DLG, editors. From Energetics to Ecosystems: the dynamics and structure of ecological systems. Dordrecht, The Netherlands: Springer. pp.1–36.
9. Hunsicker ME, Ciannelli L, Bailey KM, Buckel JA, Wilson White J, et al. (2011) Functional responses and scaling in predator–prey interactions of marine fishes: contemporary issues and emerging concepts. Ecology Letters 14: 1288–1299.
10. Kearney KA, Stock C, Aydin K, Sarmiento JL (2012) Coupling planktonic ecosystem and fisheries food web models for a pelagic ecosystem: Description and validation for the subarctic Pacific. Ecological Modelling 237–238: 43–62.
11. Gilioli G, Pasquali S, Ruggeri F (2012) Nonlinear functional response parameter estimation in a stochastic predator-prey model. Mathemathical Biosciences and Engineering 9: 75–96.
12. Mullon C, Fréon P, Cury P, Shannon L, Roy C (2009) A minimal model of the variability of marine ecosystems. Fish and Fisheries 10: 115–131.
13. Mullon C (2014) Network economics of marine ecosystems and their exploitation. Boca Raton FL, USA: CRC Press. 306 p.
14. Grimm V, Berger U, Bastiansen F, Eliassen S, Ginot V, et al. (2006) A standard protocol for describing individual-based and agent-based models. Ecological Modelling 198: 115–126.
15. Grimm V, Berger U, DeAngelis DL, Polhill JG, Giske J, et al. (2010) The ODD protocol: A review and first update. Ecological Modelling 221: 2760–2768.
16. Polovina JJ (1984) An overview of the Ecopath model. Fishbyte 2: 5–7.
17. Christensen V, Walters CJ (2004) Ecopath with Ecosim: methods, capabilities, and limitations. Ecological Modelling 172: 109–139.
18. Dulvy NK, Ellis JR, Goodwin NB, Grant A, Reynolds B, et al. (2004) Methods of assessing extinction risk in marine fishes. Fish and Fisheries 5: 255–276.
19. Hoenig JM (1983) Empirical use of longevity data to estimate mortality rates. Fisheries Bulletin US 81: 898–903.
20. Aubin J-P, Bayen A, Saint-Pierre P (2011) Viability theory: new directions. Berlin: Springer-Verlag. 803 p.
21. Shannon LJ, Moloney CL, Jarre A, Field JG (2003) Trophic flows in the southern Benguela during the 1980s and 1990s. Journal of Marine Systems 39: 83–116.
22. Yodzis P, Innes S (1992) Body size and consumer-resource dynamics. American Naturalist 139: 1151–1175.
23. Plounevez S, Champalbert G (1999) Feeding behaviour and trophic environment of *Engraulis encrasicolus* (L.) in the Bay of Biscay. Estuarine, Coastal and Shelf Science 49: 177–191.
24. Hutchings L, Barange M, Bloomer SF, Boyd AJ, Crawford RJM, et al. (1998) Multiple factors affecting South African anchovy recruitment in the spawning, transport and nursery areas. South African Journal of Marine Science 19: 211–225.
25. Bakun A (1996) Patterns in the Ocean. Ocean processes and marine population dynamics. La Jolla, CA: NOAA, California Sea Grant College System. 323 p.
26. Sydeman WJ, Hester MM, Thayer JA, Gress F, Martin P, et al. (2001) Climate change, reproductive performance and diet composition of marine birds in the southern California current system. Progress in Oceanography 49: 309–329.
27. Pinnegar J, Trenkel V, Tidd A, Dawson W, du Buit MH (2003) Does diet in Celtic sea fishes reflect prey availability? Journal of Fish Biology 63: 197–212.
28. Hilborn R, Walters CJ (1992) Quantitative fisheries stock assessment. Choice, dynamics and uncertainty. New York: Chapman and Hall, inc. 570 p.
29. Cury P, Shannon LJ, Shin Y-J (2003) The functioning of marine ecosystems: A fisheries perspective. In: M. . Sinclair and G. . Valdimarson, editors. Responsible fisheries in the marine ecosystem. FAO. pp.103–123.
30. Johannesen E, Ingvaldsen RB, Bogstad B, Dalpadado P, Eriksen E, et al. (2012) Changes in Barents Sea ecosystem state, 1970–2009: climate fluctuations, human impact, and trophic interactions. ICES Journal of Marine Science 69: 880–889.
31. Niiranen S, Blenckner T, Hjerne O, Tomczak M (2012) Uncertainties in a Baltic Sea food-web model reveal challenges for future projections. AMBIO 41: 613–625.
32. Plagányi ÉE (2007) Models for an ecosystem approach to fisheries. FAO Fisheries Technical Paper 477: 108 p.
33. Dyer ME, Frieze AM (1988) On the complexity of computing the volume of a polyhedron. SIAM Journal on Computing 17: 967–974.
34. De Martino D, Parisi V (2013) Montecarlo uniform sampling of high-dimensional convex polytopes: reducing the condition number with applications in metabolic network analysis. arXiv preprint arXiv:1312.5228.
35. Lovász L (1999) Hit-and-run mixes fast. Mathematical Programming 86: 443–461.
36. Ulanowicz RE, Holt RD, Barfield M (2014) Limits on ecosystem trophic complexity: insights from ecological network analysis. Ecology Letters 17: 127–136.
37. Kavanagh P, Newlands N, Christensen V, Pauly D (2004) Automated parameter optimization for Ecopath ecosystem models. Ecological Modelling 172: 141–149.
38. Brown JH, Gillooly JF, Allen AP, Savage VM, West GB (2004) Towards a metabolic theory of ecology. Ecology 85: 1771–1789.
39. Allen JI, Somerfield PJ, Gilbert FJ (2007) Quantifying uncertainty in high-resolution coupled hydrodynamic-ecosystem models. Journal of Marine Systems 64: 3–14.
40. Morris DJ, Speirs DC, Cameron AI, Heath MR (2014) Global sensitivity analysis of an end-to-end marine ecosystem model of the North Sea: Factors affecting the biomass of fish and benthos. Ecological Modelling 273: 251–263.
41. Grimm V, Revilla E, Berger U, Jeltsch F, Mooij WM, et al. (2005) Pattern-Oriented modeling of agent-based complex systems: lessons from ecology. Science 310: 987–991.
42. Grimm V, Railsback SF (2012) Pattern-oriented modelling: a 'multi-scope' for predictive systems ecology. Philosophical Transactions of the Royal Society B: Biological Sciences 367: 298–310.
43. Fulton EA, Link JS, Kaplan IC, Savina-Rolland M, Johnson P, et al. (2011) Lessons in modelling and management of marine ecosystems: the Atlantis experience. Fish and Fisheries 12: 171–188.
44. Hartig F, Calabrese JM, Reineking B, Wiegand T, Huth A (2011) Statistical inference for stochastic simulation models – theory and application. Ecology Letters 14: 816–827.
45. Lehuta S, Petitgas P, Mahevas S, Huret M, Vermard Y, et al. (2013) Selection and validation of a complex fishery model using an uncertainty hierarchy. Fisheries Research 143: 57–66.
46. Walters CJ, Pauly D, Christensen V (1999) Ecospace: prediction of mesoscale spatial patterns in trophic relationships of exploited ecosystems, with emphasis on the impacts of marine protected areas. Ecosystems 2: 539–554.
47. Begley J, Howell D (2004) An Overview of Gadget, the Globally applicable Area Disaggregated General Ecosystem Toolbox. ICES CM 2004/FF:13.
48. Rosindell J, Hubbell SP, He F, Harmon LJ, Etienne RS (2012) The case for ecological neutral theory. Trends in Ecology & Evolution 27: 203–208.
49. Monod J (1970) Le hasard et la nécessité. Essai sur la philosophie naturelle de la biologie moderne. Paris: Editions du Seuil.

21

An Online Database for Informing Ecological Network Models: http://kelpforest.ucsc.edu

Rodrigo Beas-Luna[1], Mark Novak[2], Mark H. Carr[1]*, Martin T. Tinker[1,3], August Black[4], Jennifer E. Caselle[4], Michael Hoban[1], Dan Malone[1], Alison Iles[2]

1 Department of Ecology and Evolutionary Biology, University of California Santa Cruz, Santa Cruz, California, United States of America, **2** Department of Integrative Biology, Oregon State University, Corvallis, Oregon, United States of America, **3** Western Ecological Research Center, United States Geological Survey, Santa Cruz, California, United States of America, **4** Marine Science Institute, University of California Santa Barbara, Santa Barbara, California, United States of America

Abstract

Ecological network models and analyses are recognized as valuable tools for understanding the dynamics and resiliency of ecosystems, and for informing ecosystem-based approaches to management. However, few databases exist that can provide the life history, demographic and species interaction information necessary to parameterize ecological network models. Faced with the difficulty of synthesizing the information required to construct models for kelp forest ecosystems along the West Coast of North America, we developed an online database (http://kelpforest.ucsc.edu/) to facilitate the collation and dissemination of such information. Many of the database's attributes are novel yet the structure is applicable and adaptable to other ecosystem modeling efforts. Information for each taxonomic unit includes stage-specific life history, demography, and body-size allometries. Species interactions include trophic, competitive, facilitative, and parasitic forms. Each data entry is temporally and spatially explicit. The online data entry interface allows researchers anywhere to contribute and access information. Quality control is facilitated by attributing each entry to unique contributor identities and source citations. The database has proven useful as an archive of species and ecosystem-specific information in the development of several ecological network models, for informing management actions, and for education purposes (e.g., undergraduate and graduate training). To facilitate adaptation of the database by other researches for other ecosystems, the code and technical details on how to customize this database and apply it to other ecosystems are freely available and located at the following link (https://github.com/kelpforest-cameo/databaseui).

Editor: Gretchen E. Hofmann, University of California- Santa Barbara, United States of America

Funding: This project was supported by the Comparative Analysis of Marine Ecosystem Organization (CAMEO) program, jointly supported by the National Science Foundation and the U.S. National Marine Fisheries Service of the National Oceanic and Atmospheric Administration (award # OCE-1041454 and OCE-10411489). Additional support for Rodrigo Beas-Luna was provided by the Consejo Nacional de Ciencia y Tecnología (www.conacyt.mx) and the University of California Institute for Mexico (www.ucmexus.ucr.edu). Funding for Open Access provided by the University of California Santa Cruz, Open Access Fund. The funders had no role in study design, data collection and analysis, decision to publish, or preparation of the manuscript.

Competing Interests: The authors have declared that no competing interests exist.

* Email: mhcarr@ucsc.edu

Introduction

Ecological network models and analyses are recognized for their value in articulating the quantitative and conceptual relationships and emergent properties of natural ecosystems, for generating plausible explanations and testable hypotheses pertaining to community structure and dynamics [1–3] and predictions regarding their responses to natural and anthropogenic perturbations [4,5]. Their importance for informing management and policies has increased markedly with the advent of ecosystem-based management (EBM) approaches (e.g., [6]). EBM requires knowledge of how the human uses of ecosystem services influence the structural (e.g., diversity, composition) and functional (e.g., productivity, nutrient cycling) attributes of ecosystems and how these attributes underpin their integrity and resilience. Quantitative ecosystem models based on species or functional group interaction networks are key tools for understanding how human activities influence ecosystems. These models allow users to forecast how entire ecosystems may respond to alternative management actions. For example, models of species interactions that describe ecosystem-wide effects of anthropogenic perturbations have proven particularly insightful for informing ecosystem-based fisheries management [7], and for understanding the effects of seasonal forcing in freshwater ecosystems [8] and carbon flux in terrestrial forests [9].

However, a critical barrier to the successful implementation of ecosystem-based models is the accessibility of the substantial data they require [10,11]. An ideal source for this data would be verifiable, comprehensive, relevant, well organized, thoroughly explained, easily updated and readily available at a single location online. Though there is a clear need for accessible online databases tailored for the development of ecological network models, few if any databases meet these criteria. Here, we describe an online interactive database with information (life history, demography, species interactions) required of many ecological network models and that fulfills these and other necessary criteria for expediting the development of these models.

Why ecological network models need databases

In a comprehensive review of ecological network models used to characterize and explore marine ecosystems, Plagányi [12] identified four general categories of models: Minimum Realistic, Individual Based, Biogeochemical, and Aggregate System Models (Table 1). These four broad categories of ecological network models illustrate the diversity of information that is required of, or can be accommodated by, the various ecological network models. Other model types, such as qualitative loop analysis [13,14] and allometric trophic network models [8] also benefit from such information. Despite differences in their assumptions and focal applications, all of these modeling approaches accommodate or require some of the same forms of information, such as knowledge of what species, life-stages or functional groups constitute an ecosystem. However, they differ in their requirements or ability to accommodate other forms of information including species' currencies (e.g., biomass, density), distributions, life history or demographic attributes, and the manner in which species interact (e.g., predation, parasitism, competition, mutualism; Table 1). For example, many ecological network models focus entirely on trophic interactions in their representation of species interactions, ignoring non-trophic interactions, such as competition for space [15] or parasitism [16]. The greater the variety of information included in a database, the greater its application across the diversity of ecological network models. Much of the same types of information are also relevant to the development of single-species population models, and are useful in non-modeling contexts. For example, including knowledge of the geographic patterns of species' life history traits and interspecific interactions can help to inform the design of experimental and observational studies, or the placement of marine reserves [17–19].

Shortcomings of existing online databases for ecological network modeling

The diversity of information required by the various kinds of ecological network models is rarely organized in a form that is useful or accessible to modelers. Several well-designed online taxon-specific databases exist that collate information on species taxonomy, phylogeny, life history traits and distribution (Table 2). However, few of these mediate with web browsers or between multiple databases, instead referring to static species-focused summaries. Fewer still translate data requests beyond species-specific searches to permit the querying of multiple species from a common functional group. Having no online database management system (DBMS), these databases preclude the integration of different functions and information in the same process to permit simultaneous access of taxonomic, life history, distribution and ecological databases [20]. Some database management systems (e.g. FishBase, Sea Life Base; Table 2) have the potential to integrate multiple databases in their queries but do not currently do so. Furthermore, database entries do not reference their datum-specific sources, leaving attribution absent or too general and difficult to reconstruct and thereby making validation and reanalysis difficult or impossible.

More generally, few existing databases housing information relevant to ecological network models also include information on species interactions. Those that do, include only the presence of the interactions without source citations or detailed description of their nature, spatial, or temporal patterns specific to those interactions. Hence, variation and uncertainties in interaction

Table 1. Comparison and examples of five categories of ecosystem models (modified from Plagany 2007).

Model type & examples	No. of nodes or functional groups	Currency	Life stages	Size structure	Spatial structure	Topology	Diet	Non-trophic interactions	Env. Data
Qualitative models Loop Analysis	<10	Biomass	A	A		Input		A	
Minimum Realistic GADGET[1]	6 to 8	Nutrients	A			Input	R		
Individual Based OSMOSE[2]	7 to 20	Biomass	A	R	A	Input	R	A	
Biogeochemical ATLANTIS	20	Nutrients	A	A	A	Input, output	R	A	R
Aggregate system EwE[3]	>40	Biomass	A		A	Input, output	R	A	A

[1]Globally Applicable Area-Disaggregated General Ecosystem Toolbox,
[2]Object-oriented simulator of marine ecosystem exploration,
[3]Ecopath with Ecosim.
"A" and "R" indicate that the model accommodates or requires information of this form, respectively. "Env" = environmental.

Table 2. List of some of the most relevant marine ecological databases and their attributes to inform ecosystem models.

Database	Subject taxa	Data type	Data visualization	Data export	Integration capabilities
Algae Base	Algae	1,2,5,6,7	x		
All about birds	Birds	1,4,5,6,7	x		
AnAge	Multiple	1,11	x		x
Catalog of Life	Multiple	2	x		x
DataMares	Multiple	2,5,6,9,10	x	x	
EOL	Multiple	1,2,3,4,5,6,7	x		x
Eurobis	Plankton	2,5,6,10	x		x
Fish Base	Fish	1,2,3,4,5,6,7,8	x		x
GoMexi	Multiple	2,8,9,10	x	x	x
ITS	Multiple	2	x	x	x
Kelpforest	Multiple	1,2,3,4,5,6,7,8,9,10,11	x	x	x
Sea Life Base	Multiple	1,2,3,4,5,6,7	x		x
Sea Net	Multiple	1,2,4,5,6,7	x		
Simon	Multiple	1,2,3,4,5,6,7	x		
WoRMS	Multiple	2	x		x

"X" indicates available function. "Integration capability" refers to ability to link with another database.
Data types: 1) Life history, 2) taxonomy, 3) biometrics, 4) behavior, 5) distribution, 6) habitats, 7) photos, 8) species interactions, 9) temporal explicit data, 10) spatial explicit data, 11) references.
Score: Sum of number of attributes valuable for data accessibility for ecosystem modelers. These attributes provide basic information for the parameterization and validation of ecosystem models.

information are difficult to obtain and remain challenging to incorporate into ecological network models.

Ecological network models for kelp forest ecosystems

Kelp forests are stands of large macroalgae of the Order Laminariales that occur on temperate and boreal rocky reefs around the world and are among the most productive and diverse ecosystems in the world (reviewed by [21–23]). These species-rich ecosystems provide many ecosystem functions, including primary production, habitat for fishes, invertebrates, mammals, and birds, and nurseries for a diversity of species (reviewed by [24–27]). Kelp forests also provide humans with many services, including carbon sequestration, shoreline protection and non-consumptive recreational activities [27,28]. In particular, they support economically and culturally significant commercial and recreational fisheries (e.g., [29,30]).

Species interactions are known to be key determinants of the structure and dynamics of kelp forests around the word such as the west coast of the United States [22,24,27,31], North Atlantic [32,33], Mexico [34], Australia and Tasmania [35] yet these are sensitive to anthropogenic and natural perturbations [35–37]. Given the importance and complexity of their species interactions, kelp forest ecosystems are strong candidates for ecosystem-based management, which greatly benefits from the use of ecological network models [26].

Only recently, a number of ecological network models have been generated for kelp forests including Espinosa-Romero [38], Ortiz [39], Brynes *et al.* [40] and Marzloff *et al.* [41], for the west coast of Canada, northern Chile, and southern California, respectively. In addition, theoretical multi-species models (not parameterized empirically), have enhanced our understanding of complex interactions in kelp forest systems [42,43] and assemblages of sessile invertebrates on temperate rocky reefs [44]. Each of these models represents local species composition and, justifiably, over-simplify the networks of kelp forest species interactions. Model-simplification can reflect a compromise between computational power, model-sensitivity, user interests, and preconceptions, but in many cases is simply a result of a lack of accessible information about life history traits and species interactions.

In the process of our development of ecological network models for the kelp forests of the eastern Pacific we found the necessary life history, demographic, and species interaction information poorly synthesized and organized and difficult to access. For these reasons, we developed an online database to collate and freely disseminate information on species life histories, demography, and species interactions. Here, we describe the development of and rationale for the database structure, and the means of accessing the information. Our goal here is to facilitate its use and describe its potential implementation for other ecosystems. That is, although the database was constructed with a focus on kelp forests, the interface, structure, utilities and functions could be easily translated for use in any other ecosystem. Moreover, because the architecture of this database is a DBMS, it can be integrated into a more comprehensive database integrating multiple ecosystems.

Methods and Results

The overarching goals of the online database, hereafter referred to as the "kelpforest database", was to create a database management system that could be conveniently populated and utilized across the community of researchers and provide users with the diversity of information required by the various types of ecosystems models. The kelpforest database consists of seven components: 1) a database management system, 2) database homepage, 3) an online data entry interface, 4) an online data entry manual, 5) graphic visualizations, 6) data export tools, and 7) a user forum for discussions, online assistance, and notification of problems. To promote and expedite adaptation of the database for modeling other ecosystems, technical information for developers is readily available, hosted at https://github.com/kelpforest-cameo/databaseui

Database management system

The database is a relational database management system that uses Structured Query Language (MySQL) and Personal Hypertext Preprocessor (PHP) languages and is hosted at the University of California Santa Cruz (http://kelpforest.ucsc.edu/). The central element of the database schema is the source (i.e. citation) of each datum entered (Figure 1). This allows all possible entries and queries to be referenced to the source of that information. This reference avoids redundant entries and promotes quality control by ensuring the legitimacy of entered data. The relational database links the various data tables of the database. Taxonomic information is linked to the Integrated Taxonomic Information System (ITIS; www.ITIS.gov) to ensure that entries are standardized (e.g., avoiding misspellings) and that taxonomic designations and synonyms are continuously updated.

Database website

The database website is created using WordPress web software (wordpress.org), providing an introduction to the database that includes its purpose, information on how to access it, and up-to-date contact information. The website hosts the other components of the database (i.e., data entry interface, visualization and export tools, user forum), and provides access to a sign-up form for users who wish to obtain data-entry privileges. Access to the data itself does not require registration.

Online data entry interface

The data entry interface allows multiple users to simultaneously enter information into the database. Access to the data entry interface requires a username and password. This username is linked to every datum entered by an individual in order to provide attribution of user contributions and to simplify quality control. A "sandbox" replica of the database and its data-entry interface allows individuals to practice entering data that will not be archived. Access to this "sandbox" does not require user registration.

The data entry interface provides links to three separate data entry forms: *nodes*, *interactions*, and *citations*. All forms are used to enter and look at information. The *nodes* form is used to enter information relevant to taxa (i.e. species, higher taxonomic units, or species groups). The *interactions* form is used to enter information characterizing interaction between nodes. The *citations* form is to enter the citation information associated with each datum that is entered. Each form contains a range of different sub forms. We therefore, first, provide an overview of each form before detailing its contents.

Within the *nodes* form, the user may list or search for existing nodes, or enter a new node. The first section of the *nodes* form indicates information that is relevant to the entire node, whereas the second section pertains to life stage-specific information. (The database distinguishes between a node's different life stages, detailed below).

The *interactions* form allows users to enter interaction information between specific life stages of two previously entered nodes. Importantly, species interactions are recorded as stage-specific observations of the interaction. That is, multiple observations of an interaction between two focal species (stages) may be recorded from different source citations or from the same citation

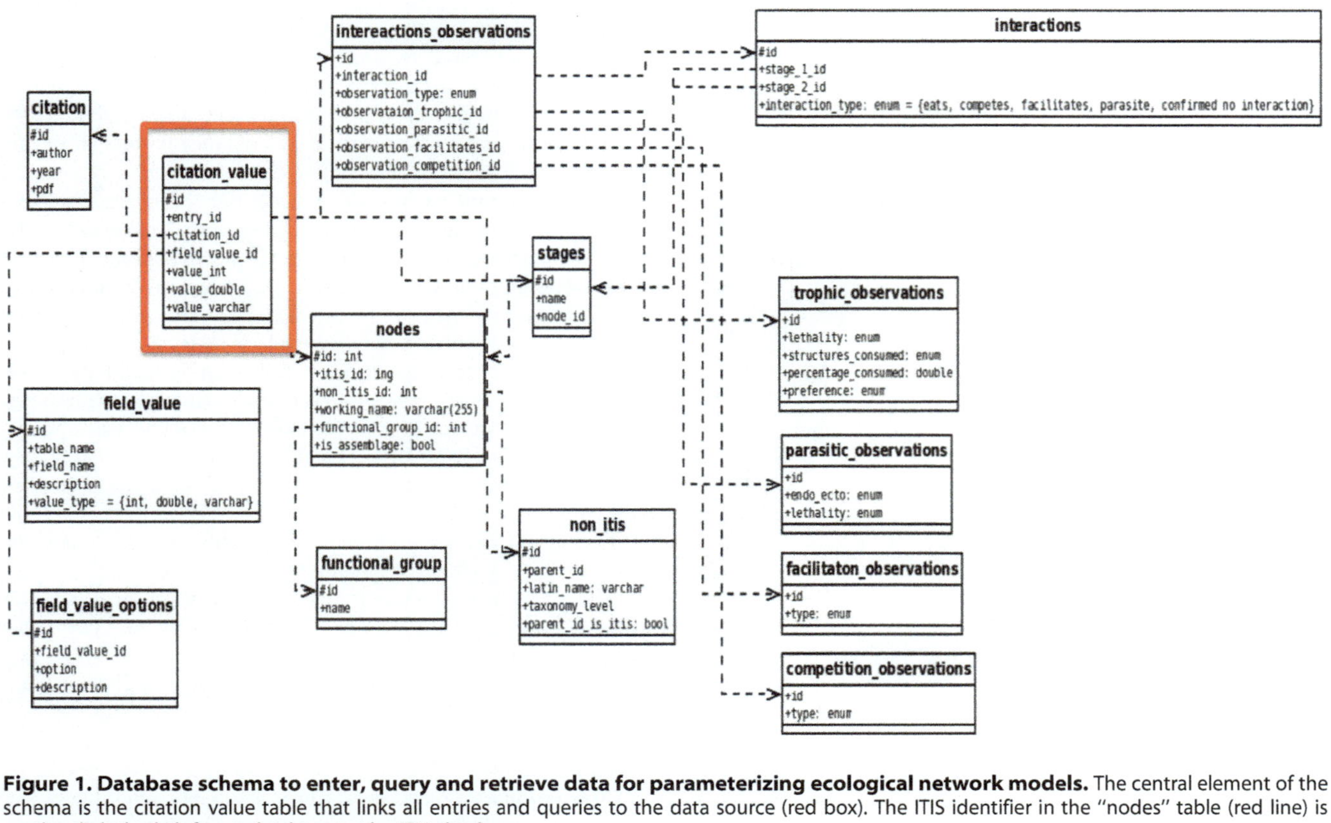

Figure 1. Database schema to enter, query and retrieve data for parameterizing ecological network models. The central element of the schema is the citation value table that links all entries and queries to the data source (red box). The ITIS identifier in the "nodes" table (red line) is used to link the kelpforest database to the ITIS database.

(pertaining, for example, to different locations or time-periods). We believe such information is key to describing the breadth, spatio-temporal variation, and uncertainty in our knowledge of species interactions.

The *citations* form allows users to enter new citations and authors to which entries are to be attributed, and to list all previously entered authors and full citations. The citation form requires users to identify the category of the source information. That is, sources from which entries have been obtained to-date are primarily from the published peer-reviewed literature, but also include unpublished reports, theses, other online databases, unpublished datasets, and qualified personal observations. The *citation* form is directly linked to the *nodes* and *interactions* forms. Every entry requires a citation. Check boxes located next to each source citation on the list of entered source citations permit data-entry users to indicate when all its pertinent information has been extracted.

Data entry fields and manual

All entry fields in both the *nodes* and *interactions* forms permit inclusion of the temporal and geographic information associated which each entry. The "time stamp" sub form for individual entry fields allows users to specify whether an entry pertains to a single time point or a window of time points at daily to annual scales. Nodes and their stage-specific interactions may be specified with a geographic location, or range of locations. Location(s) can be identified using either a map-based interactive interface or by entering a latitude and longitude. Nodes and interaction observations are thus geo-referenced across a range of spatial resolutions spanning regional, sub regional, and within sub regional scales and point locations (Figure 2). Regions and sub regions are based on recognized biogeographic sections of the eastern Pacific coast spanning from Baja, Mexico to the western Aleutian Islands. Polygons within each sub region reflect 20 km sections of the coast. Each of these standardized spatial units can be identified by the user directly on the map, or from a hierarchical legend in the mapping interface.

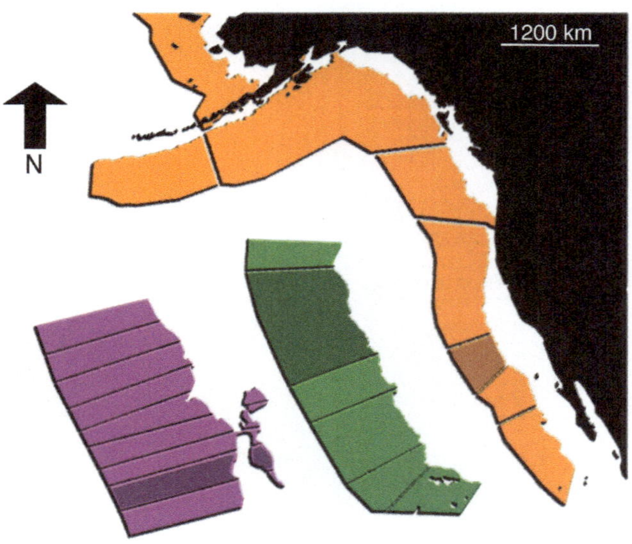

Figure 2. Map interface with delimited regions in the northeast Pacific. The interface allows users to identify location from which data were collected in the data entry process. Orange color represents regions, green sub regions and purple locations.

Each entry field also includes a comment box that allows users to clarify their input, when necessary. This is a critical element of the database. Many variables required by ecological network models are not directly available in the literature and must be calculated. The comment box allows users to describe the equations or methods that were used to derive values or standardize units from the information that was available in a given source. For example, estimates of biomass density are often derived from estimates of population size structure and density.

Data entry and standardization is facilitated by drop down menus and "mouse over" descriptions of each data entry field. In addition, the online data entry manual provides users with an overview of the database schema and the interface forms, as well as general information on data entry protocols, tips, and shortcuts.

Content

As introduced above, there are two general categories of content that may directly or indirectly inform kelp forest ecological network models: content associated with the characterization of nodes, and content describing observations of between–node interactions.

Nodes. We refer to the basic taxonomic units of the database as "nodes" rather than "species" or "taxon" because these may represent differing taxonomic resolutions (species, genera, family, etc.), or aggregated assemblages of indistinguishable taxa (e.g., phytoplankton). Each node is identified with a unique node identification number (nodeID), a common or "*working name*", scientific name, and is associated with an Integrated Taxonomic Information System (ITIS) identifier number. ITIS is an international partnership (USA, Canada and Mexico) that provides consistent and reliable information on the taxonomy and nomenclature of species in North America. Integration with the ITIS database allows nodes to be organized in a current taxonomic hierarchy and minimizes errors associated with relic synonyms and misspelled taxon names. However, the ITIS database is not complete, some taxa or assemblages found along the eastern Pacific are absent. Our database stores these nodes separately, identifying them using the working name and the ITIS id of its most resolved taxonomic level until they become available in ITIS.

Characterization of a node includes life history traits (e.g., reproductive strategy, age and size at maturity, maximum body size) and demographic information (e.g., production–biomass ratios, consumption-biomass ratios, length-weight relationships, von Bertalanffy equations, biomass). This information maybe specific to the ontogenetic stages of a node, or specified as "general" when stage-specificity is unknown. The number and types of stages may be customized for each node, with users choosing from an open-ended list of potential stages when stage-specific information is to be entered. Currently, animal stages include egg, larvae, juvenile, adult, and dead. Algae stages include sporophyte, gametophyte, and dead.

The database was initially populated with species lists from the Partnership for Interdisciplinary Studies of Coastal Oceans (PISCO– www.piscoweb.org), Reef Check California (http://reefcheck.org/rcca/rcca_home.php), Cailliet's *et al.* [45], the Monterey Bay National Marine Sanctuary Integrated Monitoring Network (SIMoN- http://sanctuarysimon.org/) and a species interaction table created by Byrnes *et al.* [46]. Many other species have since been included as a result of an intensive literature search.

Interactions. Four general categories of interactions between nodes are included in the database: trophic, competitive, facilitative, and parasitic. Individual observations for all of these interaction categories are described by their observation type (e.g., direct observation, diet analysis) and must be attributed to a source citation. Each interaction category also has entry fields particular to it. For example, trophic interactions may be described by their lethality, the structures consumed, and the percent of the consumer's diet that a particular resource represents. Similarly, parasitic interactions may be described as being endo- or ectoparasite, and by their prevalence and intensity. The interactions between two nodes are not assumed to be reciprocal.

Citations. Though most information in the database will likely continue to be extracted from the published, peer-reviewed literature, the demand for information with which to inform modeling efforts motivates a means for making it available that is faster than the rate at which it can be published. Thus, to accommodate unpublished data and personal observations, citations may refer to individuals who provide their contact information.

Data visualization

A series of static and dynamic visualization tools permit real-time access and interaction with the information contained in the database. These tools query the database in real-time to produce graphics (Figure 3) and tables of summary statistics, interaction networks, adjacency matrices, body size frequency distributions, and interaction observation maps. These utilities rely on a combination of PHP and MySQL languages and capitalize on the capabilities of D3.js (http://d3js.org), a JavaScript library that uses Hyper Text Markup Language (HTML), Scalable Vector Graphics (SVG), and Cascading Style Sheets (CSS) to create and manipulate data-driven visualizations.

Data export

Information in the kelpforest database is public and accessible to unregistered users through several export tools. These include database queries for tables and matrices containing information about nodes, interactions and citations, allowing users to download the data as comma-separated values (CSV) files (Table 3). Future additions will permit registered users to query the database directly.

Discussion

Our overarching goal in developing the kelpforest database is to provide a means for expediting the process by which information is accumulated, organized, and made accessible to those making and using ecological network models specific to temperate kelp forests. Its development has been greatly facilitated by collaborations involving federal agency scientists and academics from Canada, the United States, and Mexico. As such, we believe that with similar collaborations, its framework is applicable to any ecosystem. Our description of the structure and elements of the database is meant to inform the reader of the system's capabilities, to both motivate interest in contributing to and using the information it contains, and to suggest features to consider in the development of other databases.

In our experience to date, the online presence of the kelpforest database has been one of its most important features, allowing the research community to populate and access the database simultaneously and internationally. This has greatly enhanced the rate at which the database has been populated with entries and has facilitated communication among the kelp forest research

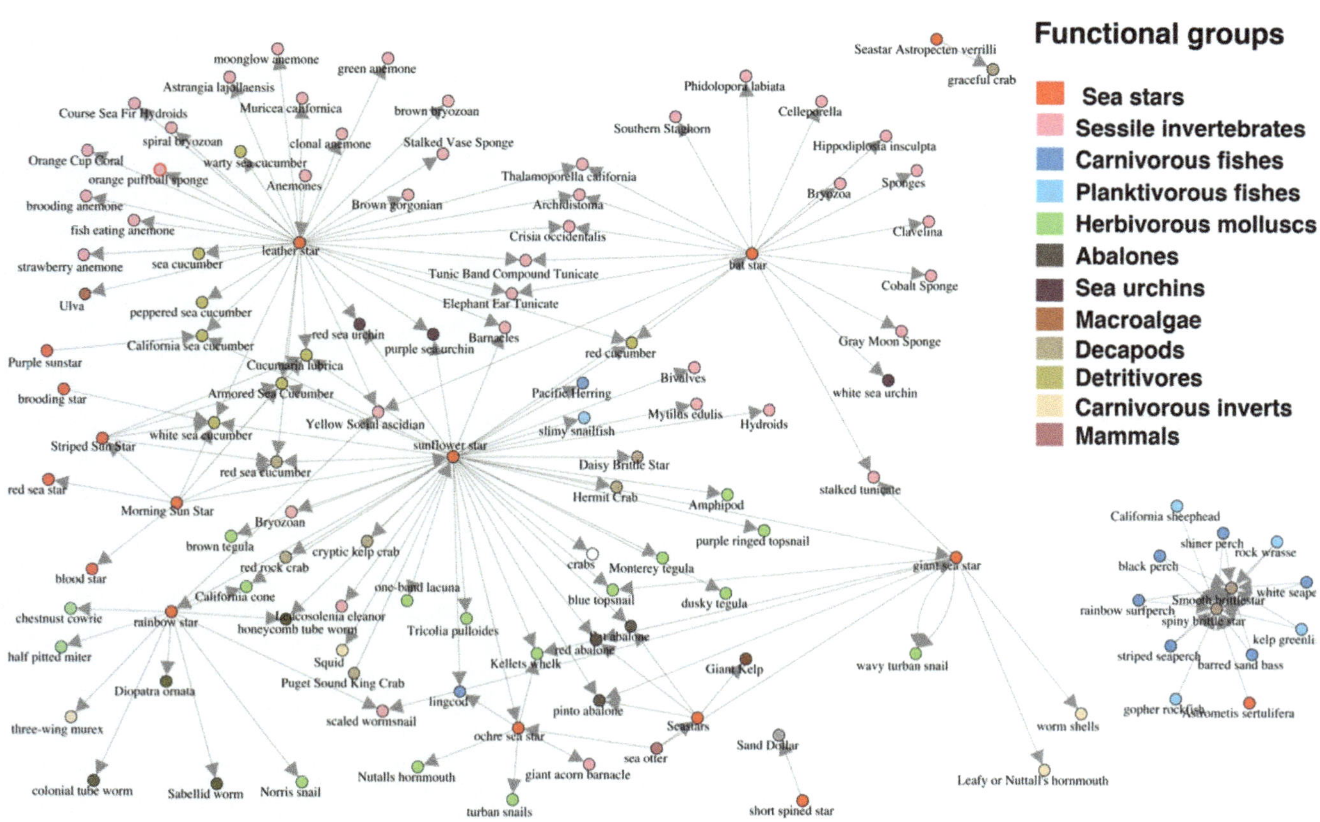

Figure 3. Example of the database visualization tool illustrating a trophic interaction network for an assemblage of kelp forest seastars, color-coded by functional group after Graham *et al.* [25].

Table 3. Example of a csv file exported from the database.

Node 1 working name	Node 1 Stage	Node 2 working name	Node 1 Stage	Type	Citations
b&y rockfish	general	crabs	general	trophic	(Larson, 1972)
b&y rockfish	general	tubesnout	general	trophic	(Quast, 1968)
b&y rockfish	general	kelp perch	general	trophic	(Robbins, 2006)
b&y rockfish	general	spiral bryozoan	general	trophic	(Perez, 1981)
b&y rockfish	general	foliate kelp crab	general	trophic	(Hines, 1982)
b&y rockfish	general	moss crab	general	trophic	(Hines, 1982)
b&y rockfish	general	sharpnose crab	general	trophic	(Hines, 1982)
b&y rockfish	general	cryptic kelp crab	general	trophic	(Hines, 1982)
b&y rockfish	general	hairy chiton	general	trophic	(Love et al. 2002)
b&y rockfish	general	monkeyface prickleback	general	trophic	(Love et al. 2002)
b&y rockfish	general	rock prickleback	general	trophic	(Love et al. 2002)
b&y rockfish	general	giant kelpfish	general	trophic	(Love et al. 2002)
b&y rockfish	larval	copepods	general	trophic	(Love et al. 2002)
b&y rockfish	juvenile	zooplankton	general	trophic	(Love et al. 2002)
b&y rockfish	adult	monkeyface prickleback	general	trophic	(Love et al. 2002)
b&y rockfish	adult	rock prickleback	general	trophic	(Love et al. 2002)
b&y rockfish	adult	northern clingfish	general	trophic	(Love et al. 2002)
b&y rockfish	adult	rockweed gunnel	general	trophic	(Love et al. 2002)
b&y rockfish	adult	fluffy sculpin	general	trophic	(Love et al. 2002)
b&y rockfish	adult	smoothhead sculpin	general	trophic	(Love et al. 2002)
b&y rockfish	adult	striped kelpfish	general	trophic	(Love et al. 2002)
b&y rockfish	adult	octopus	general	trophic	(Love et al. 2002)
b&y rockfish	adult	amphipods	general	trophic	(Love et al. 2002)
b&y rockfish	adult	polycheate worm	general	trophic	(Love et al. 2002)
b&y rockfish	general	china rockfish	juvenile	trophic	(Love et al. 2002)
b&y rockfish	general	squid	general	trophic	(Love et al. 2002)
b&y rockfish	general	octopus	general	trophic	(Hallacher & Roberts, 1985)
b&y rockfish	general	blue rockfishes	juvenile	trophic	(Hallacher & Roberts, 1985)
b&y rockfish	general	euphausiids	general	trophic	(Hallacher & Roberts, 1985)
b&y rockfish	general	caprellid amphipod	general	trophic	(Hallacher & Roberts, 1985)
b&y rockfish	general	gammarid amphipod	general	trophic	(Love et al. 2002)
b&y rockfish	general	Idotea isopods	general	trophic	(Hallacher & Roberts, 1985)
b&y rockfish	general	polycheate worm	general	trophic	(Hallacher & Roberts, 1985)
b&y rockfish	general	hydroids	general	trophic	(Hallacher & Roberts, 1985)

This table identifies some trophic interactions of the black and yellow (b&y) rockfish (Sebastes chrysomelas).

community. To date, 81 registered users across 7 institutions, the majority of whom are undergraduate and graduate students, have contributed to populating the database. Thus, this database has been used as an education and training tool for human resources from different backgrounds. Through their combined effort, the database currently contains 795 nodes and 3616 interactions based on 515 citations. That said, a critical component of the database's online nature is also the online support provided to users through the online forum, webpage, manual, and data field features described above.

A second key feature adding value to the database has been its ability to accommodate a variety of data sources, including information from the literature and existing databases, as well as user-generated values (including our own field data collection to actively fill data gaps identified by the database) and values calculated by synthesis of data in the peer-reviewed and grey literature. This has both enabled users to populate the database with their own information demands, and has made the same information immediately available to other users. Thus, the database is a clearinghouse of information on species life histories, demography and species interactions that are useful not only in the development of kelp forest ecological network models, but also for a variety of other ecological applications. The database has thereby served to inform the design of observational and experimental studies at our institutions; it has been used to train students in the use and applications of this tool, and promoted collaboration between research institutions.

Of course, few if any databases will ever collect all the relevant knowledge that has and is being obtained about kelp forest ecosystems. Databases need to be sufficiently flexible to not only accommodate new information as it is generated, but also to accommodate new kinds of information. For examples, as genetic information becomes increasingly available, the database could be modified to integrate this new information and enable users to explore the genetic basis of varying demographic relationships and species interactions and how variation in those variables contribute to patterns of genetic variability and structure and ecological-evolutionary feedbacks. To facilitate the expansion and evolution of this database and its adoption for other ecosystem databases, access to the code and technical details on how to customize this database and apply it to other ecosystems is freely available and located at the following link (https://github.com/kelpforest-cameo/databaseui).

We see the development of the kelpforest database as an important step forward toward a simpler, more organized, and more reliable integration of the collective biological knowledge of species life histories, demographics, and interactions. Our goal is to enhance the accessibility and quality of information in order to facilitate the development and use of ecological network models and inform ecosystem-based approaches to management.

Acknowledgments

The design and development of the database benefited greatly from discussions with our collaborators on this project: Jarrett Byrnes, Jim Estes, John Field, Scott Hamilton, Mike Kenner, Kevin Lafferty, Phil Levin, and Dan Reed. We are also grateful to all the students who helped to populate the database. Particularly, Jaszmin Gasca, Steven Cunningham, and Philip Ericson at University of California Santa Cruz and Stephen Gosnell, who oversaw data input at University of California Santa Barbara. We are grateful for data provided us by the Partnership for Interdisciplinary Studies of Coastal Oceans (PISCO), the Santa Barbara Coastal Long Term Ecological Research Project funded by the U.S. National Science Foundation, Reef Check California, and the Monterey Bay National Marine Sanctuary's Integrated Monitoring Network (SIMoN). Funding for Open Access provided by the UC Santa Cruz Open Access Fund. Any use of trade, product, or firm names in this publication is for descriptive purposes only and does not imply endorsement by the U.S. government or University of California. The U.S. government and University of California will not be liable for any loss of data or any other damages associated with the use of the database.

Author Contributions

Conceived and designed the experiments: RBL MHC JEC MN MTT. Performed the experiments: RBL AB MN MTT MH DM. Wrote the paper: RBL MHC MN JEC AI.

References

1. Montoya JM, Pimm SL, Solé RV (2006) Ecological networks and their fragility. Nature 442: 259–264
2. Thompson RM, Brose U, Dunne JA, Hall RO, Hladyz S, et al. (2012) Food webs: reconciling the structure and function of biodiversity. Trends Ecol Evol 27: 689–697.
3. Borrett SR, Moody J, Edelmann A (2014) The rise of Network Ecology: Maps of the topic diversity and scientific collaboration. Ecol Modell 7158: 1–17.
4. Clark JS, Carpenter SR, Barber M, Collins S, Dobson A, et al. (2001) Ecological forecasts: an emerging imperative. Science 293: 657–660.
5. Yodzis P (2001) Must top predators be culled for the sake of fisheries? Trends Ecol Evol 16: 78–84.
6. Fulton E, Link JS, Kaplan IC, Savina-Rolland M, Johnson P, et al. (2011) Lessons in modeling and management of marine ecosystems: the Atlantis experience. Fish Fish 12: 171–188.
7. Ainsworth CH, Morzaria-Luna HN, Kaplan IC, Levin PS, Fulton Ea (2012) Full compliance with harvest regulations yields ecological benefits: Northern Gulf of California case study. J Appl Ecol 49: 63–72.
8. Boit A, Martinez ND, Williams RJ, Gaedke U (2012) Mechanistic theory and modelling of complex food-web dynamics in Lake Constance. Ecol Lett 15: 594–602.
9. Morales P, Sykes MT, Prentice IC, Smith P, Smith B, et al. (2005) Comparing and evaluating process-based ecosystem model predictions of carbon and water fluxes in major European forest biomes. Glob Chang Biol 11: 2211–2233.
10. Tallis H, Levin PS, Ruckelshaus M, Lester SE, McLeod KL, et al. (2010) The many faces of ecosystem-based management: Making the process work today in real places. Mar Policy 34: 340–348.
11. Hudson L, Reuman D (2013) A cure for the plague of parameters: constraining models of complex population dynamics with allometries. Proc R Soc Biol 19.
12. Plagányi E (2007) Models for an ecosystem approach to fisheries. FAO Fish Tech Pap 477: 107.
13. Levins R (1974) The qualitative analysis of partially specified systems. Ann N Y Acad Sci 231.
14. Dambacher JM, Gaughan DJ, Rochet MJ, Rossignol PA., Trenkel VM (2009) Qualitative modeling and indicators of exploited ecosystems. Fish Fish 10: 305–322.
15. Wootton JT (2001) Local interactions predict large-scale pattern in empirically derived cellular automata. Nature 413: 841–844.
16. Lafferty KD, Allesina S, Arim M, Briggs CJ, De Leo G, et al. (2008) Parasites in food webs: the ultimate missing links. Ecol Lett 11: 533–546.
17. Gerber L, Botsford L, Hastings A, Possingham HP, Gaines SD, et al. (2003) Population models for marine reserve design: a retrospective and prospective synthesis. Ecol Appl 13: 47–64.
18. Micheli F, Halpern B, Botsford LW, Warner RR (2004) Trajectories and correlates of community change in no-take marine reserves. Ecol Appl 14: 1709–1723.
19. Caselle JE, Hamilton SL, Schroeder DM, Love MS, Standish JD, et al. (2011) Geographic variation in density, demography, and life history traits of a harvested, sex-changing, temperate reef fish. Can J Fish Aquat Sci 68: 288–303.
20. Stein LD (2003) Integrating biological databases. Nat Rev Genet 4: 337–345.
21. Mann KH (1973) Seaweeds: Their Productivity and Strategy for Growth: The role of large marine algae in coastal productivity is far more important than has been suspected. Science 182: 975–981.
22. Dayton PK (1985) Ecology of Kelp Communities. Annu Rev Ecol Syst 16: 215–245.
23. Schiel DR, Foster MS (*in press*) The Biology and Ecology of Giant Kelp Forests. UC press.
24. Schiel DR, Foster MS (1986) The structure of subtidal algal stands in temperate waters. Oceanogr Mar Biol Annu Rev 24: 265–307.
25. Graham M, Halpern B, Carr MH (2008) Diversity and Dynamics of Californian Subtidal Kelp Forests. In: McClanahan TR, Branch GM, editors. Food Webs and the Dynamics of Marine Reefs. Oxford: Oxford University Press. pp. 103–134.
26. Springer Y, Hays C, Carr MH, Mackey M (2010) Toward ecosystem-based management of marine macroalgae—The bull kelp, *Nereocystis luetkeana*. Oceanogr Mar Biol Annu Rev 48: 1–42.
27. Carr MH, Reed DC (in press) Shallow rocky reefs and kelp forests. In: H M, Zabaleta E, editors. Ecosystems of California. Berkeley: UC Press.
28. Hölker F, Beare D, Dörner H, di Natale A, Rätz HJ, et al. (2007) Comment on "Impacts of biodiversity loss on ocean ecosystem services". Science 316: 1285; author reply 1285.
29. Starr R, Cope JM, Kerr LA (2002) Trends in fisheries and fishery resources associated with the Monterey Bay National Marine Sanctuary from 1981–2000. San Diego, La Jolla, California.
30. Ling S, Johnson C (2009) Supplemental material -Overfishing reduces resilience of kelp beds to climate driven catastrophic shift. Proc Natl Acad Sci: 1–3.
31. Estes J, Palmisano J (1974) Sea Otters: Their Role in structuring nearshore communities. Science 185(4156): 1058–1060.
32. Johnson C, Mann K (1988) Diversity, patterns of adaptation, and stability of Nova Scotian kelp beds. Ecol Monogr 58: 129–154.
33. Steneck RS, Graham MH, Bourque BJ, Corbett D, Erlandson JM, et al. (2003) Kelp forest ecosystems: biodiversity, stability, resilience and future. Environ Conserv 29: 436–459.
34. Beas-Luna R, Ladah LB (2014) Latitudinal, seasonal, and small-scale spatial differences of the giant kelp, Macrocystis pyrifera, and an herbivore at their southern range limit in the northern hemisphere. Bot Mar 57: 73–83.
35. Ling SD, Johnson CR, Frusher SD, Ridgway KR (2009) Overfishing reduces resilience of kelp beds to climate-driven catastrophic phase shift. Proc Natl Acad Sci U S A 106: 22341–22345.
36. Ebeling AW, Laur DR, Rowley R (1985) Severe storm disturbances and reversal of community structure in a southern California kelp forest. Mar Biol 84: 287–294.
37. Reed DC, Rassweiler A, Carr MH, Cavanaugh KC, Malone DP, et al. (2011) Wave disturbance overwhelms top-down and bottom-up control of primary production in California kelp forests. Ecology 92: 2108–2116.
38. Espinosa-Romero MJ (2010) Towards ecosystem based management: Integrating stakeholder values in decision making and improving the representation of ecosystems in ecosystem models. Master thesis by The University of British Columbia.
39. Ortiz M, Avendaño M (2010) Trhopic mass balanced models and dynamic simulations of benthic communities from la Rinconada Marine Reserve off

northern Chile: network properties and multispecies harvest scenario assessments. Aquat Conserv Mar Freshw Ecosyst 20: 58–73.
40. Byrnes JE, Reed DC, Cardinale BJ, Cavanaugh KC, Holbrook SJ, et al. (2011) Climate-driven increases in storm frequency simplify kelp forest food webs. Glob Chang Biol 17: 2513–2524.
41. Marzloff MP, Johnson CR, Little LR, Soulié J-C, Ling SD, et al. (2013) Sensitivity analysis and pattern-oriented validation of TRITON, a model with alternative community states: Insights on temperate rocky reefs dynamics. Ecol Modell 258: 16–32.
42. Baskett ML, Salomon AK (2010) Recruitment facilitation can drive alternative states on temperate reefs. Ecology 91: 1763–1773.
43. Marzloff MP, Dambacher JM, Johnson CR, Little LR, Frusher SD (2011) Exploring alternative states in ecological systems with a qualitative analysis of community feedback. Ecol Modell 222: 2651–2662.
44. Dunstan PK, Johnson CR (2006) Linking richness, community variability, and invasion resistance with patch size. Ecology 87: 2842–2850.
45. Cailliet G, Burton E, Cope J, Kerr L, Lartson R, et al. (2000) Biological characteristics of nearshore fishes of California: A review of existing knowledge and proposed additional studies for the Pacific Ocean Interjurisdictional Fisheries Management Plan Coordination and Development Project.
46. Byrnes JE, Reed DC, Cardinale BJ, Cavanaugh KC, Holbrook SJ, et al. (2011) Climate-driven increases in storm frequency simplify kelp forest food webs. Glob Chang Biol 17: 2513–2524.

Precipitation Regime Shift Enhanced the Rain Pulse Effect on Soil Respiration in a Semi-Arid Steppe

Liming Yan[1,2], Shiping Chen[1]*, Jianyang Xia[3], Yiqi Luo[3]

1 State Key Laboratory of Vegetation and Environmental Change, Institute of Botany, Chinese Academy of Sciences, Beijing, China, **2** School of Life Sciences, Fudan University, Shanghai, China, **3** Department of Microbiology and Botany, University of Oklahoma, Norman, Oklahoma, United States of America

Abstract

The effect of resource pulses, such as rainfall events, on soil respiration plays an important role in controlling grassland carbon balance, but how shifts in long-term precipitation regime regulate rain pulse effect on soil respiration is still unclear. We first quantified the influence of rainfall event on soil respiration based on a two-year (2006 and 2009) continuously measured soil respiration data set in a temperate steppe in northern China. In 2006 and 2009, soil carbon release induced by rainfall events contributed about 44.5% (83.3 g C m^{-2}) and 39.6% (61.7 g C m^{-2}) to the growing-season total soil respiration, respectively. The pulse effect of rainfall event on soil respiration can be accurately predicted by a water status index (WSI), which is the product of rainfall event size and the ratio between antecedent soil temperature to moisture at the depth of 10 cm ($r^2 = 0.92$, $P < 0.001$) through the growing season. It indicates the pulse effect can be enhanced by not only larger individual rainfall event, but also higher soil temperature/moisture ratio which is usually associated with longer dry spells. We then analyzed a long-term (1953–2009) precipitation record in the experimental area. We found both the extreme heavy rainfall events (>40 mm per event) and the long dry-spells (>5 days) during the growing seasons increased from 1953–2009. It suggests the shift in precipitation regime has increased the contribution of rain pulse effect to growing-season total soil respiration in this region. These findings highlight the importance of incorporating precipitation regime shift and its impacts on the rain pulse effect into the future predictions of grassland carbon cycle under climate change.

Editor: Bazartseren Boldgiv, National University of Mongolia, Mongolia

Funding: This study was supported partly by the "Strategic Priority Research Program" of the Chinese Academy of Sciences, Climate Change: Carbon Budget and Relevant Issues (XDA05050402), the National Natural Science Foundation of China (31170453), the National Basic Research Program of China (973 program, 2010CB833501), and a Selected Young Scientist Program of the State Key Laboratory of Vegetation and Environment Change to SC. The funders had no role in study design, data collection and analysis, decision to publish, or preparation of the manuscript.

Competing Interests: The authors have declared that no competing interests exist.

* Email: spchen@ibcas.ac.cn

Introduction

Global precipitation has been predicted to change with increasing intra-annual variability and more frequent extreme rainfall events [1,2]. Such shifts in precipitation regime could have profound impacts on belowground carbon (C) release, especially in the arid and semiarid ecosystems [3–6]. It has been widely reported that soil respiration in grassland ecosystems can increase significantly and immediately after a rainfall event, followed by a decrease with declining soil moisture [4,7]. This pulse effect of rainfall event on soil respiration has been suggested as an important contributor to ecosystem C release in grassland ecosystems [3,8,9]. Therefore, a better understanding of how the changing precipitation regime will affect rain pulse effect is important for predicting future grassland C feedbacks to climate change.

Previous studies have summarized that variations in soil respiration are determined by several factors, including soil temperature, soil water availability and carbon substrate supply [10–12]. Yet it is not clear which factors control the pulse effect of a single rainfall event on soil respiration. There have been both laboratory [13–15] and field [7,16,17] attempts trying to identify the regulatory mechanisms of rain pulses on soil respiration. The magnitude of soil respiration response (e.g., soil carbon release) is positively correlated with rainfall event size [4,18,19]. Other studies (e.g. [20]) have observed that the antecedent soil water condition is also an important influencing factor. The pulse effect is intensified by the dry condition of the antecedent soil [20], and the effect is less obvious if the soil is wet before the rainfall event [21–23]. Soil temperature has been widely reported to regulate soil microbial activities and thus soil heterotrophic respiration in various ecosystems [22,24]. Other conditions, such as plant activity [25,26] and soil organic matter content [20,27], are also believed to influence the effects of rainfall event on soil respiration. Therefore, the predictability of rain pulse effect on soil respiration is still low and no effective approach or indicator has been developed so far.

In semi-arid ecosystems, water availability is the dominant factor regulating soil respiration [28–30]. Water availability also moderates the effects of the other factors on soil respiration such as temperature and substrate supply [30]. Water availability and its intra- and inter-annual variations are directly linked to both the intensity and the frequency of precipitation. Although IPCC [2] has reported a trend of significant change in both total amount and temporal patterns of precipitation, only a few studies have analyzed the general changes of precipitation regimes especially in semi-arid temperate grasslands [31]. Shifts in precipitation regime will alter not only the size of individual rainfall event, but also the length of the dry-spell duration and thus the antecedent soil water condition. Previous laboratory experiments have shown that both of these two factors can significantly influence the soil CO_2 release [32,33]. A growing body of works using field experiments have demonstrated that shifts in precipitation regime alone, even when the total rainfall amount does not change, can have large impacts on grassland soil CO_2 release [29,34,35] and its pulse responses to rainfall events [4,17]. However, to our knowledge, few studies have observed and quantified the dynamics of soil respiration after rainfall events [36,37]. Therefore, it is necessary to incorporate the changes in precipitation regime when evaluating the effect of rainfall event on C cycling in natural grassland ecosystems.

The area of grasslands contributed to about 40% of the total territory of China [38], and the temperate steppe in China is the third largest grassland area in the world [39]. In this study, we analyzed both a long-term (1953–2009) rainfall data set and two years (2006 and 2009) of continuously measured soil respiration data in a temperate steppe in northern China. We attempt to address the following questions: (1) What are the controlling factors of the rainfall event effect on soil respiration through the growing season in this ecosystem? (2) Has precipitation regime changed over the last ~60 yr in this semi-arid steppe? (3) How shifting precipitation regime would influence soil respiration responses to pulses in water availability?

Materials and Methods

2.1. Site description

Our field site (42°27′N, 116° 41′E), located Duolun County in the northeastern Inner Mongolia, China, belongs to Duolun Restoration Ecology Experimentation and Demonstration Station (DREEDS), Institute of Botany. No specific permissions were required for scientific researches. The site has been fenced in to exclude grazing since 2001, and no other management like mowing or fertilizing was applied. The mean altitude is about 1430 m above sea level. The vegetation was dominated by C3 grasses (e.g. *Stipa krylovii, Agropyron cristatum, Leymus chinensis*) and a semi-shrub species (*Artemisia frigida*). The soil in our study site is classified as Calcic-orthic Aridisol. The mean annual air temperature is 3.3°C and the mean annual precipitation is 377 mm, 95% (i.e. 358 mm) of which occurs during the growing season from April to October. Average air temperature was 3.13°C in 2006 and 3.09°C in 2009. Annual precipitation was 425 mm in 2006 and 185 mm in 2009, respectively.

2.2 Variable measurement

The growing season in this ecosystem usually starts from late April and ends in early October. Five PVC collars (20.3 cm in diameter and 8 cm in height) were inserted into the soil to a depth of 3 cm. The PVC collars were randomly distributed in the study site with 5 m–30 m distance between any two of them in late April, 2006, and covered an area of about 300 m^2. The site is flat so topography had little impact on the difference in measured soil respiration among replicates. Soil respiration rates were continuously measured during the growing season (from May to October) of 2006 and 2009 on a half-hour basis using an infra-red gas analyzer (LI-840, Li-Cor Inc., Lincoln, NE, USA) that was connected to five automatic measurement chambers. All living plants inside the chambers were clipped by hand weekly to exclude aboveground plant respiration. The clipped aboveground plants were left in the chambers to include CO_2 efflux from the litter decomposition. The CO_2 concentrations in the chambers were recorded in CR1000 data logger (Campbell Scientific Inc., CSI, Utah, USA), and were processed by LoggerNet 3.1 (CSI, USA). Each measurement took 120 s. The soil respiration rates were determined from the time series of soil CO_2 efflux concentrations. Since the measuring system worked well during 2006 and 2009, only small data gaps (less than 2 h) existed in two years data sets and were filled by linear interpolation method.

Soil temperatures (°C) at a depth of 10 cm and volumetric soil moisture (%) at a depth of 0–10 cm were measured using 107 soil temperature probes (CSI, USA) and CS616 soil water probes (CSI, USA), respectively. Three combinations of soil temperature and water probes were placed close (about 1 m) to the soil respiration chambers. The mean values of half-hour soil temperature and moisture were recorded in a CR1000 data logger (CSI, USA) simultaneously.

The precipitation data of the past 57 years were provided by local meteorological stations in Duolun County. The precipitation data of the growing season of 2006 and 2009 were measured by a tipping bucket rain gauge (TE525MM, CSI) at the study site.

2.3. Statistical analyses

Considering the significant impact of soil temperature on the diurnal variation of soil respiration, we analyzed precipitation and soil respiration data on a daily time step. There was a rainfall event when precipitation was recorded, and a day without any precipitation was defined as a dry day. The rainfall event size was defined as the total amount of rainfall during a rainfall event. Note that one day may have more than one rainfall event and, at the other extreme, one rainfall event may last for a few days. An uninterrupted sequence of dry days preceded and succeeded by at least one rainfall event is referred to as a dry-spell event [40,41]. We added up the total precipitation amount and calculated the frequencies of different dry-spell durations (days) in growing seasons of the past 57 years (1953–2009). The rainfall and dry-spell events were then classified into different categories based on their size (rainfall event size: 0–2, 2–5, 5–10, 10–15, 15–20, 20–30, 30–40, 40–50, or >50 mm) and duration (dry-spell duration: 0–5, 5–10, 10–20, 20–30, or >30 days).

Since the size of individual rainfall events varies greatly, the rain pulse effect on soil respiration can last from several hours to a few weeks. Thus, if the rain pulse response lasted for at least three days, we would use daily soil respiration data with the following function to estimate the effect of a rainfall event on soil respiration [7]:

$$y_t = y_0 + ate^{-bt} \quad (1)$$

where y_t is daily soil respiration rate after the rainfall event, y_0 is initial daily soil respiration before the rainfall event, t is the time (continuous but not a discrete daily time-step) after the rainfall event, and a and b are coefficients. However, if the rain pulse response lasted less than three days, we would compare the difference between post-rainfall soil respiration and the control at half-hourly time scale. The sum of all the differences over one day greater than 0 ± 0.01 (g C m^{-2} day^{-1}) was defined as rainfall event. The calculation was applied for each chamber. Based on the measurements, there were totally 47 and 45 rainfall events during the growing season of 2006 and 2009, respectively. According to their durations, the half-hourly analyses were applied to 10 and 13 rainfall events in 2006 and 2009, respectively. We further defined the time for the soil respiration response to peak after a rainfall event as the peak time (T_{peak}), and calculated the duration of the pulse ($T_{duration}$) before soil respiration decreased to 99% of the antecedent soil respiration. The rain pulse effects were estimated by the accumulative differences during this period. These analyses were conducted using Matlab (Mathworks, Natick, MA).

Linear regression analyses were used to evaluate the relationships between rain pulse response patterns (T_{peak}, $T_{duration}$, pulse effect) of soil respiration, rain size and water status index (WSI) in the growing seasons of 2006 and 2009. The water status index (WSI) was defined as,

$$WSI = \text{Rainfall event size} \times \frac{ST_{pre}}{SM_{pre}} \quad (2)$$

where ST_{pre} and SM_{pre} are the soil temperature and moisture at 10 cm depth before a rainfall event. The index thus takes both rainfall event size and antecedent condition into consideration. The development of the WSI was based on the findings from previous studies, which have shown that larger rainfall event size [4] as well as drier [23] and warmer [22] pre-event soil conditions positively affect rain pulse effect. Growing-season total soil respiration was defined as the sum of daily soil respiration from May to October in each year.

Results

3.1. Growing-season total soil respiration and relative environmental factors

Soil respiration rates showed clear seasonal variations and reached maximum values in the middle of the growing seasons in both 2006 and 2009 (Fig. 1), with the growing-season total soil respiration as 187.2±16.47 g C m^{-2} in 2006 and 155.9±4.58 g C m^{-2} in 2009. Precipitation during the growing season (May to October) was slightly above the long-term average in 2006 (403 mm) but was far below the average in 2009 (168 mm) (Fig. 1). In both years, the heaviest rainfall occurred in the middle of the growing season (late July to early August; Fig 1). Accordingly, soil moisture fluctuated through the growing season in both years, with the highest soil moisture occurred in late July in both years (Fig. 1). Soil temperature had a pronounced seasonal dynamic in both years, with the highest temperature occurring in late July of 2006 and early August of 2009 (Fig. 1). Lower seasonal mean soil temperature (16.86°C) and higher soil moisture (11.26%) were found in 2006 than those of 2009 (17.14°C and 7.81%).

3.2. Pulse effects of soil respiration and its controlling factors

Equation (1) well fitted the temporal patterns of soil respiration following heavy rainfall events, with r^2 ranging from 0.91 to 0.99 ($P<0.05$; Fig. 2a–f). In 2009, for example, daily soil respiration rates increased by 72–160% within 2 days following rainfall events (Fig. 2d–f). For small rainfall events, daily average of soil respiration could not capture the rapid responses of soil respiration to rainfall events. For example, similar to the predicted curves in Figure 3, the rainfall event effect on soil respiration peaked within several hours of the end of small rainfall events (Figure S1 in File S1). After reaching the peak, the soil respiration rate gradually returned to its antecedent level in 2.93 to 27.9 days after rainfall events of 0.4 mm to 31 mm, respectively (estimated from the fitted curve in Fig. 2). Both the magnitude and duration of the pulse effect on soil respiration (T_{peak} and $T_{duration}$) were positively correlated with rainfall event size (Fig. 3a–c). A higher sensitivity of rainfall event effect to rainfall event size was found in 2009 than in 2006 (Fig. 3c). It could be ascribed to the higher sensitivities of T_{peak} and $T_{duration}$ to rainfall event size in 2009 than in 2006 (Fig. 3a, b). It suggested that rainfall event size alone was not a good indicator for the effect of rainfall event on soil respiration in this ecosystem. When compared against WSI instead of rainfall event size, however, there were no significant differences in regressive curves between 2006 and 2009 ($P>0.1$) (Fig. 3d–f). The values of T_{peak} (about 3 days; $r^2 = 0.86$, $P<0.001$) and $T_{duration}$ (about 30 days; $r^2 = 0.94$, $P<0.001$) of the pulse responses increased logarithmically with the WSI (Fig. 3d and e). Accumulative soil respiration during the pulse processes increased linearly with the WSI (Pulse effect $= 0.1169 \times WSI$, $r^2 = 0.92$, $P<0.001$, Fig. 3f). Based on this relationship, we calculated the pulse effect of each rainfall event on soil respiration, and found the pulse effect contributed 44.5% and 39.6% of the measured growing-season total soil respiration in 2006 and 2009, respectively.

3.3. Long-term trend of the precipitation regime during 1953-2009

The mean precipitation during the growing season from 1953-2009 was 358 mm (Fig. 4b). The frequency of rainfall event occurrence decreased with increasing amount of rainfall (Fig. 4a). Long-term mean rainfall frequency with rainfall amount of 0–2 mm, 2–5 mm, 5–10 mm, 10–15 mm, 15–20 mm, 20–30 mm, 30–40 mm, 40–50 mm, and >50 mm per event was 48.4%, 21.8%, 14.0%, 7.3%, 3.8%, 2.8%, 1.0%, 0.4%, and 0.4%, respectively (Fig. 4a). We classified the duration of dry-spell into five groups (<5 days, 5–10 days, 10–20 days, 20–30 days, and > 30 days). The long-term mean occurrence frequency of the dry-spells was 71.7%, 21.4%, 5.9%, 0.8%, and 0.3% for the five groups, respectively (Fig. 4c).

During the 57 years, no clear temporal trend of the total growing-season precipitation was found (Figure S2 in File S1). The 40–50 mm ($r^2 = 0.16$, $P = 0.061$) and >50 mm ($r^2 = 0.23$, $P< 0.05$) rainfall events showed significant increasing tendencies over the years (Fig. S2k and l in File S1), while those of the smaller-sized rainfall events did not show any clear trend (Figure S2b–g in File S1). The mean dry-spell duration of the growing season increased from 1953 to 2009 ($P<0.001$; Figure S3a). During these 57 years, the frequency of dry spell events <5 days in duration significantly decreased ($r^2 = 0.28$, $P<0.001$; Figure S3b in File S1), while the dry-spell events of 5–10 days ($r^2 = 0.13$, $P = 0.003$) and 10–20 days ($r^2 = 0.11$, $P = 0.01$) occurred more frequently (Figure S3c–d in File S1). The frequency of dry-spell events longer than 20 days did not change during this time period (Figure S3e–f in File S1).

Discussion

4.1. Factors controlling rain pulse on soil respiration

In this study, significant pulse responses of soil respiration were found after rainfall events (Fig. 2), which was consistent with the results reported in other arid ecosystems [18,25]. The rain pulse induced soil respiration was 44.5% (83.3 g C m^{-2}) and 39.6% (61.7 g C m^{-2}) of the growing-season total soil respiration in 2006 and 2009, respectively. Since only few studies have quantified the contribution of rain pulse effect to growing-season total soil respiration, it is difficult to compare our results to those from other grasslands. However, some evidence show that the contribution of rain pulse to total soil respiration is lower in forest ecosystems. For example, Lee et al. [42] reported that the rainfall-induced soil CO_2 release accounted for 16–21% of the annual soil respiration in a deciduous forest in Japan, whereas Yuste et al. [43] estimated

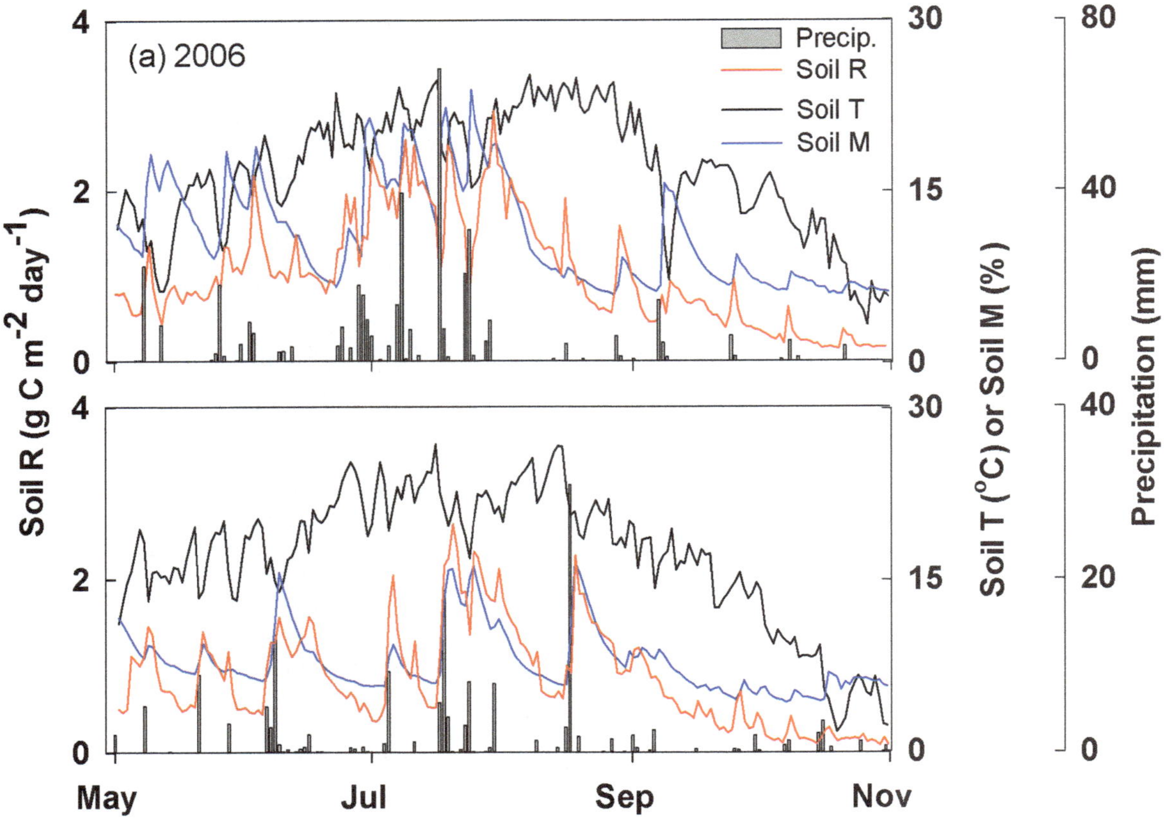

Figure 1. Seasonal patterns of precipitation (gray rectangle), soil moisture (black dashed line), soil temperature (black line) and soil respiration (gray line) during the growing seasons of 2006 (a) and 2009 (b).

that approximately 9–14% of the annual soil respiration in Belgian Campine region in Belgium was rainfall induced. The stronger rain pulse responses in grasslands could be the result of more severe water deficiency before rainfall events in grasslands and higher soil organic carbon content at the soil surface layer of grasslands than of shrublands or wood stands [36]. Therefore, the rain pulse effects on belowground C release is critical to ecosystem C balance in the semi-arid steppe ecosystem.

A rainfall event consists of at least three processes. Firstly, the percolating rainwater replaces the CO_2 in soil pore spaces. Secondly, the rainwater activates microbial activity and induces microbe-respired CO_2 in the shallow soil [44]. Thirdly, the rainwater also promotes the assimilation process by roots, which increases root respiration [3,45]. These processes have been confirmed by the positive relationship found between the rainfall event effect on soil respiration and precipitation in previous studies through field manipulative experiments [4,7,18]. Similar trends have also been found in our study. During the two growing seasons of 2006 and 2009, we observed that a larger rainfall event induced a greater pulse effect on soil respiration (Fig. 2a–c). Rather than using the limited water and resources in the soil to maintain activity, most microbes may simply become dormant in the dry soil [46]. The sudden increase in soil water availability after rains can induce microbial cell lysis or the rapid mineralization of cytoplasmic solutes and release the mineralized product into the surrounding environment to dispose of its osmolytes, which have accumulated during the dry period [27,47]. Consequently, rainfall event induced soil respiration largely comes from the decomposition of microbial cellular material. In addition, more rainfall means more water percolates to the rhizosphere and triggers root activity and respiration [3]. Therefore, the dependence of pulse effect and its duration of soil respiration on precipitation amount in this study can be explained by the greater increase in water availability under heavier rainfall event (Fig. 5).

Although the effects of rainfall events on the changes in water availability before and after rain were similar between 2006 and 2009 (Fig. 5), a significant inter-annual difference in response patterns of soil respiration to rainfall event was found (Fig. 2a–c). It indicates that some other factors rather than rainfall event size regulate the effect of rain pulse on soil respiration. For example, many studies [20,23,48] have found that the same amount of water addition had greater effects on soil respiration under drier antecedent soil condition. Soil temperature also has significant effect on soil respiration pulse [5,24,36,]. In addition, vegetation, which influences both water and C cycles, are important in regulating the effect of rain pulse on soil respiration [49,50]. In this study, we introduced the WSI to predict the effect of rain pulse on soil respiration in two growing seasons with contrasting precipitation regimes (Fig. 3d–f). It should be noted that this empirical index does not explain the underlying mechanisms of the response of soil respiration to rainfall events, which have been studied by some hourly-scale stochastic models [37]. However, it highlights the importance of precipitation regime on the pulse response of soil respiration in semi-arid areas. Since the WSI was developed from only two growing-season measurements and based on five chambers, this index needs to be tested in other sites and years before it is used as an indicator for modeling and predicting rain pulse effect on soil respiration in semi-arid grasslands.

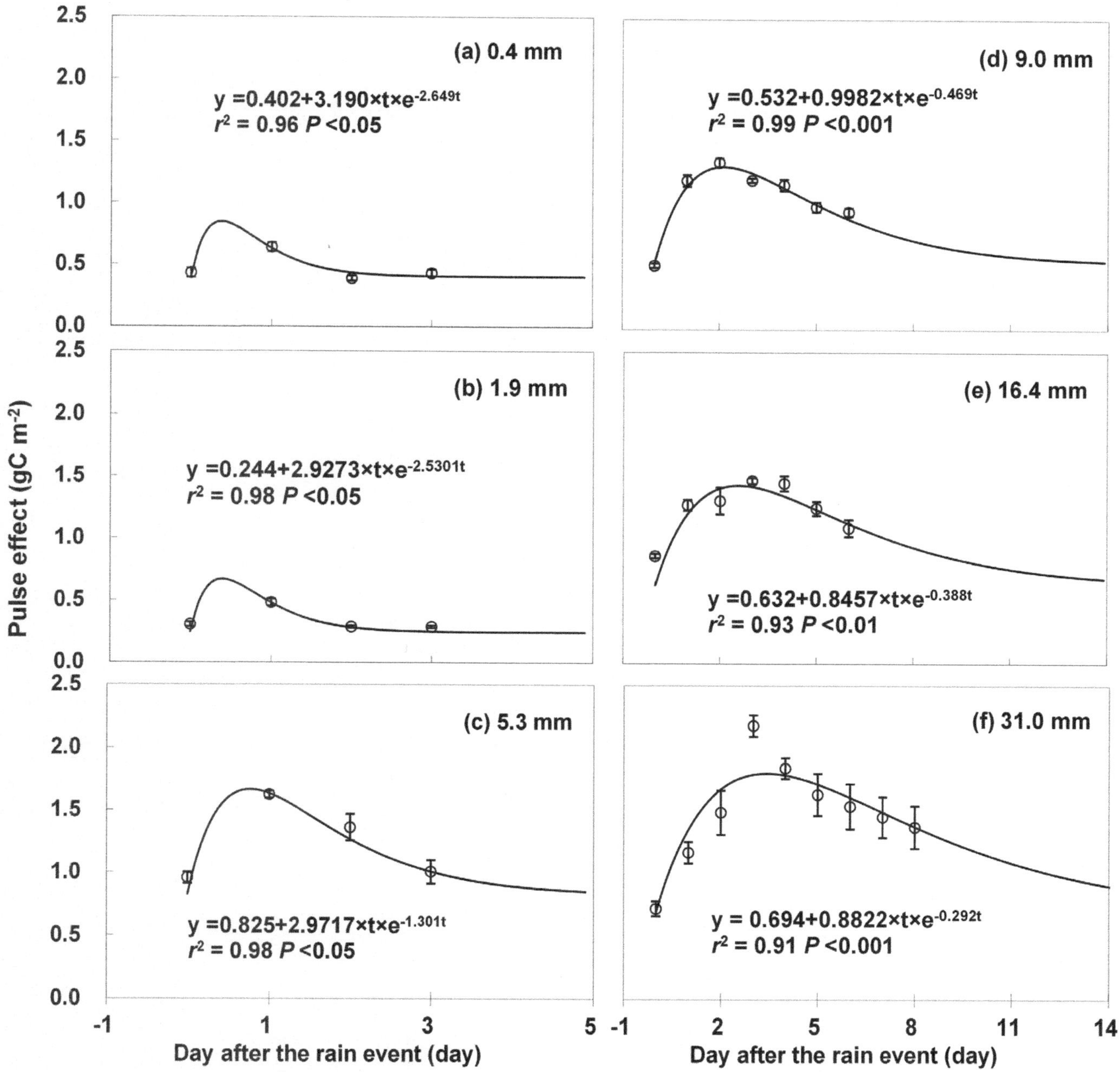

Figure 2. The response of soil respiration to rain pulses under different rain sizes. Open circles are the measured data and shown as mean ± standard deviation. Exponential equations in the form of $y=y_0+ate^{-bt}$ are used to fit the data (represented by solid lines). Rain size is shown in each panel. The 0 in the x-axis represents the day before the rainfall event.

4.2. Precipitation regime shift and its influence on soil respiration

The long-term precipitation records showed no clear trend of total precipitation amount during a growing season but a higher frequency of heavy rainfall events in our study region (Figure S2 in File S1). It was consistent with the modeling predictions of greater precipitation intensity globally in the future [1,51]. That is, more precipitation will fall in a given daily rainfall event, leading to more extreme rainfall events in the future. With similar total precipitation amount, higher frequency of heavy rainfall events means longer dry-spells during the growing season (Figure S3 in File S1). Such a shift in precipitation regime has been widely observed [40,52,53], although changes in precipitation amount would be greatly different among regions across the globe [2].

We found that a large proportion of the rain pulse effect on soil respiration was determined by the WSI, which considered the influences of rain size and antecedent soil temperature and moisture. The longer dry-spells are likely to lower the antecedent soil moisture. As a consequence, belowground C release is expected to be greater after a rainfall event compared to soils in rainfall regimes with a lower frequency of long dry-spells. The robustness of the WSI in explaining rain pulse effect on soil respiration needs to be examined in other types of grasslands. If the WSI is effective in most grassland ecosystems, it suggests not only total amount of annual precipitation but also the temporal

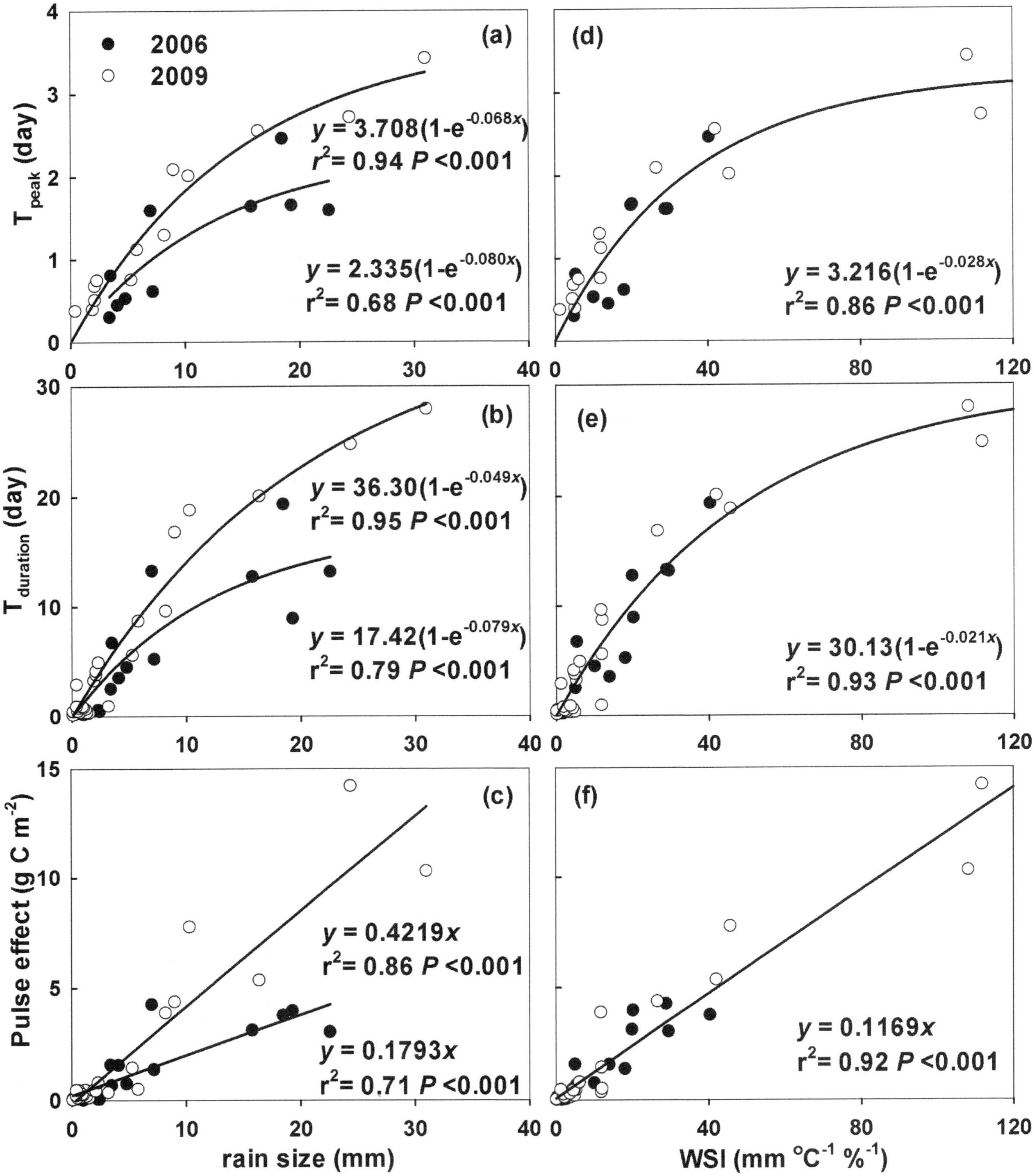

Figure 3. The relationship between dependent variables including the peak time (T_{peak}) and the duration ($T_{duration}$) of the pulses, and the pulse effects on soil respiration, and independent variables including rain size (a, b, c) and WSI (rain size × the ratio of antecedent soil temperature to soil moisture; d, e, f) in 2006 (solid circle) and 2009 (open circle), respectively.

distribution of rainfall events must be taken into consideration when predicting the belowground C release and evaluating C budget of an ecosystem undergoing climate change.

4.3. Implications for modeling soil carbon release

Currently, process-based ecosystem models are common tools for simulating and predicting the future states of terrestrial C cycle [2]. Most of these models are established at hourly or daily time

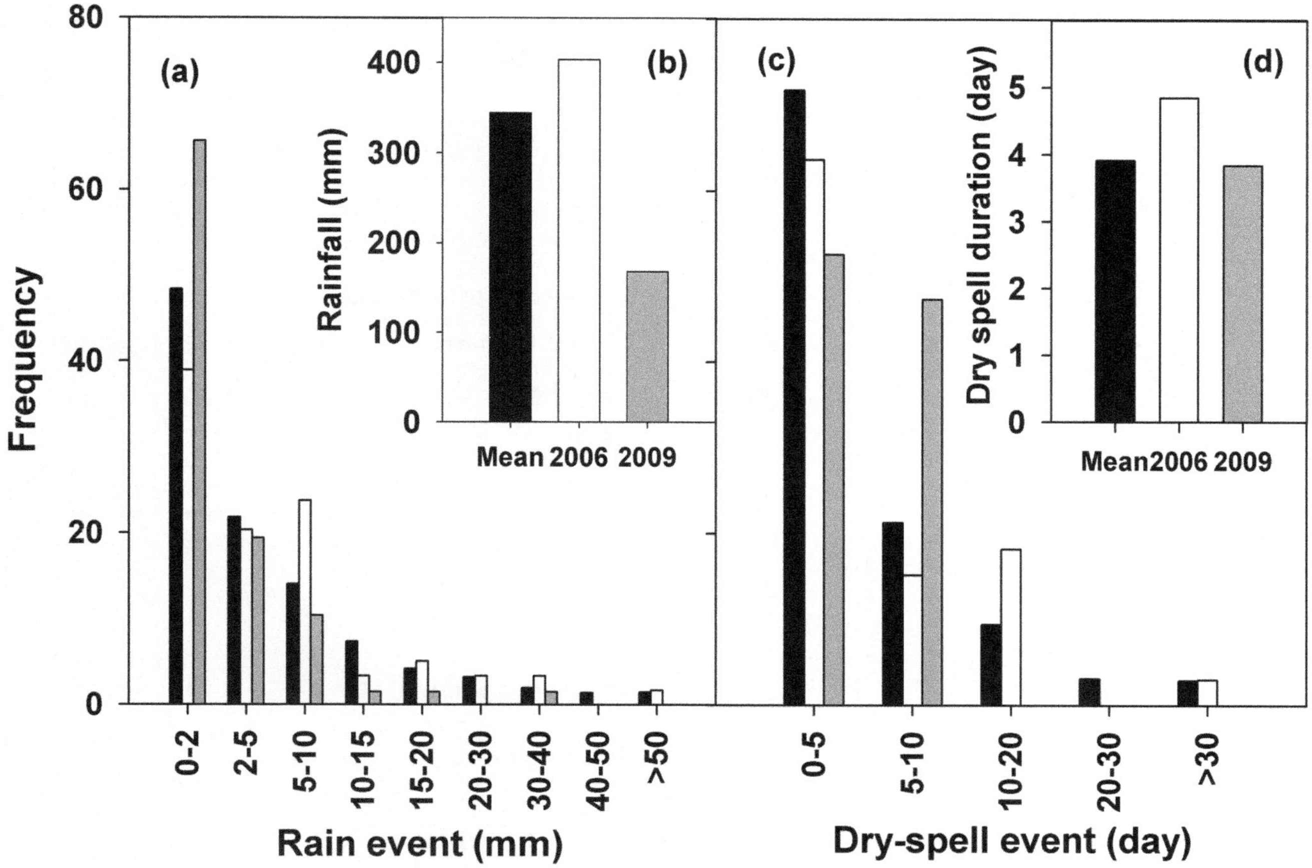

Figure 4. The frequency distribution of different class-categories of rainfall events (a) and dry-spell events (c) during the growing seasons of 2006 and 2009. The long-term mean annual precipitations from 1953 to 2009 are also represented here by black columns. The insert panels show the precipitation during the growing season (b) and the mean dry-spell duration during the growing season (d) in 2006 and 2009.

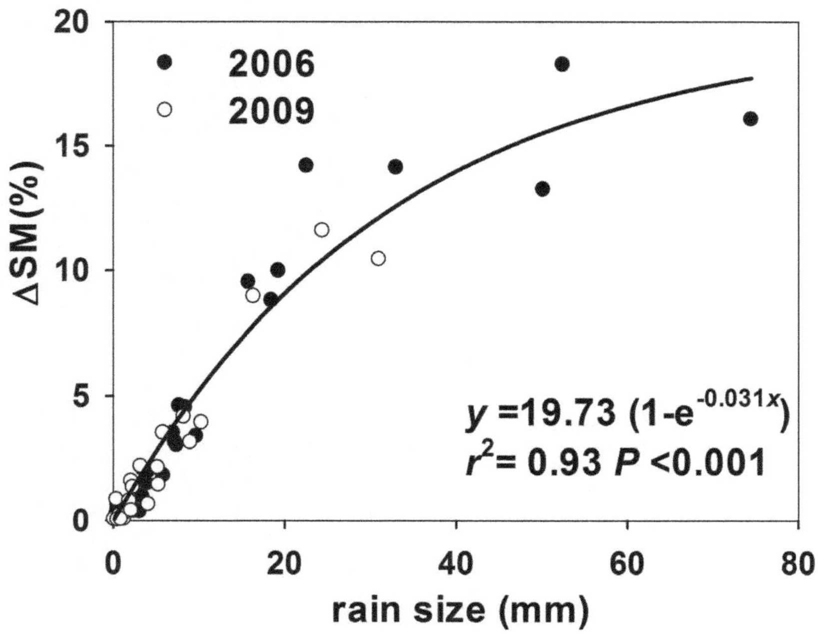

Figure 5. The relationships between the rain size and the change in water availability after a rainfall event (ΔSM; the changes in the largest volumetric soil moisture after a rainfall event from the antecedent volumetric soil moisture) in 2006 (open circle) and 2009 (solid circle) with the coefficient (r^2) and the significance level (P-value).

scales without incorporating the immediate pulse responses of soil respiration to rainfall events. Therefore, rain pulse effect on soil respiration is one of the key sources of the uncertainty in the modeled ecosystem C cycle, especially in grasslands [8,36]. In this study, we found that the rain pulse effect on soil respiration and its duration increased with the product of rain size and antecedent soil temperature-to-moisture ratio. However, the relationship between the WSI and rain pulse effect on soil respiration ($y = 0.1169 \times WSI$) as well as its duration ($y = 30.13(1\text{-}e^{-0.021x})$) was obtained from the data of only one grassland site during two growing seasons. The parameters in the regression model might vary with soil texture and soil organic matter content, which supplies C substrate for respiration [5,20,27]. Therefore, the applicability of the WSI we proposed in this study still needs to be tested in broader spatial scales and at more locations. Parameter information that constrains the relationship between the WSI and rain pulse effect on soil respiration could be useful for improving our ability to predict soil C cycle under future precipitation regimes.

There is evidence that CO_2 from carbonates may [54] or may not [55] contribute significantly to total CO_2 release when soils are moistened. Since the soil in our site is calcic, it is still unclear whether rain pulse effect will increase CO_2 release from inorganic C sources. Future study may pay more attention to the pulse effect of rainfall events on not only the decomposition of soil organic matters but also CO_2 release from inorganic C sources.

Conclusions

In this study, we presented a water status index (WSI) as a good indicator for predicting the effect of rain pulse on soil respiration in a temperate steppe in northern China. Long-term precipitation data showed a significant increase in frequency of large rainfall events and dry-spell duration with time. Although there are no measurements of soil temperature and moisture from the historical record to compare against measurements made in 2006 and 2009, the longer dry-spell may leads to larger antecedent ratio between temperature:moisture in the soils, and thus greater rain pulse effect on soil respiration. In this study, rain pulse effect contributed to about 39–44.5% of the growing-season total soil respiration, which is much larger than previous studies in this [4] and other [23] ecosystems. It indicates that the effect of rain pulses on soil respiration cannot be neglected in future climate-carbon cycle modeling. Future research could test the robustness of WSI in other regions before it is used as a simple approach to estimate the rain pulse effect on soil respiration in grassland ecosystems.

Supporting Information

File S1 Supporting figures. Figure S1, Half-hourly soil respiration rate (SR, μmol m^{-2} s^{-1}) under different small rainfall events. Rainfall size is shown in each panel. The 0 in the x-axis is the beginning time of a rainfall. **Figure S2**, The precipitation amount during the growing season (a) and the occurrence frequency of the nine class-categories of precipitation during the growing season (b–l) from 1953 to 2009. **Figure S3**, The mean dry-spell duration during the growing season (a) and the occurrence frequency of the five class-categories of the dry-spell duration during the growing season (b–f) from 1953 to 2009.

Acknowledgments

We thank Dr. Bazartseren Boldgiv and two anonymous reviewers for their valuable comments and suggestions. Thanks, Ms. Jin Chen, for polishing the English language of an early version of this manuscript.

Author Contributions

Conceived and designed the experiments: SC LY. Analyzed the data: LY SC. Contributed to the writing of the manuscript: LY SC. Conducted the filed measurement: LY. Provided constructive suggestions on data analysis: JX YL. Provided substantial comments and revised the early version of manuscript: JX YL.

References

1. Dore MHI (2005) Climate change and changes in global precipitation patterns: What do we know? Environment International 31: 1167–1181.
2. IPCC (2007) Climate change 2007: the physical science basis. In: Solomon S, Qin D, Manning M, Chen Z, Marquis M, Averyt KB, Tignor M, Miller HL, editors, Contribution of Working Group I to the Fourth Assessment Report of the Intergovernmental Panel on Climate Change. Cambridge: Cambridge University Press. pp. 30–33.
3. Huxman TE, Snyder KA, Tissue D, Leffler AJ, Ogle K, et al. (2004) Precipitation pulses and carbon fluxes in semiarid and arid ecosystems. Oecologia 141: 254–268.
4. Chen SP, Lin GH, Huang JH, Jenerette GD (2009) Dependence of carbon sequestration on the differential responses of ecosystem photosynthesis and respiration to rain pulses in a semiarid steppe. Global Change Biology 15: 2450–2461.
5. Kim DG, Vargas R, Bond-Lamberty B, Turetsky MR (2012) Effects of soil rewetting and thawing on soil gas fluxes: a review of current literature and suggestions for future research. Biogeosciences 9: 2459–2483.
6. Liu R, Cieraad E, Li Y (2013) Summer rain pulses may stimulate a CO2 release rather than absorption in desert halophyte communities. Plant and Soil 373: 799–811.
7. Liu XZ, Wan SQ, Su B, Hui DF, Luo YQ (2002) Response of soil CO_2 efflux to water manipulation in a tallgrass prairie ecosystem. Plant and Soil 240: 213–223.
8. Lee X, Wu HJ, Sigler J, Oishi C, Siccama T (2004) Rapid and transient response of soil respiration to rain. Global Change Biology 10: 1017–1026.
9. Ma S, Baldocchi DD, Hatala JA, Detto M, Yuste JC (2012) Are rain-induced ecosystem respiration pulses enhanced by legacies of antecedent photodegradation in semi-arid environments? Agricultural and Forest Meteorology 154–155: 203–213.
10. Högberg P, Nordgren A, Buchmann N, Taylor AFS, Ekblad A, et al. (2001) Large-scale forest girdling shows that current photosynthesis drives soil respiration. Nature 411: 789–792.
11. Davidson EA, Janssens IA (2006) Temperature sensitivity of soil carbon decomposition and feedbacks to climate change. Nature 440, 165–173.
12. Wan S, Norby RJ, Ledford J, Weltzin JF (2007) Responses of soil respiration to elevated CO_2, air warming, and changing soil water availability in a model old-field grassland. Global Change Biology 13: 2411–2424.
13. Griffiths E, Birch HF (1961) Microbiological Changes in Freshly Moistened Soil. Nature 189: 424.
14. Kieft TL, Soroker E, Firestone MK (1987) Microbial biomass response to a rapid increase in water potential when dry soil is wetted. Soil Biology and Biochemistry 19: 119–126.
15. Borken W, Xu YJ, Davidson EA, Beese A (2002) Site and temporal variation of soil respiration in European beech, Norway spruce, and Scots pine forests. Global Change Biology 8: 1205–1216.
16. Schimel JP, Gulledge JM, Clein-Curley JS, Lindstrom JE, Braddock JF (1999) Moisture effects on microbial activity and community structure in decomposing birch litter in the Alaskan taiga. Soil Biology and Biochemistry 31: 831–838.
17. Chen SP, Lin GH, Huang JH, He M (2008) Responses of soil respiration to simulated precipitation pulses in semiarid steppe under different grazing regimes. Journal of Plant Ecology-UK 1: 237–246.
18. Sponseller RA (2007) Precipitation pulses and soil CO_2 flux in a Sonoran Desert ecosystem. Global Change Biology 13: 426–436.
19. Vargas R, Collins SL, Thomey ML, Johnson JE, Brown RF, et al. (2012) Precipitation variability and fire influence the temporal dynamics of soil CO_2 efflux in an arid grassland. Global Change Biology 18: 1401–1411.
20. Cable JM, Ogle K, Williams DG, Weltzin JF, Huxman TE (2008) Soil texture drives responses of soil respiration to precipitation pulses in the Sonoran Desert: Implications for climate change. Ecosystems 11: 961–979.
21. Bowling DR, Grote EE, Belnap J (2011) Rain pulse response of soil CO_2 exchange by biological soil crusts and grasslands of the semiarid Colorado Plateau, United States. Journal of Geophysical Research -Biogeosciences 116: G03028, doi:10.1029/2011JG001643.

22. Carbone MS, Still CJ, Ambrose AR, Dawson TE, Williams AP, et al. (2011) Seasonal and episodic moisture controls on plant and microbial contributions to soil respiration. Oecologia 167: 265–278.
23. Wu HJ, Lee X (2011) Short-term effects of rain on soil respiration in two New England forests. Plant and Soil 338: 329–342.
24. Jager G, Bruins EH (1975) Effect of Repeated Drying at Different Temperatures on Soil Organic-Matter Decomposition and Characteristics, and on Soil Microflora. Soil Biology and Biochemistry 7: 153–159.
25. Smart DR, Penuelas J (2005) Short-term CO_2 emissions from planted soil subject to elevated CO_2 and simulated precipitation. Applied Soil Ecology 28: 247–257.
26. Vargas R, Allen MF (2008) Environmental controls and the influence of vegetation type, fine roots and rhizomorphs on diel and seasonal variation in soil respiration. New Phytologist 179: 460–471.
27. Fierer N, Schimel JP (2003) A proposed mechanism for the pulse in carbon dioxide production commonly observed following the rapid rewetting of a dry soil. Soil Science Society of America Journal 67: 798–805.
28. Riveros-Iregui DA, Emanuel RE, Muth DJ, McGlynn BL, Epstein HE, et al. (2007) Diurnal hysteresis between soil CO_2 and soil temperature is controlled by soil water content. Geophysical Research Letters 34: L17404, doi:10.1029/2007GL030938.
29. Liu WX, Zhang Z, Wan SQ (2009) Predominant role of water in regulating soil and microbial respiration and their responses to climate change in a semiarid grassland. Global Change Biology 15: 184–195.
30. Yan LM, Chen SP, Huang JH, Lin GH (2011) Water regulated effects of photosynthetic substrate supply on soil respiration in a semiarid steppe. Global Change Biology 17: 1990–2001.
31. Liu BH, Xu M, Henderson M, Qi Y (2005) Observed trends of precipitation amount, frequency, and intensity in China, 1960–2000. Journal of Geophysical Research-Atmospheres 110: D08103, doi:10.1029/2004JD004864.
32. Fierer N, Schimel JP (2002) Effects of drying-rewetting frequency on soil carbon and nitrogen transformations. Soil Biology and Biochemistry 34: 777–787.
33. Miller AE, Schimel JP, Meixner T, Sickman JO, Melack JM (2005) Episodic rewetting enhances carbon and nitrogen release from chaparral soils. Soil Biology and Biochemistry 37: 2195–2204.
34. Knapp AK, Fay PA, Blair JM, Collins SL, Smith MD, et al. (2002) Rainfall variability, carbon cycling, and plant species diversity in a mesic grassland. Science 298: 2202–2205.
35. Harper CW, Blair JM, Fay PA, Knapp AK, Carlisle JD (2005) Increased rainfall variability and reduced rainfall amount decreases soil CO_2 flux in a grassland ecosystem. Global Change Biology 11: 322–334.
36. Xu LK, Baldocchi DD, Tang JW (2004) How soil moisture, rain pulses, and growth alter the response of ecosystem respiration to temperature. Global Biogeochemical Cycles 18: GB4002.
37. Daly E, Oishi AC, Porporato A, Katul GG (2008) A stochastic model for daily subsurface CO_2 concentration and related soil respiration. Advances in Water Resources 31: 987–994.
38. Ni J (2002) Carbon storage in grasslands of China. Journal of Arid Environments 50: 205–218.
39. Lee R, Yu FF, Price KP, Ellis J, Shi PJ (2002) Evaluating vegetation phenological patterns in Inner Mongolia using NDVI time-series analysis. International Journal of Remote Sensing 23: 2505–2512.
40. Gong DY, Shi PJ, Wang JA (2004) Daily precipitation changes in the semi-arid region over northern China. Journal of Arid Environments 59: 771–784.
41. Paulo AA, Pereira LS (2006) Drought concepts and characterization: Comparing drought indices applied at local and regional scales. Water International 31: 37–49.
42. Lee MS, Nakane K, Nakatsubo T, Mo WH, Koizumi H (2002) Effects of rainfall events on soil CO_2 flux in a cool temperate deciduous broad-leaved forest. Ecology Research 17: 401–409.
43. Yuste JC, Janssens IA, Ceulemans R (2005) Calibration and validation of an empirical approach to model soil CO_2 efflux in a deciduous forest. Biogeochemistry 73: 209–230.
44. Sharkhuu A, Plante AF, Enkhmandal O, Casper BB, Helliker BR, et al. (2013) Effects of open-top passive warming chambers on soil respiration in the semi-arid steppe to taiga forest transition zone in Northern Mongolia. Biogeochemistry 115: 333–348.
45. Reynolds JF, Kemp PR, Ogle K, Fernandez RJ (2004) Modifying the 'pulse–reserve' paradigm for deserts of North America: precipitation pulses, soil water, and plant responses. Oecologia 141: 194–210.
46. Boot CM, Schaeffer SM, Schimel JP (2013) Static osmolyte concentrations in microbial biomass during seasonal drought in a California grassland. Soil Biology and Biochemistry 57: 356–361.
47. Schimel J, Balser TC, Wallenstein M (2007) Microbial stress-response physiology and its implications for ecosystem function. Ecology 88: 1386–1394.
48. Yan LM, Chen SP, Huang JH, Lin GH (2010) Differential responses of auto- and heterotrophic soil respiration to water and nitrogen addition in a semiarid temperate steppe. Global Change Biology 16: 2345–2357.
49. Kim DG, Mu S, Kang S, Lee D (2010) Factors controlling soil CO_2 effluxes and the effects of rewetting on effluxes in adjacent deciduous, coniferous, and mixed forests in Korea. Soil Biology and Biochemistry 42: 576–585.
50. Vargas R, Detto M, Baldocchi DD, Allen MF (2011) Multiscale analysis of temporal variability of soil CO_2 production as influenced by weather and vegetation. Global Change Biology 16: 1589–1605.
51. Meehl GA, Arblaster JM, Tebaldi C (2005) Understanding future patterns of increased precipitation intensity in climate model simulations. Geophysical Research Letters 32: L18719.
52. Schmidli J, Frei C (2005) Trends of heavy precipitation and wet and dry spells in Switzerland during the 20th century. International Journal of Climatology 25: 753–771.
53. New M, Hewitson B, Stephenson DB, Tsiga A, Kruger A, et al. (2006) Evidence of trends in daily climate extremes over southern and west Africa. Journal of Geophysical Research-Atmospheres 111: D14102, doi:10.1029/2005JD006289.
54. Tang JW, Baldocchi DD (2005) Spatial-temporal variation in soil respiration in an oak-grass savanna ecosystem in California and its partitioning into autotrophic and heterotrophic components. Biogeochemistry 73: 183–207.
55. Inglima I, Alberti G, Bertolini T, Vaccari FP, Gioli B, et al. (2009) Precipitation pulses enhance respiration of Mediterranean ecosystems: the balance between organic and inorganic components of increased soil CO_2 efflux. Global Change Biology 15: 1289–1301.

PERMISSIONS

All chapters in this book were first published in PLOS ONE, by The Public Library of Science; hereby published with permission under the Creative Commons Attribution License or equivalent. Every chapter published in this book has been scrutinized by our experts. Their significance has been extensively debated. The topics covered herein carry significant findings which will fuel the growth of the discipline. They may even be implemented as practical applications or may be referred to as a beginning point for another development.

The contributors of this book come from diverse backgrounds, making this book a truly international effort. This book will bring forth new frontiers with its revolutionizing research information and detailed analysis of the nascent developments around the world.

We would like to thank all the contributing authors for lending their expertise to make the book truly unique. They have played a crucial role in the development of this book. Without their invaluable contributions this book wouldn't have been possible. They have made vital efforts to compile up to date information on the varied aspects of this subject to make this book a valuable addition to the collection of many professionals and students.

This book was conceptualized with the vision of imparting up-to-date information and advanced data in this field. To ensure the same, a matchless editorial board was set up. Every individual on the board went through rigorous rounds of assessment to prove their worth. After which they invested a large part of their time researching and compiling the most relevant data for our readers.

The editorial board has been involved in producing this book since its inception. They have spent rigorous hours researching and exploring the diverse topics which have resulted in the successful publishing of this book. They have passed on their knowledge of decades through this book. To expedite this challenging task, the publisher supported the team at every step. A small team of assistant editors was also appointed to further simplify the editing procedure and attain best results for the readers.

Apart from the editorial board, the designing team has also invested a significant amount of their time in understanding the subject and creating the most relevant covers. They scrutinized every image to scout for the most suitable representation of the subject and create an appropriate cover for the book.

The publishing team has been an ardent support to the editorial, designing and production team. Their endless efforts to recruit the best for this project, has resulted in the accomplishment of this book. They are a veteran in the field of academics and their pool of knowledge is as vast as their experience in printing. Their expertise and guidance has proved useful at every step. Their uncompromising quality standards have made this book an exceptional effort. Their encouragement from time to time has been an inspiration for everyone.

The publisher and the editorial board hope that this book will prove to be a valuable piece of knowledge for researchers, students, practitioners and scholars across the globe.

LIST OF CONTRIBUTORS

Hem Nalini Morzaria-Luna and Cameron H. Ainsworth
Marine Resources Assessment Group Americas Incorporated, Seattle, Washington, United States of America

Isaac C. Kaplan and Phillip S. Levin
Conservation Biology Division, Northwest Fisheries Science Center, National Marine Fisheries Service, National Oceanic and Atmospheric Administration, Seattle, Washington, United States of America

Elizabeth A. Fulton
Division of Marine and Atmospheric Research, The Commonwealth Scientific and Industrial Research Organisation Wealth from Oceans Flagship, Hobart, Tasmania, Australia

Wenquan Zhu
State Key Laboratory of Earth Surface Processes and Resource Ecology, Beijing Normal University, Beijing, China
College of Resources Science and Technology, Beijing Normal University, Beijing, China

Guangsheng Chen
Environmental Sciences Division, Oak Ridge National Laboratory, Oak Ridge, Tennessee, United States of America

Nan Jiang, Jianhong Liu and Minjie Mou
College of Resources Science and Technology, Beijing Normal University, Beijing, China

Jeffrey J. Polovina and Phoebe A. Woodworth-Jefcoats
Pacific Islands Fisheries Science Center, NOAA Fisheries, Honolulu, Hawaii, United States of America

Jennifer Beseres Pollack
Department of Life Sciences, Texas A&M University-Corpus Christi, Corpus Christi, Texas, United States of America

David Yoskowitz and Paul A. Montagna
Harte Research Institute, Texas A&M University- Corpus Christi, Corpus Christi, Texas, United States of America

Hae-Cheol Kim
I.M. Systems Group at National Centers for Environmental Prediction, National Oceanic Atmospheric Administration, College Park, Maryland, United States of America

Elise Pendall
Department of Botany and Program in Ecology, University of Wyoming, Laramie, Wyoming, United States of America

Jana L. Heisler-White
United States Department of Agriculture - Agricultural Research Service, Rangeland Resources Research Unit and Northern Plains Area, Fort Collins, Colorado, United States of America Departments of Botany; Ecosystem Science and Management, and Program in Ecology, University of Wyoming, Laramie, Wyoming, United States of America

David G. Williams
Departments of Botany; Ecosystem Science and Management, and Program in Ecology, University of Wyoming, Laramie, Wyoming, United States of America

Feike A. Dijkstra
United States Department of Agriculture - Agricultural Research Service, Rangeland Resources Research Unit and Northern Plains Area, Fort Collins, Colorado, United States of America
Faculty of Agriculture and Environment, University of Sydney, Sydney, New South Wales, Australia

Yolima Carrillo
Department of Botany and Program in Ecology, University of Wyoming, Laramie, Wyoming, United States of America
Faculty of Agriculture and Environment, University of Sydney, Sydney, New South Wales, Australia

Jack A. Morgan and Daniel R. LeCain
United States Department of Agriculture - Agricultural Research Service, Rangeland Resources Research Unit and Northern Plains Area, Fort Collins, Colorado, United States of America

Yonggang Chi and Ruichang Shen
Key Laboratory of Ecosystem Network Observation and Modeling, Institute of Geographic Sciences and Natural Resources Research, Chinese Academy of Sciences, Beijing, China
University of Chinese Academy of Sciences, Beijing, China

Ming Xu
Key Laboratory of Ecosystem Network Observation and Modeling, Institute of Geographic Sciences and Natural Resources Research, Chinese Academy of Sciences, Beijing, China
Department of Ecology, Evolution and Natural Resources, Center for Remote Sensing and Spatial Analysis, Rutgers University, New Brunswick, New Jersey, United States of America

Qingpeng Yang
Key Laboratory of Ecosystem Network Observation and Modeling, Institute of Geographic Sciences and Natural Resources Research, Chinese Academy of Sciences, Beijing, China
Huitong Experimental Station of Forest Ecology, State Key Laboratory of Forest and Soil Ecology, Institute of Applied Ecology, Chinese Academy of Sciences, Shenyang, China

Bingru Huang
Department of Plant Biology and Pathology, Rutgers University, New Brunswick, New Jersey, United States of America

Shiqiang Wan
Key Laboratory of Plant Stress Biology, College of Life Sciences, Henan University, Kaifeng, Henan, China

Li Xu, Kun Zhang and Erkang Wang
State Key Laboratory of Electroanalytical Chemistry, Changchun Institute of Applied Chemistry, Chinese Academy of Sciences, Changchun, Jilin, China

Feng Zhang
State Key Laboratory of Electroanalytical Chemistry, Changchun Institute of Applied Chemistry, Chinese Academy of Sciences, Changchun, Jilin, China
College of Physics, Jilin University, Changchun, Jilin, China

Jin Wang
State Key Laboratory of Electroanalytical Chemistry, Changchun Institute of Applied Chemistry, Chinese Academy of Sciences, Changchun, Jilin, China
College of Physics, Jilin University, Changchun, Jilin, China
Department of Chemistry & Physics, State University of New York at Stony Brook, Stony Brook, New York, United States of America

William K. Nuttle
Eco-Hydrology, Ottawa, Canada

Christopher R. Kelble
Ocean Chemistry Division, Atlantic Oceanographic & Meteorological Laboratory, National Oceanic & Atmospheric Administration, Miami, Florida, United States of America

Dave K. Loomis
Institute for Coastal Science and Policy, East Carolina University, Greenville, North Carolina, United States of America

Susan Lovelace
JHT Incorporated, Hollings Marine Laboratory, National Centers for Coastal Ocean Science, National Oceanic and Atmospheric Administration, Charleston, South Carolina, United States of America

Peter B. Ortner
Cooperative Institute for Marine and Atmospheric Studies, Rosenstiel School of Marine and Atmospheric Science, University of Miami, Miami, Florida, United States of America

Pamela Fletcher
University of Florida Sea Grant, Gainesville, Florida, United States of America

Geoffrey S. Cook
Ocean Chemistry Division, Atlantic Oceanographic & Meteorological Laboratory, National Oceanic & Atmospheric Administration, Miami, Florida, United States of America
Cooperative Institute for Marine and Atmospheric Studies, Rosenstiel School of Marine and Atmospheric Science, University of Miami, Miami, Florida, United States of America

Jerry J. Lorenz
Tavernier Science Center, Audubon Florida, Tavernier, Florida, United States of America

Joseph N. Boyer
Center for the Environment, Plymouth State University, Plymouth, New Hampshire, United States of America

Gangsheng Wang, Melanie A. Mayes and Lianhong Gu
Climate Change Science Institute, Oak Ridge National Laboratory, Oak Ridge, Tennessee, United States of America
Environmental Sciences Division, Oak Ridge National Laboratory, Oak Ridge, Tennessee, United States of America

Christopher W. Schadt
Climate Change Science Institute, Oak Ridge National Laboratory, Oak Ridge, Tennessee, United States of America
Biosciences Division, Oak Ridge National Laboratory, Oak Ridge, Tennessee, United States of America

Hem Nalini Morzaria-Luna and Cameron H. Ainsworth
Marine Resources Assessment Group Americas Incorporated, Seattle, Washington, United States of America

Isaac C. Kaplan and Phillip S. Levin
Conservation Biology Division, Northwest Fisheries Science Center, National Marine Fisheries Service, National Oceanic and Atmospheric Administration, Seattle, Washington, United States of America

Elizabeth A. Fulton
The Commonwealth Scientific and Industrial Research Organisation Wealth from Oceans Flagship, Division of Marine and Atmospheric Research, Hobart, Tasmania, Australia

Lyne Morissette
Institut des Sciences de la Mer de Rimouski, Université du Québec à Rimouski, Rimouski, Québec, Canada
Fisheries Centre, University of British Columbia, Vancouver, British Columbia, Canada

Villy Christensen and Daniel Pauly
Fisheries Centre, University of British Columbia, Vancouver, British Columbia, Canada

Xiaohua Wei
Department of Earth and Environmental Sciences, University of British Columbia, Kelowna, British Columbia, Canada

Juan A. Blanco
Departmento de Ciencias del Medio Natural, Universidad Pública de Navarra, Pamplona, Navarra, Spain

Michael Drexler and Cameron H. Ainsworth
College of Marine Science, University of South Florida, Saint Petersburg, Florida, United States of America

Dongsheng Zhao, Shaohong Wu and Yunhe Yin
Institute of Geographical Sciences and Natural Resources Research, Chinese Academy of Sciences, Anwai, Beijing, China

Qiong Gao
Landscape Analysis and Modeling Group, National Laboratory of Earth Surface Processes and Resources Ecology, Beijing Normal University, Beijing, China

Mei Yu
Institute for Tropical Ecosystem Studies, University of Puerto Rico - Rio Piedras, San Juan, Puerto Rico, United States of America

Chan Zhou
College of Life Sciences, Liaoning University, Shenyang, China

Fanny Chenillat
Laboratoire des Sciences de l'Environnement Marin (LEMAR), CNRS/UBO/IRD/IFREMER, Institut Universitaire Européen de la Mer (IUEM), Plouzané, France
Laboratoire de Physique des Océans (LPO), CNRS/IFREMER/IRD/UBO, Université de Bretagne Occidentale, Brest, France

Pascal Rivière
Laboratoire des Sciences de l'Environnement Marin (LEMAR), CNRS/UBO/IRD/IFREMER, Institut Universitaire Européen de la Mer (IUEM), Plouzané, France

Xavier Capet
Laboratoire d'Océanographie et du Climat (LOCEAN), CNRS/ UPMC/IRD/MNHN, Institut Pierre Simon Laplace (IPSL), Paris, France

Peter J. S. Franks
Integrative Oceanography Division, Scripps Institution of Oceanography, University of California San Diego, La Jolla, California, United States of America

Bruno Blanke
Laboratoire de Physique des Océans (LPO), CNRS/IFREMER/IRD/UBO, Université de Bretagne Occidentale, Brest, France

Yaai Dang
State Key Laboratory of Soil Erosion and Dryland Farming on the Loess Plateau, Institute of Soil and Water Conservation, Northwest A&F University, Yangling, Shaanxi, China
International Center for Climate and Global Change Research, School of Forestry & Wildlife Sciences, Auburn University, Auburn, Alabama, United States of America
College of Science, Northwest A&F University, Yangling, Shaanxi, China

Wei Ren, Bo Tao, Guangsheng Chen, Chaoqun Lu, Jia Yang, Shufen Pan and Hanqin Tian
International Center for Climate and Global Change Research, School of Forestry & Wildlife Sciences, Auburn University, Auburn, Alabama, United States of America

Guodong Wang
College of Science, Northwest A&F University, Yangling, Shaanxi, China

Shiqing Li
State Key Laboratory of Soil Erosion and Dryland Farming on the Loess Plateau, Institute of Soil and Water Conservation, Northwest A&F University, Yangling, Shaanxi, China

Dan Liu, Wenwen Cai, Jiangzhou Xia, Wenjie Dong, Yang Chen, Haicheng Zhang and Wenping Yuan
State Key Laboratory of Earth Surface Processes and Resource Ecology, Beijing Normal University, Beijing, China

Guangsheng Zhou
Chinese Academy of Sciences, Institute of Botany, State Key Laboratory of Vegetation and Environmental Change, Beijing, China
Chinese Academy of Meteorological Sciences, Beijing, China

Megan L. Mobley
Department of Botany, University of Wyoming, Laramie, Wyoming, United States of America
Helga Otto Haub School of Environment and Natural Resources University of Wyoming, Laramie, Wyoming, United States of America

Matthew J. Cleary
Department of Botany, University of Wyoming, Laramie, Wyoming, United States of America

Ingrid C. Burke
Department of Botany, University of Wyoming, Laramie, Wyoming, United States of America
Helga Otto Haub School of Environment and Natural Resources, University of Wyoming, Laramie, Wyoming, United States of America
Program in Ecology, University of Wyoming, Laramie, Wyoming, United States of America

Benjamin Planque and Ulf Lindstrøm
Institute of Marine Research, Tromsø, Norway

Sam Subbey
Institute of Marine Research, Bergen, Norway

Rodrigo Beas-Luna, Michael Hoban, Dan Malone and Mark H. Carr
Department of Ecology and Evolutionary Biology, University of California Santa Cruz, Santa Cruz, California, United States of America

Mark Novak and Alison Iles
Department of Integrative Biology, Oregon State University, Corvallis, Oregon, United States of America

Martin T. Tinker
Department of Ecology and Evolutionary Biology, University of California Santa Cruz, Santa Cruz, California, United States of America
Western Ecological Research Center, United States Geological Survey, Santa Cruz, California, United States of America

August Black and Jennifer E. Caselle
Marine Science Institute, University of California Santa Barbara, Santa Barbara, California, United States of America

Liming Yan
State Key Laboratory of Vegetation and Environmental Change, Institute of Botany, Chinese Academy of Sciences, Beijing, China
School of Life Sciences, Fudan University, Shanghai, China

Shiping Chen
State Key Laboratory of Vegetation and Environmental Change, Institute of Botany, Chinese Academy of Sciences, Beijing, China

Jianyang Xia and Yiqi Luo
Department of Microbiology and Botany, University of Oklahoma, Norman, Oklahoma, United States of America

Index

O

P

R

S

T

V

W